AF411445

Breastfeeding

Breastfeeding: Short and Long-Term Benefits to Baby and Mother

Special Issue Editors

Kingsley Emwinyore Agho
Benjamin John Wheeler

MDPI • Basel • Beijing • Wuhan • Barcelona • Belgrade • Manchester • Tokyo • Cluj • Tianjin

Special Issue Editors

Kingsley Emwinyore Agho Benjamin John Wheeler
Western Sydney University University of Otago
Australia New Zealand

Editorial Office
MDPI
St. Alban-Anlage 66
4052 Basel, Switzerland

This is a reprint of articles from the Special Issue published online in the open access journal *Nutrients* (ISSN 2072-6643) (available at: https://www.mdpi.com/journal/nutrients/special_issues/breastfeeding_baby_mother).

For citation purposes, cite each article independently as indicated on the article page online and as indicated below:

LastName, A.A.; LastName, B.B.; LastName, C.C. Article Title. *Journal Name* **Year**, *Article Number*, Page Range.

ISBN 978-3-03928-921-9 (Pbk)
ISBN 978-3-03928-922-6 (PDF)

Contents

About the Special Issue Editors

Kingsley Emwinyore Agho, B.Sc., B.Ed., M.Sc., M.Eng., Ph.D., MPH, is a Senior Lecturer in Biostatistics at the School of Health Sciences, Western Sydney University, Australia. Before joining Western Sydney University, Kingsley gained research and teaching experience from the University of Sydney and the University of Newcastle. He has over 15 years' experience in the design and analysis of large epidemiological studies including longitudinal and cluster RCT data. Other key research interests include epidemiology, international health, and water management and engineering (one-to-one water saving program for high water users). Kingsley has over 180 peer-reviewed journal articles and has supervised 14 Ph.D. and 2 Masters research students and currently supervises 8 Ph.D. students.

Benjamin John Wheeler, MBCHB, DCH, CCE, FRACP, Ph.D., is a Paediatric Endocrinologist and Paediatrician working for the University of Otago and the Southern District Health Board, New Zealand. His research focuses on access to and use of new technologies for children and young people affected by diabetes, as well as factors that impact on glycemic control in diabetes. His research interests also include vitamin D and bone health during pregnancy, lactation, and infancy. He has a number of collaborations ongoing in these areas, and usually has multiple clinical trials or studies running in these areas at any one time. He is always looking for new collaborations and for new Ph.D. and other research students to join his team.

Preface to "Breastfeeding: Short and Long-Term Benefits to Baby and Mother"

Breastfeeding is the preferred method of feeding in early life. It is also one of the most cost-effective childhood survival interventions. Breastfeeding practices are important for preventing child mortality and morbidity, as well as ensuring the optimal growth, health, and development of infants. The public health benefits of breastfeeding have been well documented in the medical literature, and include the following: associations with decreased risk for early-life diseases such as otitis media, respiratory tract infection, diarrhoea, and early childhood obesity (to name but a few).

In the late 1990s, The Baby Friendly Hospital Initiative (BFHI), based on appropriate training and a system of hospital accreditation, was introduced and was proven to be an effective health systems approach to improving breastfeeding outcomes. Since the introduction of the BFHI approach, breastfeeding has improved mothers' health because breastfeeding practices burn extra calories, which help mothers to lose pregnancy weight faster and lower the risk of developing heart disease as well as breast and ovarian cancer. Breastfeeding also improves mothers' emotional health and plays an important role in child spacing, which strongly correlates with improved maternal and child survival and health.

In 2008, the World Health Organization (WHO) developed guidelines for assessing infant and young child feeding practices, including 10 indicators for assessing breastfeeding and another 5 indicators for assessing complementary feeding practices to achieve the optimal growth, development, and health of infants. These measures aim to provide public health researchers with a consistent measure for assessing breastfeeding indicators.

Although all of the above have led to an improved understanding of the benefits of breastfeeding and improved rates of breastfeeding in many locations, more research and efforts are needed if we are to continue to make gains in this area and realize the full potential of breastfeeding for both mothers and babies.

This Special Issue book includes a collection of studies on the use of novel methods to improve breastfeeding rates, and research exploring the short- and long-term benefits of breastfeeding for both the infant and mother, including technology-based approaches.

Kingsley Emwinyore Agho, Benjamin John Wheeler
Special Issue Editors

 *nutrients*

Article

Exclusive Breastfeeding Rates and Associated Factors in 13 "Economic Community of West African States" (ECOWAS) Countries

Kingsley Emwinyore Agho [1,2,*], Osita Kingsley Ezeh [1], Pramesh Raj Ghimire [1], Osuagwu Levi Uchechukwu [3], Garry John Stevens [4], Wadad Kathy Tannous [2,3,5], Catharine Fleming [1,2,3], Felix Akpojene Ogbo [2,6] and Global Maternal and Child Health Research collaboration (GloMACH) [†]

[1] School of Science and Health, Western Sydney University, Locked Bag 1797, Penrith, NSW 2571, Australia; ezehosita@yahoo.com (O.K.E.); Prameshraj@hotmail.com (P.R.G.); Catharine.Fleming@westernsydney.edu.au (C.F.)

[2] Translational Health Research Institute (THRI), School of Medicine, Western Sydney University, Campbelltown Campus, Locked Bag 1797, Penrith, NSW 2571, Australia; K.Tannous@westernsydney.edu.au (W.K.T.); F.Ogbo@westernsydney.edu.au (F.A.O.);

[3] Diabetes, Obesity and Metabolism Translational Research Unit, Western Sydney University, Campbelltown, NSW 2560, Australia; L.osuagwu@westernsydney.edu.au

[4] Humanitarian and Development Research Initiative (HADRI), School of Social sciences and Psychology, Western Sydney University, Locked Bag 1797, Penrith, NSW 2751, Australia; G.Stevens@westernsydney.edu.au

[5] School of Business, Western Sydney University, Locked Bag 1797, Penrith, NSW 2751, Australia

[6] General Practice Unit, Prescot Specialist Medical Centre, Welfare Quarters, Makurdi, Benue State 972261, Nigeria

[*] Correspondence: k.agho@westernsydney.edu.au; Tel.: +61-2-4620-3635

[†] Members are listed at the end of Acknowledgments.

Received: 19 September 2019; Accepted: 29 November 2019; Published: 9 December 2019

Abstract: Exclusive breastfeeding (EBF) has important protective effects on child survival and also increases the growth and development of infants. This paper examined EBF rates and associated factors in 13 "Economic Community of West African States" (ECOWAS) countries. A weighted sample of 19,735 infants from the recent Demographic and Health Survey dataset in ECOWAS countries for the period of 2010–2018 was used. Survey logistic regression analyses that adjusted for clustering and sampling weights were used to determine the factors associated with EBF. In ECOWAS countries, EBF rates for infants 6 months or younger ranged from 13.0% in Côte d'Ivoire to 58.0% in Togo. EBF decreased significantly by 33% as the infant age (in months) increased. Multivariate analyses revealed that mothers with at least primary education, older mothers (35–49 years), and those who lived in rural areas were significantly more likely to engage in EBF. Mothers who made four or more antenatal visits (ANC) were significantly more likely to exclusively breastfeed their babies compared to those who had no ANC visits. Our study shows that EBF rates are still suboptimal in most ECOWAS countries. EBF policy interventions in ECOWAS countries should target mothers with no schooling and those who do not attend ANC. Higher rates of EBF are likely to decrease the burden of infant morbidity and mortality in ECOWAS countries due to non-exposure to contaminated water or other liquids.

Keywords: breastfeeding; Africa; antenatal care; infants; ECOWAS; mortality

---

1. Introduction

Exclusive breastfeeding (EBF) is the best source of nutrients for the healthy growth and development of newborns, as well as a natural immunity for protection against infectious (e.g., diarrhoea) [1–4] and long-term chronic diseases [4,5]. The World Health Organization and United Nations Children's Fund (WHO/UNICEF) recommends that nursing mothers should practice EBF, defined as providing the infant human breastmilk only, and oral rehydration solution, or drops/syrups of vitamins, minerals, or medicines, when required [6]. This is because EBF is strongly correlated with a reduced burden of infant and child morbidity and mortality [7]. A recently published report on EBF indicated that approximately 12% of deaths among children under five years old could be averted annually in low- and middle-income countries if all neonates were exclusively breastfed [8]. In sub-Saharan Africa, including Economic Community of West African States (ECOWAS), approximately 42% of diarrhoea-related deaths among children younger than five years of age could be attributed to prelacteal foods or unimproved water or liquids provided to newborns [9]. Mothers who exclusively breastfeed also have a reduced risk of developing type 2 diabetes mellitus and breast and ovarian cancers [4]. ECOWAS is a political and economic union of fifteen member countries in West Africa. The members include Benin, Burkina Faso, Cabo Verde, Côte d'Ivoire, The Gambia, Ghana, Guinea, Guinea-Bissau, Liberia, Mali, Niger, Nigeria, Senegal, Sierra Leone, and Togo. ECOWAS main aim is to promote socioeconomic integration among member countries in order to improve living standards and health outcomes of the population and promote the economic growth of member states [10].

Despite the numerous evidence showing the benefits of EBF for the mother–infant pair [1,4,11], EBF has remained low across ECOWAS countries and varies widely between and within countries [12–15]. Results from the recent Demographic and Health Survey (DHS) in ECOWAS countries between 2010 and 2018 revealed that many countries in the region had EBF rates that were well below the global target (50%) required to considerably reduce infant mortality [16,17]. This low rate of EBF indicates that a substantial proportion of newborns in ECOWAS countries are fed with water and/or prelacteal foods before the age of at least six months. This will, in turn, predispose the infant to gastrointestinal infections due to intake of contaminated prelacteal [18] or complementary foods [1,3,19]. In a recent study conducted in Nigeria, one of the ECOWAS countries with the largest population and economy [10], an estimated 22,371 diarrhoea-related deaths among children under five years could have been averted in 2016 if nursing mothers had strictly adhered to appropriate breastfeeding practices, particularly EBF [20].

In ECOWAS countries (Nigeria, Ghana, Mali, and Niger), past studies conducted based on data collected between 2006 and 2012 have elucidated some factors associated with appropriate EBF practice. The factors include place of residence and birthing [12], higher maternal and paternal education [14,21–23], higher household wealth index [21,23], child's birth order, frequent (≥4) antenatal and postnatal visits [12,21,23], average size of newborns, vaginal birthing [12,24], assistance during delivery, maternal employment, and higher maternal age [12,14,21]. However, none of the previous studies had investigated EBF rates across ECOWAS countries or identified what acts as an enabler or barrier to appropriate EBF behaviours using the most recent country-specific and standardised data. Additionally, some of the previous studies were conducted at the subnational level in Nigeria and Ghana, where their findings may not inform national-level policy interventions [15,25]. Some of the previous nationwide studies used old datasets, which may not reflect the current sociodemographic and economic settings of those countries [12,23].

Understanding the current EBF rates and associated factors across ECOWAS countries is essential to key stakeholders in formulating integrated and effective health policy interventions. Furthermore, the study findings could offer credible information needed to prioritise cost-effective, evidence-based interventions that may rapidly improve EBF rates and subsequently reduce the high burden of under-five deaths in ECOWAS countries [26]. In the current United Nations Decade of Action on Nutrition (2016–2025) [16], the present study will provide policy-relevant data about EBF practice that would assist ministries of health, international agencies, and nongovernmental organizations to design programs that promote, protect, and support appropriate EBF practice in the region. This study aimed to examine EBF rates and the potential demographic, socioeconomic, and proximate factors that are associated with EBF in 13 ECOWAS countries using the combined dataset from 2010 to 2018.

2. Materials and Methods

The analyses were based on the most recent DHS dataset from 13 ECOWAS countries, which were obtained from a password-enabled Measure DHS website [27]. The DHS data were nationally representative and population-based surveys, collected by country-specific ministries of health or other relevant government-owned agencies, with technical support largely provided by Inner City Fund (ICF) International. These surveys were comparable, given the standardised nature of the data collection methods and instruments [28].

The DHS collects demographic data and population health status of people, including reproductive health, maternal and child health, mortality, nutrition, and self-reported health behaviour among adults [28]. Information was collected from eligible women, that is, all women aged 15–49 years who were either permanent residents in the households or visitors present in the households on the night before the survey. Child health information was collected from the mother based on the youngest child aged less than five years, with response rates that ranged from 96% to 99% [27]. Detailed information on the sampling design and questionnaire used is provided in the respective country-specific Measure DHS reports [27]. Our analyses were restricted to the last born child aged 0–5 months and living with the respondent, which yielded a weighted total of 19,735 infants for all 13 ECOWAS countries.

2.1. Outcome, Confounding and Exploratory Variables

EBF rate was estimated using the WHO/UNICEF definitions for assessing infant and young feeding practices in populations [29] and used by Measure DHS. EBF was measured as the proportion of infants 0–5 months of age who were fed exclusively with breast milk (but allows oral rehydration solution and drops or syrups of vitamins and medicines when required). Information on EBF was collected based on maternal recall on feeds provided to the infant in the last 24 h. EBF was categorized as "Yes" (1 = if the infant was exclusively breastfed) or "No" (0 = if the infant was not exclusively breastfed).

Previous studies conducted in sub-Saharan African countries that examined factors associated with EBF [23,30–33] played a vital role in determining the potential confounding variables for this study. The confounding variables were subdivided into four groups, and these were country and demographic factors, socioeconomic factors, access to media factors, and healthcare utilisation factors. The country variables were Benin, Burkina Faso, Côte d'Ivoire, The Gambia, Ghana, Guinea, Liberia, Mali, Niger, Nigeria, Senegal, Sierra Leone, and Togo. We considered Benin as the referenced category because it was the first country on the list of ECOWAS countries. The demographic variables were place of residence (urban or rural), mother's age, marital status, combined birth rank (the position of the youngest under-five child in the family), and birth interval (the interval between births; that is, whether there were no previous births, birth less 24 months prior, or birth more than or equal to 24 months prior), sex of baby, age of the child, and perceived size of the newborn by the mother. The socioeconomic level factors considered were maternal education, maternal work status, maternal literacy, and household wealth index variable. For the combined datasets, the household wealth index was constructed using the "hv271" variable. In the household wealth index categories, the bottom 20% of households was arbitrarily referred to as the poorest households, and the top 20% as the richest households, and was divided into poorest, poor, middle, rich, and richest. Access to media factors consists of the frequency of mothers listening to the radio, watching television, and reading newspapers or magazines. Healthcare utilisation factors were considered and included (birthplace, birth order, mode of delivery, delivery assistance, and antenatal clinic visits (ANC).

2.2. Statistical Analysis

Population-level weights were used for survey tabulation, which adjusts for a unique country-specific stratum, and clustering was used to determine the percentage, frequency count, and univariate and multivariate logistic regression of all selected characteristics. Country-specific weights were used for the Taylor series linearization method in the surveys when estimating 95% confidence intervals around the rate of EBF in each country.

For the combined dataset, sampling weight was denormalised, and a new population-level weight was created by dividing the sampling weights by the denormalised weight. We then created a unique country-specific cluster and strata because each country had individual clusters and strata in the DHS. This was done to account for the uneven country-specific population across the organisation and to avoid the effect of countries with a large population (such as Nigeria with over 175 million people in 2013) offsetting countries with a small population (such as The Gambia with about 1.8 million people in 2013) [10].

In the multivariate analyses, the factors associated were further tested by adjusted odds ratios (AOR) using hierarchical multiple logistic regression analyses as described in Table 1. The first stage (Model 1) included country and demographic factors. The second stage (Model 2) also included socioeconomic factors. The third stage (Model 3) added access to media covariates. The fourth and final stage (Model 4) added healthcare utilisation factors. The objective of this modelling strategy was to allow for a comparison of the relationship between each of the different sets of covariates in examining factors associated with EBF. All analyses were performed in Stata version 14.0 (Stata Corp, College Station, Texas, USA).

Table 1. Potential covariates used for hierarchical survey logistic regression model.

Model 1	Model 2	Model 3	Model 4
Country [1]	Country [1]	Country [1]	Country [1]
Demographic	Demographic	Demographic	Demographic
Residence	Socioeconomic	Socioeconomic	Socioeconomic
Mother's age	Household Wealth Index	Access to media	Access to media
Marital status	Work in the last 12 months	Frequency of reading newspaper or magazine	Healthcare utilization factors
Birth rank and birth interval	Maternal education	Frequency of listening to Radio	Place of delivery
Sex of baby	Maternal Literacy	Frequency of watching Television	Mode of delivery
Age of child			Type of delivery assistance
Size of baby			Antenatal Clinic visits

[1] Benin, Burkina Faso, Côte d'Ivoire, The Gambia, Ghana, Guinea, Liberia, Mali, Niger, Nigeria, Senegal, Sierra Leone or Togo.

3. Results

3.1. Characteristics of the Sample and Unadjusted Analyses for EBF

The overall EBF rate for all infants aged 0–5 months in the 13 ECOWAS countries was 31.0% between 2010 and 2018 (Figure 1). Higher EBF rates in ECOWAS countries ranged from 52.0% in Ghana to 58.0% in Togo, while Nigeria and Côte d'Ivoire had the lowest EBF rates of less than 20%. Table 2 illustrates the characteristics of the sample of infants aged 0–5 months and their unadjusted odds ratios. Overall, the number of mother–infant dyads included in our sample varied by country, with Nigeria contributing the most (20.0% of the study sample) and Togo the least (3.3%). Bivariate analyses revealed that Nigerian mothers were less likely to exclusively breastfeed their babies compared to those who resided in Benin (OR = 0.29, 95% CI: 0.22, 0.38).

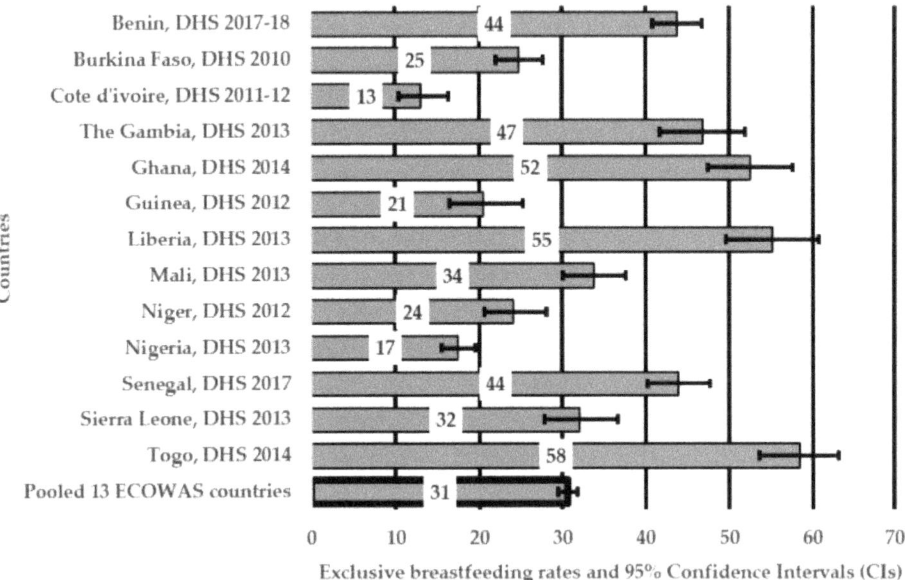

Figure 1. Exclusive breastfeeding rates of infants aged 0–5 months and corresponding 95% confidence intervals in 13 "Economic Community of West African States" (ECOWAS) countries.

Table 2. Individual, household, and community level characteristics and unadjusted odd ratios (OR) (95% CI) of exclusive breastfeeding (EBF) among infants aged 0–5 months in 13 ECOWAS countries.

Characteristic	n *	% *	%	EBF		
				OR	95% CI		p-Value
Demographic Factors						
Country						
Benin	1475	7.5	8.9	1.00			
Burkina Faso	1837	9.3	10.0	0.46	0.37	0.58	<0.001
Côte d'Ivoire	1110	5.6	5.3	0.17	0.12	0.26	<0.001
The Gambia	1422	7.2	6.6	0.99	0.64	1.52	0.946
Ghana	805	4.1	4.2	1.33	0.98	1.80	0.069
Guinea	981	5.0	4.9	0.42	0.26	0.68	<0.001
Liberia	927	4.7	4.9	1.50	1.02	2.21	0.040
Mali	1192	6.0	6.9	0.66	0.51	0.86	0.002
Niger	2196	11.1	9.0	0.46	0.32	0.64	<0.001
Nigeria	3996	20.2	20.2	0.29	0.22	0.38	<0.001
Senegal	1298	6.6	7.3	1.09	0.84	1.41	0.537
Sierra Leone	1842	9.3	7.7	0.55	0.39	0.78	0.001
Togo	655	3.3	4.2	1.81	1.37	2.39	<0.001
Residence						
Urban	6630	33.6	29.8	1.00			
Rural	13,106	66.4	70.2	0.79	0.66	0.94	0.008
Mother's age (in years)						
15–19	7215	36.6	36.3	1.00			
20–34	9267	47.0	46.0	1.01	0.88	1.17	0.876
35–49	3253	16.5	17.7	1.00	0.84	1.20	0.968
Marital status						
Currently married	18,357	93.0	93.3	1.00			
Formerly married ^	275	1.4	1.5	0.80	0.52	1.23	0.314
Never married	1103	5.6	5.1	1.10	0.79	1.53	0.574
Child Age (in months)	-	-	-	0.70	0.67	0.73	<0.001
Birth order						
First-born	3913	19.8	20.0	1.00			
2nd-4th	9579	48.5	46.6	0.93	0.79	1.09	0.355
5 or more	6243	31.6	33.4	0.70	0.59	0.83	<0.001
Preceding birth interval (n = 19,695)						
No previous birth	3913	19.8	19.7	1.00			
<24 months	1831	9.3	9.2	0.85	0.65	1.11	0.243
≥24 months	13,951	70.7	70.9	0.83	0.72	0.97	0.016

Table 2. *Cont.*

Characteristic	*n* *	% *	%	EBF		
				OR	95% CI	*p*-Value
Demographic Factors						
Combined Birth rank and birth interval						
1st birth rank	3913	19.8	19.7	1.00		
2nd/3rd birth rank, more than 2 years interval	8191	41.5	40.1	0.91	0.77 1.08	0.281
2nd/3rd birth rank, less than or equal to 2 years interval	1388	7.0	6.7	1.00	0.76 1.33	0.963
4th birth rank, more than 2 years interval	5432	27.5	29.2	0.71	0.60 0.85	<0.001
4th birth rank, less than or equal to 2 years interval	811	4.1	4.3	0.66	0.47 0.93	0.016
Sex of baby						
Male	9777	49.5	50.1	1.00		
Female	9958	50.5	49.9	1.09	0.95 1.24	0.206
Size of the baby (*n* = 19,584)						
Small	3701	18.8	18.5	1.00		
Average	8433	42.7	42.7	1.21	1.02 1.42	0.025
Large	7450	37.8	37.8	1.13	0.96 1.33	0.155
Socioeconomic factors						
Household Wealth Index						
Poorest	3252	16.5	20.2	1.00		
Poorer	3771	19.1	20.2	1.43	1.13 1.81	0.003
Middle	4071	20.6	20.4	1.45	1.16 1.82	0.001
Richer	3790	19.2	19.5	1.51	1.21 1.88	<0.001
Richest	4851	24.6	19.7	1.96	1.54 2.51	<0.001
Work in the last 12 months (*n* = 19,733)						
Not working	9252	46.9	44.2	1.00		
Working	10,481	53.1	55.8	0.92	0.80 1.07	0.288
Maternal education (*n* = 19,732)						
No formal education	12105	61.3	61.9	1.00		
Primary	3251	16.5	17.9	1.36	1.14 1.63	0.001
Secondary and above	4376	22.2	20.2	1.86	1.57 2.21	<0.001
Maternal Literacy (*n* = 19,603)						
Cannot read at all	14,935	75.7	77.6	1.00		
Able to read only part of sentences	4668	23.7	21.8	1.56	1.33 1.82	<0.001

Table 2. *Cont.*

Characteristic	*n* *	% *	%	EBF		
				OR	95% CI	*p*-Value
Access to media						
Frequency of reading a newspaper or magazine (*n* = 19,688)						
Not at all	17,770	90.0	91.6	1.00		
Less than once a week	998	5.1	4.4	1.63	1.22 2.19	0.001
At least once a week/almost every day	914	4.6	3.7	1.45	1.03 2.05	0.035
Frequency of listening to radio (*n* = 19,703)						
Not at all	6844	34.7	36.2	1.00		
Less than once a week	4957	25.1	25.2	1.18	1.00 1.41	0.054
At least once a week/almost every day	7734	39.2	37.6	1.35	1.16 1.58	<0.001
Frequency of watching television (*n* = 19,684)						
Not at all	11,558	58.6	61.0	1.00		
Less than once a week	2692	13.6	14.6	1.18	0.98 1.43	0.075
At least once a week/almost every day	5286	26.8	23.5	1.36	1.14 1.63	0.001
Healthcare utilization factors						
Place of delivery						
Home	8493	43.0	42.0	1.00		
Health facility	11,242	57.0	58.0	1.73	1.49 2.01	<0.001
Mode of delivery (*n* = 19,672)						
Noncaesarean	18,990	96.2	96.4	1.00		
Caesarean section	682	3.5	3.4	1.54	1.11 2.15	0.010
Type of delivery assistance (*n* = 15,940)						
Health professional	12,342	65.0	62.0	1.00		
Traditional birth attendant	476	2.4	2.7	0.65	0.48 0.89	0.007
Other untrained	1913	9.0	10.9	0.84	0.63 1.13	0.251
No one	1209	6.1	6.0	0.50	0.39 0.65	<0.001
Antenatal Clinic visits (*n* = 19,147)						
None	2446	12.4	12.8	1.00		
1–3	6485	32.9	34.4	2.12	1.65 2.71	<0.001
≥4	10,216	51.8	50.4	2.75	2.15 3.51	<0.001

* Weighted sample sizes and percentages; ^ divorced/separated/widowed.

Infants whose mothers were from the richer and richest households were significantly more likely to exclusively breastfeed compared to those from poor or poorer households (OR = 1.51, 95% CI: 1.21, 1.88 for richer and OR = 1.96, 95% CI: 1.54, 2.51 for richest). Mothers who watched television almost every day reported higher odds of EBF compared to those who did not watch television at all (OR = 5.44, 95% CI: 3.14, 9.43). This association was also observed among mothers who listened to the radio almost every day (OR = 3.98, 95% CI: 2.41, 6.58). Approximately 57.0% of mothers who delivered their babies in health facilities were significantly more likely to exclusively breastfeed their babies compared to those 43.0% who delivered at home (OR = 1.73, 95% CI: 1.49, 2.01).

Mothers who had caesarean delivery reported much lower proportions (3.5%) of EBF compared to those who had vaginal delivery (96.2%). More than half (51.8%) of mothers had four or more ANC visits, and those mothers who attended ANC visits reported a higher likelihood of EBF compared to those who did not attend any ANC visits (OR = 2.75, 95% CI: 2.15, 3.51) (Table 2).

3.2. Factors Associated with Exclusive Breastfeeding in Multivariate (Adjusted) Analyses

Table 3 illustrates the factors associated with EBF among infants aged 0–5 months in multivariate analyses. Among the 13 ECOWAS countries, mothers in Côte d'Ivoire and Nigeria showed the lowest likelihood to engage in EBF practice compared to those in Benin (OR = 0.18, 95% CI: 0.12, 0.28 for Côte d'Ivoire and OR = 0.33, 95% CI: 0.24, 0.45 for Nigeria). Mothers who resided in Burkina Faso, Guinea, Niger, and Sierra Leone were also less likely to engage in EBF compared to those in Benin (Model 4; OR = 0.49, 95% CI: 0.37, 0.64 for Burkina Faso, OR = 0.47, 95% CI: 0.26, 0.84 for Guinea, OR = 0.50, 95% CI: 0.32, 0.79 for Niger, and OR = 0.62, 95% CI: 0.40, 0.97 for Sierra Leone). Liberian and Togolese mothers were more likely to practice EBF compared to their counterparts in Benin (Model 4; OR = 1.65, 95% CI: 1.03, 2.66 for Liberia and OR = 1.52, 95% CI: 1.04, 2.21 for Togo) (Table 3).

Mothers who perceived their babies to be of average birth size were more likely to engage in EBF compared to those who perceived their babies to be of small birth size (OR = 1.22, 95% CI: 1.00, 1.50). Mothers who lived in rural areas and those who had female infants reported higher odds of EBF compared to their counterparts (OR = 1.25, 95% CI: 1.01, 1.55 for rural and OR = 1.17, 95% CI: 1.00, 1.37 for female infants) (Table 3).

Higher maternal age (35–49 years) and education (primary and above education) were associated with increased odds of EBF practice compared to young maternal age (15–19 years) and no maternal education, respectively (OR = 1.50, 95% CI: 1.12, 1.99 for maternal age, and OR = 1.32, 95% CI: 1.06, 1.65 for primary and OR = 1.83, 95% CI: 1.18, 2.82 for secondary and above education). Mothers who made ANC visits had higher odds of EBF practice compared to those who did not make any ANC visits (OR = 1.52, 95% CI: 1.14, 2.03 for 1–3 ANC visits and OR = 1.66, 95% CI: 1.24, 2.23 for four or more ANC visits) (Table 3).

Increasing age of the infant was the only common factor associated with a lower likelihood of EBF in ECOWAS countries (see Table 3) and across all ECOWAS countries (see Supplementary Tables S1 and S2).

Table 3. Adjusted OR (95% CI) of factors associated with exclusive breastfeeding among infants aged 0–5 months in 13 ECOWAS countries.

Characteristic	Model 1				Model 2				Model 3				Model 4			
	aOR	95% CI		p-Value	aOR	95% CI		p-Value	aOR	95% CI		p-Value	aOR	95% CI		p-Value
Demographic factors																
Benin	1.00				1.00				1.00				1.00			
Burkina Faso	0.46	0.36	0.58	<0.001	0.46	0.36	0.59	<0.001	0.46	0.35	0.59	<0.001	0.49	0.37	0.64	≤0.001
Côte d'Ivoire	0.16	0.11	0.25	<0.001	0.17	0.11	0.26	<0.001	0.17	0.11	0.26	<0.001	0.18	0.12	0.28	≤0.001
The Gambia	1.06	0.70	1.60	0.795	0.97	0.63	1.48	0.872	0.97	0.64	1.47	0.878	0.89	0.56	1.40	0.618
Ghana	1.43	1.01	2.04	0.047	1.17	0.77	1.77	0.462	1.16	0.76	1.78	0.493	1.15	0.73	1.80	0.557
Guinea	0.39	0.25	0.61	<0.001	0.40	0.24	0.65	<0.001	0.40	0.24	0.65	<0.001	0.47	0.26	0.84	0.010
Liberia	1.76	1.19	2.61	0.005	1.72	1.15	2.57	0.008	1.73	1.15	2.59	0.008	1.65	1.03	2.66	0.039
Mali	0.64	0.49	0.85	0.002	0.76	0.55	1.04	0.083	0.76	0.55	1.04	0.086	0.81	0.57	1.13	0.211
Niger	0.46	0.32	0.67	<0.001	0.45	0.31	0.66	<0.001	0.45	0.31	0.65	<0.001	0.50	0.32	0.79	0.003
Nigeria	0.28	0.22	0.37	<0.001	0.26	0.20	0.34	<0.001	0.26	0.20	0.34	<0.001	0.33	0.24	0.45	<0.001
Senegal	1.17	0.88	1.56	0.279	1.11	0.81	1.51	0.518	1.12	0.81	1.54	0.497	1.08	0.77	1.52	0.665
Sierra Leone	0.56	0.38	0.83	0.004	0.53	0.36	0.78	0.001	0.54	0.37	0.78	0.001	0.62	0.40	0.97	0.037
Togo	2.00	1.48	2.68	<0.001	1.82	1.34	2.47	<0.001	1.58	1.12	2.23	0.009	1.52	1.04	2.21	0.031
Residence																
Urban	1.00				1.00				1.00				1.00			
Rural	0.98	0.81	1.18	0.838	1.24	1.01	1.51	0.035	1.24	1.01	1.51	0.037	1.25	1.01	1.55	0.042
Mother's age																
15–19 years	1.00				1.00				1.00				1.00			
20–34 years	1.23	1.03	1.48	0.022	1.20	1.00	1.45	0.051	1.22	1.01	1.46	0.041	1.17	0.96	1.42	0.121
35–49 years	1.50	1.15	1.95	0.003	1.46	1.12	1.90	0.005	1.47	1.13	1.92	0.004	1.50	1.12	1.99	0.006
Marital status																
Currently married	1.00				1.00				1.00				1.00			
Formerly married ^	0.80	0.50	1.30	0.370	0.82	0.51	1.33	0.427	0.83	0.51	1.34	0.441	0.78	0.45	1.36	0.386
Never married	0.81	0.56	1.17	0.260	0.72	0.49	1.06	0.096	0.73	0.50	1.07	0.111	0.75	0.50	1.13	0.167
Combined Birth rank and birth interval																
1st birth rank	1.00				1.00				1.00				1.00			
2nd/3rd birth rank, more than 2 years interval	0.88	0.73	1.06	0.186	0.96	0.78	1.17	0.688	0.95	0.78	1.17	0.651	0.93	0.75	1.16	0.542
2nd/3rd birth rank, less than or equal to 2 years interval	0.97	0.71	1.32	0.843	1.03	0.76	1.40	0.858	1.03	0.76	1.39	0.861	0.85	0.60	1.20	0.361
4th birth rank, more than 2 years interval	0.61	0.48	0.78	<0.001	0.75	0.57	0.97	0.031	0.74	057	0.97	0.030	0.76	0.56	1.03	0.074
4th birth rank, less than or equal to 2 years interval	0.64	0.44	0.93	0.019	0.75	0.50	1.01	0.146	0.74	0.50	1.10	0.141	0.81	0.52	1.28	0.358
Sex of baby																
Male	1.00				1.00				1.00				1.00			
Female	1.12	0.98	1.29	0.102	1.13	0.98	1.30	0.094	1.13	0.98	1.31	0.082	1.17	1.00	1.37	0.046
Child Age (in months)	0.67	0.64	0.69	<0.001	0.66	0.64	0.69	<0.001	0.66	0.64	0.69	<0.001	0.67	0.64	0.71	<0.001
Size of the baby																
Small	1.00				1.00				1.00				1.00			
Average	1.26	1.05	1.51	0.013	1.23	1.02	1.48	0.032	1.23	1.02	1.48	0.030	1.22	1.00	1.50	0.050
Large	1.22	1.01	1.47	0.039	1.18	0.98	1.43	0.081	1.19	0.98	1.43	0.078	1.18	0.95	1.45	0.135

Table 3. *Cont.*

Characteristic	Model 1			Model 2				Model 3				Model 4			
	aOR	95% CI	p-Value	aOR	95% CI		p-Value	aOR	95% CI		p-Value	aOR	95% CI		p-Value
Socioeconomic factors															
Household Wealth Index															
Poorest				1.00				1.00				1.00			
Poorer				1.39	1.04	1.84	0.024	1.39	1.04	1.85	0.026	1.34	0.99	1.80	0.056
Middle				1.35	1.04	1.75	0.025	1.35	1.04	1.76	0.026	1.13	0.86	1.50	0.369
Richer				1.32	1.03	1.68	0.027	1.31	1.02	1.68	0.035	1.14	0.87	1.48	0.340
Richest				1.46	1.09	1.95	0.010	1.44	1.05	1.99	0.025	1.20	0.86	1.68	0.272
Work in the last 12 months															
Not working				1.00				1.00				1.00			
Working				1.00	0.85	1.17	0.962	0.99	0.84	1.16	0.911	1.01	0.85	1.20	0.899
Maternal education															
No education				1.00				1.00				1.00			
Primary				1.22	0.99	1.51	0.058	1.22	0.99	1.51	0.058	1.32	1.06	1.65	0.015
Secondary and above				1.79	1.12	2.86	0.014	1.84	1.15	2.93	0.011	1.83	1.18	2.82	0.006
Maternal Literacy															
Cannot read at all				1.00				1.00				1.00			
Able to read only part of sentences				0.97	0.64	1.47	0.873	0.95	0.62	1.46	0.826	1.00	0.67	1.48	0.983
Access to media															
Frequency of reading newspaper or magazine															
Not at all								1.00							
Less than once a week								1.08	0.76	1.53	0.655	1.03	0.73	1.46	0.860
At least once a week/Almost every day								0.83	0.53	1.30	0.404	0.83	0.53	1.31	0.420
Frequency of listening to Radio															
Not at all								1.00							
Less than once a week								0.99	0.82	1.19	0.878	0.91	0.75	1.11	0.362
At least once a week/Almost every day								1.01	0.84	1.22	0.916	0.98	0.81	1.19	0.846
Frequency of watching Television															
Not at all								1.00							
Less than once a week								1.03	0.83	1.27	0.800	1.02	0.81	1.28	0.883
At least once a week/Almost every day								1.02	0.83	1.27	0.543	0.93	0.74	1.17	0.551

Table 3. *Cont.*

Characteristic	Model 1			Model 2			Model 3			Model 4			
	aOR	95% CI	*p*-Value	aOR	95% CI	*p*-Value	aOR	95% CI	*p*-Value	aOR	95% CI		*p*-Value
Healthcare utilization factors													
Place of delivery													
Home										1.00			
Health facility										1.09	0.84	1.43	0.508
Mode of delivery													
Non-caesarean										1.00			
Caesarean section										1.24	0.84	1.83	0.270
Type of delivery assistance													
Health professional										1.00			
Traditional birth attendant.										0.80	0.53	1.21	0.285
Other untrained										1.00	0.71	1.41	0.993
No one										0.81	0.59	1.10	0.178
Antenatal Clinic visits													
None										1.00			
1–3										1.52	1.14	2.03	0.004
≥4										1.66	1.24	2.23	0.001

^ Divorce/separated/widowed.

4. Discussion

Our study indicated that the overall rate of EBF among infants aged 0–5 months of life in 13 ECOWAS countries was at a suboptimum level between 2010 and 2018 compared to the global target of 50% by 2025 [17]. There is a need for further improvement in order to gain the full benefits of EBF as 10 out of the 13 ECOWAS countries had EBF rates between 13% and 47%. This study found that place of residence (rural), maternal education (primary or secondary), mother's age (35–49 years), and antenatal care visit (1–3 or ≥4), increased the likelihood of EBF in 13 ECOWAS countries. Baby size at birth (average) and child's gender (female) were associated with EBF.

The present study showed that nursing mothers who resided in rural areas were more likely to exclusively breastfeed their newborns for less than six months compared with their urban counterparts. This finding is consistent with previous studies conducted in Ethiopia [34], Egypt [35], Malaysia [36], and Lebanon [37]. This finding, however, contradicts earlier reports from Nigeria [30] and Bangladesh [38], which indicated a negative association between rural residence and EBF. An increased likelihood of the association between EBF and rural residence noted in the current study (even after accounting for maternal employment status) could be attributed to a high number of nonworking rural women (rural: 68% vs. urban: 32%). This can, in turn, allow unlimited time for nonworking rural women to breastfed their babies exclusively. Most working nursing mothers reside in urban areas, and labour law concerning maternity leave duration in most ECOWAS countries remains a serious challenge for promotion, protection, and support of EBF practice [13,39]. Fewer than six months of maternity leave are granted to nursing mothers in many ECOWAS countries; for example, in Ghana (12 weeks) [40], Nigeria (16 weeks) [39], and Senegal (14 weeks) [41]. This often leads to early weaning of EBF among working nursing mothers. In addition to the short duration of maternity leave, inadequate facilities and unallocated official time for breastfeeding in the workplace also limit EBF [42]. Urban regions (e.g., urban slums) with potentially disadvantaged communities may deserve special focus with targeted interventions for improving EBF practice in these ECOWAS countries.

The odds of EBF practice were higher among nursing mothers who had primary or higher education, in comparison to their counterparts who had no formal education. This is contrary to previous studies conducted in Ethiopia [34] and Bangladesh [38], where mothers who had no formal education were more likely to report EBF practice. Nevertheless, other studies conducted in sub-Saharan African countries [21,23,33] have also suggested that higher maternal education was significantly associated with EBF practice. Educated mothers are more knowledgeable in utilizing maternal health care services appropriately, including ANC [43], postnatal care (PNC) [44], and institutional delivery [45], which remains an important initiative to scale up EBF participation [46]. To add, past studies have suggested that no formal education among mothers was associated with both underutilization of ANC [47,48] and PNC [49] services.

Furthermore, older mothers (≥35 years) reported a higher likelihood of EBF practice compared with younger mothers. This result is not consistent with previous studies [35,36,50], which indicated that younger mothers were more likely to maintain EBF practice than older mothers. The difference in outcomes may be linked to maternal age classification, characteristics adjusted for, and diverse populations. However, our finding is in line with previous studies conducted in Bangladesh [38] and six low- and middle-income countries [51] that used similar data and maternal age limits. An explanation for the higher odds of EBF noted among older mothers may be their experience in child-rearing, better knowledge of breastfeeding, and being well informed of the benefits of breastfeeding to both newborns and mothers. Additionally, older mothers may also have more time for EBF due to decreased paid job opportunities compared with younger mothers.

Mothers who had one or more ANC visits to the health institution prior to childbirth showed higher odds of EBF practice compared to those who did not attend any ANC visits. This finding is similar to cross-sectional studies carried out in sub-Saharan Africa [23,30] and Bangladesh [38]. The increased odds of EBF in the current study may be attributed to breastfeeding counselling received through the Baby-Friendly Hospital Initiative (BFHI) program during the ANC visits, which may have improved mothers' knowledge and benefits of EBF practice. This is supported by previous evidence from regional Egypt and Ethiopia [35,52], which showed that breastfeeding counselling during ANC was positively related to EBF practice.

This study found that female infants were also associated with higher EBF rates, and this finding is similar to the research conducted in Nigeria, which found that female infants were 2.13 times more likely to be exclusively breastfed than male infants [30]. A study conducted by Johns Hopkins Children's Center revealed that the protective effect of EBF was higher in female infants than male infants [53]. Even though our result on female babies was positively related to EBF practice, caution needs to be exercised in concluding this finding and would warrant further investigation to better understand the impact of gender on EBF in ECOWAS countries. Furthermore, our study also showed that increasing infant age was the only common factor associated with a lower likelihood of EBF in all ECOWAS countries. This finding has been reported in India, where EBF practice decreases faster with increasing infant age in Indian regions with higher socioeconomic status women [54]. Country-specific decline in EBF associated with increasing infant age has not been well studied in ECOWAS countries, and thus, would need further assessment to inform targeted interventions. Reasons why other factors were not significant across the ECOWAS countries varied, including methodological (e.g., sample size) and/or qualitative factors (e.g., cultural norms).

Babies perceived by their mothers to be of average size at birth had an increased likelihood of EBF practice compared to babies perceived by their mothers to be of small size at birth. This finding was similar to a cross-sectional study carried out in Timor-Leste in 2014, which showed that babies perceived to be of a small size at birth by their mothers were less likely to be EBF [55]. Furthermore, a cross-sectional study done in the United States in 2011 reported that very low birth weight (a proxy for babies perceived to be of a small size at birth) was significantly related to a lower likelihood of EBF practice compared to babies born at normal or above-normal birth weights [56]. The significantly lower chance of EBF practice for newborns perceived to be of a small size by their mothers in ECOWAS countries may be attributed to obstacles related to breastfeeding small-sized babies, such as poor sucking and illness (e.g., hypoglycaemia), leading to babies being kept away from their mothers [57], resulting in early introduction of complementary foods and drinks. Although caesarean birthing was associated with non-EBF practice in the bivariate analyses, this association was not significant in multivariate analyses. The possible reasons why caesarean birth was related to non-EBF have been described elsewhere [12,24,58], including a limited number of BFHI-certified maternal health centres, or health practitioners' limited knowledge of appropriate EBF immediately post-birth [59,60]. The BFHI was introduced in 1990 by WHO/UNICEF to promote, protect, and support breastfeeding [61] and has been shown to be successful in increasing EBF participation worldwide [62,63].

The main strengths of our study were the nationally representative sample [27], the comprehensive data on standard infant feeding indicators, and the appropriate adjustments for sampling design in the analysis. There are five limitations of note in this study: Firstly, the subjectivity of information in recalling the time of EBF, especially among mothers who have older infants. Secondly, measurement error and recall bias by mothers may have underestimated or overestimated our effect sizes because information concerning some of the study variables was obtained from mothers whose youngest child may have been up to five years of age. Thirdly, findings in the study cannot be regarded as causal because they were obtained with a cross-sectional design. Fourthly, the study was unable to account for other study and confounding factors, including information on the health status of mothers, partner support, smoking status and alcohol intake, and prenatal breastfeeding intention as documented elsewhere [59,60]. Lastly, the 24 h recall approach was used for EBF investigation, and it is possible that day-to-day variability in nutritional intake may have affected EBF estimates.

5. Conclusions

In summary, the rates of EBF in 10 out of the 13 ECOWAS countries were below the WHO global nutrition target of 50% for breastfeeding. The low rates of EBF in ECOWAS countries may be contributing to higher infant and child morbidity and mortality in each country, given that it is a strong indicator for infant and child growth and survival. At the community level, "baby-friendly" initiative needs to be implemented as a trial in order to empower the community for better nutrition and health improvement of children, as well as to test its effectiveness. At the individual level, the practice of giving water and other liquids is a significant contributor to the decrease in EBF in Africa [18,64]. Awareness must be spread to inform mothers that their babies do not require water and liquids until the completion of the recommended time of six months. These are crucial messages that could be spread by community-level campaigns, including peer support community intervention for young mothers who are vulnerable and have lower rates of EBF. However, further improvement is needed in order to gain the full benefits of breastfeeding practices. Our results suggest that breastfeeding interventions should target urban residents, working mothers, mothers who do attend antennal care, those who give birth to male infants, and mothers who perceive their babies to be small at birth.

Supplementary Materials: The following are available online at http://www.mdpi.com/2072-6643/11/12/3007/s1, Table S1: Adjusted OR (95% CI) of factors associated with Exclusive breastfeeding among infant aged 0-5 months in Benin, Burkina Faso, Côte d'Ivoire, Gambia, Ghana, Guinea, and Liberia, Table S2: Adjusted OR (95% CI) of factors associated with Exclusive breastfeeding among infant aged 0–5 months in Mali, Niger, Nigeria, Senegal, Sierra Leone, and Togo.

Author Contributions: Conceptualization, K.E.A. and F.A.O.; methodology, K.E.A. and O.K.E.; software, K.E.A.; validation, O.L.U., F.A.O., and O.K.E.; formal analysis, K.E.A. and P.R.G.; investigation, O.K.E. and P.R.G.; resources, W.K.T. and C.F.; data curation, F.A.O., P.R.G., and O.L.U.; writing—original draft preparation, K.E.A. and O.K.E.; writing—review and editing, O.L.U., W.K.T., and C.F.; visualization, G.J.S. and O.L.U.; supervision, K.E.A., W.K.T., and G.J.S.; project administration, F.A.O., P.R.G., and K.E.A.

Funding: This research received no external funding.

Acknowledgments: The authors are grateful to Measure DHS, ICF International, Rockville, Marylands, USA for providing the data for this study. GloMACH: Members are Kingsley E Agho, Felix A Ogbo, Thierno Diallo, Osita K. Ezeh, Osuagwu L Uchechukwu, Pramesh R Ghimire, Blessing Akombi, Paschal Ogeleka, Tanvir Abir, Abukari I Issaka, Kedir Yimam Ahmed, Rose Victor, Deborah Charwe, Abdon Gregory Rwabilimbo, Daarwin Subramanee, Mehak Mehak, Nilu Nagdev and Mansi Dhami.

Conflicts of Interest: The authors declare no conflict of interest. K.E.A. is a Guest Editor for the Nutrients Special Issue on "Breastfeeding: Short- and Long-Term Benefits to Baby and Mother" and did not have any role in the journal review and decision-making process for this manuscript.

References

1. Ogbo, F.A.; Agho, K.; Ogeleka, P.; Woolfenden, S.; Page, A.; Eastwood, J.; Global Child Health Research Interest Group. Infant feeding practices and diarrhoea in sub-Saharan African countries with high diarrhoea mortality. *PLoS ONE* **2017**, *12*, e0171792. [CrossRef] [PubMed]

2. Ogbo, F.A.; Page, A.; Idoko, J.; Claudio, F.; Agho, K.E. Diarrhoea and suboptimal feeding practices in Nigeria: Evidence from the national household surveys. *Paediatr. Perinat. Epidemiol.* **2016**, *30*, 346–355. [CrossRef] [PubMed]

3. Ogbo, F.A.; Nguyen, H.; Naz, S.; Agho, K.E.; Page, A. The association between infant and young child feeding practices and diarrhoea in Tanzanian children. *Trop. Med. Health* **2018**, *46*, 2. [CrossRef] [PubMed]

4. Victora, C.G.; Bahl, R.; Barros, A.J.D.; França, G.V.A.; Horton, S.; Krasevec, J.; Murch, S.; Sankar, M.J.; Walker, N.; Rollins, N.C.; et al. Breastfeeding in the 21st century: Epidemiology, mechanisms, and lifelong effect. *Lancet* **2016**, *387*, 475–490. [CrossRef]

5. Mannan, H. Early Infant Feeding of Formula or Solid Foods and Risk of Childhood Overweight or Obesity in a Socioeconomically Disadvantaged Region of Australia: A Longitudinal Cohort Analysis. *Int. J. Environ. Res. Public Health* **2018**, *15*, 1685. [CrossRef] [PubMed]

6. WHO. *Indicators for Assessing Infant and Young Child Feeding Practices I*; World Health Organization: Geneva, Switzerland, 2008.

7. Black, R.E.; Victora, C.G.; Walker, S.P.; Bhutta, Z.A.; Christian, P.; de Onis, M.; Ezzati, M.; Grantham-McGregor, S.; Katz, J.; Martorell, R.; et al. Maternal and child undernutrition and overweight in low-income and middle-income countries. *Lancet* **2013**, *382*, 427–451. [CrossRef]

8. Debes, A.K.; Kohli, A.; Walker, N.; Edmond, K.; Mullany, L.C. Time to initiation of breastfeeding and neonatal mortality and morbidity: A systematic review. *BMC Public Health* **2013**, *13*, S19. [CrossRef]
9. Wardlaw, T.; Salama, P.; Brocklehurst, C.; Chopra, M.; Mason, E. Diarrhoea: Why children are still dying and what can be done. *Lancet* **2010**, *375*, 870–872. [CrossRef]
10. Economic Community of West African States (ECOWAS). 2019. Available online: https://www.ecowas.int/ (accessed on 23 October 2019).
11. Lamberti, L.M.; Walker, C.L.F.; Noiman, A.; Victora, C.; Black, R.E. Breastfeeding and the risk for diarrhea morbidity and mortality. *BMC Public Health* **2011**, *11*, S15. [CrossRef]
12. Ogbo, F.A.; Agho, K.E.; Page, A. Determinants of suboptimal breastfeeding practices in Nigeria: Evidence from the 2008 demographic and health survey. *BMC Public Health* **2015**, *15*, 259. [CrossRef]
13. Ogbo, F.A.; Page, A.; Idoko, J.; Claudio, F.; Agho, K.E. Have policy responses in Nigeria resulted in improvements in infant and young child feeding practices in Nigeria? *Int. Breastfeed. J.* **2017**, *12*, 9. [CrossRef] [PubMed]
14. Otoo, G.E.; Lartey, A.A.; Pérez-Escamilla, R. Perceived incentives and barriers to exclusive breastfeeding among Periurban Ghanaian women. *J. Hum. Lact.* **2009**, *25*, 34–41. [CrossRef] [PubMed]
15. Agunbiade, O.M.; Ogunleye, O.V. Constraints to exclusive breastfeeding practice among breastfeeding mothers in Southwest Nigeria: Implications for scaling up. *Int. Breastfeed. J.* **2012**, *7*, 5. [CrossRef] [PubMed]
16. United Nations General Assembly. *The UN Decade of Action on Nutrition: Working together to Implement the Outcomes of the Second International Conference on Nutrition*; United Nations: New York, NY, USA, 2016.
17. World Health Organisation. *Global Nutrition Targets 2025: Policy Brief Series (WHO/NMH/NHD/14.2)*; World Health Organisation: Geneva, Switzerland, 2014.
18. Agho, K.E.; Ogeleka, P.; Ogbo, F.A.; Ezeh, O.K.; Eastwood, J.; Page, A. Trends and predictors of prelacteal feeding practices in Nigeria (2003–2013). *Nutrients* **2016**, *8*, 462. [CrossRef]
19. Dhami, M.V.; Ogbo, F.A.; Osuagwu, U.L.; Ugboma, Z.; Agho, K.E. Stunting and severe stunting among infants in India: The role of delayed introduction of complementary foods and community and household factors. *Glob. Health Action* **2019**, *12*, 1638020. [CrossRef]
20. Ogbo, F.A.; Okoro, A.; Olusanya, B.O.; Olusanya, J.; Ifegwu, I.K.; Awosemo, A.O.; Ogeleka, P.; Page, A. Diarrhoea deaths and disability-adjusted life years attributable to suboptimal breastfeeding practices in Nigeria: Findings from the global burden of disease study 2016. *Int. Breastfeed. J.* **2019**, *14*, 4. [CrossRef]
21. Ogbo, F.A.; Eastwood, J.; Page, A.; Efe-Aluta, O.; Anago-Amanze, C.; Kadiri, E.A.; Ifegwu, I.K.; Woolfenden, S.; Agho, K.E. The impact of sociodemographic and health-service factors on breast-feeding in sub-Saharan African countries with high diarrhoea mortality. *Public Health Nutr.* **2017**, *20*, 3109–3119. [CrossRef]
22. Ogunlesi, T.A. Maternal socio-demographic factors influencing the initiation and exclusivity of breastfeeding in a Nigerian semi-urban setting. *Matern. Child Health J.* **2010**, *14*, 459–465. [CrossRef]
23. Yalçin, S.S.; Berde, A.S.; Yalçin, S. Determinants of Exclusive Breast Feeding in sub-Saharan Africa: A Multilevel Approach. *Paediatr. Perinat. Epidemiol.* **2016**, *30*, 439–449. [CrossRef]
24. Ogbo, F.A.; Page, A.; Agho, K.E.; Claudio, F. Determinants of trends in breast-feeding indicators in Nigeria, 1999–2013. *Public Health Nutr.* **2015**, *18*, 3287–3299. [CrossRef]
25. Maduforo, A.N.; Ubah, N.C.; Obiakor-okeke, P.N. The practice of exclusive breastfeeding by lactating women in Owerri metropolis, Imo State, Nigeria. *Glob. Adv. Res. J. Med. Med. Sci.* **2013**, *2*, 13–19.
26. Haidong, W.; Abajobir, A.A.; Abate, K.H.; Abbafati, C.; Abbas, K.M.; Abd-Allah, F.; Abera, S.F.; Abraha, H.N.; Abu-Raddad, L.J.; Abu-Rmeileh, N.M.E.; et al. Global, regional, and national under-5 mortality, adult mortality, age-specific mortality, and life expectancy, 1970–2016: A systematic analysis for the Global Burden of Disease Study 2016. *Lancet* **2017**, *390*, 1084–1150.
27. Measure DHS. The Demographic and Health Survey Program. 2019. Available online: https://dhsprogram.com/publications/index.cfm (accessed on 26 October 2019).
28. Corsi, D.J.; Neuman, M.; Finlay, J.E.; Subramanian, S.V. Demographic and health surveys: A profile. *Int. J. Epidemiol.* **2012**, *41*, 1602–1613. [CrossRef] [PubMed]
29. WHO; UNICEF, USAID, AED, UCDAVIS, IFPRI. *Indicators for Assessing Infant and Young Child Feeding Practices Part 2*; World Health Organization: Geneva, Switzerland, 2008.
30. Agho, K.E.; Dibley, M.J.; Odiase, J.I.; Ogbonmwan, S.M. Determinants of exclusive breastfeeding in Nigeria. *BMC Pregnancy Childbirth* **2011**, *11*, 2. [CrossRef]

31. Victor, R.; Baines, S.K.; Agho, K.E.; Dibley, M.J. Determinants of breastfeeding indicators among children less than 24 months of age in Tanzania: A secondary analysis of the 2010 Tanzania Demographic and Health Survey. *BMJ Open* **2013**, *3*, e001529. [CrossRef]

32. Aidam, B.A.; Pérez-Escamilla, R.; Lartey, A.; Aidam, J. Factors associated with exclusive breastfeeding in Accra, Ghana. *Eur. J. Clin. Nutr.* **2005**, *59*, 789–796. [CrossRef]

33. Tariku, A.; Alemu, K.; Gizaw, Z.; Muchie, K.F.; Derso, T.; Abebe, S.M.; Yitayal, M.; Fekadu, A.; Ayele, T.A.; Alemayehu, G.A.; et al. Mothers' education and ANC visit improved exclusive breastfeeding in Dabat Health and Demographic Surveillance System Site, northwest Ethiopia. *PLoS ONE* **2017**, *12*, e0179056. [CrossRef]

34. Asfaw, M.M.; Argaw, M.D.; Kefene, Z.K. Factors associated with exclusive breastfeeding practices in Debre Berhan District, Central Ethiopia: A cross sectional community based study. *Int. Breastfeed. J.* **2015**, *10*, 23. [CrossRef]

35. El Shafei, A.M.H.; Labib, J.R. Determinants of exclusive breastfeeding and introduction of complementary foods in rural Egyptian communities. *Glob. J. Health Sci.* **2014**, *6*, 236. [CrossRef]

36. Tan, K.L. Factors associated with exclusive breastfeeding among infants under six months of age in peninsular Malaysia. *Int. Breastfeed. J.* **2011**, *6*, 2. [CrossRef]

37. Batal, M.; Boulghourjian, C.; Abdallah, A.; Afifi, R. Breast-feeding and feeding practices of infants in a developing country: A national survey in Lebanon. *Public Health Nutr.* **2006**, *9*, 313–319. [CrossRef]

38. Hossain, M.; Islam, A.; Kamarul, T.; Hossain, G. Exclusive breastfeeding practice during first six months of an infant's life in Bangladesh: A country based cross-sectional study. *BMC Pediatr.* **2018**, *18*, 93. [CrossRef] [PubMed]

39. Ogbo, F.A. *Infant and Young Child Feeding Practices in Nigeria: Epidemiology and Policy Implications*; Centre for Health Research, School of Medicine, Western Sydney University: Sydney, Australia, 2016.

40. Stumbitz, B.; Kyei, A.; Lewis, S.; Lyon, F. *Maternity Protection and Workers with Family Responsibilities in the Formal and Informal Economy of Ghana Practices, Gaps and Measures for Improvement*; International Labour Organization: Geneva, Switzerland, 2017.

41. International Labour Organization. Senegal—Maternity Protection-2011. Available online: https://www.ilo.org/dyn/travail/travmain.sectionReport1?p_lang=en&p_structure=3&p_year=2011&p_start=1&p_increment=10&p_sc_id=2000&p_countries=SN&p_print=Y (accessed on 26 October 2019).

42. Atabay, E.; Moreno, G.; Nandi, A.; Kranz, G.; Vincent, I.; Assi, Ti.; Winfrey, E.M.V.; Earle, A.; Raub, A.; Heymann, S.J. Facilitating working mothers' ability to breastfeed: Global trends in guaranteeing breastfeeding breaks at work, 1995–2014. *J. Hum. Lact.* **2015**, *31*, 81–88. [CrossRef] [PubMed]

43. Ogbo, F.A.; Dhami, M.V.; Ude, E.M.; Senanayake, P.; Osuagwu, U.L.; Awosemo, A.O.; Ogeleka, P.; Akombi, B.J.; Ezeh, O.K.; Agho, K.E. Enablers and Barriers to the Utilization of Antenatal Care Services in India. *Int. J. Environ. Res. Public Health* **2019**, *16*, 3152. [CrossRef] [PubMed]

44. Agho, K.E.; Ezeh, O.K.; Issaka, A.I.; Enoma, A.I.; Baines, S.; Renzaho, A.M.N. Population attributable risk estimates for factors associated with non-use of postnatal care services among women in Nigeria. *BMJ Open* **2016**, *6*, e010493. [CrossRef]

45. Moyer, C.A.; Mustafa, A. Drivers and deterrents of facility delivery in sub-Saharan Africa: A systematic review. *Reprod. Health* **2013**, *10*, 1. [CrossRef]

46. Ogbo, F.A.; Page, A.; Idoko, J.; Agho, K.E. Population attributable risk of key modifiable risk factors associated with non-exclusive breastfeeding in Nigeria. *BMC Public Health* **2018**, *18*, 247. [CrossRef]

47. Agho, K.E.; Ezeh, O.K.; Ogbo, F.A.; Enoma, A.I.; Raynes-Greenow, C. Factors associated with inadequate receipt of components and use of antenatal care services in Nigeria: A population-based study. *Int. Health* **2018**, *10*, 172–181. [CrossRef]

48. Joshi, C.; Torvaldsen, S.; Hodgson, R.; Hayen, A. Factors associated with the use and quality of antenatal care in Nepal: A population-based study using the demographic and health survey data. *BMC Pregnancy Childbirth* **2014**, *14*, 1. [CrossRef]

49. Neupane, S.; Doku, D.T. Determinants of time of start of prenatal care and number of prenatal care visits during pregnancy among Nepalese women. *J. Community Health* **2012**, *37*, 865–873. [CrossRef]

50. Seid, A.M.; Yesuf, M.E.; Koye, D.N. Prevalence of Exclusive Breastfeeding Practices and associated factors among mothers in Bahir Dar city, Northwest Ethiopia: A community based cross-sectional study. *Int. Breastfeed. J.* **2013**, *8*, 14. [CrossRef] [PubMed]

51. Patel, A.; Bucher, S.; Pusdekar, Y.; Esamai, F.; Krebs, N.F.; Goudar, S.S.; Chomba, E.; Garces, A.; Pasha, O.; Saleem, S.; et al. Rates and determinants of early initiation of breastfeeding and exclusive breast feeding at 42 days postnatal in six low and middle-income countries: A prospective cohort study. *Reprod. Health* **2015**, *12*, S10. [CrossRef] [PubMed]

52. Mekuria, G.; Edris, M. Exclusive breastfeeding and associated factors among mothers in Debre Markos, Northwest Ethiopia: A cross-sectional study. *Int. Breastfeed. J.* **2015**, *10*, 1. [CrossRef] [PubMed]

53. More Girls than Boys Benefit from Breastfeeding, Hopkins Children's Research Shows. Available online: https://www.eurekalert.org/pub_releases/2008-06/jhmi-mgt052908.php (accessed on 15 February 2015).

54. Ogbo, F.A.; Dhami, M.V.; Awosemo, A.O.; Olusanya, B.O.; Olusanya, J.; Osuagwu, U.L.; Ghimire, P.R.; Page, A.; Agho, K.E. Regional prevalence and determinants of exclusive breastfeeding in India. *Int. Breastfeed. J.* **2019**, *14*, 20. [CrossRef] [PubMed]

55. Khanal, V.; da Cruz, J.L.N.B.; Karkee, R.; Lee, H.A. Factors associated with exclusive breastfeeding in Timor-Leste: Findings from demographic and health survey 2009–2010. *J. Nutr.* **2014**, *6*, 1691–1700. [CrossRef] [PubMed]

56. Jones, J.R.; Kogan, M.D.; Singh, G.K.; Dee, D.L.; Grummer-Strawn, L.M. Factors associated with exclusive breastfeeding in the United States. *Pediatrics* **2011**, *128*, 1117–1125. [CrossRef]

57. Mulready-Ward, C.; Sackoff, J. Outcomes and factors associated with breastfeeding for <8 weeks among preterm infants: Findings from 6 states and NYC, 2004–2007. *Matern. Child Health J.* **2013**, *17*, 1648–1657. [CrossRef]

58. Tawiah-Agyemang, C.; Kirkwood, B.R.; Edmond, K.; Bazzano, A.; Hill, Z. Early initiation of breast-feeding in Ghana: Barriers and facilitators. *J. Perinatol.* **2008**, *28*, 46–52. [CrossRef]

59. Ogbo, F.A.; Eastwood, J.; Page, A.; Arora, A.; McKenzie, A.; Jalaludin, B.; Tennant, E.; Miller, E.; Kohlhoff, J.; Noble, J.; et al. Prevalence and determinants of cessation of exclusive breastfeeding in the early postnatal period in Sydney, Australia. *Int. Breastfeed. J.* **2017**, *12*, 16. [CrossRef]

60. Ogbo, F.A.; Ezeh, O.K.; Khanlari, S.; Naz, S.; Senanayake, P.; Ahmed, K.Y.; McKenzie, A.; Ogunsiji, O.; Agho, K.; Page, A.; et al. Determinants of exclusive breastfeeding cessation in the early postnatal period among culturally and linguistically diverse (CALD) Australian mothers. *Nutrients* **2019**, *11*, 1611. [CrossRef]

61. World Health Organisation; United Nation Childen Education Fund. *Baby-Friendly Hospital Initiative Revised, Updated and Expanded for Integrated Care*; WHO Document Production Services: Geneva, Switzerland, 2009.

62. Pérez-Escamilla, R. Evidence based breast-feeding promotion: The Baby-Friendly Hospital Initiative. *J. Nutr.* **2007**, *137*, 484–487. [CrossRef] [PubMed]

63. Abrahams, S.W.; Labbok, M.H. Exploring the impact of the Baby-Friendly Hospital Initiative on trends in exclusive breastfeeding. *Int. Breastfeed. J.* **2009**, *4*, 1. [CrossRef] [PubMed]

64. Khanal, V.; Adhikari, M.; Sauer, K.; Zhao, Y. Factors associated with the introduction of prelacteal feeds in Nepal: Findings from the Nepal Demographic and Health Survey 2011. *Int. Breastfeed. J.* **2013**, *8*, 9. [CrossRef] [PubMed]

Article

A Moderated Mediation Model of Maternal Perinatal Stress, Anxiety, Infant Perceptions and Breastfeeding

Jessica P. Riedstra * and Nicki L. Aubuchon-Endsley

Department of Psychology, Idaho State University, Pocatello, ID 83209, USA; aubunick@isu.edu
* Correspondence: riedjess@isu.edu

Received: 19 November 2019; Accepted: 27 November 2019; Published: 6 December 2019

Abstract: This study examined a moderated mediation model of relations among maternal perinatal stress/anxiety, breastfeeding difficulties (mediator), misperceptions of infant crying (moderator), and maternal breastfeeding duration to understand risk factors for early breastfeeding termination. It was hypothesized that more breastfeeding difficulties would mediate the relation between greater prenatal stress/anxiety and shorter breastfeeding duration, and that perceptions of response to infant crying as spoiling would moderate the relation between more breastfeeding difficulties and reduced breastfeeding duration. Additionally, it was hypothesized that participants who breastfed through 6 months would demonstrate less postnatal stress/anxiety and there would be a positive relation between fewer breastfeeding difficulties and less postnatal stress/anxiety through 6 months. Participants included 94 expectant mothers at 33–37 weeks gestation and 6 months (±2 weeks) postpartum. Greater prenatal anxiety was associated with shorter breastfeeding duration. Results presented are the first to document negative relations between prenatal (as opposed to postnatal) anxiety and breastfeeding duration (as opposed to frequency or other indicators) in a U.S. sample. Future studies should seek to replicate findings in a more diverse sample and compare findings from clinical and non-clinical samples. Studies may also wish to explore the effects of anxiety prevention/intervention on breastfeeding duration.

Keywords: perinatal; maternal stress; maternal anxiety; infant crying; breastfeeding

---

1. Introduction

Despite efforts to improve breastfeeding rates, about half of mothers discontinue breastfeeding by 6 months, and two-thirds by 12 months [1]. Breastfed children and mothers who breastfeed benefit from improved physical and psychological health and maternal perinatal stress and anxiety may impact breastfeeding behaviors [2–11]. About 10% of pregnant women and 13% of women who recently gave birth have a mental disorder (most commonly depression and anxiety) [12]. While perinatal depression is heavily researched, stress and anxiety receive considerably less attention and remain less understood [13–17]. Furthermore, limited research has examined these psychological experiences among medically underserved perinatal women (e.g., those with limited access to physical and mental health services).

Feelings of stress (i.e., reaction to a stressor) and anxiety (i.e., emotion not necessarily tied to a specific event or stressor) can be defined as two separate, though similar, constructs [18]. Literature suggests that women may experience perinatal stress and/or anxiety; therefore, it is important to measure them separately. For this project, subjective measures of stress and anxiety were utilized given the practical utility for perinatal health providers. Regarding anxiety, worldwide estimates suggest 7%–60% of women experience high levels of general anxiety during pregnancy and 5%–33% in the postpartum period. Furthermore, 3%–39% of women experience an anxiety disorder, as do 4%–20% of postpartum women [15]. In addition, 17% of Idaho women experience some form of anxiety 3 months prior to pregnancy [19]. Research regarding the prevalence of psychosocial stress during pregnancy is limited, and suggests that during the antenatal period, 6% of women reported high stress, 78% low to moderate stress, and 16% no stress [20]. In 2010, 70% of women reported a stressful life event within the year prior to their infant's birth [21]. With regard to Idaho, 21% of women described experiencing high prenatal stress (i.e., three or more stressful life events in the 12 months prior to delivery) [19]. Therefore, women in this region may experience more pregnancy stressful life events than the national average, making perinatal samples from these understudied regions important to consider. Many of these women also have poorer access to healthcare, thereby enhancing the potential impact and outcomes of stress-related difficulties during pregnancy [22].

1.1. Breastfeeding Difficulties

Numerous research findings support relations between stress or anxiety and lower breastfeeding rates [23–31]. Breastfeeding difficulties also may play a role (e.g., mediation) in breastfeeding behaviors. Early in the postpartum period, breastfeeding concerns or perceptions, like difficulty latching or low milk supply, are a significant source of maternal stress [32]. The Centers for Disease Control (CDC) Pregnancy Risk Assessment Monitoring System (PRAMS) is perhaps the most comprehensive and widely used assessment of U.S. women's experiences during and 2–6 months after pregnancy, including breastfeeding. It is utilized to examine difficulties initiating and continuing breastfeeding, including: Difficulties for mother, infant, and with time management, among others [33]. Women discontinue breastfeeding due to physical pain and discomfort (e.g., cracked nipples, biting, and scratching); uncertainty of milk supply and latching issues; personal, professional, social, and physical difficulties; beliefs regarding infant preferences; and lack of confidence [9,11,34,35]. Several breastfeeding difficulties may influence breastfeeding initiation, frequency, and duration, however, the relation between stress and anxiety and breastfeeding difficulties remains unexplored.

1.2. Infant Crying

Considering the potential relation between breastfeeding difficulties and behaviors, perceptions of infant crying may serve as a moderating factor. Crying can be disruptive and challenging for families, including experiencing guilt, stress, frustration, social isolation, and strained relationships [36]. Many mothers use breastfeeding as a method to console crying infants, and breastfeeding to comfort is a strong predictor of longer partial breastfeeding duration [37]. However, women who perceive responding to their infant's crying cues with breastfeeding as spoiling may experience a more robust relation between increased breastfeeding difficulties and decreased breastfeeding behaviors [38,39]. The compounded effect of breastfeeding difficulties and a decreased response to infant crying with breastfeeding may serve to decrease breastfeeding duration, such that women may avoid difficulties and the risk of spoiling their infants by removing feeding to console. Therefore, maternal perceptions of spoiling an infant via breastfeeding to console crying behavior were investigated in the present study.

Taken together, little, if any, research has focused on women from medically underserved, U.S. samples wherein breastfeeding resources and support may be limited. Few studies have examined both anxiety and stress, and their impact on specific breastfeeding behaviors during the prenatal and postpartum periods. To our knowledge, no research has evaluated mechanisms by which perinatal stress, anxiety, breastfeeding difficulties, perceptions of responding to infant crying, and breastfeeding behaviors may be related. Additionally, research to date has demonstrated variability in when and how peripartum stress and/or anxiety is measured. Thus, the present study used psychometrically sound measures at two time points to explore whether prenatal (i.e., during third trimester) or postnatal (i.e., up to 6 months postpartum) stress or anxiety is more predictive of breastfeeding behavior. The study also sought to examine the role of breastfeeding difficulties as a potential mediator of relations between prenatal stress and anxiety and breastfeeding duration, and perceptions of infant crying as a moderator between breastfeeding difficulties and duration. Additionally, the study sought to examine univariate relations between postnatal maternal stress, anxiety, and breastfeeding difficulties, relevant covariates, and breastfeeding duration. Understanding risk factors that contribute to decreased breastfeeding behavior may prove beneficial for infants and mothers, which may assist in prevention/intervention research. The following models and hypotheses were used to fill these gaps in the extant literature.

1.3. Proposed Models and Hypotheses

Given the relation between breastfeeding difficulties and duration, it is plausible that breastfeeding difficulties mediate the relation between stress and anxiety and breastfeeding duration [34,35]. Furthermore, a mother's perception of responding to infant crying with breastfeeding as spoiling may interact with breastfeeding difficulties to further negatively impact breastfeeding duration. Therefore, high scores on measures of maternal misperceptions of infant crying may moderate, or strengthen, the mediated relation between anxiety and stress, breastfeeding difficulties, and breastfeeding duration. Given these predicted relations, two moderated mediation models were proposed: Hypothesis 1 consisted of a moderated mediation model with prenatal stress as the predictor, breastfeeding difficulties as the mediator, perceptions of infant crying as spoiling as the moderator along the b path, and breastfeeding duration through 6 months as the outcome variable. Hypothesis 2 proposed the same model, however the predictor was prenatal anxiety. To examine the impact of breastfeeding experiences on postnatal stress and anxiety, two further relations were hypothesized. Specifically, it was hypothesized that participants who breastfed through 6 months postpartum would demonstrate lower levels of postnatal stress (Hypothesis 3) and anxiety (Hypothesis 4) than those who discontinued prior to 6 months. We also hypothesized that among women who breastfeed through 6 months, there would be a positive relation between fewer breastfeeding difficulties and lower levels of postnatal stress (Hypothesis 5) and anxiety (Hypothesis 6). As breastfeeding difficulties appear related to shorter breastfeeding duration, women who continue breastfeeding at 6 months may experience fewer difficulties and report lower levels of stress and anxiety.

2. Methods

2.1. Participants

Data were collected as part of a larger longitudinal research project examining maternal health and infant development. The project was approved by the Idaho State University Human Subjects Committee (protocol 4191, entitled "Infant Development and Healthy Outcomes in Mothers Study"). Data were collected for 125 prenatal participants and 96 participants at 6 months postpartum (See Figure 1).

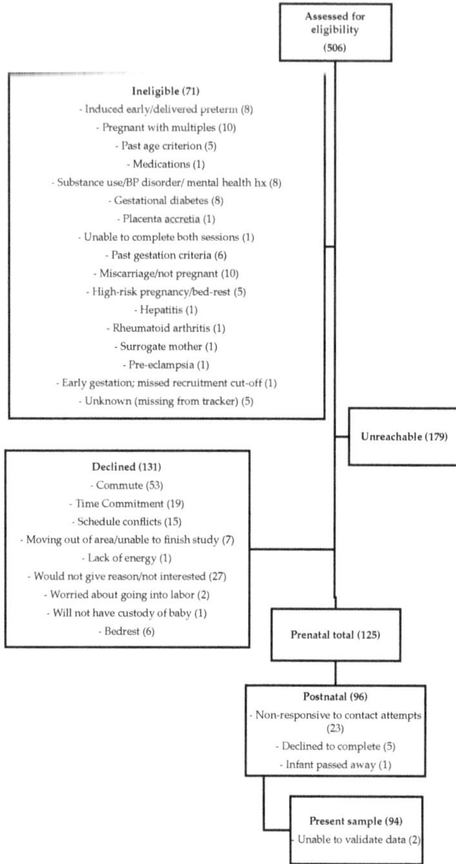

Figure 1. Response rate throughout recruitment and each data collection time point. Bolded numbers in parentheses indicate total sample size at each stage and non-bolded numbers in parentheses indicate the number of participants who withdrew for the reason listed.

2.2. Recruitment

Participants were recruited throughout southeastern Idaho and screened for eligibility via brief phone calls. During the third trimester, prospective participants met individually with a trained research assistant (RA) and completed written informed consent, if eligible. Exclusion criteria included mothers with: More than one baby; certain health conditions (e.g., gestational diabetes, pre-eclampsia, and toxemia) that could impact endocrine functioning; certain behavioral or physical health diagnoses/symptoms (e.g., Schizophrenia, Bipolar Disorder, HIV, and AIDS); or chronic use of recreational substances (e.g., marijuana, or cocaine), medications from FDA categories D and X with documented detrimental fetal effects, or alcohol (>40 drinks) during pregnancy. Mothers had to be at least 18 years of age to provide consent, and 35 years of age or younger given that infertility, chromosomal abnormalities, miscarriage, stillbirth, and multiple pregnancies occur at greater rates after age 35 and all of these factors are related to other exclusion criteria [40]. Participants were recruited from medical centers (e.g., hospitals, clinics), family programs and service centers, childcare facilities, local businesses, schools/university, libraries, recreational centers, the local tribal reservation, community organizations and professionals (e.g., doulas, La Leche League, midwives), and through a variety of mediums (e.g., newsletters, social media).

2.3. Quantitative Analyses

Based on Fritz and MacKinnon's mediation model simulations, a sample size of 71 is required to achieve a power of 0.80 in a mediation model assessed via bias-corrected bootstrapping with medium effect sizes ($d = 0.39$) for a and b paths [41]. Similarly, a priori power analyses utilizing *GPower* suggest that a sample size of 85 is sufficient for the most complex analyses tested in Hypothesis 1 and 2 [41,42]. In *GPower*, a least squares linear multiple regression with three predictors and one outcome variable and up to three covariates was used. This included a medium effect size ($f^2 = 0.15$) for R squared change in steps 1 and 2 after covariates then predictors are added into the model. Analyses also included a two-tailed p-value of 0.05 and power of 0.80. Medium effects size parameters were used based upon prior research findings on maternal affect and breastfeeding behaviors [14]. Two participants' data on the Perinatal Anxiety Screening Scale (PASS) were deemed to be likely inaccurate due to extremely low scores, which indicated that participants did not attend to measure items. Therefore, they were removed from the dataset, resulting in a final sample size of 94. The PROCESS macro v2.16, Model 14 was used to test moderated mediation Hypotheses 1 and 2 [43]. The macro uses bias-corrected bootstrapping to assess statistical significance of direct and indirect effects, which maximizes power and is robust against Type II error and violations of normality. These methods require a smaller sample size to attain sufficient power for the conditional indirect effect with recent simulation studies supporting that sample sizes <100 have power ≥0.80 with a and b paths of medium effect size [44]. The macro quantifies the effect of V on the indirect effect of X on Y through M, with significance determined via bias-corrected confidence intervals [43]. Independent samples t-tests were used to test Hypotheses 3 and 4, and Pearson's product–moment correlations for Hypotheses 5 and 6. Power analyses utilizing *GPower* suggested a sample size of 82 to achieve a medium effect (i.e., 0.3) and a power of 0.80 [42]. Prenatal and postnatal PASS scores were positively skewed and transformed using a natural log function.

2.4. Measures

2.4.1. Stress

The Perceived Stress Scale (PSS) is a 14-item self-report measure developed to assess the degree to which respondents find their lives "unpredictable, uncontrollable, and overloading," or components central to experiencing stress [45] (p. 387). In two college and one smoking cessation sample, the PSS demonstrated Cronbach's $\alpha = 0.84-0.86$, test-retest reliability after 2 days of 0.85, and test-retest reliability after 6 days of 0.55. The College Student Life-Event Schedule and PSS correlated between 0.17–0.49 in all samples [46]. The PSS-10 ($\alpha = 0.74$) and PSS-4 ($\alpha = 0.79$) have demonstrated good reliability in pregnancy samples [47] and a recent, large ($n = 2847$) validation study suggested that the PSS-14 has similar factor structure, internal consistency, and convergent and divergent validity than the PSS-10 in two prenatal samples [48]. The PSS demonstrated good reliability at prenatal and 6-month postnatal time points in the current sample (Cronbach's $a = 0.80$ and 0.78, respectively).

2.4.2. Anxiety

The PASS is a 31-item self-report questionnaire developed to screen for a broad range of anxiety symptoms in perinatal women [49]. The PASS total score is statistically significantly correlated with the Depression Anxiety Stress Scale anxiety (Pearson's product–moment $r = 0.78$, $p < 0.01$) and stress (Pearson's product–moment $r = 0.81$, $p < 0.01$) scales, anxiety scale of the Edinburg Postnatal Depression Scale (Pearson's product–moment $r = 0.74$, $p < 0.01$) and the State–Trait Anxiety Inventory (STAI; State Pearson's product–moment $r = 0.75$, $p < 0.01$; Trait Pearson's product–moment $r = 0.83$, $p < 0.01$)), which support the measure's convergent validity [50,51]. A correlation between PASS scores in subsamples of antenatal and postnatal women ($n = 35$) was 0.74. Internal consistency of the total PASS score in the present sample was high at prenatal ($a = 0.95$) and postnatal ($a = 0.93$) time points.

2.4.3. Breastfeeding difficulties

Breastfeeding difficulties were measured as part of the 6-Month Infant Dietary Questionnaire, a measure created utilizing similar item content from the PRAMS. The 7-item self-report questionnaire measures infant feeding behaviors since birth, including formula feeding, breastfeeding, consuming solid foods, and difficulties breastfeeding. Breastfeeding difficulties were measured with a yes/no question asking whether difficulties have been present, at any time and with any frequency, during breastfeeding. Mothers answering "yes" were asked to specify which difficulties they experienced including maternal-related difficulties (i.e., expressing milk, soreness, or fatigue), difficulties an infant may experience (i.e., latching, sucking, not obtaining enough milk, or not interested in feeding), difficulties with time management (i.e., returning to work and unable to pump), difficulties with insufficient environment (i.e., discomfort feeding in public or around others at home), or others (e.g., open-ended question). Difficulties were quantified as the number endorsed (ranging from 0 to 10 or more). Frequency of difficulties were provided as percentages (See Results).

2.4.4. Infant crying behavior

The Infant Crying Questionnaire-Revised (ICQ-R) is a 43-item self-report measure of maternal perceptions of infant crying [52]. The postnatal version includes 20 items that assess mothers' perceptions of their babies' crying behavior. The ICQ-R consists of five scales, and the present study utilized the Spoiling scale as an index of maternal perceptions of response to infant crying as spoiling. The Spoiling scale has demonstrated internal consistency of Cronbach's $\alpha = 0.70$ [52]. Scores for the three items related to infant spoiling are totaled. The Spoiling scale in the present sample demonstrated Cronbach's $a = 0.76$.

2.4.5. Breastfeeding duration

Breastfeeding duration was measured as part of the 6-Month Infant Dietary Questionnaire [53] via a single open-ended item asking participant how many months, weeks, or days they breastfed. All data were quantified as days.

2.5. Procedures

The prenatal visit included an interview regarding participants' current and past pregnancy-related information, brief health history, and sociodemographic characteristics (socioeconomic status, ethnicity, race, age), and electronic self-report questionnaires including the PSS and PASS. Participants were reimbursed $30 for completing the prenatal session. A mental health resource list was provided if participants endorsed any critical items or reported experiencing distress. RAs scheduled participant's 6-month postnatal session 1 month after their due date and sent reminders to ensure sessions occurred as scheduled. During the 6-month session, mothers completed interviews regarding their and their baby's health, the 6-Month Infant Dietary Questionnaire, and the following electronic self-report measures: PSS, PASS, and ICQ-R postpartum version.

3. Results

3.1. Sample Characteristics

Data were collected from adult mothers ($M_{AGE} = 27.29$ years, $SD_{AGE} = 4.02$ years) during the third trimester ($M_{GESTATION} = 34.25$ weeks, $SD_{GESTATION} = 1.22$ weeks), and 6 months postpartum ($M_{AGE} = 6.05$ months, $SD_{AGE} = 2.13$ months). The largest percentage of participants identified as White/Caucasian (94%), married (84%), and members of the Church of Jesus Christ of Latter-day Saints (LDS; 62%), with annual household incomes between $50,000–$74,999 (29%), having completed a standard college or university degree (i.e., BS/BA; 38%), employed (60%), and having birthed one other child (30%; see Table 1). Additionally, 14% qualified as living in a rural area [22]. Chi-square tests of independence revealed that there were no significant differences in race ($\chi^2(2) = 3.119$, $p = 0.210$), marital status ($\chi^2(1) = 1.407$, $p = 0.236$), employment status ($\chi^2(1) = 3.185$, $p = 0.074$) or education status ($\chi^2(20) = 12.951$, $p = 0.879$) between participants who chose to return for the 6 month session and those who did not. Independent samples *t*-tests also revealed no differences in income ($t(217) = -0.289$, $p = 0.773$), prenatal stress ($t(123) = -0.924$, $p = 0.358$), prenatal anxiety ($t(123) = -1.289$, $p = 0.200$), and parity ($t(123) = 0.187$, $p = 0.187$) between time points. Religious preference data were only collected at 6 months.

Table 1. Sample sociodemographic description.

Race (Categories Not Mutually Exclusive)	*n*	%
White/Caucasian	88	94
Black/African American	2	2
Native Hawaiian or other Pacific Islander	2	2
American Indian/Alaska Native	1	1
Hispanic/Latino	13	14
Asian	1	1
Other	5	5
Relationship Status		
Single/never married	9	10
Married	79	84
Divorced	1	1
Committed relationship	3	3
Engaged	2	2
Employment Status		
Employed	56	60
Not currently working	38	40
Highest Degree of Education		
Partial high school	2	2
High school	13	14
Partial college	33	35
Standard college or university	36	38
Graduate training with a degree	10	11
Income ($)		
<5000	1	1
5000–9000	2	2
10,000–19,000	14	15
20,000–29,000	16	17
30,000–39,000	12	13
40,000–49,000	9	10
50,000–74,999	27	29
75,000–99,999	7	7
≥100,000	6	6
Religious Preference		
Agnostic	3	3
Assembly of God	2	2
Atheist	2	2
Baptist	2	2
Catholic	4	4

Table 1. *Cont.*

Race (Categories Not Mutually Exclusive)	n	%
Lutheran	2	2
Methodist	1	1
Church of Jesus Christ of Latter-day Saints	58	62
Non-denominational	10	11
Pentecostal	1	1
Presbyterian	1	1
Other	12	13
Prefer not to answer	9	10
Parity		
No other child	39	41
One other child	28	30
Two other children	12	13
Three other children	6	6
Four other children	3	3
5 other children	5	5
6 other children	1	1

Note. $n = 94$. Categories for race and religious preference were not mutually exclusive.

3.2. Correlations

Pearson correlations between primary predictor and outcome variables (See Table 2) supported significant relations between stress and anxiety at both time points. None of the potential covariates (number of maternal prenatal, labor/delivery, or postnatal health conditions, infant physical health conditions through 6 months of age, socioeconomic status (SES) measured by the Hollingshead Four Factor Index of Social Status, and parity) significantly correlated with predictor and outcome variables and were not included in moderated mediation models [54]. SES was related to postnatal stress ($r = -0.312$, $p = 0.002$) and anxiety ($r = -0.275$, $p = 0.007$), colic was related to postnatal stress ($r = 0.215$, $p = 0.04$), and postnatal maternal illness was related to postnatal anxiety ($r = 0.257$, $p = 0.05$). Therefore, these variables were controlled for in follow-up analyses for Hypotheses 3 and 4.

Table 2. Correlations among primary study variables.

	PASS (Prenatal)	PSS (Prenatal)	PASS (6 M)	PSS (6 M)	Bf diff.	Spoiling	Duration
PASS (prenatal)	1						
PSS (prenatal)	0.686 **	1					
PASS (6 M)	0.608 **	0.506 **	1				
PSS (6 M)	0.483 **	0.578 **	0.487 **	1			
Bf difficulties	0.036	0.019	0.109	0.027	1		
Spoiling	0.218 *	0.178	0.222 *	0.169	−0.028	1	
Duration	−0.258 *	−0.121	−0.155	−0.130	−0.105	−0.102	1

Note. PASS = Perinatal Anxiety Screening Scale; PSS = Perceived Stress Scale; 6 M = postnatal 6-month time point; Bf = breastfeeding; Spoiling = response to infant crying with breastfeeding as spoiling; Duration = breastfeeding duration. * $p < 0.05$. ** $p < 0.01$.

3.3. Descriptive Statistics

Sixty five percent of participants reported breastfeeding at 6 months postpartum, with 86% having initiated breastfeeding since giving birth. The average PSS score was 19.59 (SD = 6.65) prenatally and 20.23 (SD = 5.96) at the 6-month postnatal visit, out of 56 points. Participants' average PASS score was 16.73 (SD = 11.99) prenatally and 14.46 (SD = 9.45) postnatally out of 93 points. Both PASS average scores fall in the minimal anxiety range. On average, participants ICQ-R Spoiling score was 5.62 (SD = 2.10) out of 15. Participants reported an average of 0.52 (SD = 0.96) difficulties with breastfeeding.

Specifically, 19% endorsed difficulties for themselves, 17% endorsed difficulties for their babies, 11% endorsed difficulties with time management, 2% endorsed an insufficient environment, and 3% noted other difficulties, although their responses were found to be iterations of the previously mentioned answer choices. Participants reported an average of 0.83 (SD = 0.94) prenatal maternal illnesses, 0.64 (SD = 0.80) postnatal maternal illnesses, 1.62 (SD = 0.49) complications during labor or delivery, and 4.83 (SD = 2.09) infant illnesses from birth to 6 months postpartum. The average duration of breastfeeding through 6 months was 138 days (SD = 69.13 days) or 4.4 months.

Hypothesis 1. *The moderated mediation model proposed in Hypothesis 1 was not statistically significant (See Table 3).*

Hypothesis 2. *The moderated mediation model proposed in Hypothesis 2 was also not statistically significant (See Table 3). However, the model with prenatal anxiety as the predictor revealed that prenatal anxiety significantly predicted breastfeeding duration (b = −25.253, t(93) = −2.325, SE = 10.860, p = 0.022).*

Table 3. Prenatal moderated mediation findings.

	F	R^2	S/A → bf diff	Bf Difficulties → Duration	S/A → Duration	Interaction (Bf Difficulties × Spoiling)
			b/SE	*b*/SE	*b*/SE	*b*/SE
Stress	0.747	0.033	0.003/0.015	−10.416/24.829	−1.068/1.103	0.503/4.201
Anxiety	1.889	0.078	0.052/0.149	−6.633/24.299	−25.253 */10.860	0.069/4.109

Note. Bf = breastfeeding; Spoiling = response to infant crying with breastfeeding as spoiling; Duration = breastfeeding duration; SE = standard error; S/A = stress/anxiety. Arrows indicate direction of prediction. * $p < 0.05$.

An independent samples *t*-test was conducted to determine the difference in breastfeeding duration between the 14 participants (approximately 15% of mothers) who were at or above the prenatal PASS cut-off of 26 and those below. All 96 participants were used, as natural log transformations were only necessary for postnatal PASS scores. The two groups were statistically significantly different (t(94) = −3.90, p < 0.001, d = 1.044), with mothers who were above the clinical cut score for prenatal anxiety breastfeeding 72.58 days or 2.5 months less in duration. Of the mothers with clinically elevated prenatal anxiety, only 4 (29%) breastfed through 6 months, while 56 of the 82 (68%) mothers who did not have clinically elevated prenatal anxiety continued breastfeeding at 6 months. Due to small sample sizes, results should be replicated before drawing strong conclusions.

Hypothesis 3. *We conducted an independent samples t-test to compare women who breastfed through 6 months to women who did not on postnatal stress (Hypothesis 3). Results suggests no significant differences in postnatal stress (t(92) = −0.711, p = 0.479). As follow-up analyses, we examined the relation between breastfeeding duration as a continuous variable, and postnatal stress and anxiety, while controlling for covariates. An analysis of covariance (ANCOVA) suggested that the relationship between stress and breastfeeding duration was not statistically significant (F(13) = 0.753, p = 0.706), though SES was significantly related to postnatal stress (F(1) = 7.865, p = 0.006).*

Hypothesis 4. *We conducted an independent samples t-test to compare women who breastfed through 6 months to women who did not on postnatal anxiety (Hypothesis 4). Results suggests no significant differences in postnatal anxiety (t(92) = −1.369, p = 0.174). An additional ANCOVA suggested that the relationship between greater postnatal anxiety and longer breastfeeding duration was statistically significant (F(13) = 2.385, p = 0.009), even after controlling for postnatal maternal illness and SES.*

Hypothesis 5. *Utilizing a two-tailed Pearson correlation among women who continued to breastfeed through 6 months postpartum, breastfeeding difficulties were not significantly related to postnatal stress (Hypothesis 5, r − −0.156, p − 0.114).*

Hypothesis 6. *An additional two-tailed Pearson correlation also suggested that breastfeeding difficulties were not significantly related to postnatal anxiety among women who continued to breastfeed through 6 months postpartum, (Hypothesis 6, r = −0.137, p = 0.144).*

4. Discussion

The purpose of the present project was to investigate a novel mediator and moderator in the relation between prenatal and postnatal anxiety and stress and breastfeeding duration among a medically underserved sample. While none of the moderated mediation models were significant, results indicated that greater prenatal anxiety was significantly related to shorter breastfeeding duration. This finding replicates previous literature in non-U.S. samples suggesting that high levels of trait anxiety in early and mid-pregnancy were associated with decreased breastfeeding rates [25]. Furthermore, many women who experience anxiety during pregnancy breastfeed for less than 6 months [26]. Researchers speculate that poor psychosocial health during pregnancy may be an indicator of poorer overall health, which may influence breastfeeding behaviors [25,26].

Prior literature suggests that prenatal anxiety predicts shorter exclusive breastfeeding at 1 month postpartum, but only prenatal depression is related to breastfeeding duration [55]. A literature review revealed mixed findings regarding prenatal anxiety and breastfeeding behaviors [56]. While there are no associations between prenatal anxiety and breastfeeding initiation, a relation does exist between high levels of prenatal anxiety and a decrease in breastfeeding intention and exclusivity. The relation between prenatal anxiety and breastfeeding is still poorly understood and necessitates further research [56]. Researchers report possible explanations for this relation, including that trait anxiety may interfere with oxytocin release, which stimulates the milk-ejection reflex [56]. State anxiety may result in heightened levels of cortisol and glucose, potentially decreasing milk volume and resulting in decreased breastfeeding duration. Additionally, mothers who experience prenatal anxiety about parenting may avoid challenging parenting behaviors, such as breastfeeding, leading to decreased maternal-infant interaction and decreased breastfeeding.

Few, if any studies have examined and found that various forms of prenatal anxiety, including state, trait, and perinatal, influence breastfeeding duration. Thus, the present project provides novel and important findings regarding correlates of breastfeeding duration. Future research should investigate whether prenatal state, strait, or perinatal anxiety is most related to breastfeeding duration. While the present study provides further support for the relation between heightened prenatal anxiety and decreased breastfeeding duration, the interactive biopsychosocial mechanisms by which this relation exists remains elusive and should be further explored (e.g., stress hormone systems, immune functioning, maternal-infant bonding, and sociocultural understanding and support for breastfeeding). Our analyses also suggest a positive relation between postnatal anxiety and breastfeeding duration, which further indicates the importance of continuing to examine mechanisms of action (e.g., mediating role of breastfeeding difficulties) for these complex and poorly understood relations. Additionally, the positive relation between increased SES and increased postnatal stress may be due to factors like occupation, which warrants investigation.

Correlation analyses suggest that stress and anxiety are related within and over time. Moreover, prenatal and postnatal anxiety were also significantly related to perceptions of infant spoiling. As previous literature suggests, excessive infant crying can be challenging for mothers, and this may lead to heightened feelings of stress and anxiety [36]. If mothers misperceive their response to infant cries with breastfeeding as spoiling, they may decrease breastfeeding behaviors [38,39]. Subsequently, mothers' stress and anxiety may increase if they notice a decrease in their milk supply due to decreased feeding behavior, or due to infant crying behavior that they are not addressing. Additional studies are needed to further investigate these relations.

Our study variables were grounded in strong theory and largely assessed by psychometrically sound measures, and conditional process analysis allowed for a robust and multivariate examination of prenatal stress and anxiety. However, the present study also included limitations, including the way in which we measured breastfeeding difficulties/duration and perinatal stress, homogenous sample demographics (e.g., White), relatively lower power for Hypotheses 5 and 6, and a correlational design. Possible future directions include creating comprehensive and well-validated measures of breastfeeding behaviors, difficulties, and psychosocial experiences, tracking breastfeeding duration and anxiety at multiple time points and beyond 6 months postpartum, examining other breastfeeding outcomes (e.g., initiation or frequency), studying more heterogeneous samples, exploring additional risk and reliance factors, and examining perceptions of infant crying more broadly. Future research should continue to explore possible explanations for decreased breastfeeding duration among women with psychological distress, as these relations are still not fully understood. Focus in each of these areas will lead to a richer understanding of maternal experiences in the perinatal period, informing future intervention and policy.

Practical Implications

The current study's findings hold important implications. Although our sample was a relatively low-risk, healthy community sample, we still found a significant relation between greater prenatal anxiety and shorter breastfeeding duration. Furthermore, even minimal levels of prenatal anxiety are related to breastfeeding duration, thus pregnant women with low levels of anxiety may benefit from screening and intervention. Future prevention and intervention research should focus on the nature of maternal prenatal anxiety, including women who experience relatively low levels of anxiety as well as clinical samples. Understanding the characteristics of prenatal anxiety may help inform more specific targets for psychosocial prevention, intervention, and support.

Follow-up analysis using the PASS cut-off score of 26, demonstrated that mothers who experienced clinically significant levels of prenatal anxiety on average breastfed for 72.58 days, approximately 2.5 months, less than mothers without clinically significant levels of prenatal anxiety (i.e., 148 days or almost 5 months). The magnitude of this effect was quantified by a large Cohen's *d* effect size as well, indicating a strong statistical effect ($d = 1.044$). A review of the literature suggests that even a 2.5-month difference in breastfeeding duration is clinically meaningful, including numerous implications in maternal and infant physical health (e.g., infant risk of infectious diseases) [57,58]. Of note, only 4 of the 14 mothers who were elevated on the prenatal PASS breastfed through 6 months, while 56 of the 82 mothers who did not have clinically elevated prenatal anxiety continued breastfeeding at 6 months. This suggests that the effects of prenatal anxiety on breastfeeding duration may play a robust and critical role within the first few months, such that this is a time period when breastfeeding difficulties should be assessed in greater detail and breastfeeding resources, support, and interventions may be most efficacious. As the two groups in the present analyses were not proportional (i.e., only 14 mothers met the threshold for clinically significant prenatal anxiety), findings should be replicated with larger sample sizes. Nonetheless, these findings indicate that maternal prenatal anxiety may be clinically significantly related to breastfeeding duration, and subsequently, health outcomes for mother-infant dyads. While the present study examined several possible covariates in relation to these variables, breastfeeding duration and PASS scores were related even after controlling for maternal illness and SES. Future studies should consider examining the role of social support among other factors that may influence broad relations between maternal affect and breastfeeding outcomes.

5. Conclusions

The present project sought to better understand the relations among perinatal stress, anxiety, breastfeeding difficulties, perceptions of responding to infant crying, and breastfeeding duration in an underserved sample. Previous cross-cultural research has demonstrated some relations among these variables, however there is inconsistency in these findings. Furthermore, this is the first study to our knowledge to examine these variables collectively through moderated mediation models. The present project also fills a gap in the literature by exploring the unique experiences of U.S. women, specifically in Idaho. Study results revealed a statistically and clinically significant relation between clinical levels of prenatal anxiety and decreased breastfeeding duration, highlighting that even low levels of prenatal anxiety should not be overlooked in research or clinical settings.

Author Contributions: J.P.R., the first and corresponding author contributed to the present project in conceptualization, data curation, formal analysis, investigation, visualization, and writing. N.L.A.-E. contributed in conceptualization, data curation, formal analysis, funding acquisition, methodology, project administration, resources, supervision, visualization, and writing.

Funding: This research was funded by Idaho State University via internal grants from the Departments of Psychology and Physical/Occupational Therapy, College of Arts & Letters, and Office of Research.

Acknowledgments: We would like to thank Idaho State University for their assistance in funding this project. Additionally, we would like to acknowledge the research assistants in the Perinatal Psychobiology Lab who assisted with data collection, and thank our many participants for their contribution to science.

Conflicts of Interest: The authors declare no conflict of interest. The funders had no role in the design of the study; in the collection, analyses, or interpretation of data; in the writing of the manuscript, or in the decision to publish the results.

References

1. U.S. Department of Health and Human Services. Centers for Disease Control and Prevention, National Center for Chronic Disease Prevention and Health Promotion. Available online: https://www.cdc.gov/breastfeeding/data/reportcard.htm (accessed on 6 June 2017).
2. Cook, C. Oxytocin and prolactin suppress cortisol responses to acute stress in both lactating and non-lactating sheep. *J. Dairy Res.* **1997**, *64*, 327–339. [CrossRef] [PubMed]
3. Freeman, M.E.; Kanyicska, B.; Lerant, A.; Nagy, G. Prolactin: Structure, function, and regulation of secretion. *Physiol Rev.* **2000**, *80*, 1523–1631. [CrossRef] [PubMed]
4. Heinrichs, M.; Neumann, I.; Ehlert, U. Lactation and stress: Protective effects of breast-feeding in humans. *Stress* **2002**, *5*, 195–203. [CrossRef] [PubMed]
5. Mezzacappa, E.S. Breastfeeding and maternal stress response and health. *Nutr. Rev.* **2004**, *62*, 261–268. [CrossRef]
6. Tu, M.T.; Lupien, S.J.; Walker, C. Multiparity reveals the blunting effect of breastfeeding on physiological reactivity to psychological stress. *J. Neuroendocrinol.* **2006**, *18*, 494–503. [CrossRef]
7. Uvnäs Moberg, K.; Prime, D.K. Oxytocin effects in mothers and infants during breastfeeding. *Infant* **2013**, *9*, 201–206.
8. Johnson, K. Maternal-infant bonding: A Review of literature. *Int. J. Childbirth Educ.* **2013**, *28*, 17–22.
9. Phillips, K.F. First-time breastfeeding mothers: Perceptions and lived experiences with breastfeeding. *Int. J. Childbirth Educ* **2011**, *26*, 17–20.
10. Prendergast, E.; James, J. Engaging mothers: Breastfeeding experiences recounted (EMBER). A pilot study. *Breastfeed. Rev.* **2016**, *24*, 11–19.
11. Schmied, V.; Barclay, L. Connection and pleasure, disruption and distress: Women's experience of breastfeeding. *J. Hum. Lact.* **1999**, *15*, 325–334. [CrossRef]
12. World Health Organization. Available online: http://www.who.int/mental_health/maternal-child/maternal_mental_health/en/ (accessed on 5 May 2017).
13. Dunkel Schetter, C.; Tanner, L. Anxiety, depression and stress in pregnancy: Implications for mothers, children, research, and practice. *Curr. Opin. Psychiatry* **2012**, *25*, 141–148. [CrossRef]
14. Fallon, V.; Groves, R.; Halford, J.G.; Bennett, K.M.; Harrold, J.A. Postpartum anxiety and infant-feeding outcomes. *J. Hum. Lact.* **2016**, *32*, 740–758. [CrossRef] [PubMed]
15. Leach, L.S.; Poyser, C.; Fairweather-Schmidt, K. Maternal perinatal anxiety: A Review of Prevalence and correlates. *Br. J. Clin. Psychol.* **2017**, *21*, 4–19. [CrossRef]

16. Leight, K.L.; Fitelson, E.M.; Weston, C.A.; Wisner, K.L. Childbirth and mental disorders. *Int. Rev. Psychiatry* **2010**, *22*, 453–471. [CrossRef]

17. O'Hara, M.; Wisner, K. Perinatal mental illness: Definition, description and aetiology. *Best Pract. Res. Clin. Obstet Gynaecol.* **2013**, *28*, 3–12. [CrossRef]

18. American Psychiatric Association. *Diagnostic and Statistical Manual of Mental Disorders*, 5th ed.; American Psychiatric Association: Washington, DC, USA, 2013.

19. Idaho Department of Health and Welfare. Available online: http://healthandwelfare.idaho.gov/Health/VitalRecordsandHealthStatistics/HealthStatistics/PregnancyRiskAssessmentTrackingSystem/tabid/915/Default.aspx (accessed on 5 May 2017).

20. Woods, S.; Melville, J.; Guo, Y.; Fan, M.; Gavin, A.; Woods, S.M.; Gavin, A. Psychosocial stress during pregnancy. *Am. J. Obstet. Gynecol.* **2010**, *202*, 61. [CrossRef]

21. Burns, E.R.; Farr, S.L.; Howards, P.P. Stressful life events experienced by women in the year before their infants' births–United States, 2000–2010. *Morb. Mortality Weekly Report* **2015**, *64*, 247–251.

22. Rural Health Information Hub. Available online: https://www.ruralhealthinfo.org/data-explorer?id=209&state=ID (accessed on 1 September 2019).

23. Britton, J.R. Postpartum anxiety and breastfeeding. *J. Reprod. Med.* **2007**, *52*, 689–695.

24. Doulougeri, K.; Panagopoulou, E.; Montgomery, A. The impact of maternal stress on initiation and establishment of breastfeeding. *J. Nanosci. Nanotechnol.* **2013**, *19*, 162–167. [CrossRef]

25. Insaf, T.Z.; Fortner, R.T.; Pekow, P.; Dole, N.; Markenson, G.; Chasan-Taber, L. Prenatal stress, anxiety and depressive symptoms as predictors of intention to breastfeed among Hispanic women. *J. Women's Health* **2011**, *20*, 1183–1192. [CrossRef]

26. Kehler, H.L.; Chaput, K.H.; Tough, S.C. Risk factors for cessation of breastfeeding prior to six months postpartum among a community sample of women in Calgary, Alberta. *Can J Public Health* **2009**, *100*, 376–380. [CrossRef] [PubMed]

27. Li, J.; Kendall, G.E.; Henderson, S.; Downie, J.; Landsborough, L.; Oddy, W.H. Maternal psychosocial well-being in pregnancy and breastfeeding duration. *Acta Paediatr.* **2008**, *97*, 221–225. [CrossRef] [PubMed]

28. O'Brien, M.; Buikstra, E.; Hegney, D. The influence of psychological factors on breastfeeding duration. *J. Adv. Nurs.* **2008**, *63*, 397–408. [CrossRef]

29. Papinczak, T.A.; Turner, C.T. An analysis of personal and social factors influencing initiation and duration of breastfeeding in a large Queensland maternity hospital. *Breastfeed. Rev.* **2000**, *8*, 25–33. [PubMed]

30. Ystrom, E. Breastfeeding cessation and symptoms of anxiety and depression: A longitudinal cohort study. *BMC Pregnancy Childbirth* **2012**, *12*, 36–41. [CrossRef]

31. Zhu, P.; Hao, J.; Jiang, X.; Huang, K.; Tao, F. New insight into onset of lactation: Mediating the negative effect of multiple perinatal biopsychosocial stress on breastfeeding duration. *Breastfeed. Med.* **2013**, *8*, 151–158. [CrossRef]

32. Henshaw, E.J.; Fried, R.; Siskind, E.; Newhouse, L.; Cooper, M. Breastfeeding self- efficacy, mood, and breastfeeding outcomes among primiparous women. *J. Hum. Lact.* **2015**, *31*, 511. [CrossRef]

33. Centers for Disease Control and Prevention. Available online: https://www.cdc.gov/prAm.s/pdf/questionnaire/phase-7-standard-questions-508.pdf (accessed on 5 May 2017).

34. Purdy, I.B. Social, cultural, and medical factors that influence maternal breastfeeding. *Issues Ment Health Nurs.* **2010**, *31*, 365–367. [CrossRef]

35. Ertem, I.O.; Votto, N.; Leventhal, J.M. The timing and predictors of the early termination of breastfeeding. *Pediatrics* **2001**, *107*, 543. [CrossRef]

36. Long, T.; Johnson, M. Living and coping with excessive infantile crying. *J. Adv. Nurs.* **2001**, *34*, 155–162. [CrossRef]

37. Howard, C.R.; Lanphear, N.; Lanphear, B.P.; Eberly, S.; Lawrence, R.A. Parental responses to infant crying and colic: The effect on breastfeeding duration. *Breastfeed. Med.* **2006**, *1*, 146–155. [CrossRef] [PubMed]

38. Bell, S.M.; Ainsworth, M.S. Infant crying and maternal responsiveness. *Child Dev.* **1972**, *43*, 1171–1190. [CrossRef] [PubMed]

39. Mathews, M.; Leerkes, E.; Lovelady, C.; Labban, J. Psychosocial predictors of primiparous breastfeeding initiation and duration. *J. Hum. Lact.* **2014**, *30*, 480–487. [CrossRef] [PubMed]

40. The American College of Obstetrics and Gynecology. Available online: https://www.acog.org/Patients/FAQs/Having-a-Baby-After-Age-35-How-Aging-Affects-Fertility-and-Pregnancy?IsMobileSet=false (accessed on 2 November 2019)

41. Fritz, M.S.; Mackinnon, D.P. Required sample size to detect the mediated effect. *Psychol. Sci.* **2007**, *18*, 233–239. [CrossRef] [PubMed]

42. Faul, F.; Erdfelder, E.; Lang, A.-G.; Buchner, A. G *Power 3: A flexible statistical power analysis program. for the social, behavioral, and biomedical sciences. *Behav. Res. Methods* **2007**, *39*, 175–191. [CrossRef] [PubMed]

43. Hayes, A.F. *Introduction to Mediation, Moderation, and Conditional Process Analysis: A Regression-Based Approach*, 2nd ed.; The Guilford Press: New York, NY, USA, 2013.

44. Yzerbyt, V.; Muller, D.; Batailler, C.; Judd, C. New recommendations for testing indirect effects in the mediational models: The need to report and test component paths. *J. Personal. Soc. Psychol.* **2018**, *115*, 929–943. [CrossRef] [PubMed]

45. Cohen, S.; Kamarck, T.; Mermelstein, R. A global measure of perceived stress. *J. Health Soc. Behav.* **1983**, *4*, 385. [CrossRef]

46. Levine, M.; Perkins, D.V. Tailor making life events scale. In Proceedings of the 88th Annual Convention of the American Psychological Association, Montreal, QC, Canada, 1–5 September 1980.

47. Lee, E. Review of the psychometric evidence of the perceived stress scale. *Asian Nurs. Res.* **2012**, *6*, 121–127. [CrossRef]

48. Yokokura, A.; Silva, A.; Fernandes, J.; Del-Ben, C.; Figueiredo, F.; Barbieri, M.; Bettiol, H. Perceived Stress Scale: Confirmatory factor analysis of the PSS14 and PSS10 versions in two samples of pregnant women from the BRISA cohort. *Cad. Saude Pub.* **2017**, *33*, e00184615. [CrossRef]

49. Somerville, S.; Dedman, K.; Hagan, R.; Oxnam, E.; Wettinger, M.; Byrne, S.; Page, A. The Perinatal Anxiety Screening Scale: Development and preliminary validation. *Arch. Women's Ment. Health* **2014**, *17*, 443–454. [CrossRef]

50. Cox, J.L.; Holden, J.M.; Sagovsky, R. Detection of postnatal depression: Development of the 10-item Edinburgh Postnatal Depression Scale. *Br. J. Psychiatry* **1987**, *150*, 782–786. [CrossRef] [PubMed]

51. Spielberger, C.D.; Gorsuch, R.L.; Lushene, R.; Vagg, P.R.; Jacobs, G.A. *Manual for the State-Trait. Anxiety Inventory*; Consulting Psychologists Press: Palo Alto, CA, USA, 1983.

52. Haltigan, J.; Leerkes, E.; Burney, R.; O'Brien, M.; Supple, A.; Calkins, S. The Infant Crying Questionnaire: Initial factor structure and validation. *Infant Behav. Dev.* **2012**, *35*, 876–883. [CrossRef] [PubMed]

53. Fein, S.B.; Labiner-Wolfe, J.; Shealy, K.; Li, R.; Chen, J.; Grummer-Strawn, L.M. Infant feeding practices study II: Study methods. *Pediatrics* **2008**, *122* (Suppl. 2), S28–S35. [CrossRef] [PubMed]

54. Hollingshead, A.B. *Four-Factor Index of Social Status*; Yale University: New Haven, CT, USA, 1975.

55. Mehta, U.J.; Siega-Riz, A.M.; Herring, A.H.; Adair, L.S.; Bentley, M.E. Pregravid body mass index, psychological factors during pregnancy and breastfeeding duration: Is there a link? *Matern. Child Nutr.* **2012**, *8*, 423–433. [CrossRef]

56. Fallon, V.; Bennett, K.M.; Harrold, J.A. Prenatal anxiety and infant feeding outcomes. *J. Hum. Lact.* **2016**, *32*, 53. [CrossRef]

57. Dieterich, C.; Felice, J.; O'Sullivan, E.; Rasmussen, K. Breastfeeding and health outcomes for the mother-infant dyad. *Pediatr. Clin. N. Am.* **2013**, *60*, 31–48. [CrossRef]

58. Duijts, L.; Jaddoe, V.V.; Moll, H.A.; Hofman, A. Prolonged and exclusive breastfeeding reduces the risk of infectious diseases in infancy. *Pediatrics* **2010**, *126*, E18–E25. [CrossRef]

 *nutrients* 

Article

Human Milk Omega-3 Fatty Acid Composition Is Associated with Infant Temperament

Jennifer Hahn-Holbrook [1,*], Adi Fish [1] and Laura M. Glynn [2]

[1] Department of Psychology, University of California, Merced, 5200 North Lake Rd,
Merced, CA 95343, Canada; afish@ucmerced.edu

[2] Department of Psychology, Chapman University, Orange, CA 92866, USA; lglynn@chapman.edu

* Correspondence: jhahn-holbrook@ucmerced.edu

Received: 15 October 2019; Accepted: 27 November 2019; Published: 4 December 2019

Abstract: There is growing evidence that omega-3 (n-3) polyunsaturated fatty-acids (PUFAs) are important for the brain development in childhood and are necessary for an optimal health in adults. However, there have been no studies examining how the n-3 PUFA composition of human milk influences infant behavior or temperament. To fill this knowledge gap, 52 breastfeeding mothers provided milk samples at 3 months postpartum and completed the Infant Behavior Questionnaire (IBQ-R), a widely used parent-report measure of infant temperament. Milk was assessed for n-3 PUFAs and omega-6 (n-6) PUFAs using gas-liquid chromatography. The total fat and the ratio of n-6/n-3 fatty acids in milk were also examined. Linear regression models revealed that infants whose mothers' milk was richer in n-3 PUFAs had lower scores on the negative affectivity domain of the IBQ-R, a component of temperament associated with a risk for internalizing disorders later in life. These associations remained statistically significant after considering covariates, including maternal age, marital status, and infant birth weight. The n-6 PUFAs, n-6/n-3 ratio, and total fat of milk were not associated with infant temperament. These results suggest that mothers may have the ability to shape the behavior of their offspring by adjusting the n-3 PUFA composition of their milk.

Keywords: breastfeeding; breast milk; temperament; fatty acids; LC-PUFA; omega-3; omega-6; DHA; AA; children; early life nutrition

1. Introduction

Early life nutrition plays a foundational role in brain development [1–3]. In recent decades, research has shown that exposure to human milk and the variation in its composition contribute meaningfully to children's behavior, cognition, and disease risk [4–6]. The American Academy of Pediatrics recommends that human milk be the sole source of infant nutrition for the first 6 months of life [7], a sensitive period characterized by rapid brain development [2,3]. One key nutritional factor that is present in human milk and that is necessary for an optimal brain development during this period are omega-3 (n-3) polyunsaturated fatty-acids (PUFAs) [8,9].

Human milk contains relatively high levels of n-3 PUFAs, which are essential to visual, motor, and cognitive development [8,9]. Among the 11 n-3 PUFAs, the three most important and prevalent in human milk are alpha-linolenic acid (ALA), eicosapentaenoic acid (EPA), and docosahexaenoic acid (DHA) [5,10]. ALA is the most common n-3 PUFA in human milk [10] (and the adult diet) and can be converted into EPA and DHA [11]. DHA is the most abundant n-3 PUFA in the central nervous system in mammals and forms the structural matrix of grey matter and retinal membranes [9,12]. EPA is used to produce eicosanoids, signaling molecules that play numerous roles, including reducing inflammation in the body and the brain [13,14]. The U.S. Department of Health and Human Services 2015–2020 Dietary Guidelines for Americans recommend that all adults, particularly pregnant and breastfeeding women, consume 8 ounces of seafood per week, providing approximately 250 mg of

EPA and DHA per day [15]. Breast-fed infants have significantly higher levels of n-3 PUFAs in their plasma lipids at three months than do infants given formula lacking n-3 fortification [16]. An autopsy study showed that the rate of accumulation of DHA is approximately 5.0 mg a day in the brains of breast-fed infants versus 2.3 mg a day in infants fed non-DHA fortified formula [17].

Numerous studies have shown that deficiencies in n-3 PUFAs are related to mood and anxiety disorders [18,19]. In adults, n-3 supplementation has benefits for the prevention and treatment of major depression [20], bipolar disorder [21], and anxiety disorders [22]. Much less is known about how early variations in the exposure to n-3 PUFAs impact the mood and behavior of children. However, animal models suggest that early n-3 exposure can have a lasting impact on offspring temperament and behavioral phenotypes. For example, feeding pregnant rats diets deficient in n-3 PUFAs increases anxiety-like behavior in rat pups, and upregulates anxiogenic-related glucocorticoid receptors in the frontal cortex, hypothalamus, and hippocampus [23]. These pre-clinical studies are consistent with correlational studies in human children. For example, in a diverse study of 255 women, eating a diet higher in n-3 relative to n-6 fatty acids during pregnancy buffered Black infants against the detrimental effects of maternal stress on infant regulatory capacities [24]. Plasma DHA levels have also been found to negatively correlate with depressive symptoms in children and adolescents with bipolar disorder [25].

No studies, however, could be found examining whether the n-3 PUFA composition of milk influences the infant mood or anxiety in humans, although several studies have linked cortisol levels in milk to infant temperament [26–28]. For example, one study showed that higher levels of cortisol in milk predicted an enhanced performance on the autonomic stability cluster on the Neonatal Behavioral Assessment Scale in neonates [27]. Studies have also found that cortisol levels in mothers' milk are positively correlated with a negative affectivity and fear reactivity in humans [26,28]. Most of what we know about lactational programing comes from animal models, and findings are generally consistent with the notion that milk is an important early moderator of the infant phenotype [29–31]. For example, in a study of Rhesus Macaques, heavier mothers with more reproductive experience were able to supply more calories to their infant through milk [32]. Moreover, infants whose mothers supplied more calories through milk in the early postnatal period showed higher activity levels and a greater confidence in a stressful setting later in infancy [32]. In sum, research has neglected the question of whether the fatty acid composition of milk influences infant temperament in humans.

To fill this gap, the current study tests whether the n-3 PUFA composition of mothers' milk is associated with the temperament of their infants. To address this question, milk samples were collected from 52 mothers of 3-month old infants and assessed for n-3 PUFA levels using gas-liquid chromatography. Mothers also completed the Rothbart Revised Infant Behavior Questionnaire (IBQ-R) [33], a widely-used parental-report instrument that assesses three broad dimensions of infant temperament (negative affectivity, orienting/regulation, and surgency/extraversion). Given that n-3 PUFAs and higher n-3/n-6 ratios have been found to be protective against mood disorders and anxiety symptoms [20], we predicted that n-3 PUFAs and the n-3/n-6 ratio in milk would be inversely correlated with the negative affectivity dimension of the IBQ-R. n-3 PUFAs have also been linked to enhanced executive functioning in children [9]. Therefore, we predicted that milk n-3 PUFAs levels would be positively correlated with scores on the orienting/regulation temperament dimension. We had no predictions regarding the surgency/extraversion temperament dimension. Exploratory analyses are also presented; they test whether the levels of n-6 PUFAs, total PUFAs or total milk fat concentrations are associated with infant temperament.

2. Methods

2.1. Participants

Fifty-two breastfeeding mothers and their three-month-old infants were enrolled in a large longitudinal study of early development from a medical center in Southern California. To be eligible to participate in the study, participants had to be over the age of 18, English-speaking, with a singleton intrauterine pregnancy. Women were ineligible for this study if they used alcohol, tobacco, illicit drugs, had cervical or uterine abnormalities, used medications that impacted the endocrine function, had a diagnosis of a disease influencing the neuroendocrine function, or had an infant admittance to the Neonatal Intensive Care Unit at birth because of compromised health (e.g., intrauterine growth restriction and respiratory distress syndrome).

After mothers gave informed consent to participate in the study, they were asked to come into the laboratory at 3 months postpartum (M = 3.01 months, SD = 0.25) to provide milk samples and fill out a survey regarding their infants' temperament. This study was approved by the Institutional Review Board at the University of California, Irvine (ethics approval code HS# 2002−2441, first approved: January 31, 2003; renewed most recently: 25 February 2019).

2.2. Determination of Milk Fatty Acid Levels

Mothers were asked to empty the contents of one breast with an electric breast pump into a sterile plastic container (Medela, Inc., McHenry, IL, USA). In an effort to ensure that the mother's milk donation for the study would not adversely impact infant nutrition, providing milk samples was made an optional part of this larger longitudinal study, and mothers were only asked to empty the contents of one breast, leaving milk in the other breast for the infant. Before the collection, the mothers were asked to clean the breast and nipple area with an antibacterial wipe and wait until the area dried (to leave open the possibility to examine the microbial composition of the milk for later research). The milk samples were then immediately aliquoted into polypropylene tubes and stored at 70 °C until they were assessed for fatty acids. The time of day of the milk collection was recorded and modestly correlated with the n-3 fatty acid composition of the milk (Pearson's $r = -0.344$, $p = 0.015$). The time of day of the sample collection was not related to any of the other fatty acids' composite variables (Pearson's r ranged from -0.175 to -0.008; $ps > 0.153$).

The fatty acid composition of the breast milk was analyzed by gas-liquid chromatography in prepared fatty acid methyl esters. The total lipids were extracted from the milk by a method described previously [34], using tridecanoin as the internal standard [35]. The thawed milk samples were shaken vigorously and saponified in 6% ethanolic KOH for 1 h at 37 °C. The mixture was extracted twice with hexane (20 min on rotator), and spun to separate the phases. The hexane layers were then combined and evaporated under a gentle stream of nitrogen to reduce lipid peroxidation. The fatty acid methyl esters were prepared by heating the reconstituted samples in 12% boron trifluoride-methanol for 10 min at 100 °C in tightly sealed tubes. The derivatized samples were separated with a Hewlett-Packard gas chromatograph equipped with an Omegawax 250 capillary column and isothermal oven.

The total of the n-3 PUFAs was computed by summing the total fats from ALA, EPA, DHA, and a rare n-3 PUFA detected in milk, Eicosatrienoic acid (ETE). The total of the n-6 fatty acids was computed by summing the total fats from Linolelaidic acid, Arachidic acid, Linoleic acid, Linolenic acid, Dihomo-gamma-linolenic acid (DGLA), Arachidonic acid (AA), and Eicosadienoic Acid. The n-6/n-3 PUFA ratio was computed by dividing the total fats from n-6 PUFAs by the total fats from n-3 PUFAs. Higher ratios represent milk that has more n-6 fatty acids relative to n-3s. The total PUFA concentration in milk was determined by summing the levels of n-3 and n-6 PUFAs.

2.3. Infant Temperament

To assess the infant temperament, mothers were asked to fill out the Rothbart Revised Infant Behavior Questionnaire (IBQ-R) during their laboratory visit [33]. The IBQ-R includes 191 specific questions addressing concrete behaviors such as, "During a peek-a-boo game, how often did the baby smile?" and "How often during the last week did the baby startle to a sudden or loud noise?" To prevent errors in recall, the scale only asks mothers about recently occurring events, using a 7-point Likert scale (1-never to 7-always). The IBQ-R measures three broad dimensions of temperament: negative affectivity, surgency/extraversion, and orienting/regulation. The negative affectivity dimension is created by averaging the scores across four subscales assessing infant sadness, fear, falling reactivity, and distress to limitations. The orienting/regulation dimension is created by averaging across four subscales of mothers' ratings of infants' cuddliness/affiliation, low intensity pleasure, duration of orienting, and soothability. The surgency/extraversion dimension is a composite averaged from six subscales assessing infant approach, vocal reactivity, high intensity pleasure, smiling and laughter, activity level, and perceptual sensitivity. This widely used parental-report instrument has been shown to be reliable across parental reports [33] and to correlate well with behavioral observations of infants [36]. The IBQ-R has been validated for use for infants aged 3 to 12 months, scores are relatively stable over the first year of life [36], and the scale takes approximately 30 min to complete [33]. As part of the larger longitudinal study, mothers also completed the IBQ-R at 6 months. See Table S1 in the Supplemental Materials for exploratory analyses testing the association between the milk fatty acid composition at 3 months and the infant temperament at 6 months.

2.4. Demographic and Health Information

Various demographic and health measures were tested as potential covariates. Maternal reports of race/ethnicity, age, education level, income, parity, marital status, and breastfeeding exclusivity were collected during a structured interview. A medical record review was conducted to assess the birth outcomes, including the infant sex, gestational age at delivery, and birth weight and length. The early pregnancy BMI (at 15 weeks) and weight gain during pregnancy (weight in pounds at 37 weeks-15 weeks) were collected during laboratory visits as part of the larger longitudinal study. At 3 months, the infant weight was assessed using a digital scale (Midmark, Versailles, OH, USA), and the height was determined while the child lay in a supine position. The child BMI percentiles (BMIP) at birth and 3 months were calculated using an SPSS macro that fits the child's height and weight to standard WHO growth curves and generates a child's BMI z-score standardized for age and sex [37,38]. For ease of interpretation, these z-scores were converted to percentiles.

2.5. Statistical Analysis Strategy

First, preliminary analyses were performed to check that the variables were normally distributed. Outliers that were greater than 3 standard deviations above or below the mean were winsorized to bring them to within 2 standard deviations before the analysis. We planned to log-transform any skewed variables; however, after winsorization, all variables were normally distributed. Second, we sought to identify potential confounds by testing whether demographic or health characteristics were associated with fatty acid concentrations in milk. Linear regression models were run with various demographic (age, income, education, marital status, race/ethnicity and infant sex) and health (maternal BMI, weight gain in pregnancy, gestational age at birth, exclusive breastfeeding status, and infant BMIP at birth and at 3 months) covariates entered simultaneously to predict n-3, n-6, n-6/n-3 ratios, and the total milk fat. Demographic or health variables that were associated with any one of the milk fatty acid composite variables with a *p*-value < 0.10 were included as covariates in the subsequent analysis. For the primary analysis, a multivariate linear regression was used to test whether the n-3, n-6, n-6/n-3 ratios, total PUFAs, or total milk fat predicted the negative affectivity or orienting/regulation dimension of the IBQ-R, adjusting for potential confounds. We also ran an exploratory analysis to see whether the fatty acid composition of milk was associated with the surgency/extraversion factor. If there was a significant association, we then used a follow-up linear regression analysis to test: (i) which specific sub-scales that made up the IBQ-R temperament dimension were significantly related to the fatty acid composite, and (ii) which specific fatty acid type (e.g., DHA vs. ALA) predicted the IBQ-R temperament dimension. Finally, we tested whether infant sex or exclusive breastfeeding moderated any observed significant association between the fatty acid levels and IBQ-R dimension. A moderation analysis was carried out by creating cross products between the potential moderator (infant sex, exclusive breastfeeding) and fatty acid levels, and the cross-products were then included in a linear regression model with the constituent variables and covariates. All analyses were performed in SPSS version 21.0. The findings were considered statistically significant if the *p*-values were under 0.05. The effect sizes (Standardized Betas or β) are also provided for all analyses, regardless of the significance level.

3. Results

3.1. Preliminary and Descriptive Analyses

The demographic and health information of the samples is presented in Table 1. The fatty acids composition of the milk samples are presented in Table 2. Several milk samples were more than three standard deviations higher than the mean in terms of n-3 PUFA levels ($n = 1$), EPA levels ($n = 2$), ALA levels ($n = 1$), and ETE levels ($n = 1$) and were winsorized to within 2 standard deviations of the mean before the statistical analysis.

Table 1. Sample characteristics and their association with the fatty acid composition of the milk.

	Mean/%	Standard Deviation	Range	Omega-3	Omega-6	Omega-6/3 Ratio	Total Fat
				Stand. β	Stand. β	Stand. β	Stand. β
Maternal Characteristics							
Maternal age	29.67	4.86	19.2–39.9	**0.407** [t]	0.285	−0.066	0.254
Education	2.70	1.051	0–4	0.113	−0.014	−0.167	−0.060
Household Income	69,489	33,980	25k–105k	0.008	−0.186	−0.286	−0.181
Pre-Pregnancy BMI	24.17	5.98	16.4–47.4	−0.102	−0.086	−0.045	−0.026
Pregnancy Weight Gain (lbs)	35.60	13.27	9.00–71.00	0.103	0.221	0.168	0.243
Parity (% Primiparous)	48.1%		1–4	0.085	−0.053	−0.123	−0.076
Exclusive Breastfeeding	63.27%			0.133	0.168	0.027	0.260
% Married	75%			−0.223	0.341	**0.593** *	0.223
Race/Ethnicity							
% White	58.3%						
% Latina	18.8%			0.006	0.118	0.117	0.220
% Asian	10.4%			0.057	−0.006	−0.052	0.015
% Multi-Ethnic/Other	12.5%			−0.152	−0.011	0.170	−0.145
Infant Characteristics							
Birth Weight (grams)	3465.49	386.874	2470–4220	0.104	0.405	**0.449** [t]	0.215
Gestational Age at Birth (weeks)	39.6822	1.12	37.1–42.3	−0.093	−0.070	0.090	−0.147
BMIP at Birth	56.92%	0.30	0.34–98%	−0.114	−0.125	−0.175	−0.164
BMIP at 3 mos	46.25%	0.27	0.02–97%	0.164	0.101	−0.034	−0.013
% Female Infants	48%			0.117	0.181	0.035	0.120

Note: Race/ethnicity was dummy coded to create contrasts to compare Hispanic, Asian, and Multi-Ethnic/Other groups to whites. Married was coded as 1 = Married, 0 = Not married; Infant sex was coded as Female = 2, Male = 1. Bolded coefficients indicate those with *p*-values < 0.10 = t and *p*-values < 0.05 = *. BMIP = Body Mass Index Percentile.

See Table 1 for the results of the linear regression analysis testing for associations between sociodemographic and health factors and the milk fatty acid composition. Married mothers had significantly higher n-6/n-3 ratios in their milk than mothers who were not married ($p = 0.010$). Older mothers tended to have higher levels of n-3 PUFAs in their milk ($p = 0.060$). Babies with a higher birth weight also tended to have higher n-6/n-3 ratios in their milk ($p = 0.086$). Infant sex, gestational age at delivery, and infant BMIP at birth or 3 months were not associated with the milk fatty acid composition. Likewise, mothers' pre-pregnancy BMI, weight gain in pregnancy, income, education, exclusive breastfeeding status, parity and race/ethnicity did not predict fatty acid levels in milk. Therefore, only the maternal age, marital status, and infant birth weight were included as covariates in the subsequent linear regression models.

Table 2. Milk fatty acid concentrations.

	IUPAC Name	Mean	Standard Deviation	Range
		(ug/mL)	(ug/mL)	(ug/mL)
Omega-3 Fatty Acids				
Linolenic acid (ALA)	(9Z,12Z,15Z)-octadeca-9,12,15-trienoic acid	13.80	6.15	4.26–42.25
Eicosatrienoic acid (ETE)	(11Z,14Z,17Z)-icosa-11,14,17-trienoic acid	0.74	1.02	0.02–7.00
Eicosapentaenoic acid (EPA)	(5Z,8Z,11Z,14Z,17Z)-icosa-5,8,11,14,17-pentaenoic acid	1.09	1.16	0.06–7.00
Docosahexaenoic acid (DHA)	(4Z,7Z,10Z,13Z,16Z,19Z)-docosa-4,7,10,13,16,19-hexaenoic acid	2.72	1.91	0.77–10.71
Omega-6 Fatty Acids				
Linolelaidic acid	(9E,12E)-octadeca-9,12-dienoic acid	1.16	0.30	1.00–3.08
Arachidic acid	(1^{13}C)icosanoic acid	194.60	72.78	65.12–398.72
Linoleic acid	(9Z,12Z)-octadeca-9,12-dienoic acid	0.54	.06	0.51–0.89
Linolenic acid	(9Z,12Z,15Z)-octadeca-9,12,15-trienoic acid	2.26	1.71	0.14–8.79
Dihomo-gamma-linolenic acid (DGLA)	(8Z,11Z,14Z)-icosa-8,11,14-trienoic acid	5.41	2.21	1.41–10.55
Arachidonic acid (AA)	(5Z,8Z,11Z,14Z)-icosa-5,8,11,14-tetraenoic acid	1.17	0.94	0.12–4.92
Eicosadienoic Acid	(11E,14E)-icosa-11,14-dienoic acid	3.72	1.78	1.04–9.01
Composite Fatty Acid Variables				
Total Omega-3		18.36	7.98	6.66–55.64
Total Omega-6		208.85	77.88	70.11–433.86
Omega-6/Omega-3 Ratio		12.12	4.36	4.86–27.14
Total Milk PUFAs		227.21	83.57	83.58–489.50
Total Milk Fat		942.51	346.76	398–2301

The average infant score on the surgency/extraversion factor was 3.92 (SD = 0.88), the negative affectivity factor was 2.99 (SD = 0.56), and the orienting regulation factor was 5.02 (SD = 0.62). These mean scores are similar to those reported in the original validation study that had a larger sample size [33].

3.2. Primary Analysis

See Table 3 for the results of the multivariate linear regression analyses testing the association between the fatty acid composite variables and the three primary temperament dimensions of the IBQ-R. These analyses revealed that only higher levels of n-3 PUFA levels predicted lower scores on the negative affectivity dimension (see Figure 1). This association remained statistically significant after adjustment for the total milk fat content ($\beta = -0.443$, $p = 0.023$) and when all covariates were removed from the model ($\beta = -0.335$, $p = 0.015$). In addition, controlling for the time of day of the milk collection did not change the significant association between n-3 fatty acids and negative affectivity.

Table 3. The association between the milk fatty acid composition and the infant temperament assessed with multivariate linear regression models.

	Negative Affectivity	Orienting/Regulation	Surgency/Extraversion
	Standardized β (*p*-Value)	Standardized β (*p*-Value)	Standardized β (*p*-Value)
Omega-3	−0.352 (0.020) *	−0.014 (0.927)	−0.108 (0.479)
Omega-6	−0.249 (0.106)	0.131 (0.394)	0.053 (0.727)
Omega-6/3 ratio	0.134 (0.387)	0.159 (0.297)	0.179 (0.236)
Total PUFAs	−0.266 (0.083)	0.127 (0.408)	0.042 (0.784)
Total Milk Fat	−0.124 (0.401)	−0.004 (0.981)	−0.039 (0.791)

Note: All coefficients are statistically adjusted for the maternal age, mother's marital status, and infant birth weight.
* $p < 0.05$.

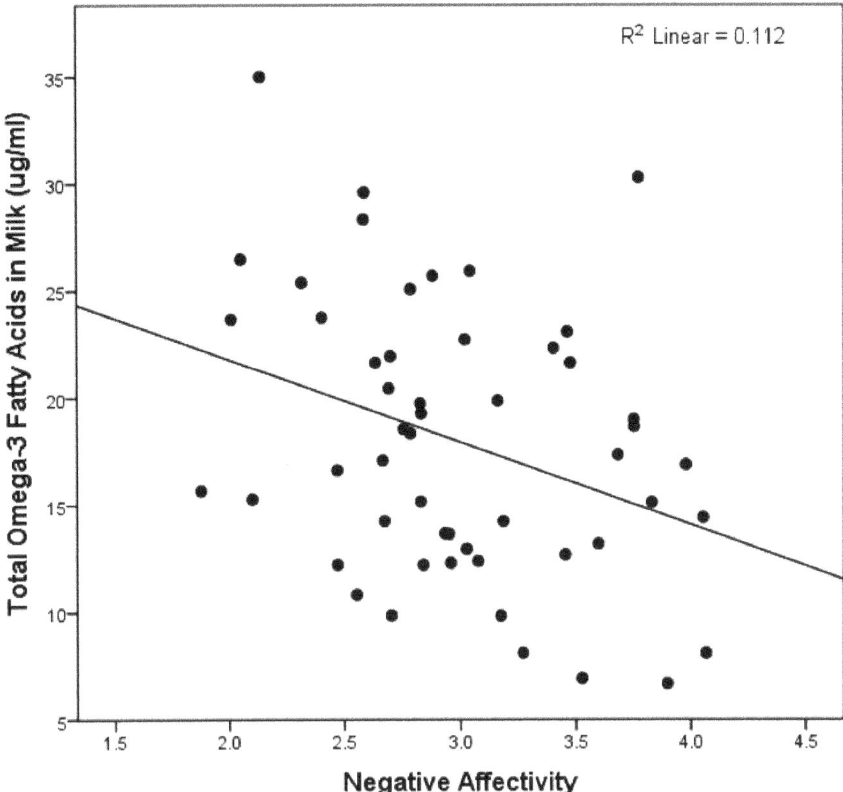

Figure 1. The scatterplot of the total omega-3 fatty acids in the milk and the infant negative affectivity.

Because the milk n-3 PUFA levels predicted a negative affectivity, we performed a follow-up analysis to test how the n-3 PUFAs predicted each of the four subscales that comprised this dimension of the IBQ-R. Higher n-3 PUFAs were associated with less sadness ($\beta = -0.349, p = 0.021$) and less distress to limitations ($\beta = -365, p = 0.017$). Milk n-3s were not significantly related to infant fear ($\beta = -0.187, p = 0.219$) or falling reactivity ($\beta = -0.087, p = 0.584$), although these effects were in the same direction as those observed for sadness and limitations to distress.

In addition, given that n-3 PUFAs predicted negative affectivity, we tested whether specific subtypes of n-3 PUFAs (ALA, EPA, DHA, or ETE) were, independently, related to negative affectivity. Mothers with higher ALA in their milk reported that their infants had significantly less negative affectivity ($\beta = -0.368, p = 0.015$). Although they did not reach the criterion for statistical significance, the DHA, EPA, and ETE levels were all negatively associated with negative affectivity (DHA: $\beta = -0.178, p = 0.229$; EPA: $\beta = -0.251, p = 0.095$; ETE: $\beta = -0.103, p = 0.472$).

3.3. Moderation Analysis

Neither the infant sex, nor the breastfeeding exclusivity, significantly moderated the association between the milk n-3 PUFA levels and negative affectivity (sex: $\beta = -0.170, p = 0.780$; breastfeeding exclusivity: $\beta = 0.975, p = 0.071$).

4. Discussion

The present study found evidence to support a link between the n-3 PUFA composition of human milk and infant temperament. Specifically, infants whose mothers had higher levels of n-3 PUFAs in their milk were reported to display lower levels of negative affectivity at 3 months than did infants whose mothers had lower levels of n-3 PUFAs in their milk. To the extent that negative affectivity in infancy is associated with a higher risk for internalizing disorders in later childhood [39,40], n-3 PUFAs in milk may represent a novel developmental origins pathway through which environmental exposures influence the risk for later mental health disorders. To our knowledge, this is the first study to report an association between the variation in the n-3 PUFAs levels in milk and the temperament in humans. One previous study did report that there was no difference in the temperaments of infants who were randomly assigned to receive formula supplemented with DHA (vs. non-DHA-supplemented formula) [41] (see [42] for similar null results with rhesus macaques). However, formula supplementation studies are not directly comparable to studies of human milk, since n-3 PUFAs are metabolized differently in human milk than they are in formula [43], and since the absorption of fatty acids from formula is highly dependent on the specific type of oil from which the n-3s were harvested [44]. Still, the results of this study should be considered preliminary until they can be replicated in larger studies.

There are several potential physiological pathways through which n-3 PUFAs in milk could influence infant affect regulation and temperament. Most directly, n-3 PUFAs and their derivatives help to regulate neurotransmission, and animal studies suggest that deficiencies increase the serotonin 2 (5-HT$_2$) and decrease dopamine 2 (D$_2$) receptor density in the frontal cortex [45]. The upregulation of 5-HT$_2$ and downregulation of D$_2$ in the brain have been implicated in mood disorders [46]. Less directly, n-3 PUFAs might influence infant negative affectivity through inflammatory pathways [18,19,47]. n-3 PUFAs, and EPA especially, have anti-inflammatory properties and reduce the levels of pro-inflammatory cytokines [18,19,47]. In adults, we know from animal and human experimental studies that pro-inflammatory cytokines can readily cross the blood–brain barrier and induce sickness behavior characterized by depressive-like symptoms such as low mood, social withdrawal, and anxiety [48]. The extent to which n-3s PUFAs regulate inflammatory processes in early development is unknown, but future research could investigate if n-3 PUFAs' levels in milk are related to inflammatory processes in infants and if, by doing so, they indirectly influence negative affectivity.

It was notable that, of the three primary types of n-3 PUFAs in milk, only higher levels of ALA were significantly associated with negative affectivity. The DHA and EPA levels were also inversely correlated with negative affectivity, but not to a statistically significant degree. It would be premature for readers to conclude, however, that only ALA (and not DHA or EPA) in milk influences infant temperament given that all three of the primary n-3 PUFAs were negatively correlated with negative affectivity, with effect sizes in the small-to-medium range. Future studies that combine the methods used here with direct assessments of infants' plasma levels of ALA, DHA, and EPA are necessary to address the question of whether it is n-3 PUFAs in general, or one subtype of n-3 PUFA in particular, that influences infant temperament.

In their strongest form, these results could have several potential policy implications. First, the maternal diet can influence the levels of n-3 PUFAs in milk and, by doing so, may influence infant temperament. For example, breast milk EPA and DHA concentrations are closely linked to maternal dietary EPA and DHA intake [49], and randomized control trials have shown that mothers given n-3 PUFA supplements show corresponding increases in the n-3 PUFA composition of their milk [50,51]. Unfortunately, this study did not include information on maternal diets, and only two mothers in our study reported taking omega-3 PUFA supplements, so we were not able to test this hypothesis. Future studies could examine whether increasing the dietary intake of n-3 PUFAs in breastfeeding women would lead to a reduced negative affectivity in infants. Second, considering that negative affectivity in infancy predicts internalizing disorders (i.e., anxiety/depression) in later childhood [39,40], perhaps supplementing breastfeeding mothers with n-3 PUFAs could help ameliorate the risk for later mental health problems. While this possibility is only hypothetical, maternal diets during pregnancy and lactation are modifiable and present promising targets for the prevention of adverse developmental outcomes in children [52,53].

Although this study had several strengths, including the direct assessment of the fatty acid composition of milk and the carefully characterized cohort, these results should be considered in the context of several limitations. First, we relied on maternal reports of infant temperament, which may be biased. Despite this, the IBQ-R has in previous studies been found to correlate highly with more objective measures of infant temperament (e.g., behavioral observations) [36]. Moreover, the association between the reports of mothers and other caregivers is generally high [33]. Still, future research would benefit from including additional object measures of infant temperament and behavior. Second, this study was correlational and cannot rule out the possibility that a confounding factor may explain the observed association between the n-3 PUFA levels in milk and infant temperament. To address this possibility, we tested whether a number of demographic (e.g., income, education) and health (e.g., maternal BMI, weight gain in pregnancy) factors predicted the fatty acid composition of milk. Only the marital status, maternal age, and infant birth weight predicted the fatty acid composition of milk, and statistically adjusting for these covariates did not negate the association between n-3 PUFA levels and negative affectivity in this study. Regardless, there are numerous bio-active factors in milk, such as cortisol, which have been shown to correlate with total milk fat [54] and negative affectivity in infants [26]. However, given that we found no association between the total fat or n-6 fatty acid content of milk and infant temperament, we believe that the n-3 composition of milk and cortisol levels in milk likely represent distinct lactational programming pathways. Furthermore, the association between n-3 PUFA levels in milk and negative affectivity remains statistically significant after adjusting for the total fat content of the milk, suggesting that it is n-3 PUFAs specifically, and not richer milk generally, that shapes infant temperament. Finally, this study did not include information on the last time mothers had breastfed or on the total milk volume that mothers expressed. We were unable to identify any studies that tested how the time since the last feeding or the volume of milk production impacts the omega-3 fatty acid composition of milk. Future research is needed to determine whether these factors are associated with LC PUFAs and could therefore have influenced these results.

Findings from this study add to the growing body of research demonstrating that the variation in the composition of mothers' milk plays an important role in shaping the development of offspring. Specifically, we found that milk that was richer in n-3 PUFAs was associated with reduced sadness and distress in infants. These new data are consistent with previous research linking n-3 PUFAs to an improved mood and mental health in adults [18–22]. We hope that the exciting research taking place in the field of lactational programming will open up new possibilities for preventing mood disturbances.

Supplementary Materials: The following are available online at http://www.mdpi.com/2072-6643/11/12/2964/s1, Table S1: Exploratory analysis of the association between the fatty-acid concentration of milk at 3-months and negative affectivity at 6-months (*n* = 44).

Author Contributions: Methodology, investigation, project administration, funding acqusition L.M.G., formal analysis, J.H.-H. and A.F., writing—original draft, J.H.-H., writing—review and editing, J.H.-H., A.F., L.M.G.

Funding: This research was supported by grants from the National Institutes of Health (HD-40967 to LMG and NS-41298) and by a grant from the University of California, Irvine (CORCLR to LMG). The breast pumps were donated by Medela.

Acknowledgments: The authors thank the families who participated in this project and the staff at the UCI Women and Children's Health and Well-Being project for their excellent work.

Conflicts of Interest: The authors declare no conflict of interest.

References

1. Prado, E.L.; Dewey, K.G. Nutrition and brain development in early life. *Nutr. Rev.* **2014**, *72*, 267–284. [CrossRef] [PubMed]
2. Keunen, K.; Van Elburg, R.M.; Van Bel, F.; Benders, M.J. Impact of nutrition on brain development and its neuroprotective implications following preterm birth. *Pediatr. Res.* **2015**, *77*, 148. [CrossRef] [PubMed]
3. Cusick, S.E.; Georgieff, M.K. The role of nutrition in brain development: The golden opportunity of the "first 1000 days". *J. Pediatr.* **2016**, *175*, 16–21. [CrossRef] [PubMed]
4. Belfort, M.B.; Anderson, P.J.; Nowak, V.A.; Lee, K.J.; Molesworth, C.; Thompson, D.K.; Doyle, L.W.; Inder, T.E. Breast milk feeding, brain development, and neurocognitive outcomes: A 7-year longitudinal study in infants born at less than 30 weeks' gestation. *J. Pediatr.* **2016**, *177*, 133–139. [CrossRef]
5. Martin, C.; Ling, P.R.; Blackburn, G. Review of infant feeding: Key features of breast milk and infant formula. *Nutrients* **2016**, *8*, 279. [CrossRef]
6. Innis, S.M.; Gilley, J.; Werker, J. Are human milk long-chain polyunsaturated fatty acids related to visual and neural development in breast-fed term infants? *J. Pediatr.* **2001**, *139*, 532–538. [CrossRef]
7. Eidelman, A.I.; Schanler, R.J. Breastfeeding and the use of human milk. *Pediatrics* **2012**, *5*, 323–324.
8. Birch, D.G.; Birch, E.E.; Hoffman, D.R.; Uauy, R.D. Retinal development in very-low-birth-weight infants fed diets differing in omega-3 fatty acids. *Investig. Ophthalmol. Vis. Sci.* **1992**, *33*, 2365–2376.
9. Innis, S.M. Dietary omega 3 fatty acids and the developing brain. *Brain. Res.* **2008**, *1237*, 35–43. [CrossRef]
10. Jensen, R.G. Lipids in human milk. *Lipids* **1999**, *34*, 1243–1271. [CrossRef]
11. Brenna, J.T.; Salem, N., Jr.; Sinclair, A.J.; Cunnane, S.C. α-Linolenic acid supplementation and conversion to n-3 long-chain polyunsaturated fatty acids in humans. *Prostaglandins Leukot. Essent. Fat. Acids* **2009**, *80*, 85–91. [CrossRef] [PubMed]
12. Giusto, N.; Salvador, G.; Castagnet, P.; Pasquare, S.; de Boschero, M.I. Age-associated changes in central nervous system glycerolipid composition and metabolism. *Neurochem. Res.* **2002**, *27*, 1513–1523. [CrossRef] [PubMed]
13. Meydani, S.N. Effect of (n-3) polyunsaturated fatty acidson cytokine production and their biologic function. *Nutrition* **1996**, *12*, S8–S14. [CrossRef]
14. Endres, S.; Ghorbani, R.; Kelley, V.E.; Georgilis, K.; Lonnemann, G.; Van Der Meer, J.W.; Cannon, J.G.; Rogers, T.S.; Klempner, M.S.; Weber, P.C. The effect of dietary supplementation with n—3 polyunsaturated fatty acids on the synthesis of interleukin-1 and tumor necrosis factor by mononuclear cells. *N. Engl. J. Med.* **1989**, *320*, 265–271. [CrossRef] [PubMed]
15. U.S Department of Health and Human Services. *2015–2020 Dietary Guidelines for Americans*, 8th ed.; U.S Department of Health and Human Services: Washington, DC, USA, 2015.
16. Decsi, T.; Kelemen, B.; Minda, H.; Burus, I.; Kohn, G. Effect of type of early infant feeding on fatty acid composition of plasma lipid classes in full-term infants during the second 6 months of life. *J. Pediatr. Gastroenterol. Nutr.* **2000**, *30*, 547–551. [CrossRef] [PubMed]
17. Cunnane, S.C.; Francescutti, V.; Brenna, J.T.; Crawford, M.A. Breast-fed infants achieve a higher rate of brain and whole body docosahexaenoate accumulation than formula-fed infants not consuming dietary docosahexaenoate. *Lipids* **2000**, *35*, 105–111. [CrossRef] [PubMed]
18. Su, K.P.; Matsuoka, Y.; Pae, C.U. Omega-3 polyunsaturated fatty acids in prevention of mood and anxiety disorders. *Clin. Psychopharmacol. Neurosci.* **2015**, *13*, 129. [CrossRef]
19. Su, K.P. Biological Mechanism of Antidepressant Effect of Omega–3 Fatty Acids: How Does Fish Oil Act as a 'Mind-Body Interface'? *Neurosignals* **2009**, *17*, 144–152. [CrossRef]
20. Lin, P.Y.; Huang, S.Y.; Su, K.P. A meta-analytic review of polyunsaturated fatty acid compositions in patients with depression. *Biol. Psychiatry* **2010**, *68*, 140–147. [CrossRef]
21. Stoll, A.L.; Severus, W.E.; Freeman, M.P.; Rueter, S.; Zboyan, H.A.; Diamond, E.; Cress, K.K.; Marangell, L.B. Omega 3 fatty acids in bipolar disorder: A preliminary double-blind, placebo-controlled trial. *Arch. Gen. Psychiatry* **1999**, *56*, 407–412. [CrossRef]
22. Buydens-Branchey, L.; Branchey, M.; Hibbeln, J.R. Associations between increases in plasma n-3 polyunsaturated fatty acids following supplementation and decreases in anger and anxiety in substance abusers. *Prog. Neuro-Psychopharmacol. Biol. Psychiatry* **2008**, *32*, 568–575. [CrossRef] [PubMed]

23. Bhatia, H.S.; Agrawal, R.; Sharma, S.; Huo, Y.-X.; Ying, Z.; Gomez-Pinilla, F. Omega-3 fatty acid deficiency during brain maturation reduces neuronal and behavioral plasticity in adulthood. *PLoS ONE* **2011**, *6*, e28451. [CrossRef] [PubMed]
24. Brunst, K.J.; Enlow, M.B.; Kannan, S.; Carroll, K.N.; Coull, B.A.; Wright, R.J. Effects of prenatal social stress and maternal dietary fatty acid ratio on infant temperament: Does race matter? *Epidemiology* **2014**, *4*, 1000167. [PubMed]
25. Clayton, E.H.; Hanstock, T.L.; Hirneth, S.J.; Kable, C.J.; Garg, M.L.; Hazell, P.L. Long-chain omega-3 polyunsaturated fatty acids in the blood of children and adolescents with juvenile bipolar disorder. *Lipids* **2008**, *43*, 1031–1038. [CrossRef] [PubMed]
26. Grey, K.R.; Davis, E.P.; Sandman, C.A.; Glynn, L.M. Human milk cortisol is associated with infant temperament. *Psychoneuroendocrinology* **2013**, *38*, 1178–1185. [CrossRef]
27. Hart, S.; Boylan, L.M.; Border, B.; Carroll, S.R.; McGunegle, D.; Lampe, R.M. Breast milk levels of cortisol and Secretory Immunoglobulin A (SIgA) differ with maternal mood and infant neuro-behavioral functioning. *Infant Behav. Dev.* **2004**, *27*, 101–106. [CrossRef]
28. Nolvi, S.; Uusitupa, H.M.; Bridgett, D.J.; Pesonen, H.; Aatsinki, A.K.; Kataja, E.L.; Korja, R.; Karlsson, H.; Karlsson, L. Human milk cortisol concentration predicts experimentally induced infant fear reactivity: Moderation by infant sex. *Dev. Sci.* **2018**, *21*, e12625. [CrossRef]
29. Hinde, K. Lactational programming of infant behavioral phenotype. In *Building Babies*; Springer: Berlin/Heidelberg, Germany, 2013; pp. 187–207.
30. Dettmer, A.M.; Murphy, A.M.; Guitarra, D.; Slonecker, E.; Suomi, S.J.; Rosenberg, K.L.; Novak, M.A.; Meyer, J.S.; Hinde, K. Cortisol in neonatal mother's milk predicts later infant social and cognitive functioning in rhesus monkeys. *Child Dev.* **2018**, *89*, 525–538. [CrossRef]
31. Petrullo, L.; Hinde, K.; Lu, A. Steroid hormone concentrations in milk predict sex-specific offspring growth in a nonhuman primate. *Am. J. Hum. Biol.* **2019**, *31*, e23315. [CrossRef]
32. Hinde, K.; Capitanio, J.P. Lactational programming? Mother's milk energy predicts infant behavior and temperament in rhesus macaques (Macaca mulatta). *Am. J. Primatol. Off. J. Am. Soc. Primatol.* **2010**, *72*, 522–529. [CrossRef]
33. Gartstein, M.A.; Rothbart, M.K. Studying infant temperament via the revised infant behavior questionnaire. *Infant Behav. Dev.* **2003**, *26*, 64–86. [CrossRef]
34. Francois, C.A.; Connor, S.L.; Bolewicz, L.C.; Connor, W.E. Supplementing lactating women with flaxseed oil does not increase docosahexaenoic acid in their milk. *Am. J. Clin. Nutr.* **2003**, *77*, 226–233. [CrossRef] [PubMed]
35. Jacobson, J.L.; Jacobson, S.W.; Muckle, G.; Kaplan-Estrin, M.; Ayotte, P.; Dewailly, E. Beneficial effects of a polyunsaturated fatty acid on infant development: Evidence from the Inuit of Arctic Quebec. *J. Pediatr.* **2008**, *152*, 356–364. [CrossRef] [PubMed]
36. Worobey, J.; Blajda, V.M. Temperament ratings at 2 weeks, 2 months, and 1 year: Differential stablity of activity and emotionality. *Dev. Psychol.* **1989**, *25*, 257. [CrossRef]
37. World Health Organization. *WHO Child Growth Standards: Length/Height-for-Age, Weight-for-Age, Weight-for-Length, Weight-for-Height and Body Mass Index-for-Age: Methods and Development*; World Health Organization: Geneva, Switzerland, 2006.
38. Cole, T.J.; Green, P.J. Smoothing reference centile curves: The LMS method and penalized likelihood. *Stat. Med.* **1992**, *11*, 1305–1319. [CrossRef]
39. Gartstein, M.A.; Putnam, S.P.; Rothbart, M.K. Etiology of preschool behavior problems: Contributions of temperament attributes in early childhood. *Infant Ment. Health J.* **2012**, *33*, 197–211. [CrossRef]
40. Rende, R.D. Longitudinal relations between temperament traits and behavioral syndromes in middle childhood. *J. Am. Acad. Child Adolesc. Psychiatry* **1993**, *32*, 287–290. [CrossRef]
41. Auestad, N.; Halter, R.; Hall, R.T.; Blatter, M.; Bogle, M.L.; Burks, W.; Erickson, J.R.; Fitzgerald, K.M.; Dobson, V.; Innis, S.M. Growth and development in term infants fed long-chain polyunsaturated fatty acids: A double-masked, randomized, parallel, prospective, multivariate study. *Pediatrics* **2001**, *108*, 372–381. [CrossRef]
42. Champoux, M.; Hibbeln, J.R.; Shannon, C.; Majchrzak, S.; Suomi, S.J.; Salem Jr, N.; Higley, J.D. Fatty acid formula supplementation and neuromotor development in rhesus monkey neonates. *Pediatr. Res.* **2002**, *51*, 273. [CrossRef]

43. Carnielli, V.P.; Verlato, G.; Pederzini, F.; Luijendijk, I.; Boerlage, A.; Pedrotti, D.; Sauer, P. Intestinal absorption of long-chain polyunsaturated fatty acids in preterm infants fed breast milk or formula. *Am. J. Clin. Nutr.* **1998**, *67*, 97–103. [CrossRef]

44. de Souza, C.O.; Leite, M.E.Q.; Lasekan, J.; Baggs, G.; Pinho, L.S.; Druzian, J.I.; Ribeiro, T.C.M.; Mattos, Â.P.; Menezes-Filho, J.A.; Costa-Ribeiro, H. Milk protein-based formulas containing different oils affect fatty acids balance in term infants: A randomized blinded crossover clinical trial. *Lipids Health Dis.* **2017**, *16*, 78. [CrossRef] [PubMed]

45. Delion, S.; Chalon, S.; Hérault, J.; Guilloteau, D.; Besnard, J.C.; Durand, G. Chronic dietary α-linolenic acid deficiency alters dopaminergic and serotoninergic neurotransmission in rats. *J. Nutr.* **1994**, *124*, 2466–2476. [CrossRef] [PubMed]

46. Meltzer, H. Serotonergic dysfunction in depression. *Br. J. Psychiatry* **1989**, *155*, 25–31. [CrossRef]

47. Stahl, L.A.; Begg, D.P.; Weisinger, R.S.; Sinclair, A.J. The role of omega-3 fatty acids in mood disorders. *Curr. Opin. Investig. Drugs* **2008**, *9*, 57–64.

48. Eisenberger, N.I.; Berkman, E.T.; Inagaki, T.K.; Rameson, L.T.; Mashal, N.M.; Irwin, M.R. Inflammation-induced anhedonia: Endotoxin reduces ventral striatum responses to reward. *Biol. Psychiatry* **2010**, *68*, 748–754. [CrossRef]

49. Francois, C.A.; Connor, S.L.; Wander, R.C.; Connor, W.E. Acute effects of dietary fatty acids on the fatty acids of human milk. *Am. J. Clin. Nutr.* **1998**, *67*, 301–308. [CrossRef]

50. Weseler, A.R.; Dirix, C.E.; Bruins, M.J.; Hornstra, G. Dietary arachidonic acid dose-dependently increases the arachidonic acid concentration in human milk. *J. Nutr.* **2008**, *138*, 2190–2197. [CrossRef]

51. Makrides, M.; Neumann, M.; Gibson, R.A. Effect of maternal docosahexaenoic acid (DHA) supplementation on breast milk composition. *Eur. J. Clin. Nutr.* **1996**, *50*, 352–357.

52. O'Neil, A.; Itsiopoulos, C.; Skouteris, H.; Opie, R.S.; McPhie, S.; Hill, B.; Jacka, F.N. Preventing mental health problems in offspring by targeting dietary intake of pregnant women. *BMC Med.* **2014**, *12*, 208. [CrossRef]

53. Greer, F.R.; Sicherer, S.H.; Burks, A.W. Effects of early nutritional interventions on the development of atopic disease in infants and children: The role of maternal dietary restriction, breastfeeding, timing of introduction of complementary foods, and hydrolyzed formulas. *Pediatrics* **2008**, *121*, 183–191. [CrossRef]

54. Hinde, K.; Skibiel, A.L.; Foster, A.B.; Del Rosso, L.; Mendoza, S.P.; Capitanio, J.P. Cortisol in mother's milk across lactation reflects maternal life history and predicts infant temperament. *Behav. Ecol.* **2014**, *26*, 269–281. [CrossRef] [PubMed]

 *nutrients*

Article

Validation of the Breastfeeding Score—A Simple Screening Tool to Predict Breastfeeding Duration

Hanne Kronborg * and Michael Væth

Department of Public Health, Faculty of Health, Aarhus University, 8000 Aarhus, Denmark; vaeth@ph.au.dk
* Correspondence: hk@ph.au.dk; Tel.: +45-30233626

Received: 7 October 2019; Accepted: 18 November 2019; Published: 21 November 2019

Abstract: Easy to use screening tools to identify mothers in risk of early breastfeeding cessation are needed. The purpose was to validate a revised version of the breastfeeding score, consisting of four questions addressing completed education, earlier breastfeeding duration, self-efficacy, and sense of security not knowing the exact amount of milk the baby ingests. We used two cohorts from 2004 ($n = 633$) and 2017 ($n = 579$) to explore the predictive validity of the breastfeeding score to identify mothers at risk of breastfeeding cessation within the first 17 weeks postpartum. The analyses included sensitivity and specificity, clinically relevant cut-points, and calibrations plots. A cut-point ≥5 points identified 61% of first-time and 42% of multiparous mothers in the validation cohort 2017 to be at risk of early breastfeeding cessation with a sensitivity and specificity of 80% and 60% for first-time, and 69% and 82% for multiparous, respectively. The corresponding numbers in the 2004 cohort were almost identical. The area under the receiver operating characteristic (ROC) curves were 0.77 and 0.78 and the calibration plots showed good agreement for the two cohorts. The breastfeeding score indicated good ability to discriminate between mothers at risk of early exclusive breastfeeding cessation. The simple form of the tool makes it easy to use in daily practice.

Keywords: breastfeeding; infant; parity; maternal behavior; educational status; self-efficacy; risk factors; prognosis; sensitivity and specificity; practice

---

1. Introduction

Breastfeeding has been shown to be beneficial for the health of both the mother and infant, not only in the postnatal period, but also later in life [1]. However, breastfeeding rates are still below the recommended in western societies [2]. One third of new mothers experience that establishment of breastfeeding is complicated by early breastfeeding problems [3,4]. From ongoing Cochrane reviews, we know that professional support in the early postpartum period helps mothers to successfully establish breastfeeding [5]. The challenge is to identify the mothers at risk of early breastfeeding cessation and thus in need of early support. Usable screening tools have to be simple to use in practice and reliable to discriminate between mothers.

Successful breastfeeding depends on multiple factors [6–8], and a tool to predict breastfeeding duration should therefore include different kinds of indicators. The challenge is still to establish a link from the risk profile of predictors for early cessation of breastfeeding to a practical tool (index) providing valid information for professionals to initiate preventive efforts. Consensus has not yet been reached on which indicators or indexes are the most reliable. Numerous postpartum screening tools developed during the last decades include behavioral and observational factors connected to breastfeeding [9]; others include more self-reported psychosocial factors [10]. So far, the tools have primarily been evaluated for construct and content validity and psychometric properties, and, to a lesser extent, for their predictive validity [10–13]. Moreover, a number of tools are comprehensive and contain a large number of variables, making them less useful in practice [10–13]. The Breastfeeding

Self-Efficacy Scale-Short Form (BSES-SF), including 14 items measuring perceived self-efficacy towards breastfeeding, has been the tool that has shown the best ability to predict breastfeeding success [12,14]. Still, the question remains whether it is possible to produce a reliable self-reported tool with fewer questions to predict early breastfeeding cessation.

In 2007, we introduced a breastfeeding score, developed on the basis of data from a cohort of breastfeeding mothers from 1999 and validated using cohort data from 2004 [15]. This breastfeeding score was based on four simple questions and showed a strong prediction of the breastfeeding cessation in the first four months postpartum. Thus, the tool complied with demands of both simplicity and reliability. The aim of the present study was to validate a revised version of the breastfeeding score using new and expanded data.

2. Materials and Methods

The validation process followed the method suggested by Altman et al. [16–18]. The breastfeeding score was first tested in a temporal validation on a second dataset collected independently of the first dataset which was used in the development process. In the present study, we also validated the revised version of the breastfeeding score on a third dataset of mothers from different geographical areas where we had no exclusion criteria.

2.1. Three Cohorts Included in the Derivation and Validation Process

The breastfeeding score was obtained from the derivation cohort collected in 1999 [19], which included 471 Danish-speaking mothers who had given birth to a single child with a gestational age ≥37 weeks. The temporal validation was performed on a cohort of 723 mothers from 2004 [20], which constituted the comparison group from a randomized trial with the same inclusion criteria and data obtained from the same area as the derivation cohort. Detailed description of the cohorts can be found elsewhere [19,20].

The external validation cohort from 2017 (n = 612) consisted of the comparison group in a community-based randomized trial [21,22]. The aim of the trial was to test the effect of the Newborn Behavioral Observation method with the comparison group treated as usual. The health visitor recruited mothers from the beginning of January 2017 until the end of January 2018 at the first home visit one to two weeks after birth. There were no exclusion criteria, except exclusion of mothers who were unable to manage own legal affairs. The validation cohort from 2017 thus included first-time mothers, premature births, twin births, postpartum depressed mothers, and mothers with a different ethnic background than Danish. The study was performed in four Danish municipalities, representing several geographical regions of Denmark [22]. The comparison groups were chosen as validation cohorts to avoid impact from the interventions studied in the randomized trials.

2.2. Data Collection

Data in all three cohorts were collected from self-reported questionnaires. Questions included in the breastfeeding score were identical in wording and were collected approximately two weeks postpartum. The duration of exclusive breastfeeding was defined as only ingesting mother's milk according to indicators for assessing breastfeeding practices from the WHO [23]. Data on breastfeeding duration were collected by the local health visitors in the derivation cohort and self-reported in the validation cohorts. The follow-up periods were 17 and 26 weeks for the 1999 and 2004 cohorts, and nine months for the 2017 cohort. Data were obtained by mail in the first two cohorts, and by a web-based system in the 2017 cohort.

2.3. Ethical Approval

Permission to conduct the cohort studies in 1999 and 2004 was obtained from the Science Ethic Committee for the Counties of Ringkoebing, Ribe and Soenderjylland (ref. 2047-99) and (ref. 2480-03), and the Danish Data Protection Agency (j.nr. 2013-41-1866, 1344) and (j.nr. 2003-41-3306), respectively.

According to the cohort study 2017 the Central Denmark Region Committee on Health Research Ethics found no need for ethical approval (ref.no.172/2016) as there was no biomedicine involved in the project. Written consent was obtained from all participating mothers. Approval was obtained from the Danish Data Protection Agency (ref.no. 62908/2016).

2.4. Risk Factors Included in the Breastfeeding Score

Originally, the breastfeeding score included the following four risk factors: duration of schooling, previous breastfeeding experience, self-efficacy concerning breastfeeding, and mother's confidence not knowing the exact amount of milk the baby ingests when breastfeeding [15,19]. These factors represent information within sociodemographic, pre and perinatal as well as psychosocial domains.

In the current validation, we chose to change the sociodemographic variable from years of schooling to level of education. Originally, we used the distinction between having completed primary or secondary school (high school) [19]. Today, more than 70% finish secondary school in Denmark, and the level of further education is thus more informative. Furthermore, we changed the variable name from confidence to sense of security (not knowing the exact amount of milk the baby ingests when breastfeeding). The variable name was changed primarily to achieve a more descriptive translation of the content of the question and secondly to make a clearer distinction from the self-efficacy variable. The wording of the question (in Danish) was not changed. Thus, the current validation included the following four risk factors: Completed education measured at levels: none, short, skilled, theoretical bachelor, master. Earlier breastfeeding duration measured in weeks: none for primipara, 0–5, 6–17, >17 for multipara. Self-efficacy measured on a five-point Likert scale from very certain to very uncertain. Sense of security not knowing the exact amount of milk the baby ingests measured on a five-point Likert scale from very secure to very unsecure. Questions and their categorization appear from Table 1.

2.5. Statistical Methods

Initially, descriptive analyses of the risk factors in the three cohorts were performed, and Chi^2 tests were used to compare the derivation and validation cohorts, as seen in Table 1. Next, we used Cox regression analysis to assess the influence of the four risk factors on exclusive breastfeeding cessation in the first 17 weeks postpartum in the derivation cohort. For comparison, data from the two validation cohorts were analysed in the same way. The hazard ratios (HR) estimated the ratio of the cessation rate of exclusive breastfeeding for a given category of the risk factor relative to the cessation rate for the reference category. Integer scores roughly proportional to the log (HR) estimates in the derivation cohort were subsequently obtained for each category of the four risk factors. Finally, for each mother, the breastfeeding score was computed as the sum of the scores for the categories to which she belonged, providing a possible risk score of 0–12 points, as seen in Table 2. The scoring thus obtained was subsequently validated on the two independently collected dataset to evaluate the robustness of the prediction of the breastfeeding score as suggested by Altman et al. [16,17].

Table 1. Characteristics of mothers in the derivation and validation cohorts with complete information on the four questions included in the breastfeeding score.

Characteristics	Derivation Cohort 1999	Validation Cohort 2004	p Value *	Validation Cohort 2017	p Value *
	n = 391	n = 633		n = 579	
	n (%)	n (%)		n (%)	
Sociodemographic Factors					
Maternal age: How old are you?					
17–23	36 (9)	43 (7)		34 (6)	
24–31	245 (63)	391 (62)	0.32	323 (56)	0.002
>32	109 (28)	196 (31)		222 (38)	
Vocational Education: Which vocational education have you completed?					
None–short-skilled	256 (65)	373 (59)	0.04	195 (34)	<0.001
Theoretical bachelor–master	135 (35)	260 (41)		384 (66)	
Smoking: Do you smoke now after giving birth?					
Yes	78 (20)	113 (18)	0.40	33 (6)	<0.001
No	313 (80)	520 (82)		546 (94)	
Perinatal factors					
Parity: How many children have you given birth to?					
Primipara	157 (40)	257 (41)	0.91	276 (48)	0.02
Multipara	233 (60)	376 (59)		303 (53)	
Previous breastfeeding experience: How many weeks did you breastfeed your previous child without giving it anything else?					
None, primipara	157 (40)	257 (41)		276 (48)	
0–5 weeks	31 (8)	40 (6)	0.09	70 (12)	<0.001
6–17 weeks	104 (27)	137 (22)		83 (14)	
>17 weeks	99 (25)	199 (31)		150 (26)	
Psychosocial factors					
Self-efficacy with respect to breastfeeding: How certain are you that you can complete four months of exclusive breastfeeding					
Medium to very uncertain (1–3)	131 (33)	153 (24)	<0.001	189 (33)	0.78
Certain to very certain (4–5)	260 (67)	480 (76)		390 (67)	
Sense of security not knowing the exact amount of milk the baby ingests: I feel fine about not knowing exactly how much milk my child gets when being breastfed					
Is not true (1–2)	181 (46)	240 (38)	0.01	246 (42)	0.24
Is fairly–exactly true (3–5)	210 (54)	393 (62)		333 (58)	

Note: p value * from a comparison with derivation cohort.

Table 2. Risk factors associated with breastfeeding cessation in the first 17 weeks after delivery.

Risk Factors	Prognostic Index						
	Derivation Cohort 1999			Validation Cohort 2004		Validation Cohort 2017	
	Log HR	95% CI	Scores	Log HR	95% CI	Log HR	95% CI
Completed Vocational Education							
None–short–skilled	0.32	−0.04–0.69	1	0.51	0.24–0.79	0.24	0.00–0.48
Theoretical bachelor–master	ref.	ref.	0	ref.	ref.	ref.	ref.
Previous breastfeeding experience:							
None, primipara	0.89	0.25–1.53	3	1.02	0.62–1.42	0.49	0.11–0.87
0–5 weeks	1.79	1.05–2.52	6	1.49	0.96–2.03	0.85	0.39–1.31
6–17 weeks	1.14	0.49–1.79	4	0.77	0.33–1.21	0.78	0.34–1.22
>17 weeks	ref.	ref.	0	ref.	ref.	ref.	ref.
Self-efficacy with respect to breastfeeding							
Uncertain	0.82	0.48–1.17	3	0.98	0.70–1.25	1.11	0.85–1.38
Certain	ref.	ref.	0	ref.	ref.	ref.	ref.
Sense of security not knowing the exact amount of milk the baby ingests							
Insecure	0.72	0.35–1.08	2	0.51	0.24–0.77	0.71	0.46–97
Secure	ref.	ref.	0	ref.	ref.	ref.	ref.

Cox regression analysis of data from 391 mothers with complete information in the derivation cohort, and assigned scores included in the breastfeeding score for each category of the four risk factors. For comparison, the corresponding Cox regression analyses of data from 633 and 579 mothers in the validation cohorts.

The two validation cohorts were used to investigate the breastfeeding score's ability to predict maternal risk of early breastfeeding cessation defined as cessation of exclusive breastfeeding before the end of week 17. Subsequently, we used receiver operating characteristic (ROC) curves to explore the diagnostic validity of the breastfeeding score and to identify a suitable cut-point, as seen in Figure 1. Moreover, the validity of the breastfeeding score was evaluated by calibration plots, as seen in Figure 2, and analyses, describing numbers and proportions of mothers who stopped breastfeeding within 17 weeks according to their breastfeeding score, as seen in Table 3.

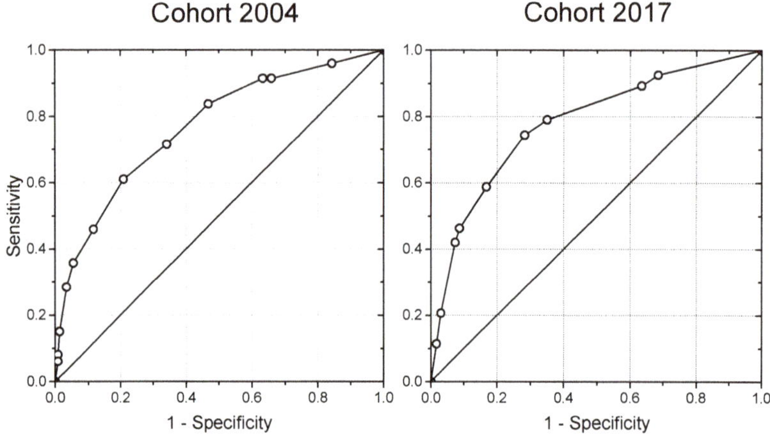

Figure 1. Receiver operating characteristic (ROC) curves for the breastfeeding scores in the validation 2004 (n = 633) and 2017 (n = 579) cohorts.

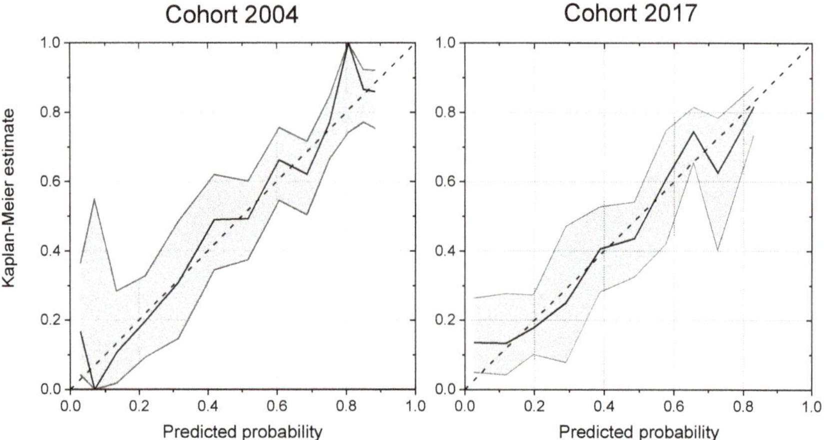

Figure 2. Calibrations plots with 95% confidence intervals for the breastfeeding scores in the validation cohorts 2004 (n = 633) and 2017 (n = 579).

Table 3. Primiparous and multiparous mothers in the validation cohorts divided into risk groups according to their breastfeeding scores. Numbers and proportions of mothers who stopped breastfeeding within 17 weeks in each risk group.

Breastfeeding Score	Validation Cohort 2004			Validation Cohort 2017		
	n (%)	No. Stopped	% (95% CI)	n (%)	No. Stopped	% (95% CI)
Primipara						
0–4 points	103 (40)	32	31 (22–41)	109 (39)	29	27 (19–36)
5–19 points	154 (60)	92	60 (0.52–0.68)	167 (61)	113	68 (60–75)
Multipara						
0–4 points	222 (59)	38	17 (12–23)	177 (58)	43	24 (18–31)
5–19 points	154 (41)	84	55 (46–63)	126 (42)	96	76 (68–83)

3. Results

In the derivation and validation cohorts, we had 391 (83%), 633 (88%), and 579 (95%) mothers, respectively, with complete information on the included risk factors in the breastfeeding score. Among mothers in these cohorts, 194 (41%), 246 (39%), and 281 (49%) had stopped exclusive breastfeeding within 17 weeks after birth. A comparison of the distribution of the risk factors in the three cohorts showed that the educational level changed over time with an increasing proportion of mothers with a long-term education. Moreover, the age of the mothers at birth increased and fewer were smoking. Earlier breastfeeding duration among multipara only showed minor differences, whereas mothers in the validation cohort 2017 had reported similar levels of self-efficacy and sense of security towards breastfeeding as mothers in the derivation cohort 1999; mothers in the validation cohort 2004 reported slightly higher levels, as seen in Table 1.

Table 2 presents results of the Cox regression analysis in the three cohorts with log hazard ratios quantifying risk of breastfeeding cessation for each category relative to the reference category of the four risk factors. Scores assigned to each category of the four risk factors from the derivation cohort are also shown.

The ROC curves, seen in Figure 1, illustrate the association between sensitivity and specificity for different cut-points of the breastfeeding score, when mothers in the validation cohorts had been allocated points depending on their replies concerning the risk factors early after birth. The areas under the ROC curves were 0.77 (CI: 0.73–0.81) and 0.78 (CI: 0.74–0.82), respectively, for the validation cohorts from 2004 and 2017. A detailed calculation for the sensitivity and specificity with associated cut-points showed that a cut-point ≥5 gave the highest sensitivity and specificity. A cut-point ≥5 points predicted that 61% of first-time mothers and 42% of multiparous mothers in the validation cohort 2017 had a high risk of early breastfeeding cessation before 17 weeks with a sensitivity and specificity of 80% and 60% for first-time and 69% and 82% for multiparous mothers, respectively. The corresponding numbers for a cut-point ≥5 in the validation cohort 2004 were almost identical, pointing out that 60% of first-time and 41% of multiparous mothers were at risk; sensitivity and specificity were 74% and 53% for first-time mothers and 69% and 72% for multiparous mothers, respectively.

In both validation cohorts, the calibration plot with confidence intervals, as seen in Figure 2, showed a good agreement between the observed and predicted probability of breastfeeding cessation before the end of week 17 postpartum. From Table 3, it appears that 80% of first-time mothers and 69% of multiparous mothers with a risk score of 5–19 points stopped exclusive breastfeeding before 17 weeks in the validation cohort 2017. The corresponding numbers for mothers with a risk score of 5–19 points in the validation cohort 2004 were 68% and 70%, respectively.

4. Discussion

Validation of the revised version of the breastfeeding score in a general population of Danish mothers confirmed its ability to identify new mothers at risk of early breastfeeding cessation.

The breastfeeding score combines four risk factors to a simple score. The score showed a high ability to discriminate between mothers with different risks and had a good predictive performance.

The updating of the breastfeeding score with a replacement of the sociodemographic risk factor from completed primary or secondary school to completed level of further education included recalibration of the relative weight of the four risk factors included in the breastfeeding score, as suggested by Moons et al. [18]. The use of retrospective data when testing the performance of the breastfeeding score made it possible not only to compare observed events of early breastfeeding cessation with the prediction from the breastfeeding score, but also to assess the ability of the breastfeeding score to discriminate and calibrate between risk groups and thereby assess the clinical usefulness of the score [18]. Self-reported data on the risk factors used to validate the breastfeeding score were collected early after birth independently of collection of data on the outcome of exclusive breastfeeding. Although mothers generally recall their breastfeeding duration quite accurately [24], the collection of data on breastfeeding duration by health professionals must be considered more reliable, which was the case in the derivation cohort. The transferability of the breastfeeding score was supported by the validation of the cohort from 2017, consisting of an unselected study population of new mothers living in different geographical areas of Denmark, with a diverse population of mothers reflecting the Danish background population according to age and level of education [25].

The four included risk factors in the breastfeeding score represented different domains of the multifactorial structure of breastfeeding. Level of completed education contributed with information on maternal sociodemographic background, well-known to influence breastfeeding duration [6]. Previous breastfeeding experience contributed with information on the pre and perinatal period, including first time motherhood and the success of breastfeeding for the multiparous mother expressed by the duration of exclusive breastfeeding of the previous child. Mothers tend to breastfeed for just as long as or a little longer than they did with a previous child [26]. Self-efficacy and sense of security concerning breastfeeding contributed with information on the psychosocial aspects of breastfeeding, which have been shown to be essential predictors of breastfeeding duration [7,19,27]. The general self-efficacy towards breastfeeding expressed the mothers' confidence in overcoming any difficulties and achieving success [28]. Thereby, the predictor expressed maternal ability and strengths to overcome early problems and continue breastfeeding. Sense of security towards breastfeeding gathered information about the insecurity many mothers experience when they start breastfeeding. Concerns may be related to having enough milk [29] and getting used to a responsive feeding style with an infant-led approach [30,31]. Breastfeeding implies to learn how to read the infant's cues on hunger and satiety to have a sense of security because the amount of milk the infant ingests is not visible. The inclusion of predictors from different domains of the multifactorial construct of breastfeeding may explain why the breastfeeding score maintains its predictive ability over time in different populations of new mothers.

The breastfeeding score indicated a high predictive value. The Self-efficacy Scale-Short Form (BSES-SF) has been found to predict the exclusive breastfeeding to 16 weeks in smaller study samples [12,14]. A direct comparison of the predictive values of the two screening tools are not possible, as we have not been able to find a comprehensive validation of the BSES-SF. The cut-point of five revealed a good ability to discriminate between subgroups by pointing to 60% of first-time mothers and 40% of multiparous mothers to be at risk of early breastfeeding cessation. The outcome of exclusive breastfeeding was determined based on mothers who stopped breastfeeding within 17 weeks, because this is the minimum recommended duration of exclusive breastfeeding according to the European Committees on Nutrition ESPGHAN and EFSA [32,33]. The relatively high proportion of mothers at risk of early breastfeeding cessation reflected the high proportion of mothers who generally stop exclusive breastfeeding before four months [1]. Cochrane reviews have repeatedly shown positive effects of supporting mothers in the breastfeeding act [5]. The breastfeeding score may ensure that the intervention is addressed to the insecure mothers who wish to breastfeed by pointing to mothers at risk of early cessation and in need for early support. From a public health perspective, it is especially important to identify the insecure first-time mothers and support them in establishing breastfeeding, considering the association between breastfeeding duration of the first and later infants [27].

From a practical point of view, the breastfeeding score is easy to use with a simple scoring system, which does not require technical assistance. The breastfeeding score aims to identify women with increased need for additional support when establishing breastfeeding. A short questionnaire can be used to obtain the breastfeeding score as it is based on self-reported questions. An example of a questionnaire is enclosed in Table 4. Because the breastfeeding score is self-reported, any health professional with an interest in identifying mothers in need of additional support when establishing breastfeeding can use the score. The time after discharge to home is the optimal time to screen the mother, since at that time she has experiences with breastfeeding and is therefore able to answer the questions more reliably in relation to her experiences.

Table 4. Practical application of the breastfeeding score. The written postpartum questionnaire for mothers.

The purpose of the questions is to give you the breastfeeding guidance that meet your needs *Just enter a number or cross in the box that best suits your situation*		
Basic information Which education have you completed?	❑ ❑	none / short / skilled bachelor / master
Your experiences with breastfeeding This is my first child	❑	yes
If you have given birth earlier, how many weeks did you breastfeed your previous child without giving it anything else than your milk?	❑	weeks
Your expectations to breastfeeding How certain are you that you can complete four months of exclusive breastfeeding?	❑ ❑ ❑ ❑ ❑	very uncertain very certain
How does the following apply to you? I feel fine about not knowing exactly how much milk my baby gets when being breastfed?	❑ ❑ ❑ ❑ ❑	is not true exactly true

The breastfeeding score is an addition to the health professional's own assessment. The health professional often knows the educational level of the mother in advance, but less about the mothers' experiences with and expectations to breastfeeding. When used in a Danish setting, health visitors describe the questions included in the screening tool as suitable to initiate and open a dialogue about how mothers experience breastfeeding and create basis for further guidance on areas of importance. Moreover, the simple scoring system connected to the answers of the questions makes it easy to calculate the mother's score. A practical application with a key for reading the score is enclosed in Table 5. A cut-point of five will identify the first-time mother who expresses a low level of self-efficacy or sense of security towards breastfeeding, and, likewise, the multiparous mother with previous short breastfeeding experience. To prove these benefits, the breastfeeding score needs further process evaluation among health professionals using the breastfeeding score in their daily practice in different settings.

Table 5. Practical application of the breastfeeding score. Key for reading the score.

When the mother has completed the questionnaire, her answers are converted into a total score. The individual questions are awarded points according to the system below.		
Completed vocational education	none / short / skilled	1
	bachelor / master	0
Previous breastfeeding experience	None, first-time mother	3
	0–5 weeks	6
	6–17 weeks	4
	>17 weeks	0
Self-efficacy with respect to breastfeeding	Medium to very uncertain, 1–3	3
	Certain to very certain, 4–5	0
Sense of security not knowing the exact amount of milk the baby ingests	Is not true, 1–2	2
	Is fairly–exactly true, 3–5	0

The points are added up to a total score
Overall score: 0–5 = mothers are offered standard care
Overall score: 6–19 = mothers are offered additional support during establishment of breastfeeding

The validation of the breastfeeding score was carried out in Denmark, a country with a well-educated population and a long tradition for initiating breastfeeding after birth and for breastfeeding for a relatively long period of time [34]. There is no reason to believe the questions included in the breastfeeding score will be understood or perceived differently in other western societies with a culture similar to the Danish; however, additional studies are needed to confirm the prognostic value of using the breastfeeding score in other settings.

5. Conclusions

The breastfeeding score, consisting of four risk factors representing the sociodemographic, perinatal, and psychosocial domains of breastfeeding, showed a high predictive value and a good ability to discriminate between mothers at risk of exclusive breastfeeding cessation before 17 weeks postpartum. Both the temporal and external validation of the breastfeeding score supported the predictive value of the four questions asked shortly after birth. The construction with a simple score connected to the answers of the four questions makes the breastfeeding score easy to use in daily practice.

Author Contributions: H.K. conceived the study idea and wrote the original draft of the article. Both authors were involved in data analysis, writing, reviewing and editing of the article. Both authors have read and approved the final version

Funding: The cohort study from 1999 received no specific grant from any funding agency or commercial sectors. The cohort study only received funding from the non-for-profit public sector: The Public Health Research Foundation established by the counties of Ringkoebing and Ribe in Denmark. The cohort study from 2004 was funded by The Health Insurance Foundation (J.nr. 2003B045 www.helsefonden.dk), and the Lundbeck Foundation (J.nr. FP10 www.lundbeckfonden.dk). The cohort study 2017 was funded by TrygFonden's Centre for Child Research (grant no. 18480 http://childresearch.au.dk) and the following foundations: Det Obelske Familiefond (grant no. 27951 http://obel.com), Lauridsen Fonden (grant no. 32948 www.lauritzenfonden.com), and the Innovation Fund Denmark (grant no. 5155-00001B https://innovationsfonden.dk). All funding in the three studies came from non-profit centers or foundations, and none of the funders had any role in the design, analysis or writing of this article.

Acknowledgments: The authors thank the families and health visitors who participated in the studies and thereby made the present study possible. The authors also wish to thank Ingeborg H. Kristensen, PhD, Department of Public Health at Aarhus University, for valuable and indispensable work in conjunction with the 2017 cohort study.

Conflicts of Interest: The authors declare no conflict of interest.

References

1. Bahl, R.; Barros, A.J.; França, G.V.; Horton, S.; Krasevec, J.; Murch, S.; Sankar, M.J.; Walker, N.; Rollins, N.C. Lancet Breastfeeding Series Group. Breastfeeding in the 21st century: Epidemiology, mechanisms, and lifelong effect. *Lancet* **2016**, *387*, 475–490.
2. Rollins, N.C.; Bhandari, N.; Hajeebhoy, N.; Horton, S.; Lutter, C.K.; Martines, J.C.; Piwoz, E.G.; Richter, L.M.; Victora, C.G. Lancet Breastfeeding Series Group. Why invest, and what will it take to improve breastfeeding practices? *Lancet* **2016**, *387*, 491–504. [CrossRef]
3. Kronborg, H.; Vaeth, M. How effective breastfeeding technique and pacifier use related to breastfeeding problems and breastfeeding duration? *Birth* **2009**, *36*, 34–42. [CrossRef] [PubMed]
4. Wagner, E.A.; Chantry, C.J.; Dewey, K.G.; Nommsen-Rivers, L.A. Breastfeeding concerns at 3 and 7 days postpartum and feeding status at 2 months. *Pediatrics* **2013**, *132*, e865–e875. [CrossRef]
5. McFadden, A.; Gavine, A.; Renfrew, M.J.; Wade, A.; Buchanan, P.; Taylor, J.L.; Veitch, E.; Rennie, A.M.; Crowther, S.A.; Neiman, S.; et al. Support for healthy breastfeeding mothers with healthy term babies. *Cochrane Database Syst. Rev.* **2017**, *28*, CD001141. [CrossRef]
6. Cohen, S.S.; Alexander, D.D.; Krebs, N.F.; Young, B.E.; Cabana, M.D.; Erdmann, P.; Hays, N.P.; Bezold, C.P.; Levin-Sparenberg, E.; Turini, M.; et al. Factors Associated with Breastfeeding Initiation and Continuation: A Meta-Analysis. *J. Pediatr.* **2018**, *203*, 190–196. [CrossRef]
7. De Jager, E.; Broadbent, J.; Fuller-Tyszkiewicz, M.; Skouteris, H. The role of psychosocial factors in exclusive breastfeeding to six months postpartum. *Midwifery* **2014**, *30*, 657–666. [CrossRef]
8. Colombo, L.; Crippa, B.L.; Consonni, D.; Bettinelli, M.E.; Agosti, V.; Mangino, G.; Bezze, E.N.; Mauri, P.A.; Zanotta, L.; Roggero, P.; et al. Breastfeeding Determinants in Healthy Term Newborns. *Nutrients* **2018**, *10*, 48. [CrossRef]
9. Sartorio, B.T.; Coca, K.P.; Marcacine, K.O.; Abuchaim, É.S.V.; Abrão, A.C.F.V. Breastfeeding assessment instruments and their use in clinical practice. *Rev. Gaucha Enferm.* **2017**, *38*, e64675.
10. Casal, C.S.; Lei, A.; Young, S.L.; Tuthill, E.L. A Critical Review of Instruments measuring Breastfeeding Attitudes, Knowledge, and Social Support. *J. Hum. Lact.* **2017**, *33*, 21–47. [CrossRef]
11. Moran, V.H.; Dinwoodie, K.; Bramwell, R.; Dykes, F. A critical analysis of the content of the tools that measure breast-feeding interaction. *Midwifery* **2000**, *16*, 260–268. [CrossRef] [PubMed]
12. Tuthill, E.L.; McGrath, J.M.; Graber, M.; Cusson, R.M.; Young, S.L. Breastfeeding Self-efficacy: A Critical Review of Available Instruments. *J. Hum. Lact.* **2016**, *32*, 35–45. [CrossRef] [PubMed]
13. Ho, Y.J.; McGrath, J.M. A review of the psychometric properties of breastfeeding assessment tools. *J. Obstet. Gynecol. Neonatal Nurs.* **2010**, *39*, 386–400. [CrossRef] [PubMed]
14. Wutke, K.; Dennis, C.L. The reliability and validity of the Polish version of the Breastfeeding Self-Efficacy Scale-ShortForm: Translation and psychometric assessment. *Int. J. Nurs. Stud.* **2007**, *44*, 1439–1446. [CrossRef] [PubMed]
15. Kronborg, H.; Vaeth, M.; Olsen, J.; Iversen, L.; Harder, I. Early breastfeeding cessation: Validation of a prognostic breastfeeding score. *Acta Paediatr.* **2007**, *96*, 688–692. [CrossRef]
16. Altman, D.G.; Royston, P. What do we mean by validating a prognostic model? *Stat. Med.* **2000**, *19*, 453–473. [CrossRef]
17. Altman, D.G.; Vergouwe, Y.; Royston, P.; Moons, K.G. Prognosis and prognostic research: Validating a prognostic model. *BMJ* **2009**, *28*, b605. [CrossRef]
18. Moons, K.G.M.; Altman, D.G.; Vergouwe, Y.; Royston, P. Prognosis and prognostic research: Application and impact of prediction models in clinical practice. *BMJ* **2009**, *338*, b606. [CrossRef]
19. Kronborg, H.; Vaeth, M. The influence of psychosocial factors on the duration of breastfeeding. *Scand. J. Public Health* **2004**, *32*, 210–216. [CrossRef]
20. Kronborg, H.; Vaeth, M.; Olsen, J.; Iversen, L.; Harder, I. Effect of early postnatal breastfeeding support: A cluster-randomized community based trial. *Acta Paediatr.* **2007**, *96*, 1064–1070. [CrossRef]
21. Kristensen, I.H.; Vinter, M.; Nickell, I.K.; Kronborg, H. Health visitors' competences before and after implementing the newborn behavioral observations (NBO) system in a community setting: A cluster randomised study. *Public Health Nurs.* **2019**. [CrossRef] [PubMed]

22. Kristensen, I.H.; Kronborg, H. What are the effects of supporting early parenting by enhancing parents' understanding of the infant? Study protocol for a cluster-randomized community-based trial of the Newborn Behavioral Observation (NBO) method. *BMC Public Health* **2018**, *18*, 832. [CrossRef] [PubMed]

23. World Health Organisation (WHO). *Indicators for Assessing Breastfeeding Practices*; Division of Child Health and Development: Geneva, Switzerland, 1991; WHO/CDD/SER/91.14, Corr. 1; Available online: https://www.who.int/maternal_child_adolescent/documents/cdd_ser_91_14/en/ (accessed on 2 October 2019).

24. Natland, S.T.; Andersen, L.F.; Nilsen, T.I.; Forsmo, S.; Jacobsen, G.W. Maternal recall of breastfeeding duration twenty years after delivery. *BMC Med. Res. Methodol.* **2012**, *12*, 1471–2288. [CrossRef] [PubMed]

25. Statistics Denmark. Live Birth by Municipality, Age of Mothers, and Sex of Child. Available online: https://www.statistikbanken.dk/FODIE (accessed on 29 October 2019).

26. Bai, D.L.; Fong, D.Y.; Tarrant, M. Previous breastfeeding experience and duration of any and exclusive breastfeeding among multiparous mothers. *Birth* **2015**, *42*, 70–77. [CrossRef] [PubMed]

27. Kronborg, H.; Foverskov, E.; Væth, M.; Maimburg, R.D. The role of intention and self-efficacy on the association between breastfeeding of first and second child, a Danish cohort study. *BMC Pregnancy Childbirth* **2018**, *18*, 454. [CrossRef] [PubMed]

28. Bandura, A. Theoretical Perspectives and The Nature and Structure of Self-Efficacy. In *Self-Efficacy the Exercise of Control*; Freeman, W.H. and Company: New York, NY, USA, 1997; pp. 5–42.

29. Li, R.; Fein, S.B.; Chen, J.; Grummer-Strawn, L.M. Why mothers stop breastfeeding: mothers' self-reported reasons for stopping during the first year. *Pediatrics* **2008**, *122*, S69–S76. [CrossRef]

30. Brown, A.; Arnott, B. Breastfeeding duration and early parenting behaviour: The importance of an infant-led, responsive style. *PLoS ONE* **2014**, *9*, e83893. [CrossRef]

31. Little, E.E.; Legare, C.H.; Carver, L.J. Mother infant Physical Contact Predicts Responsive Feeding among U.S. Breastfeeding Mothers. *Nutrients* **2018**, *10*, 1251. [CrossRef]

32. Agostoni CDecsi TFewtrell, M. Complementary feeding: A commentary by the ESPGHAN Committee on Nutrition. *J. Pediatr. Gastroenterol. Nutrients* **2008**, *46*, 99–110. [CrossRef]

33. European Food Safety Authority, EFSA. *Scientific Opinion on the Appropriate Age for Introduction of Complementary Feeding of Infants*; EFSA Panel on Dietetic Products, NDA Nutrition and Allergies: Parma, Italy, 2019; Available online: http://www.efsa.europa.eu/en/efsajournal/doc/1423.pdf (accessed on 2 October 2019).

34. Kronborg, H.; Foverskov, E.; Væth, M. Breastfeeding and introduction of complementary food in Danish infants. *Scand. J. Public Health* **2015**, *43*, 138–145. [CrossRef]

Article

Is There an Association between Breastfeeding and Dental Caries among Three-Year-Old Australian Aboriginal Children?

Dandara G. Haag [1,2], Lisa M. Jamieson [1], Joanne Hedges [1] and Lisa G. Smithers [2,*]

1   Indigenous Oral Health Unit, Australian Research Centre for Population Oral Health,
    The University of Adelaide, Adelaide, SA 5005, Australia; dandara.haag@adelaide.edu.au (D.G.H.);
    lisa.jamieson@adelaide.edu.au (L.M.J.); joanne.hedges@adelaide.edu.au (J.H.)
2   School of Public Health, The University of Adelaide, Adelaide, SA 5005, Australia
*   Correspondence: lisa.smithers@adelaide.edu.au; Tel.: +61-8-83130546

Received: 29 September 2019; Accepted: 14 November 2019; Published: 18 November 2019

Abstract: An unresolved question about breastfeeding is its effect on caries, in particular, early childhood caries (ECC). In secondary analyses of data from an ECC intervention, we describe breastfeeding among Aboriginal children and associations between breastfeeding and ECC. Breastfeeding (duration and exclusivity to six months) was grouped into mutually exclusive categories. ECC was observed by a calibrated dental professional. Outcomes were prevalence of ECC (% decayed, missing, and filled teeth in the primary dentition (% dmft>0)) and caries severity (mean number of decayed, missing, and filled surfaces (mean dmfs)) in children aged three years. Analyses were adjusted for confounding. Multiple imputation was undertaken for missing information. Of 307 participants, 29.3% were never breastfed, 17.9% exclusively breastfed to six months, and 9.3% breastfed >24 months. Breastfeeding >24 months was associated with higher caries prevalence (adjusted prevalence ratio (PR$_a$) 2.06 (95%CI 1.35, 3.13, *p*-value = 0.001) and mean dmfs (5.22 (95% CI 2.06, 8.38, *p*-value = 0.001), compared with children never breastfed. Exclusive breastfeeding to six months with breastfeeding <24 months was associated with 1.45 higher caries prevalence (95% CI −0.92, 2.30, *p*-value = 0.114) and mean dmfs 2.04 (−0.62, 4.71, *p*-value = 0.132), compared with never breastfeeding. The findings are similar to observational studies on breastfeeding and caries but not with randomized controlled trials of breastfeeding interventions. Despite attending to potential biases, inconsistencies with trial evidence raises concerns about the ability to identify causal effects of breastfeeding in observational research.

Keywords: breastfeeding; caries; childhood; Aboriginal

1. Introduction

Breastfeeding has many well-established benefits for both mother and baby, and is recommended by governments, scientific, and health institutions all over the world. The World Health Organization (WHO) recommends that infants are exclusively breastfed to age six months, and that breastfeeding continues to two years of age [1]. Individual countries may vary slightly in their recommendations, for example, the Australian government recommends exclusive breastfeeding to six months and continued breastfeeding to up to and beyond one year [2]. Despite these recommendations, there is little recent evidence on adherence to these guidelines by marginalized communities, such as Aboriginal and Torres Strait Islander peoples. Breastfeeding is culturally supported by Aboriginal and Torres Strait Islanders and is encouraged by local Aboriginal Community Controlled Health Service workers and midwifes.

An unresolved question about breastfeeding is its effect on caries, in particular, early childhood caries (ECC). ECC is defined as the presence of one or more decayed (cavitated or not), missing (due to dental caries), or filled tooth surfaces in any primary tooth among children aged <6 years old [3]. In recent years, there have been a number of systematic reviews reporting the effect of breastfeeding on ECC that have arrived at slightly differing conclusions. One of the first systematic reviews published by Tham et al. [4] included 63 papers published up until 2014. Tham et al. concluded that breastfeeding protected against ECC but breastfeeding beyond 12 months of age increased the risk of ECC. Cui et al. suggested similar findings from a search up to 2015 that included 35 studies but called for more studies to strengthen the evidence [5]. While it seems that breastfeeding in the early postnatal period is protective, the time at which the association reverses and breastfeeding becomes a risk factor for ECC is unclear. Earlier this year, Moynihan et al. published a systematic review that investigated this further [6], where they examined studies that compared the effects of breastfeeding ≥12 months compared to <12 months on ECC, and studies of breastfeeding ≥24 months compared to <24 months. For the 21 studies comparing breastfeeding <12 vs. ≥12 months, a null association was observed. The evidence from the eight studies comparing breastfeeding ≥24 months compared to <24 months varied widely, with some studies reporting strong positive associations with ECC, but studies were too small to generate a precise effect estimate and a null association was inferred.

We previously collected breastfeeding and ECC information from Australian Aboriginal children and their families who participated in a randomized controlled trial of an early childhood caries intervention. In Australia, oral health inequalities between Aboriginal and non-Aboriginal people start to emerge in childhood. According to the Australian National Child Oral Health Study (NCOHS) conducted between 2012 and 2014, the prevalence of untreated dental caries in the primary dentition among Indigenous children aged 5–10 years old was 44% compared to 26% among their non-Indigenous counterparts [7]. This difference represents a 70% increase in the prevalence of untreated dental caries, and it was larger than the inequalities observed according to other important sociodemographic factors, such as household income, parental education, and residential location. Furthermore, not only the prevalence of untreated dental caries is higher among Indigenous children, but they also have more severe disease. Indigenous Australian children aged 5 to 10 years old have, on average, almost three times the mean number of decayed tooth surfaces observed among their non-Indigenous counterparts (3.4 vs. 1.2). This means that Indigenous children are more likely to be exposed to the consequences of dental caries, such as dental pain, treatment under general anesthesia, poorer concentration in a classroom setting, and higher school absenteeism. Therefore, understanding potential drivers of dental caries at an early age in this population is the first step to developing culturally appropriate interventions to manage dental disease in its early stages, and to instituting effective preventive measures to improve oral health in adulthood. Here, we conduct a secondary (observational) analyses of these data with two aims: (1) To describe patterns of breastfeeding among Australian Aboriginal children, and (2) to examine the association between different categories of breastfeeding on three-year-old ECC outcomes.

2. Materials and Methods

2.1. Participants and Study Design

The current study comprises a secondary analysis using data from an early childhood caries intervention that involved 448 women who identified as being pregnant with an Aboriginal and/or Torres Strait Islander child between February 2010 and May 2011 [8].

Recruitment was mainly through the antenatal clinics at South Australian hospitals and through Aboriginal Community Controlled Health Organizations. The study was a 2-arm parallel, outcome assessor-blinded, randomized controlled trial that aimed to assess if an intervention involving dental care to mothers during pregnancy, application of fluoride varnish to the teeth of children, anticipatory guidance, and motivational interviewing reduced prevalence of dental disease among Aboriginal children in South Australia at follow-up ages of 24 and 36 months [9,10]. The intervention took place during pregnancy and when children were aged 6, 12, and 18 months for the early intervention group, and when children were aged 24, 30, and 36 months for the delayed intervention group. Three-year follow-up data were collected from November 2014 to February 2016. All information was collected at participants' households or at their ACCHO. Children with any information collected at baseline, 2,- and 3-year-old follow-up were included in the study ($n = 307$).

All participants provided written informed consent. Both the original trial and the follow-up studies were conducted in accordance with the Declaration of Helsinki, and the protocol was approved by the Ethics Committee of The University of Adelaide Human Research Ethics Committee (Project code: H-057-2010), the Aboriginal Health Council of South Australia (Project code: 04-09-362), the South Australian Department for Health, including the human research ethics committees of participating South Australian hospitals (Flinders Medical Centre (Project code: 435-10), Lyell McEwin Hospital (Project code: 2010-160), and Women's and Children's Hospital (Project code: REC2322/11/13)).

2.2. Breastfeeding (Exposure)

Breastfeeding information was collected when children were aged 24 months by asking mothers the duration of breastfeeding and the period of exclusive breastfeeding. Patterns of breastfeeding were categorized according to two definitions based on previous studies examining the effect of breastfeeding on dental caries (definition A) [4,6] and following the WHO recommendations (definition B) [1]. Definition A allows us to compare the risks of ECC for breastfeeding over 12 months with other durations of breastfeeding, whereas definition B aligns with the WHO's international breastfeeding recommendations. Definition A consisted of four categories including children who were never breastfed, breastfed for <12 months, breastfed for 12–23 months, and breastfed for 24 months or longer. Definition B consisted of the following four categories of breastfeeding; never breastfed, breastfed for <24 months with exclusive breastfeeding until 6 months, breastfed for <24 months with no exclusive breastfeeding until 6 months, and breastfed for 24 months or longer.

2.3. Oral Health (Outcomes)

The primary outcome was the prevalence of ECC, which is the proportion of children with ≥1 decayed teeth (with or without a break in the tooth surface), ≥1 missing teeth due to dental caries, and ≥1 filled teeth) at 3 years of age (% dmft>0). The primary dentition comprises 20 teeth, which normally erupt in the oral cavity between 6 and 24 months of age. Dental decay was considered since its initial stages, which consists of white areas on the surface of the tooth. We also examined the severity of ECC by assessing the mean number of tooth surfaces with decayed, missing, and filled surfaces (mean dmfs). Oral examinations were conducted by calibrated and masked examiners following a standardized protocol to record dental disease experience. Children were examined in the knee-to-knee position on their mother's lap, as is considered appropriate for this age group. Teeth were dried with cotton pads before examination. The light source was a fiber-optic light and standard infection control procedures were followed. Only visual criteria were used to assess diagnoses. Any children identified as having untreated dental decay were referred to the South Australian Dental Service (SADS).

2.4. Confounding

Confounding factors included a range of variables that share a causal association with both exposures and outcomes and were assessed at baseline. These included smoking and alcohol consumption in pregnancy (yes/no), child sex, main source of household income (job vs. government social welfare support), maternal education (secondary school or less vs. trade/technical and university), parity (first time mother vs. others), maternal age (continuous), maternal ethnicity (Aboriginal vs. non-Aboriginal), and a national indicator of socioeconomic position calculated from residential postcode (the index of relative socioeconomic advantage and disadvantage (IRSAD) [11]. Additionally, we adjusted for randomization group because breastfeeding occurred during the intervention, even though there were no differences in breastfeeding duration between the treatment groups.

2.5. Statistical Analysis

In order to address potential bias due to missing data, we undertook multiple imputation by chained equations under the assumption that data were missing at random conditional on the observed data [12]. Twenty imputed datasets were generated from imputation models containing all the variables included in the final regression analyses. Imputed values were generated for smoking in pregnancy (missing $n = 1$), alcohol consumption in pregnancy (missing $n = 1$), main source of household income at baseline (missing $n = 1$), IRSAD (missing $n = 3$), breastfeeding duration (missing $n = 3$), exclusive breastfeeding duration (missing $n = 9$), and mean number of tooth surfaces with dental caries (missing $n = 1$). All other variables were fully observed. Analyses were conducted by combining all imputed datasets following Robin's rules on a final sample of $n = 307$ children who had some information collected at baseline, 2-, and 3-year-old follow-up.

Descriptive statistics were computed for the response sample, complete cases, and imputed sample, and included the distribution of participants according to exposures (breastfeeding), confounding factors (listed above), and outcomes (prevalence of ECC and mean dmfs). Generalized linear models with a log-Poisson link function and robust standard errors were used to estimate the prevalence ratios [13] and their 95% confidence intervals (CI) of dental caries according to breastfeeding patterns (definitions A and B). Next, adjusted PRs and their respective 95% CIs were assessed after the inclusion of the abovementioned confounding variables in the models. Unadjusted and adjusted and linear regressions were used to estimate the association between breastfeeding patterns and mean dmfs.

Analyses were carried out using STATA 15.0.

3. Results

The sample consisted of 307 children who had some information collected at baseline, 2-, and 3-year-old follow-up. Sample characteristics are shown in Table 1. Most children were born to mothers who identified as Aboriginal (82.7%). On average, mothers were 24.7 years old at baseline. Just over half of the children (52.1%) were boys and 39.4% were the first child. Four in five children (82.4%) lived in households where government support was the main source of income and over two-thirds of mothers (71.7%) had high school or less educational attainment. Over half of the children (55.3%) lived in the most disadvantaged IRSAD quintile. Smoking in pregnancy was observed among half of the sample, whereas 10.8% of mothers reported alcohol consumption during pregnancy. One in three children were never breastfed (29.2%), whereas half of the sample (49.9%) were breastfed for less than 12 months. The proportion of children breastfed for 12–23 months and 24 or more months was 11.5% and 9.3%, respectively. One in four children were exclusively breastfed until six months (23.8%) (data not shown). At three years of age, one-third had experienced dental caries, and the mean number of decayed, missed, or filled tooth surfaces was 3.2 (Table 1).

Table 1. Characteristics of the study sample.

	Response Sample %	Complete Cases % (*n* = 292)	Imputed Sample % (*n* = 307)
Exposure: Breastfeeding patterns			
Breastfeeding pattern A			
Never breastfed	29.3	29.8	29.2
Breastfed for <12 months	49.8	51.7	49.9
Breastfed for 12 to 23 months	11.4	11.3	11.5
Breastfed for 24 months or more	9.5	7.2	9.3
Breastfeeding pattern B (WHO recommendations)			
Never breastfed	29.3	29.8	29.2
Breastfed for <24 months, exclusively until 6	17.9	18.5	18.2
Breastfed for <24 months, not exclusively until 6	43.3	44.5	43.2
Breastfed for 24 months or more	9.5	7.2	9.3
Outcomes at child age 3 years			
Any decayed, missed or filled tooth (% dmft>0)	32.0	31.9	32.1
Mean number of decayed, missed or filled tooth surfaces (mean dmfs)	3.1	3.2	3.2
Potential confounding factors			
Child sex (male)	52.1	52.1	52.1
Mean maternal age at childbirth	24.7	24.6	24.7
Main household income source			
Job	17.6	18.2	17.6
Government support	82.4	81.9	82.4
Maternal education at birth			
High school or less	71.7	71.2	71.7
TAFE or university	28.3	28.8	28.3
Index of Relative Socioeconomic Advantage and Disadvantage			
1st quintile (most disadvantaged)	55.4	55.1	55.3
2nd quintile	17.3	16.8	17.4
3rd quintile	21.1	23.6	21.0
4th quintile	2.6	2.7	2.6
5th quintile (most advantaged)	1.6	1.7	1.6
Alcohol consumption in pregnancy (yes)	10.8	10.6	10.8
Smoking in pregnancy (yes)	50.1	50.7	50.0
Parity (first child)	39.4	40.1	39.4
Maternal Ethnicity (Aboriginal and/or Torres Strait Islander)	82.7	82.5	82.7
Child received the intervention (early)	49.5	49.0	49.5

WHO: World Health Organization.

Figure 1 displays the prevalence of dental caries experience (% dmft>0) and mean dmfs according to breastfeeding patterns (definition A). In comparison with children who were never breastfed, the prevalence of ECC was lower among those who were breastfed <12 months (33.1% vs. 25.7%), although both groups exhibited a similar mean dmfs (2.3 vs. 3.0). Over two-thirds of the children breastfed for 12 to 23 months had dental caries (37.0%). The highest prevalence (56.4%) and severity (6.6 mean dmfs) of disease was observed among children breastfed for 24 months.

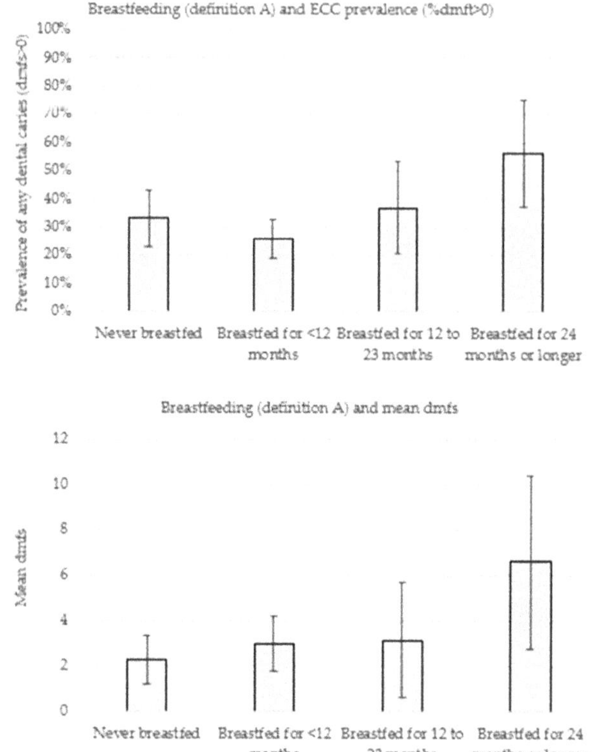

Figure 1. Prevalence of early childhood caries (ECC) and mean number of decayed, missing, and filled teeth (mean dmfs) according to breastfeeding patterns (definition A).

Figure 2 displays the prevalence of dental caries experience (% dmft>0) and mean dmfs according to breastfeeding patterns (definition B). While a higher prevalence of dental caries was observed among children breastfed for less than 24 months with exclusive breastfeeding until six months (33.1%) in comparison with those breastfed for the same duration but without exclusive breastfeeding (25.7%), both groups exhibited a similar mean dmfs (3.0 vs. 3.1).

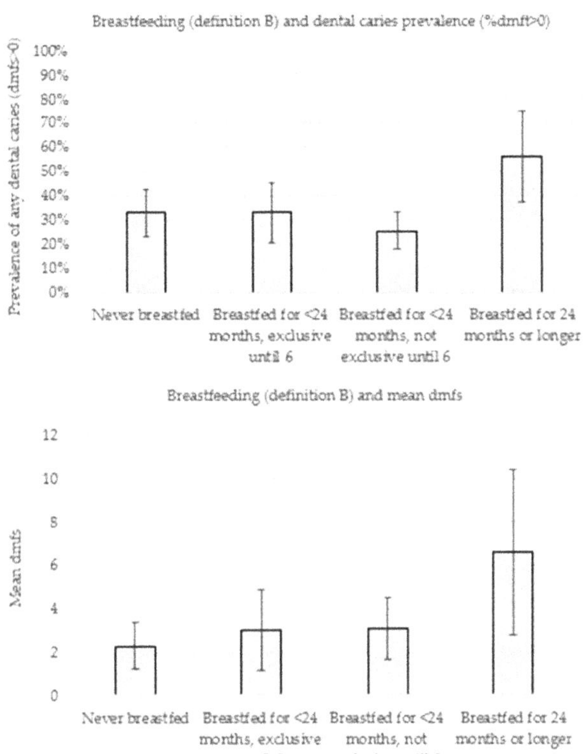

Figure 2. Prevalence of early childhood caries (ECC) and mean number of decayed, missing, and filled teeth (mean dmfs) according to breastfeeding patterns (definition B).

Table 2 displays the unadjusted and adjusted associations between breastfeeding patterns (definitions A and B) with % dmft>0 and mean dmfs. For definition A, the point estimate for prevalence of % dmft>0 was 6% lower among children who were breastfeeding <12 months compared with no breastfeeding (PR = 0.94 (95%CI 0.63; 1.41)). Breastfeeding for 12 to 23 months and breastfeeding for two years or more were associated with a 47% (PR = 1.47 (95%CI 0.85; 2.54)) and two-fold higher ECC prevalence (PR = 2.06 (95%CI 1.35; 3.13)), respectively. In comparison with children who were never breastfed, adjusted models for the mean dmfs showed higher dmfs among children breastfed for less than 12 months (β = 1.31 (95%CI-0.66; 3.30)), those breastfed for 12 to 13 months (β = 2.24 (95%CI –0.74; 5.24)), and among those breastfed for 24 months or longer (β = 5.22 (95%CI 2.06; 8.39). For definition B, breastfeeding up to 23 months and exclusively until six months was associated with a 45% (PR = 1.45 (95%CI 0.92; 2.30)) higher ECC prevalence and a higher dmfs (β = 2.04 (95%CI –0.62; 4.71)).

Table 2. Associations between breastfeeding patterns with dental caries prevalence and severity.

	Any Decayed, Missed or Filled Tooth (% dmft> 0)		Number of Decayed, Missed or Filled Tooth Surfaces (Mean dmfs)	
	Unadjusted PR (95%CI)	Adjusted PR (95%CI)	Unadjusted β (95%CI)	Adjusted β (95%CI)
Breastfeeding (definition A)				
Never breastfed	Ref	Ref	Ref	Ref
Breastfed for <12 months	0.77 (0.52; 1.17)	0.94 (0.63; 1.41)	0.72 (−1.02; 2.65)	1.31 (−0.66; 3.30)
Breastfed for 12 to 23 months	1.11 (0.66; 1.89)	1.47 (0.85; 2.54)	0.88 (−2.00; 3.76)	2.24 (−0.74; 5.24)
Breastfed for 24 months or longer	1.70 (1.09; 2.64)	2.06 (1.35; 3.13)	4.31 (1.19; 7.43)	5.22 (2.06; 8.39)
Breastfeeding (definition B)				
Never breastfed	Ref	Ref	Ref	Ref
Breastfed for <24 months, exclusively until 6	1.00 (0.52; 1.17)	1.45 (0.92; 2.30)	0.70 (−1.85; 3.24)	2.04 (−0.62; 4.71)
Breastfed for <24 months, not exclusively until 6	0.77 (0.66; 1.89)	0.90 (0.59; 1.36)	0.77 (−1.23; 2.77)	1.26 (−0.79; 3.31)
Breastfed for 24 months or longer	1.70 (1.09; 2.64)	2.06 (1.35; 3.13)	4.31 (1.19; 7.43)	5.22 (2.05; 8.39)

Prevalence ratios obtained from Poisson regressions. Beta coefficients obtained from linear regressions. Models adjusted for child: Child sex, household main income source, maternal education at childbirth, index of relative socioeconomic advantage and disadvantage, alcohol consumption in pregnancy, smoking in pregnancy, parity, maternal ethnicity, and whether the child received the intervention (early, delayed).

4. Discussion

The results presented here are secondary, observational analyses from a sample of Aboriginal families in South Australia who were enrolled in an early childhood caries intervention. They show that that around one-third of children are never breastfed, half are breastfed for less than 12 months, and one-fifth are breastfed for longer than 12 months (breastfeeding definition A). Applying the WHO breastfeeding recommendations in breastfeeding definition B shows that almost one-fifth of children are breastfed exclusively for six months. For our aim of investigating observational associations between breastfeeding and oral health, these results suggest that breastfeeding for 24 months or longer is associated with twice the prevalence of ECC and four-fold higher severity of caries (indicated by mean dmfs). Both duration and exclusivity of breastfeeding may be important, but evidence from the current study is stronger for the duration of breastfeeding. Evidence for duration comes from the breastfeeding pattern A results, where the adjusted point estimates suggest monotonically increasing risks of both % dmft>0 and dmfs from 12–23 and then ≥24 months of breastfeeding. Breastfeeding pattern B provides some suggestive observational evidence of an increased risk of caries with exclusive breastfeeding to six months, although there are major challenges with disentangling duration and exclusivity (discussed further below). Additionally, the modest sample size has likely contributed to imprecision in estimating these effects. Notably, this sample has poor oral health with 25% prevalence of ECC even though children were only three years old, which has contributed to why we have been able to detect these associations at such a young age.

It is important to emphasize that any potential effects of breastfeeding on caries can be mitigated by better oral hygiene and access to oral healthcare, and these should be the targets of public policy, particularly since oral health is a major cause of problems (such as hospitalizations, extractions) in children. Maternal-child health nurses encourage breastfeeding and with appropriate training they might present an opportunity for teaching caregivers to begin clean deciduous teeth as soon as they erupt. The promotion of wide-spread water fluoridation is also important in reducing oral health inequalities, particularly between Indigenous and non-Indigenous Australian children [7].

Compared with Australian national statistics on breastfeeding the prevalence, data reported here appear lower. For example, national data indicate that around 80% of Aboriginal children were ever breastfed [14], whereas 70% of children in the current study were reportedly ever breastfed. The national data were collected in 2014–2015, at a similar time to our study, so temporal shifts in breastfeeding cannot explain this discrepancy. This is probably not due to selection bias as, compared with birth registry data, we enrolled around half of all eligible Aboriginal children born in the state over the recruitment period, and families who were not enrolled or dropped out are likely to be more marginalized. While it is difficult to pinpoint exactly why we found lower breastfeeding outcomes compared with national averages, we speculate that this may be because our participants are more disadvantaged than families included in the national data. The high level of disadvantage is indicated in Table 1, where half of the participants in our study are in the most disadvantaged quintile for socioeconomic position (on a national scale), half smoked during pregnancy, and about 80% received government income support. The state of South Australia has a more disadvantaged socioeconomic profile than other mainland Australian states. However, many other (non-socioeconomic) issues around breastfeeding should also be considered. Such factors might include mothers being very time-poor, impediments to breastfeeding due to mental health problems (postnatal depression, anxiety), low self-confidence in their ability to breastfeed, and lack of partner support. Taking a strengths-based view, since only one-third of women did not breastfeed, this indicates this is an area of cultural strength. Contributors to breastfeeding success are other key women involved in perinatal care and social networks of new mothers', which may improve self-confidence in their ability to successfully initiate and maintain breastfeeding for longer. Social context and culturally safe birthing programs are important influences on breastfeeding among Aboriginal women [15,16].

The debate about whether breastfeeding has a causal role in ECC remains, with discussions on the evidence from the three systematic reviews on this topic vary. Our results appear to confirm concepts raised in previous systematic reviews that the risk of ECC increases with breastfeeding beyond 12 months and possibly with exclusive breastfeeding [4,6]. We were surprised to find suggestive evidence that exclusive breastfeeding to six months (among children breastfed for less than 24 months) was linked to higher caries, as first teeth only erupt around six months. It is possible that the exclusive breastfeeding to six months is an indicator of other related factors, such as mothers being more committed to breastfeeding and more likely to breastfeed for longer. Support for this idea comes from the fact that the duration of breastfeeding among children exclusively breastfed to six months was 2.5 times longer than the duration observed among children who were not exclusively breastfed to six months (10 vs. 4 months). It is clearly difficult to separate out effects of exclusive breastfeeding from the effect of duration of breastfeeding in observational research, and for this reason we look to the literature for evidence from specific breastfeeding randomized controlled trials (RCTs). The standout in this field is the large Promotion of Breastfeeding Intervention Trial (PROBIT), trial which is a cluster-RCT where intervention hospitals implemented the baby-friendly hospital initiative, and this resulted in increased breastfeeding duration and exclusivity [13]. Follow up of caries data in the PROBIT participants indicated no differences in caries outcomes between the groups, despite having more women in the intervention group breastfeeding for longer durations and higher proportions of women exclusively breastfeeding to six months of age [17]. Another cluster RCT of an intervention that increased exclusive breastfeeding to six months conducted in Uganda also indicated no effect of caries outcomes at five years of age [18]. Observational research on breastfeeding is notoriously confounded and one way forward might be to apply different study designs that can help overcome such problems, as has been done for the effects of breastfeeding on intelligence quotient and obesity [19].

We approached this study with causal inference in mind. With respect to study validity, we tried to minimize confounding by adjusting for multiple variables that were identified through content knowledge and literature. One issue relevant to this field is whether to adjust for confounding by other aspects of the child's diet such as sugary foods and beverages. In causal inference theory, confounding is considered to be a common cause of the exposure and outcome [20]. While there is no doubt that the sugar content of the diet has a direct effect on caries, it is not a direct cause of initiating breastfeeding. Any reduction in breastfeeding will lead to consumption of other foods and therefore sugary foods may be a mediator on the breastfeeding → diet → caries pathway. Factors further back in the causal chain, such as parental education, influence breastfeeding and diet, a phenomenon occasionally referred to as M-bias in the epidemiological literature. Our approach was to adjust for many of these more distal factors to block the influence of these cofounders. It is also worth noting that the participants were involved in a health promotion trial and therefore their breastfeeding behaviors may have been different had they not been involved in the trial. We used multiple imputation to address bias due to missing information, which is uncommon in this field [4,6]. Some effect estimates have wide confidence intervals which is likely due to the sample size, which was limited to the original trial. Increasing the sample size was not possible because the study is estimated to have enrolled around half of the local Aboriginal and Torres Strait Islander children in the state of South Australia at the time of recruitment.

5. Conclusions

There is room to improve breastfeeding prevalence, duration, and exclusivity among mothers of Aboriginal children in South Australia, and this could have flow on effects of improving other health outcomes. Regarding ECC, the observational analyses presented here suggest that breastfeeding for longer and more exclusively may be linked to poorer outcomes. Despite attempting to address sources of bias, doubt is raised about interpreting this observational research as causal.

Author Contributions: Conceptualization, D.G.H. and L.G.S.; data curation, L.M.J., J.H., and L.G.S.; formal analysis, D.G.H.; funding acquisition, L.M.J. and J.H.; methodology, D.G.H. and L.G.S.; project administration, L.M.J.; supervision, L.G.S.; writing—original draft, D.G.H. and L.G.S.; writing—review and editing, D.G.H., L.M.J., J.H., and L.G.S..

Funding: This research was funded by the National Health and Medical Research Council of Australia, grant number 627350.

Acknowledgments: The authors acknowledge the support of the Baby Teeth Talk study participants and staff including the Aboriginal Reference Group (Chaired by Cathy Leane); support of staff from the South Australian Aboriginal Community-Controlled Health Organisations. We thank Colgate Palmolive for donations of toothpaste, toothbrushes, dental floss and disclosing tables that were offered as part of a 'thank you' package to participants and their families.

Conflicts of Interest: The authors declare no conflict of interest. Neither the funders nor Colgate Palmolive had any role in the design of the study; in the collection, analyses or interpretation of data; in the writing of the manuscript, or in the decision to publish the results.

References

1. World Health Organization. Breastfeeding. Available online: https://www.who.int/topics/breastfeeding/en/ (accessed on 27 September 2019).

2. National Health and Medical Research Council. *Infant Feeding Guidelines*; National Health and Medical Research Council: Canberra, Australia, 2013.

3. Wyne, A.H. Early childhood caries: Nomenclature and case definition. *Community Dent. Oral Epidemiol.* **1999**, *27*, 313–315. [CrossRef] [PubMed]

4. Tham, R.; Bowatte, G.; Dharmage, S.C.; Tan, D.J.; Lau, M.X.; Dai, X.; Allen, K.J.; Lodge, C.J. Breastfeeding and the risk of dental caries: A systematic review and meta-analysis. *Acta Paediatr.* **2015**, *104*, 62–84. [CrossRef] [PubMed]

5. Cui, L.; Li, X.; Tian, Y.; Bao, J.; Wang, L.; Xu, D.; Zhao, B.; Li, W. Breastfeeding and early childhood caries: A meta-analysis of observational studies. *Asia Pac. J. Clin. Nutr.* **2017**, *26*, 867–880. [CrossRef] [PubMed]

6. Moynihan, P.; Tanner, L.M.; Holmes, R.D.; Hillier-Brown, F.; Mashayekhi, A.; Kelly, S.A.M.; Craig, D. Systematic Review of Evidence Pertaining to Factors That Modify Risk of Early Childhood Caries. *JDR Clin. Transl. Res.* **2019**, *4*, 202–216. [CrossRef] [PubMed]

7. Do, L.G.; Ha, D.H.; Roberts-Thomson, K.F.; Jamieson, L.; Peres, M.A.; Spencer, A.J. Race- and income-related inequalities in oral health in Australian children by fluoridation status. *JDR Clin. Trans. Res.* **2018**, *3*, 170–179. [CrossRef] [PubMed]

8. Merrick, J.; Chong, A.; Parker, E.; Roberts-Thomson, K.; Misan, G.; Spencer, J.; Broughton, J.; Lawrence, H.; Jamieson, L. Reducing disease burden and health inequalities arising from chronic disease among Indigenous children: An early childhood caries intervention. *BMC Public Health* **2012**, *12*, 323. [CrossRef] [PubMed]

9. Jamieson, L.; Smithers, L.; Hedges, J.; Parker, E.; Mills, H.; Kapellas, K.; Lawrence, H.P.; Broughton, J.R.; Ju, X. Dental Disease Outcomes Following a 2-Year Oral Health Promotion Program for Australian Aboriginal Children and Their Families: A 2-Arm Parallel, Single-blind, Randomised Controlled Trial. *E Clin. Med.* **2018**, *1*, 43–50. [CrossRef] [PubMed]

10. Jamieson, L.M.; Smithers, L.G.; Hedges, J.; Aldis, J.; Mills, H.; Kapellas, K.; Lawrence, H.P.; Broughton, J.R.; Ju, X. Follow-up of an Intervention to Reduce Dental Caries in Indigenous Australian Children: A Secondary Analysis of a Randomized Clinical Trial. *JAMA Netw. Open* **2019**, *2*, e190648. [CrossRef] [PubMed]

11. Australian Bureau of Statistics. Information Paper: An Introduction to Socio-Economic Indexes for Areas (SEIFA). 2006. Available online: http://www.ausstats.abs.gov.au/Ausstats/subscriber.nsf/0/D729075E079F9FDECA2574170011B088/$File/20390_2006.pdf (accessed on 27 September 2019).

12. Bhaskaran, K.; Smeeth, L. What is the difference between missing completely at random and missing at random? *Int. J. Epidemiol.* **2014**, *43*, 1336–1339. [CrossRef] [PubMed]

13. Kramer, M.S.; Chalmers, B.; Hodnett, E.D.; Sevkovskaya, Z.; Dzikovich, I.; Shapiro, S.; Collet, J.-P.; Vanilovich, I.; Mezen, I.; Ducruet, T.; et al. Promotion of Breastfeeding Intervention Trial (PROBIT): A Randomized Trial in the Republic of Belarus. *J. Am. Med. Assoc.* **2001**, *285*, 413–420. [CrossRef] [PubMed]

14. Australian Government Department of Health. *Australian National Brestfeeding Strategy 2018 and Beyond: Draft for Public Consultation*; Australian Government Department of Health: Canberra, Australia, 2018; pp. 1–47.

15. Hartz, D.L.; Blain, J.; Caplice, S.; Allende, T.; Anderson, S.; Hall, B.; McGrath, L.; Williams, K.; Jarman, H.; Tracy, S.K. Evaluation of an Australian Aboriginal model of maternity care: The Malabar Community Midwifery Link Service. *Women Birth* **2019**, *32*, 427–436. [CrossRef] [PubMed]

16. Brown, S.; Stuart-Butler, D.; Leane, C.; Glover, K.; Mitchell, A.; Deverix, J.; Francis, T.; Ah Kit, J.; Weetra, D.; Gartland, D.; et al. Initiation and duration of breastfeeding of Aboriginal infants in South Australia. *Women Birth* **2019**, *32*, e315–e322. [CrossRef] [PubMed]

17. Kramer, M.S.; Vanilovich, I.; Matush, L.; Bogdanovich, N.; Zhang, X.; Shishko, G.; Muller-Bolla, M.; Platt, R.W. The effect of prolonged and exclusive breast-feeding on dental caries in early school-age children. New evidence from a large randomized trial. *Caries Res.* **2007**, *41*, 484–488. [CrossRef] [PubMed]

18. Birungi, N.; Fadnes, L.T.; Okullo, I.; Kasangaki, A.; Nankabirwa, V.; Ndeezi, G.; Tumwine, J.K.; Tylleskar, T.; Lie, S.A.; Astrom, A.N. Effect of Breastfeeding Promotion on Early Childhood Caries and Breastfeeding Duration among 5 Year Old Children in Eastern Uganda: A Cluster Randomized Trial. *PLoS ONE* **2015**, *10*, e0125352. [CrossRef] [PubMed]

19. Smithers, L.G.; Kramer, M.S.; Lynch, J.W. Effect of breastfeeding on obesity and intelligence: Causal insights from different study designs. *JAMA Pediatrics* **2015**, *169*, 707–708. [CrossRef] [PubMed]

20. Glymour, M.M.; Greenland, S. Causal diagrams. In *Modern Epidemiology*, 3rd ed.; Rothman, K.J., Greenland, S., Lash, T.L., Eds.; Lippincott Williams & Wilkins: Philadelphia, PA, USA, 2008; pp. 183–209.

 *nutrients*

Article

DNA Methylation Signatures of Breastfeeding in Buccal Cells Collected in Mid-Childhood

Veronika V. Odintsova [1,2,3,*], Fiona A. Hagenbeek [1,2], Matthew Suderman [4],
Doretta Caramaschi [4], Catharina E. M. van Beijsterveldt [1], Noah A. Kallsen [5], Erik A. Ehli [5],
Gareth E. Davies [5], Gennady T. Sukhikh [3], Vassilios Fanos [6], Caroline Relton [4], Meike Bartels [1,2],
Dorret I. Boomsma [1,2] and Jenny van Dongen [1,2,*]

[1] Department of Biological Psychology, Vrije Universiteit Amsterdam, 1081 BT Amsterdam, The Netherlands; f.a.hagenbeek@vu.nl (F.A.H.); t.van.beijsterveldt@vu.nl (C.E.M.v.B.); m.bartels@vu.nl (M.B.); di.boomsma@vu.nl (D.I.B.)

[2] Amsterdam Public Health Research Institute, 1081 BT Amsterdam, The Netherlands

[3] Kulakov National Medical Research Center for Obstetrics, Gynecology and Perinatology, Moscow 101000, Russia; inter_otdel@mail.ru

[4] MRC Integrative Epidemiology Unit, Bristol Medical School, Population Health Science, University of Bristol, Bristol BS8 1TH, UK; matthew.suderman@bristol.ac.uk (M.S.); d.Caramaschi@bristol.ac.uk (D.C.); caroline.relton@bristol.ac.uk (C.R.)

[5] Avera Institute for Human Genetics, Sioux Falls, SD 57101, USA; Noah.Kallsen@avera.org (N.A.K.); Erik.Ehli@avera.org (E.A.E.); Gareth.Davies@avera.org (G.E.D.)

[6] Neonatal Intensive Care Unit, Department of Surgical Sciences, AOU and University of Cagliari, 09121 Cagliari, Italy; vafanos@tin.it

* Correspondence: v.v.odintsova@vu.nl (V.V.O.); j.van.dongen@vu.nl (J.v.D.); Tel.: +31–20–5983570 (J.v.D.)

Received: 15 October 2019; Accepted: 12 November 2019; Published: 17 November 2019

Abstract: Breastfeeding has long-term benefits for children that may be mediated via the epigenome. This pathway has been hypothesized, but the number of empirical studies in humans is small and mostly done by using peripheral blood as the DNA source. We performed an epigenome-wide association study (EWAS) in buccal cells collected around age nine (mean = 9.5) from 1006 twins recruited by the Netherlands Twin Register (NTR). An age-stratified analysis examined if effects attenuate with age (median split at 10 years; $n_{<10}$ = 517, mean age = 7.9; $n_{>10}$ = 489, mean age = 11.2). We performed replication analyses in two independent cohorts from the NTR (buccal cells) and the Avon Longitudinal Study of Parents and Children (ALSPAC) (peripheral blood), and we tested loci previously associated with breastfeeding in epigenetic studies. Genome-wide DNA methylation was assessed with the Illumina Infinium MethylationEPIC BeadChip (Illumina, San Diego, CA, USA) in the NTR and with the HumanMethylation450 Bead Chip in the ALSPAC. The duration of breastfeeding was dichotomized ('never' vs. 'ever'). In the total sample, no robustly associated epigenome-wide significant CpGs were identified ($\alpha = 6.34 \times 10^{-8}$). In the sub-group of children younger than 10 years, four significant CpGs were associated with breastfeeding after adjusting for child and maternal characteristics. In children older than 10 years, methylation differences at these CpGs were smaller and non-significant. The findings did not replicate in the NTR sample (n = 98; mean age = 7.5 years), and no nearby sites were associated with breastfeeding in the ALSPAC study (n = 938; mean age = 7.4). Of the CpG sites previously reported in the literature, three were associated with breastfeeding in children younger than 10 years, thus showing that these CpGs are associated with breastfeeding in buccal and blood cells. Our study is the first to show that breastfeeding is associated with epigenetic variation in buccal cells in children. Further studies are needed to investigate if methylation differences at these loci are caused by breastfeeding or by other unmeasured confounders, as well as what mechanism drives changes in associations with age.

Keywords: breastfeeding; EWAS; DNA methylation; twins; EPIC; NTR; ALSPAC

---

1. Introduction

Early life environmental influences are associated with development and disease. One mechanism that has been hypothesized to account for the association is through the "epigenetic programming" of genes, some of which may act to confer plasticity to developmental processes [1,2]. Nutrition in early life may play a crucial role in modulating gene expression [3,4]. It has been hypothesized that nutrition-induced epigenetic variation may result in different development trajectories and may be associated with metabolic and immune development during critical periods of early life [5–8]. Epigenetic mechanisms are a key element in understanding the developmental origins of later life disease risk [9–11]. One of the best studied epigenetic mechanisms is DNA methylation—the modification of a cytosine base, usually at CpG dinucleotides, with a methyl group, which regulates gene expression and seems to already be sensitive to nutrition during the prenatal period [12–17].

In humans, prenatal maternal famine has been associated with long-term DNA methylation changes that are still observable in middle-aged individuals [18,19]. The early postnatal period is also believed to be a critical period at which permanent long-term changes may be induced by environmental exposures that affect the child's susceptibility to chronic disease [20]; however, whether similar long-term changes in the human epigenome may be induced in this period is unexplored. The first nutrition period, including breastfeeding, has long-term effects on children [3]. Human breast milk has a unique composition that differs from other lactating animals and that is quite impossible to reproduce in artificial production; in fact, breast milk contains a unique mixture of microorganisms (the microbiome) [21], metabolites [22], multipotent stem cells [23], growth factors [24] and other components that render it unique and individualized for each newborn [25]. Moreover, breast milk varies its composition according to the lactation period and circadian rhythm, and it even varies from the start till the end of one feeding [26]. The benefits of breastfeeding on the health of children are widely described [27–29] and may involve the transmission of nutrients, hormones, and antibodies from mother to child [24,30,31]. These benefits also include the process of interaction and attachment of a child to their mother, although the effect of bonding may also be achieved through formula feeding [32,33].

Balanced newborn feeding is the basis for the adequate growth and development in childhood and beyond [25,28,34]. Breast milk is important for sensory, neurological and cognitive development [35–37], especially in preterm infants [38], but its effects on cognition are confounded by maternal education [39]. The association between breastfeeding duration and a lower risk of infectious diseases, obesity, cancer, coronary heart disease, some allergies, autoimmune disease, diabetes mellitus, inflammatory bowel diseases, and metabolic syndrome at later age has been widely studied [27,40–49]. Even though a protective effect of breastfeeding in hypertension, diabetes, obesity, and metabolic syndrome has not been evident in some large studies [30,50], the benefits of breastfeeding are well recognized, and the understanding of the biological mechanisms of its influence is of interest. Previous studies on the association between breastfeeding and child outcomes have used a variety of definitions of breastfeeding, including: never vs. ever breastfeeding, breastfeeding duration which has been assessed as a continuous measure in months or weeks, or as a categorical variable (e.g., long vs. short). Some studies have examined the percentage of breastfed meals, exclusive breastfeeding duration (when a child is exclusively breastfed until formula feed and/or solid foods are introduced). In a meta-analysis by Victora et al. [27], breastfeeding never vs. ever was associated with a reduction in sudden infant deaths, a reduction of acute otitis during the first few years of life, a protection against allergic rhinitis in children <5 years, a higher child Intelligence Quotient (IQ) of about three IQ points. More versus less breastfeeding has been associated with major protection against diarrhea morbidity, a reduction in severe respiratory infection, and effects on deciduous teeth. Exclusive breastfeeding has been associated with a strong protective result against infectious disease and allergic rhinitis in children <5 years. A dose–response association with duration of breastfeeding was found for a higher IQ and a decreased risk for overweight and obesity.

Epigenome-wide association studies (EWASs) can offer new insights into DNA–environment interactions in determining child development and health [51]. To our knowledge, seven association studies (two candidate gene studies, two EWASs with breastfeeding as covariate and three EWASs of breastfeeding) have been performed on breastfeeding and DNA methylation in humans to date (see Table S1). A candidate gene study of Obermann-Borst et al. found a negative association between the duration of breastfeeding (in categories) and methylation level in blood cells from 120 children (mean age 1.4 years) at seven CpG sites in the promoter of the *LEP* gene; a hormone that regulates energy homeostasis [52]. A suggestive positive association of the methylation level of 2201 CpGs and a negative association of the methylation level of 2075 CpGs with the duration of breastfeeding (continuous measure in weeks) were reported in blood samples from 37 infants (mean age 25.7 months) [53]. These CpGs were annotated to genes predominantly involved in the control of cell signaling systems, the development of anatomical structures and cells, and the development and function of the immune and central nervous systems [53]. The impact of breastfeeding duration (continuous measure in months) on DNA methylation patterns in 200 children (mean age 11.6 years) was suggested in a study of asthma [54]. An EWAS of Sherwood et al. on exclusive breastfeeding supported the findings of Obermann-Borst et al. at a later stage in childhood (10 years, $n = 297$) but not in young adulthood (18 years, $n = 305$) [55]. This suggests that methylation changes induced by breastfeeding may change with time and may be more evident at an early age. Similarly, it has been observed that associations between DNA methylation and maternal smoking and birthweight attenuate during childhood [56,57]. Nevertheless, a long-lasting modulating effect of breastfeeding (continuous measure in weeks) on the effects of methylation quantitative trait loci (mQTLs) for CpG sites at the 17q21 locus, where the *IL4R* (interleukin-4) gene is located, has been suggested at age 18 ($n = 245$) [58]. Not having been breastfed has been associated with an increase in methylation of the promoter of the tumor suppressor gene *CDKN2A* (cyclin dependent kinase inhibitor 2A) in premenopausal breast tumors of 639 women (mean age of 57.6 years) [59]. In a more recent EWAS study, breastfeeding (dichotomized as never vs. ever) was associated with changes in the *TTC34* gene at age 7 ($n = 640$), which were still evident in adolescence ($n = 709$) [60]. These previous epigenetic studies of breastfeeding were often conducted with relatively small samples (average sample size = 307, range = 37–640). In all studies, DNA was extracted from peripheral blood [52–55,58], or from tumor tissues in adults [59].

We aimed to carry out an EWAS of breastfeeding in 1006 children around nine years of age recruited by the Netherlands Twin Register (NTR) based on buccal cell DNA and a replication analysis of loci previously associated with breastfeeding in aforementioned epigenetic studies. Buccal samples typically consist of a large proportion of epithelial cells, which might serve as a surrogate tissue for other ectodermal tissues, including the brain [61,62]. Buccal samples also consist of a smaller proportion of leukocytes [63]. To date, few EWASs have been performed on buccal DNA. As some studies have suggested that the effects of early life exposures, including breastfeeding [55–58], may fade away during childhood, we also performed an EWAS on younger children (age <10 years; where 10 corresponds to the median age of the sample) and compared effect sizes in this group with effect sizes in children older than 10 years. We applied a median split of the sample by age to achieve equal sample sizes in both groups. We hypothesized that if effects of breastfeeding attenuate with age, associations would be strongest in the younger age group. We performed replication in an independent buccal-cells DNA methylation dataset from the NTR ($n = 98$) and in a blood-DNA-methylation dataset from the Avon Longitudinal Study of Parents and Children (ALSPAC) ($n = 938$). We also examined the correlation between methylation levels of twins for the significant CpGs associated with breastfeeding. We hypothesized that the equal exposure to breastfeeding of co-twins should cause resemblance in their methylation profiles.

2. Materials and Methods

2.1. Overview

We carried out an EWAS in the NTR in 1006 children from 496 complete pairs and 14 twins from incomplete pairs with DNA methylation in buccal cells, testing 787,711 methylation sites ($\alpha = 0.05/787771 = 6.34 \times 10^{-8}$). The EWAS analyses in different age groups were performed following the median-split of the sample by age (age <10 years: $n = 517$, age range = 5–9; age $\geq$ 10 years: $n = 489$, age range = 10–12). Two models were applied with different covariates (Model 1 = "basic model" and Model 2 = "adjusted model") (see Table S2). Epigenome-wide significant results were taken forward, after checking effects of outliers, for a replication analysis in an independent sample, consisting of 98 NTR children with DNA methylation in buccal cells and in 938 ALSPAC children with DNA methylation in peripheral blood cells. Lastly, we performed a follow-up analysis of the results of previous studies (3859 CpGs) in the discovery cohort (NTR) in the total sample ($n = 1006$) and the younger age group ($n = 517$). A flowchart of the analyses is provided in Figure 1.

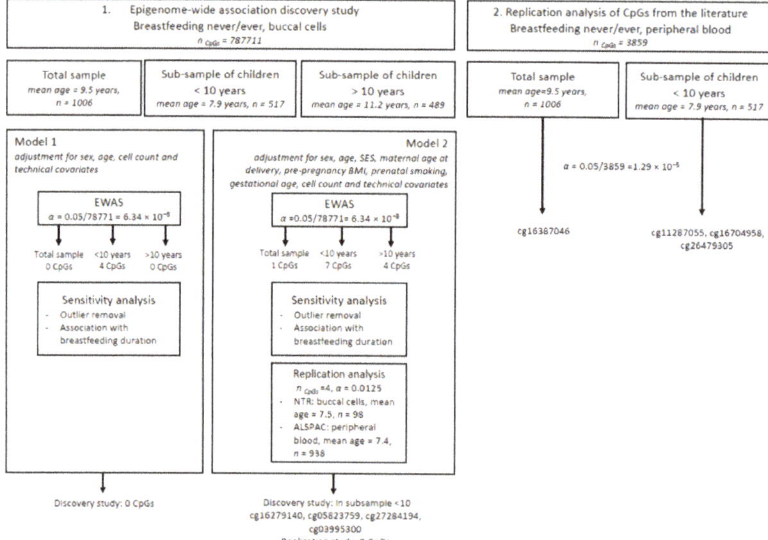

Figure 1. Flowchart of analyses.

2.2. Subjects and Samples

2.2.1. Discovery Study

The subjects were enrolled in the NTR [64] a few weeks to months after birth. Informed consent was obtained from parents. Data on breastfeeding collected around the age of 2 years of the children and good quality DNA methylation data around 9 years of age (mean = 9.5, standard deviation (SD) = 1.89, range = 5–12) were available for 1006 children. The dataset included 51.6% girls and 86% monozygotic twins. This study is embedded in a larger project on childhood aggression: Aggression in Children: Unraveling gene-environment interplay to inform Treatment and InterventiON strategies (ACTION) [65,66]. From the population-based NTR, the ACTION study identified twins who at least once scored higher or lower on a sum score for aggression [67,68].

2.2.2. NTR Replication Study

An independent group of children from the NTR for whom EPIC array methylation data were previously described [63] were included as replication study. This cohort was also embedded in the ACTION project and were comprised of 98 monozygotic twins with available data on breastfeeding duration and DNA methylation from buccal swabs (mean age = 7.5 years, SD = 2.4, age range = 1–10). The NTR studies were approved by the Central Ethics Committee on Research Involving Human Subjects of the VU University Medical Centre, Amsterdam, an Institutional Review Board certified by the U.S. Office of Human Research Protections (IRB number IRB00002991 under federal-wide Assurance-FWA00017598; IRB/institute codes, NTR 03–180).

2.2.3. ALSPAC Replication Study

Data came from the ALSPAC [69,70], a population-based birth cohort. All pregnant women living in the geographical area of Avon (UK) with expected delivery date between 1 April 1991 and 31 December 1992 were invited to participate. Approximately 85% of the eligible population was enrolled, totaling 14,541 pregnant women who gave informed and written consent. The study website contains details of all the data that are available through a fully searchable data dictionary and variable search tool (http://www.bris.ac.uk/alspac/researchers/data-access/data-dictionary/). Ethical approval for the study was obtained from the ALSPAC Ethics and Law Committee and the Local Research Ethics Committees. Of these, 938 children had information on breastfeeding duration and DNA methylation from peripheral blood cells measured on the Infinium HumanMethylation450 BeadChip (mean age = 7.4, SD = 0.13, age range = 7.1–8.8) within the Accessible Resource for Integrated Epigenomics Studies (ARIES) project [71].

2.3. Phenotype Data

In the NTR, breastfeeding was assessed in a questionnaire sent to mothers 2 years after the twins were born. Mothers were separately asked about breastfeeding for the first and second born twin. There were six answer categories: 'no', 'less than 2 weeks', '2–6 weeks', '6 weeks to 3 months', '3–6 months', and 'more than 6 months', For the main analyses, breastfeeding as exposure was recoded into 2 categories: 'ever' and 'never', No information on exclusive and mixed breastfeeding was available. The duration of breastfeeding was used in a secondary analysis based on the 6 categories from the original questionnaire coded from 0 (no) to 5 (more than 6 months). Socio-economic status (SES) was determined in two ways (depending on the version of the survey): (1) SES was obtained from a full description of the occupation of the parents and was subsequently coded according to the Standard Classification of Occupations [72]; (2) SES was obtained by the Erikson-Goldthorpe-Portocarero (EPG)-classification scheme [73], combined with information on parental education. The self-reported maternal pre-pregnancy weight just before pregnancy (in kilograms) divided by the square of height (in meters) was used to obtain the maternal pre-pregnancy body mass index (BMI) (weight/height^2). Maternal smoking during pregnancy was reported by mothers for three trimesters of pregnancy and was coded as non-smoking if the mother did not smoke during the entire pregnancy and smoking if the mother smoked at least during one trimester [74]. The mode of conception was classified in three groups: naturally conceived, conceived through stimulation, and conceived via in vitro fertilization (IVF) or intracytoplasmic sperm injection (ICSI) [75]. The mode of delivery was assessed as vaginal delivery, caesarean section (planned and urgent), and vaginal delivery with urgent intervention with forceps or vacuum extraction. Apgar scores at the 1st and 5th minute were presented in 3 conventional categories 0–6 (low), 7–9 (intermediate), and 10 (high) [76]. Gestational age, birthweight, parental age at birth were categorized in groups for descriptive statistics and treated as continuous z-scores in the analyses.

ALSPAC Replication Study

Breastfeeding was assessed via a questionnaire sent to mothers when the study children were approximately 6 months old. Breastfeeding was coded as 'ever' or 'never'. Information on sample characteristics and covariates was obtained by questionnaires completed by the mother during pregnancy and from medical birth records. Socio-economic status was determined by the highest level of maternal education (grouped as follows: certificate of secondary education or not, vocational, O-level, A-level, or university degree). Maternal characteristics included age at birth, pre-pregnancy height and weight, and smoking during pregnancy (any or none). Gestational characteristics included caesarean delivery, gestational age and birthweight.

2.4. DNA Sample Collection

2.4.1. NTR (Discovery and Replication Study)

DNA samples were collected from buccal swabs, as described previously [77]. In short, 16 cotton mouth swabs were individually rubbed against the inside of the cheek on 2 days (morning and evening) and placed in four separate 15 mL conical tubes containing 0.5 mL of a Sodium Chloride-Tris-Ethylenediaminetetraacetic acid (STE) buffer (100 mM sodium chloride, 10 mM Tris hydrochloride (pH 8.0), and 10 mM ethylenediaminetetraacetic acid) with proteinase K (0.1 mg/mL) and sodium dodecyl sulfate (SDS) (0.5%) per swab. Individuals were asked to refrain from eating or drinking 1 h prior to sampling. High molecular weight genomic DNA was extracted by standard DNA extraction techniques. The DNA samples were quantified using the Quant-iT PicoGreen dsDNA Assay Kit (ThermoFisher Scientific, Waltham, MA, USA).

2.4.2. ALSPAC Replication Study

The ALSPAC blood-based DNA methylation profiles were generated at age 7 as part of the ARIES [71], a subsample of approximately 1000 mother-child pairs from the ALSPAC study.

2.5. DNA-Methylation Measurements

2.5.1. NTR Discovery Study

The genome-wide DNA methylation in buccal cells [68] was assessed with the Infinium MethylationEPIC BeadChip (Illumina, San Diego, CA, USA) [78] by the Human Genotyping facility (HugeF) of ErasmusMC, the Netherlands (http://www.glimdna.org/). Five hundred nanograms of genomic DNA from buccal swabs were bisulfite treated using the ZymoResearch EZ DNA Methylation kit (Zymo Research Corp, Irvine, CA, USA). The quality control of the methylation data are described in detail elsewhere [68]. In brief, the quality control (QC) and normalization of the methylation data were performed using a pipeline developed by the Biobank-based Integrative Omics Study (BIOS) consortium [79], which includes sample quality control using the R package MethylAid [80] and probe filtering and functional normalization as implemented in the R package DNAmArray. We previously and successfully applied this pipeline in a pilot study of EPIC array data from buccal samples [63], which was used as replication sample in the current paper. MethylAid was applied with the default EPIC array-specific quality filter thresholds for EPIC arrays. The R package omicsPrint [81] was used to call genotypes based on methylation probes and to verify sample relationships based on those single nucleotide polymorphisms (SNPs) (e.g., the zygosity of twins and samples from the same individual). We checked for sample mismatches between methylation data and genotype data by computing the correlation between SNP genotypes called by omicsPrint based on methylation probes and genotypes based on genome-wide SNP arrays. DNAmArray and meffil [82] were used to identify sex mismatches (both packages identified the same mismatches).

Functional normalization was performed based on five control probe principal components (PCs). The following probe filters were applied: Probes were set to missing (NA) in a sample if they had an intensity value of exactly zero, detection *p*-value > 0.01, or bead count <3. Probes were excluded from all samples if they mapped to multiple locations in the genome, if they overlapped with an SNP or insertion/deletion (INDEL), or if they had a success rate <0.95 across samples. Annotations of ambiguous mapping probes (based on an overlap of at least 47 bases per probe) and probes where genetic variants (SNPs or INDELS) with a minor allele frequency >0.01 in Europeans overlap with the targeted CpG or single base extension site (SBE) were obtained from Pidsley et al. [83]. After probe filtering, the success rate of probes for each sample was checked: All samples had a success rate above 0.95 after removal of low-performing samples detected by MethylAid. Only the autosomal methylation sites were analyzed, leaving 787,711 out of 865,859 sites for analysis.

2.5.2. NTR Replication Study

Genome-wide DNA methylation in buccal cells was assessed with the Infinium MethylationEPIC BeadChip (Illumina, San Diego, CA, USA) [78] by the Avera Institute for Human Genetics. The quality control, processing and normalization of the data were performed with the same pipeline as for the NTR discovery cohort, a pipeline which has previously been described in detail [63].

2.5.3. ALSPAC Replication Study

DNA methylation wet laboratory procedures, preprocessing analyses, and quality control were performed at the University of Bristol, as previously described [71]. DNA methylation outliers were identified as those three times the inter-quartile range from the nearest of the first and third quartiles. Outliers were replaced with missing values.

2.6. Cellular Proportions

2.6.1. NTR Discovery and Replication Study

Cellular proportions were predicted with hierarchical epigenetic dissection of intra-sample-heterogeneity (HepiDISH) with the RPC method (reduced partial correlation), as described by Zheng et al. [84] and implemented in the R package HepiDISH. HepiDISH is a cell-type deconvolution algorithm that was specifically developed for estimating cellular proportions in epithelial tissues based on genome-wide methylation profiles and that makes use of reference DNA methylation data from epithelial cells, fibroblast and seven leukocyte subtypes. This was applied to the data after data QC and normalization.

2.6.2. ALSPAC Replication Study

Cell count estimates were estimated from DNA methylation profiles using a deconvolution algorithm [85] and included in statistical models to adjust for cell count heterogeneity.

2.7. Data Analyses

2.7.1. Associations between Breastfeeding and Pre- and Perinatal Factors

The association between covariates and breastfeeding was tested in the discovery cohort using a generalized estimating equations (GEE) model that accounted for the correlation structure within families. Six breastfeeding duration categories were included as the continuous outcome variable. The predictors consisted of variables previously described as covariates in EWASs of maternal smoking and birth weight [56,57], i.e., parental SES, maternal smoking during pregnancy (yes/no), gestational age (*z*-scores), maternal age at birth (*z*-score), maternal pre-pregnancy BMI (*z*-scores), cell counts of epithelial cells and natural killer (NK) cells, child's sex, and child's age at DNA-methylation (see Table S3). This analysis was performed in SPSS version 25.

2.7.2. EWAS

Discovery Study

The association between DNA methylation level and breastfeeding was tested using a generalized estimating equations (GEE) model that accounted for the correlation structure within families with DNA methylation β-value as the outcome variable (see Table S2). All analyses in the NTR were performed with GEE models, which were fitted with the R package 'gee.' The following settings were used: Gaussian link function (for continuous outcome variables), 100 iterations, and the 'exchangeable' option to account for the correlation structure within families. To examine and adjust (where applicable) for the inflation of test statistics, the R package bacon was used [86]. We first fitted a basic model with the following predictors: breastfeeding, sex, age at sample collection, the percentages of epithelial and natural killer cells, EPIC array row and bisulfite sample plate (using dummy coding). For the primary EWAS, the breastfeeding outcome was dichotomized (yes/no). Secondly, we fitted a model in which we adjusted for additional covariates (SES, maternal smoking, mother's age at birth, mother's BMI at pregnancy, and gestational age). These covariates were selected on the base of adjustments done in recent meta-analyses of maternal smoking and birthweight [56,57]. We also considered the effects of the duration of breastfeeding by evaluating the same model with the original six categories (not-breastfed; <2 weeks; 2–6 weeks; 6 weeks to 3 months; >6 months). Thirdly, we repeated the EWAS in children younger than 10 years and older than 10 years. To this end, we split up the sample into two groups (median age split) to have two equally-sized groups: children younger than 10 years and children older than 10 years. Epigenome-wide significance was assessed following Bonferroni correction for the number of methylation sites tested ($\alpha = 0.05/787{,}711 = 6.34 \times 10^{-8}$).

We carefully checked if epigenome-wide significant associations were influenced by outliers. As a sensitivity analysis, we repeated the association analysis in the discovery sample for significant CpGs without outliers defined using the Tukey method [87], in which an outlier is any value greater than the upper quartile plus three-times the interquartile range or any value lower than the lower quartile minus three-times the interquartile ranges.

Twin Correlations

The correlation between DNA methylation levels of monozygotic (MZ) and dizygotic (DZ) twins was computed for CpGs that were significant in the discovery study and which were robust to outliers. The methylation beta-values were adjusted for the set of covariates as included in the basic EWAS model with a linear model, and residuals were saved. Next, the MZ and DZ twin correlations were computed on the residuals.

Replication

CpGs that were significant in the discovery study and which were robust to outliers were selected for replication. The basic and adjusted models were applied as in the discovery sample: in the NTR, the GEE model had the same covariates; in the ALSPAC, cell count estimates for peripheral blood were included instead of the epithelium cells count; the sex, age of the child at blood collection around age 7 and the same covariates from the adjusted Model 2 were included; no siblings were included. In the ALSPAC, DNA methylation variation due to technical artefacts or unknown confounders were handled by including 20 surrogate variables generated from the DNA methylation data using the 'sva' R package [88], and associations were tested with linear models implemented by the limma package [89].

2.8. Methylation Data Annotation

To examine previously reported associations for epigenome-wide significant CpGs associated with breastfeeding, we looked them up in the EWAS atlas (https://bigd.big.ac.cn/ewas/tools; accessed on 3 October 2019 [90]) and the EWAS catalog (http://www.ewascatalog.org; accessed on 3 October 2019).

To analyze the possible function of CpGs, we searched for overlaps with mQTLs in a study that analyzed EPIC array data from 102 buccal samples [63] to identify the associated SNPs, and we looked up these SNPs and genes in the genome-wide association study (GWAS) catalog (https://www.ebi.ac.uk/gwas/; accessed on 4 October 2019).

2.9. Overlap with Previous Findings

The follow-up analyses were done for CpGs that were previously associated with breastfeeding in EWASs or candidate gene studies. These included all CpGs on the EPIC array that were annotated to the genes *LEP* (n_{CpGs} = 23) [52–55,58], *IL4R* (n_{CpGs} = 37) [58], *CDKN2A* (n_{CpGs} = 32) [59], 1 CpG from the study of Hartwig et al. [60], and 4297 suggestive CpGs from the study of Naumova et al. [53]. In total, this resulted in a list of 4370 CpGs, of which 3859 CpGs were present (after QC) in our data. For these CpGs, we performed a look-up analysis in the total sample (n = 1006) and in the subsample younger than 10 years (n = 517) of the discovery NTR cohort. Significance was assessed after Bonferroni correction for the number of CpGs tested (α = 0.05/3858 = 1.29 × 10^{-5}).

3. Results

3.1. Descriptive Statistics

Descriptive statistics are presented in Table 1. The total discovery sample consisted of 1006 children (mean age 9.5, SD = 1.89); 73.7% of the children were breastfed with different duration, and 26.3% never received breastfeeding. Most of the children were breastfed at least one month: 189 (25.5%) were breastfed for 2 weeks to 1.5 months, 181 (24.4%) were breastfed for 1.5–3 months, 148 (20%) were breastfed for 2–6 months, and 148 (20%) were breastfed more than 6 months. The subsample of children younger than 10 years old included 517 twins (mean age 7.9, SD = 1.1) (see Table S4), and the group older than 10 years included 489 twins (mean age = 11.16, SD = 0.72) (see Table S5). The majority of twin pairs was concordant for breastfeeding (99.4%). Among the discordant twin pairs were three pairs, of which one twin was not breastfed while their co-twin received breastfeeding and three pairs of which the co-twins experienced different durations of breastfeeding.

Table 1. Early life characteristics and breastfeeding in the Netherlands Twin Register (NTR) discovery sample (n = 1006).

	Breastfeeding Never (n = 265)		Breastfeeding Ever (n = 741)		Total	
	n	%	n	%	n	%
Sex						
male	138	52.1%	349	47.1%	487	48.4%
female	127	47.9%	392	52.9%	519	51.6%
Zygosity						
Monozygotic (MZ)	226	85.3%	613	82.7%	839	83.4%
Dizygotic (DZ)	39	14.7%	128	17.3%	167	16.6%
Chorionicity						
MCMA	4	3.5%	21	7.3%	25	6.2%
MCDA	61	53.0%	144	50.3%	205	51.1%
DCDA	50	43.5%	121	42.3%	171	42.6%
Gestational Age (Weeks)						
Mean (SD)	36.2	(22.2)	35.8	(25.9)	35.9	(25.1)
≤ 32	12	4.8%	61	8.4%	73	7.5%
33–36	128	51.4%	359	49.4%	487	49.9%
≥ 37	109	43.8%	306	42.1%	415	42.6%
Mother's Age at Birth (Years)						
Mean (SD)	31.9	(4.5)	31.2	(4.2)	31.4	(4.3)
19–29	76	29.0%	288	39.0%	364	36.4%
30–39	175	66.8%	435	58.9%	610	61.0%
>40	11	4.2%	15	2.0%	26	2.6%
Mother's BMI Before Pregnancy						
Mean (SD)	24.3	(4.0)	24.2	(4.11)	24.2	(4.1)
<25	149	61.3%	470	66.1%	619	64.9%
25–29	70	28.8%	169	23.8%	239	25.1%
30–39	24	9.9%	65	9.1%	89	9.3%
>40	0	0.0%	7	1.0%	7	0.7%

Table 1. *Cont.*

	Breastfeeding Never (*n* = 265)		Breastfeeding Ever (*n* = 741)		Total	
	n	%	*n*	%	*n*	%
Father's Age at Birth (Years)						
Mean (SD)	33.2	(4.4)	33.9	(5.4)	33.7	(5.2)
20–29	53	22.0%	146	20.3%	199	20.8%
30–39	163	67.6%	482	67.1%	645	67.3%
>40	25	10.4%	90	12.5%	115	12.0%
Mode of Conception						
naturally	227	92.7%	623	86.5%	850	88.1%
stimulated	4	1.6%	26	3.6%	30	3.1%
IVF/ICSI	14	5.7%	71	9.9%	85	8.8%
Maternal Smoking						
no smoking	205	86.1%	631	92.9%	836	91.2%
smoking	33	13.9%	48	7.1%	81	8.8%
Parental SES						
low skill level	0	0.0%	8	1.2%	8	0.9%
lower secondary educational level	30	11.9%	41	6.1%	71	7.6%
upper secondary education level	99	39.3%	203	30.0%	302	32.5%
higher vocational level	89	35.3%	234	34.6%	323	34.8%
scientific level	34	13.5%	191	28.2%	225	24.2%
Mode of Delivery						
vaginal	141	56.6%	416	57.1%	557	57.0%
caesarean planned	43	17.3%	97	13.3%	140	14.3%
urgent intervention (forceps, vacuum extraction)	20	8.0%	75	10.3%	95	9.7%
urgent caesarean section	45	18.1%	140	19.2%	185	18.9%
Birth Weight						
Mean (SD)	2435.7	(444.8)	2394.6	(558)	2405	(531.7)
<1500	8	3.2%	52	7.1%	60	6.2%
1500–2500	123	49.8%	338	46.4%	461	47.3%
>2500	116	47.0%	338	46.4%	454	46.6%
Apgar Score at 1st Minute						
0–6	17	12.3%	48	12.8%	65	12.7%
7–9	103	74.6%	290	77.5%	393	76.8%
10	18	13.0%	36	9.6%	54	10.5%
Apgar Score at 5th Minute						
0–6	1	0.8%	14	3.9%	15	3.1%
7–9	41	31.1%	130	36.3%	171	34.9%
10	90	68.2%	214	59.8%	304	62.0%
Breastfeeding Duration						
no	265	100.0%	0		265	26.3%
less than 2 weeks			75	10.1%	75	7.5%
2–6 weeks			189	25.5%	189	18.8%
6 weeks to 3 months			181	24.4%	181	18.0%
3–6 months			148	20.0%	148	14.7%
more than 6 months			148	20.0%	148	14.7%

Descriptive statistics of children included in the study. MCMA = monochorionic monoamniotic; MCDA = monochorionic diamniotic; DCDA = dichorionic diamniotic; SD = standard deviation; BMI = body mass index; IVF = in vitro fertilization; ICSI = intracytoplasmic sperm injection; and SES = socio-economic status.

Breastfeeding duration (six categories) was significantly positively associated with socio-economic status (β = 0.353, standard error (SE) = 0.09, and p = 0.0004) and inversely associated with maternal smoking (β = −0.733, SE = 0.32, and p = 0.02) and gestational age (β = −0.194, SE = 0.09, and p = 0.03) (see Table S3). Breastfeeding was not significantly correlated with maternal pre-pregnancy BMI (β = −0.117, SE = 0.75, and p = 0.119), maternal age at delivery (β = −0.016, SE = 0.09, and p = 0.86), and the cell composition of buccal swabs (β = 1.27, SE = 1.2, and p = 0.25 for count of epithelial cells; and β = 14.78, SE = 9.8, and p = 0.13 for count of natural killer cells).

3.2. Association Analysis Findings

First, genome-wide DNA-methylation analyses were performed to test the association between breastfeeding (never/ever) and DNA-methylation level while adjusting for sex, age at DNA sample collection, estimated proportion of epithelial cells, the estimated proportion of natural killer cells, the row of the sample on the chip, and the bisulfite plate (model 1). No epigenome-wide significant sites were identified (see Table S8). Genome-wide EWAS test statistics showed no inflation (see Figure S1). DNA methylation was also not associated with the duration of breastfeeding (see Table S9).

In the EWAS in children younger than 10 years with the same basic model, we identified four epigenome-wide significant CpGs (see Table S10): cg25178826 (β = −0.026, SE = 0.003, and $p = 8.04 \times 10^{-12}$) is located in the 5′ untranslated region (5′UTR) region of *PRLR* (prolactin receptor) gene, cg12087956 (β = −0.03, SE = 0.005, and $p = 1.8 \times 10^{-8}$) is in the gene body of *CDAN1* (codanin 1), cg24192772 (β = −0.024, SE = 0.004, and $p = 2.5 \times 10^{-8}$) is in the gene body of *FOXK2* (forkhead box K2), and cg10142656 (β = −0.02, SE = 0.004, and $p = 6.28 \times 10^{-8}$) is mapped to the transcription start site (TSS1500) of *TRMT10B* (tRNA methyltransferase 10B). These four CpGs were not strongly associated with breastfeeding duration (see Table S11). The plots of the methylation values for these four CpGs revealed several extreme methylation values in the data (see Figure S2 and Table S10), and the association with breastfeeding disappeared after outlier removal: cg25178826 β = −0.00006, SE = 0.001, and p = 0.98; cg12087956 β = 0.00004, SE = 0.0009, and p = 0.5; cg24192772 β = 0.0009, SE = 0.001, and p = 0.15; and cg10142656 β = 0.0001, SE = 0.0008, and p = 0.13 (see Table S10).

Next, we performed EWASs of breastfeeding never/ever with adjustments for SES, maternal age at delivery, maternal pre-pregnancy BMI, maternal prenatal smoking, and gestational age (Model 2). One CpG, cg22491379, was significant in the total sample (β = −0.007, SE = 0.001, and $p = 1.3 \times 10^{-9}$) (see Figure 2a, Figure S3, Table 2 and Table S12), and seven CpGs were significant in the children younger than 10 years (see Table 2 and Table S14). The Manhattan plot for the EWAS in children younger than 10 years showed a peak on chromosome 21, which contains a cluster of CpGs just below the epigenome-wide significance threshold (see Figure 2d). Results from the same analysis for breastfeeding duration are shown in Tables S13 and S15.

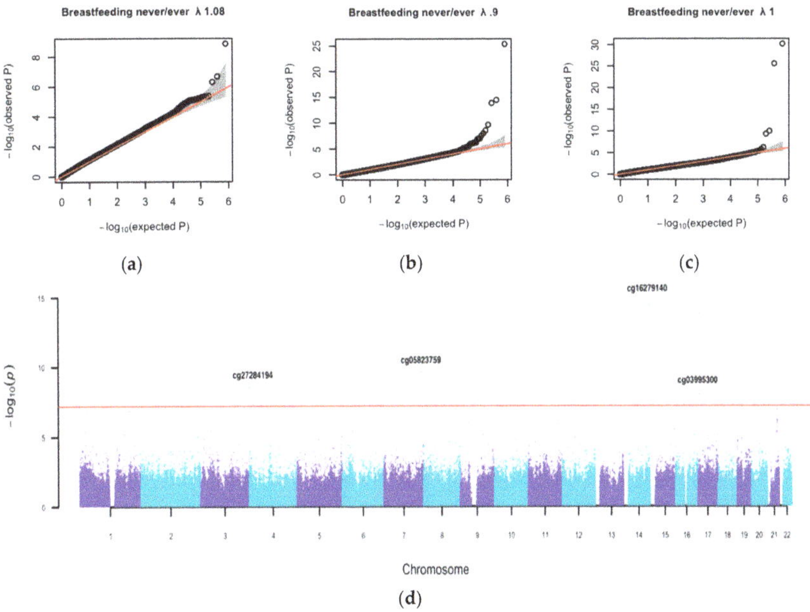

Figure 2. Quantile-Quantile (QQ) plots: (**a**) in the total sample; (**b**) in the subsample of children younger than 10 years old; and (**c**) in the subsample of children older than 10 years old. (**d**) Manhattan plot shows the epigenome-wide association study results of breastfeeding 'never'/'ever' in children younger than 10 years old. Covariates included sex, age, socio-economic status (SES), maternal age at delivery, maternal pre-pregnancy BMI, maternal prenatal smoking, gestational age, EPIC array row and bisulfite sample plate, and cell composition. In the Manhattan plot, the red line represents the Bonferroni threshold (6.34×10^{-8}). Given CpG names indicate significant loci after removal of outliers.

Table 2. Summary of epigenome-wide significant CpGs from the discovery epigenome-wide association study (EWAS) of DNA signatures of breastfeeding.

cgID	Chromosome	Position	Gene	Gene Region	Discovery Study			Discovery Study Without Outliers		
					Estimate	SE	*p*-Value	Estimate	SE	*p*-Value
Basic Model (1). Sub-Sample < 10 years (n = 517)										
cg25178826	chr5	35165447	*PRLR*	5′UTR	−0.026	0.004	8.04×10^{-12}	-6.04×10^{-5}	0.001	0.98
cg12087956	chr15	43022167	*CDAN1*	Body	−0.031	0.005	1.18×10^{-8}	4.92×10^{-5}	0.001	0.50
cg24192772	chr17	80536920	*FOXK2*	Body	−0.024	0.004	2.52×10^{-8}	9.17×10^{-4}	0.001	0.15
cg10142656	chr9	37753047	*TRMT10B*	TSS1500	−0.019	0.004	6.28×10^{-8}	9.98×10^{-5}	0.0007	0.14
Adjusted Model (2). Discovery Sample (n = 1006)										
cg22491379	chr2	120553625	*PTPN4*	5′UTR	−0.007	0.001	1.30×10^{-9}	−0.005	0.001	5.78×10^{-3}
Adjusted Model (2). Sub-Sample < 10 Years (n = 517)										
cg03463465	chr6	164143581			0.360	0.034	4.51×10^{-26}	−0.003	0.001	0.01
cg07670516	chr17	5019840	*ZNF232*	5′UTR	0.249	0.032	1.40×10^{-14}	0.006	0.014	0.65
cg20820810	chr11	71850130	*FOLR3*	Body	−0.300	0.054	2.82×10^{-8}	−0.001	0.001	0.21
cg16279140	chr14	103981749			−0.411	0.052	3.50×10^{-15}	no outliers		
cg05823759	chr7	149646627			0.205	0.032	2.35×10^{-10}	no outliers		
cg27284194	chr4	1044797			0.638	0.107	2.90×10^{-9}	no outliers		
cg03995300	chr17	5019989	*ZNF232*	5′UTR	0.229	0.040	1.02×10^{-8}	no outliers		

α = 0.05/787,711 = 6.34×10^{-8}. Basic Model 1 included breastfeeding coded as 'never' and 'ever' with adjustments for sex, age at DNA methylation, the count of epithelial cells, the count of natural killer cells, EPIC array row and bisulfite sample plate. Adjusted Model (2) included in additional to basic model covariates: SES, maternal smoking during pregnancy, maternal age at birth, maternal pre-pregnancy BMI, and gestational age. In bold: CpGs selected for replication.

Genome-wide EWAS test statistics from the analyses of Model 2 (in the total sample and in the children younger than 10 years) showed no inflation (see Figure 2a,b). For four out of the eight significant CpGs, the association disappeared after removal of outliers (see Figures S4 and S5). The four CpGs that were unaffected by outliers were selected for a replication analysis (see Figure S6): cg16279140 ($\beta = -0.4$, SE $= 0.05$, and $p = 3.5 \times 10^{-15}$), cg05823759 ($\beta = 0.2$, SE $= 0.032$, and $p = 2.35 \times 10^{-10}$), cg27284194 ($\beta = 0.64$, SE $= 0.12$, and $p = 2.9 \times 10^{-9}$), and cg03995300 ($\beta = 0.23$, SE $= 0.04$, and $p = 1.2 \times 10^{-8}$) (see Table S14). All four CpGs were significant in children younger than 10 years.

We next examined the association with breastfeeding in children older than 10 (Model 2). In this analysis, the genome-wide test statistics showed mild inflation (lambda $= 1.11$). After adjusting for inflation, four CpGs remained epigenome-wide significant, but none were significant after outlier removal. The four significant CpGs in the children younger than 10 years had smaller effects, and three CpGs showed inverse directions of effect in children older than 10: cg16279140 ($\beta = 0.04$, SE $= 0.02$, and $p = 0.04$), cg05823759 ($\beta = -0.002$, SE $= 0.02$, and $p = 0.92$), cg27284194 ($\beta = 0.18$, SE $= 0.06$, and $p = 0.002$), and cg03995300 ($\beta = -0.003$, SE $= 0.02$, and $p = 0.89$) (see Table S16). The quantile-quantile plots (see Figure 2b,c) suggested a stronger association signal for breastfeeding in children younger than 10 years than in the older group. We computed the correlations between the regression coefficients (i.e., the methylation difference associated with breastfeeding) for the top 100 CpGs from each age group. These correlations were weak ($r = 0.25$ and $p = 0.10$ for the top 100 CpG from the EWAS in children <10 years and from the EWAS in children >10 years). This suggests that methylation profiles associated with breastfeeding are different in children younger than 10 years and older than 10 years.

We computed the twin correlations in the MZ and DZ pairs for the significant CpGs, as we assumed that equal exposure to breastfeeding should cause similarity in methylation level of twins. The correlations of methylation levels of all four sites were very high in MZ twins and almost twice as high as in DZ twins: cg16279140 rMZ $= 0.906$, rDZ $= 0.505$; cg05823759 rMZ $= 0.953$, rDZ $= 0.609$; cg27284194 rMZ $= 0.972$, rDZ $= 0.462$; and cg03995300 rMZ $= 0.902$, rDZ $= 0.453$. This pattern suggests that there are heritable influences on DNA methylation at these CpGs, as previously shown by other studies of DNA methylation heritability.

3.3. Replication Analysis of Our Findings in Other Samples

Both replication datasets were comparable to the discovery subsample of children younger than 10 years old (see Tables S4, S6, and S7). As expected, the singleton children (ALSPAC) had a higher gestational age (~39 weeks vs. ~35.7), a higher birthweight (~3500 g vs. ~2400 g) and a lower proportion of caesarean deliveries (~9% vs. ~36.8%). We performed a replication analysis for significant CpGs unaffected by outliers from the adjusted Model 2 of the discovery analysis with the same set of covariates (see Table 3).

In the NTR replication cohort, for one CpG, the direction of effect changed (cg03995300, *ZNF232*, $\beta = -0.050$, SE $= 0.033$, and $p = 0.13$), and three other CpGs had the same direction of effect but were not significant: cg16279140 ($\beta = -0.03$, SE $= 0.04$, and $p = 0.43$), cg05823759 ($\beta = 0.03$, SE $= 0.033$, and $p = 0.32$), and cg27284194 ($\beta = 0.07$, SE $= 0.054$, and $p = 0.21$) (see Table 3, Figure S7).

In the ALSPAC, the most important difference to note were the different tissues used for DNA methylation profiling (peripheral blood) and the platform (the Illumina Infinium HumanMethylation450 Beadchip [Illumina, San Diego, CA, USA]). Because of the platform difference, roughly half of the CpG sites included in the discovery DNA methylation profiles were measured in the ALSPAC. In the adjusted Model 2, two of the four most strongly associated CpG sites were not included on the ALSPAC platform (cg16279140 and cg05823759), nor were any other sites within 1000 base pairs (bp) of these sites. The other two sites, cg27284194 and cg03995300, were included, but neither association was replicated ($p > 0.19$ and $p > 0.77$, respectively) in the adjusted model (see Table 3). An overall lack of replication was confirmed by low correlations between effect estimates in the discovery and the ALSPAC. In particular, we identified the 100 most strongly associated CpG sites in the discovery that were also present in the ALSPAC DNA methylation profiles. The correlation of effect estimates for these sites between studies was low (Pearson's $r = 0.13$ with $p = 0.2$ for the adjusted model).

Table 3. Summary of association between breastfeeding and significant in the discovery study CpGs in replication.

cgID	Direction of Effect in Discovery Study <10 Years	NTR Replication Study ($n = 98$)			ALSPAC Replication Study ($n = 938$)		
		Estimate	SE	*P*-Value [a]	Estimate	SE	*P*-Value
cg16279140	−	−0.0326	0.0412	0.43	NA	NA	NA
cg05823759	+	0.0329	0.0332	0.32	NA	NA	NA
cg27284194	+	0.0668	0.0542	0.21	0.0047	0.016	0.77
cg03995300	+	−0.0502	0.0334	0.13	0.0140	0.011	0.19

Adjusted Model 2 was used. [a] $\alpha = 0.05/4 = 0.01$. NA = not available on the 450k platform.

3.4. Replication Analysis of Findings from Previous EWAS

Of the 3858 CpGs from previous literature, four CpGs were associated with breastfeeding in our sample ($\alpha = 1.29 \times 10^{-5}$, see Table 4). One site was significant in the total sample (see Table S17): cg16387046 is located on chromosome 12 in the *MUCL1* (mucin like 1) gene ($\beta = 0.027$, SE = 0.005, and $p = 4.9 \times 10^{-7}$). Three sites were significant in the children younger than 10 years (see Table S18): cg16704958 ($\beta = 0.009$, SE = 0.002, and $p = 8.03 \times 10^{-6}$) and cg11287055 ($\beta = 0.056$, SE = 0.01, and $p = 4.9 \times 10^{-6}$), located on chromosome 21 in the *VPS26C* (endosomal protein sorting factor C; previous name *DSCR3*) gene, and cg26479305 ($\beta = 0.338$, SE = 0.07, and $p = 1.11 \times 10^{-5}$), located on chromosome 12 in the *ATG101* (autophagy related 101, previous name *C12orf44*) gene. All these CpGs were previously reported in the association study by Naumova et al. [53] that was carried out in infants around two years old (mean age 25.7 months) with DNA methylation profiling in peripheral blood with Illumina EPIC BeadChip (Illumina, San Diego, CA, USA). The direction of association was positive for all four CpGs in both studies. CpGs located in/nearby *LEP* [52,55], *IL4R* [58] and *CDKN2A* [59], previously discussed as candidate genes for association with breastfeeding, were not significant in our study.

Table 4. Replication of CpGs from previous literature.

cgID	Chromosome	Position	Gene	Gene Region	ESTIMATE	SE	*P*-Value
			Total Sample ($n = 1006$)				
cg16387046	chr12	55248207	*MUCL1*	TSS200	0.027	0.005	4.93×10^{-7}
			Sub-Sample <10 ($n = 517$)				
cg11287055	chr21	38630234	*VPS26C (DSCR3)*	Body	0.056	0.012	4.93×10^{-6}
cg16704958	chr21	38630728	*VPS26C (DSCR3)*	Body	0.009	0.002	8.03×10^{-6}
cg26479305	chr12	52470979	*ATG10 (C12orf44)*	3'UTR	0.338	0.077	1.11×10^{-5}

$\alpha = 0.05/3859 = 1.29 \times 10^{-5}$. The table shows all CpGs that were previously reported to be associated with breastfeeding and that were significantly replicated by our study.

3.5. Methylation Data Annotation

For the significant CpGs from discovery and follow-up studies, we looked up the associations with nearby genetic variants (mQTLs) in the results from a previously published mQTL study of buccal-derived DNA in monozygotic twins [63]. Four CpGs were associated with mQTLs (cg27284194, associated with SNPs on chromosome 4: 927973–1039876; cg16279140, associated with SNPs on chromosome 14: 103823243–104186876; cg16387046, associated with SNPs on chromosome 12: 54830518–54896008; and cg26479305, associated with SNPs on chromosome 12: 52492131). Additionally, no significant mQTLs were found for four CpGs (cg03995300, cg05823759, cg11287055, cg16704958).

Next, we compared our results against all previously associated traits in EWASs (the EWAS atlas and the EWAS catalog) and GWASs (the GWAS catalog). One CpG (cg03995300) is mapped to *ZNF232* and has been previously associated with prenatal maternal smoking, sex, and ancestry [56,91,92]. The *ZNF232* locus was associated with a family history of Alzheimer's disease in a GWA meta-analysis [93]. The intergenic CpG on chromosome 4 (cg27284194) has been previously associated with infertility [94,95]. In GWASs, the region 4: 927973–1039876 (harboring mQTLs for cg27284194) has been associated with a number of traits and diseases, including bone mineral density [96–100] and several metabolomic characteristics such as blood protein [101,102] and triglyceride [103] levels (See Table S19). The intergenic CpG on chromosome 14 (cg16279140) did not appear in previous EWASs. In GWASs, SNPs on chromosome 14: 103823243–104186876 have been linked to amino acids levels as biomarkers of metabolic disorders [104] and other blood metabolites [105], and some disease and addictions including multisite chronic pain [106], metabolite changes in chronic kidney disease [107], bipolar disorder [108], alcohol consumption [109] and risk-taking behavior [110] (See Table S19). The intergenic CpG on chromosome 7, cg05823759, has not been previously identified in an EWAS and is not associated with mQTLs.

Four CpGs that were replicated from previous studies on breastfeeding exposure have been reported in association with other traits and diseases. The *ATG10 (C12orf44)* gene, where cg26479305 is located, has been related to the autophagy pathway and cellular senescence in an EWAS [111]. It is related to regulators of inflammation (circulating cytokines and growth factors) [112], hematological traits in GWASs [113], and prenatal arsenic exposure [114].

CpGs (cg11287055, cg16704958) mapped to the *VPS26C (DSCR3)* have not been reported in other EWASs. *VPS26C* is a component of the retriever complex, which plays a role in cell surface processes such as cell migration, cell adhesion, nutrient supply and cell signaling [115].

4. Discussion

We aimed to study DNA methylation in buccal cells in mid-childhood in association with breastfeeding as an exposure. Associations were tested in the total sample and in groups stratified by age. The age distribution allowed us to analyze whether effects of breastfeeding can be attenuated with age, as this has been observed for other exposures in peripheral blood, including maternal smoking [56], birthweight [57] and breastfeeding [55].

We did not find evidence for robust epigenome-wide significant associations in the total sample (1006 children, age 5–12), but we did observe associations in the younger age group (before 10 years) that did not appear in the group of children of 10–12 years old. This suggests that epigenetic alterations in certain genomic regions might be associated with nutritional differences in the early postnatal life that tend to fade across childhood. However, since the CpGs were not replicated in the small, buccal-cells replication NTR dataset and in the large blood-cells replication dataset from the ALSPAC, further studies are required to follow up on these findings. It also remains to be examined if the associations reflect a causal effect of breastfeeding on methylation and whether these methylation differences might influence developmental trajectories and later-life health of the child.

The observation that associations between breastfeeding and methylation at some CpGs are age-dependent may be explained by age-related changes in DNA methylation that have been reported in several studies [116,117], including children in the age range of our study [118–122]. Similarly, Sherwood et al. [55] observed that breastfeeding is associated with the DNA methylation of the *LEP* gene at age 10 but not at age 18. Previous studies have identified DNA methylation signatures of prenatal nutrition in peripheral tissues at various life stages, even in adulthood. DNA methylation signatures in many genomic regions have been identified in middle-aged individuals who were exposed to the 1944/45 Dutch hunger winter at the time of conception, and many of these sites have been found to be related to growth, developmental processes and metabolism [18,19]. It is unclear if the effects of early life postnatal exposures can have similar long-term effects to prenatal exposures in humans, as most studies of early life influences on the epigenome have focused on prenatal exposures or have not examined long-term effects. None of the genes reported in studies of prenatal malnutrition (*SMAD7, CDH23, INSR, RFTN1, CPT1A, KLF13* [19], *IGF2* [18], and *LEP* [123]) were present among the top CpGs associated with breastfeeding in our study. Several explanations can be proposed. First, postnatal nutrition and prenatal nutrition exposures might differently influence the DNA methylation. Second, DNA methylation signatures induced by early nutrition exposure can be different across examined tissues (buccal cells in our study versus peripheral blood in previous famine studies). Third, the prenatal famine study examined the effects of an extreme exposure during the prenatal period.

The presence of associations with breastfeeding initiation (breastfeeding never vs. ever), and the absence of findings with breastfeeding duration suggests that the association depends on the exposure to breast milk rather than its duration. This could potentially be in line with previous findings that have shown that effects of exposure on DNA methylation occur only when the individual is exposed in certain sensitive life periods [118]. Our finding of stronger epigenetic associations with breastfeeding ever vs. never than with the duration of breastfeeding is also in line with studies providing evidence that any breastfeeding has stronger biological effects than the duration of breastfeeding, e.g., the impact of first maternal milk (colostrum) on immunoglobulins and further neonatal health, especially for small for gestational age and low-birthweight infants [22,24].

We observed associations at a total of eight CpG sites: cg16279140, chromosome 14, intergenic; cg05823759, chr7, intergenic; cg27284194, chromosome 4, intergenic; cg03995300, chromosome 17, *ZNF232*; cg16387046, chromosome 12, *MUCL1*; cg11287055, chromosome 21, *VPS26C* (*DSCR3*); cg16704958, chromosome 21, *VPS26C* (*DSCR3*); and cg26479305, chromosome 12, *ATG10* (*C12orf44*). Methylation levels at these CpGs are potentially influenced by both environmental influences, such as breastfeeding, and by genetic variation, although an alternative interpretation could be that the genomic regions of these CpGs affect breastfeeding function. Unfortunately, no GWAS on breastfeeding is available to verify this suggestion. It remains to be examined whether methylation differences induced by breastfeeding have an effect on the aforementioned traits. Interestingly, associations with breastfeeding have been previously reported in epidemiological studies for bone mineral content and bone mineral density [124] and metabolite profiles [125]; outcomes that have been associated with SNPs in the regions detected in our study as differentially methylated between breastfed and non-breastfed children. It remains to be examined whether methylation differences induced by breastfeeding have an effect on these traits.

Since the role of nutrition is systemic, it has been assumed that biomarkers must be present in different tissues. To study the epigenetic mechanisms of breastfeeding, previous studies have examined DNA from peripheral blood [52–55,58] and also from tumor tissue [59]. Buccal epithelium is of interest because it offers a non-invasive way of biosample collection for an epigenetic analysis. A number of studies have demonstrated the potential of buccal cells to study DNA methylation [126]. We previously showed that, although there is some correlation between DNA methylation in buccal and blood cells ($n = 22$, age = 18 years), it is low for most CpGs interrogated by the Illumina 450k array [127]. The methylation levels of two of the CpG sites that were included in the ALSPAC replication study were highly correlated in buccal and blood cells: cg27284194 ($r = 0.864$) and cg03995300 ($r = 0.782$). In spite of this, the associations with breastfeeding observed in the NTR were not replicated in the ALSPAC. We observed some overlap with the study of Naumova et al. [53] in peripheral blood that used the EPIC array, and these sites had medium correlations between DNA methylation in buccal and blood cells (cg16387046 $r = 0.690$, cg11287055 $r = 0.413$, and cg16704958 $r = 0.328$).

We observed that breastfeeding-associated CpGs show strong correlations in MZ twins and larger correlations in MZ twins compared to DZ twins. As the large majority of twin pairs in our study were almost always concordant for breastfeeding, it is expected that twins of both types show resemblance for DNA methylation levels at these sites. The larger correlation in MZ compared to DZ twins in our study suggests that these sites are also subject to heritable influences. In line with this observation, some of these CpGs are associated with mQTLs. It should be noted that breastfeeding and lactation itself are heritable traits. In previous twin studies, the heritability of initiation of breastfeeding ranged from 49% [128] to 70% [129]. DNA-methylation profiles may be influenced by several early life factors. In our study, almost all early environmental factors of interest were shared by twins: SES, maternal age at birth, maternal pre-pregnancy BMI, and maternal smoking during pregnancy and gestational age.

Some CpGs that did not reach epigenome-wide significance can have potential for further epigenetic studies of breastfeeding. The CpG cg22491379, located in the *PTPN4* on chromosome 2, was discovered in total sample of children of 5–12 years old and, after outlier removal, had suggestive significance ($p = 5.78 \times 10^{-3}$). *PTPN4* was listed in suggestive regions associated with breast morphology (breast size) [130]. Some top CpGs in the discovery study (non-adjusted model) are associated with genes that are involved in growth and metabolism. The *PRLR* (prolactin receptor) gene, located on chromosome 5, is involved in prolactin receptor activity and the growth hormone receptor signaling pathway. Growth hormone binds to the prolactin receptor, this being the basis of induction of lactation by growth hormone [131]. The *FOXK2* gene, located on chromosome 17, is a member of forkhead box transcription factors and is involved in glucose metabolism, aerobic glycolysis and autophagy [132], thus playing a role in metabolic reprogramming towards aerobic glycolysis [133]. It is associated lean body mass [134] and has been found to be hypomethylated in CpG islands in obese patients' adipose tissues [135]. We observed a cluster of associations on chromosome 21, each individual association just below epigenome-wide significance. The *VPS26C* (*DSCR3*) gene is located on chromosome 21 and, as mentioned earlier, contains two CpGs previously associated with breastfeeding in blood [53].

The strengths of our study include the use of the buccal epithelium methylome to investigate breastfeeding for the first time, sample size, and the availability of replication data, including data from another tissue (blood). Furthermore, our exposure, breastfeeding, was assessed very shortly after it occurred, reducing the risk of measurement error and recall bias. Both the discovery and replication samples have been thoroughly phenotyped and extensively studied. Our study also has limitations. First, the findings in the discovery study were not replicated in the cohort of children with buccal cell DNA methylation of the same age, possibly due to the small size of this cohort (98 monozygotic twins). Second, the discovery sample was included in a study of aggressive behavior, and not primarily examined for the purposes of research on breastfeeding. Third, we did not have longitudinal DNA methylation data to measure the stability of effects; however, we examined the possible attenuation of breastfeeding effects through an age-stratified analysis. Fourth, there are currently no datasets available on gene expression in buccal cells; thus, we could not examine relationships between DNA methylation and transcription. A strength of the current analysis is that it used the Illumina EPIC array, which has much greater coverage than the 450k array. Importantly, some of the top CpGs identified in the discovery EWAS were all novel EPIC probes. The replication analysis in the ALSPAC on DNA methylation data from peripheral blood, however, used the 450k array and therefore did not permit look-up of the same CpGs. The findings of this study require further replication in cohorts with buccal epithelium and other tissues to improve our understanding of breastfeeding-associated methylation changes in different tissues, and the possible utility as a biomarker of early life nutrition. Finally, a difficulty in studies of breastfeeding is that breast milk composition is unique in each mother. Lactation is influenced not only by genetic variation but also by many environmental factors such as the mother's nutrition, lifestyle, level of stress, and attachment to the child [24,25,136–139]. Future epigenetic studies of breastfeeding could stratify the breastfeeding sample on the basis of criteria of breast milk composition. This might be more informative for predicting a child's outcome but might also be a better indicator of a mother's well-being. The value of our research is that it combined breastfeeding, prenatal characteristics, and methylation data. In the future, results from studies that will integrate epigenomic data with genomics, transcriptomics, and metabolomics may be used to develop prediction models of long-term outcomes in child development and health. Understanding of the mechanisms associated with breastfeeding will help to develop interventions to improve children's health and reduce risk of chronic disease by supporting breastfeeding or optimizing infant nutrition when breastfeeding is not possible.

5. Conclusions

Our study provided a first indication that breastfeeding as an early life environmental factor may be associated with epigenetic variation in buccal cells in children. The findings point at new candidate loci influenced by breastfeeding. Future studies are needed to investigate if the DNA methylation signatures are caused by breastfeeding or by other unmeasured confounders (including a genetic predisposition to give or receive breastfeeding and other aspects of prenatal or postnatal diet), whether they are influenced by percentage of breastfed meals, exclusive breastfeeding duration, breastmilk composition, etc., and what age-related mechanisms drive changes in the association between breastfeeding and methylation.

Supplementary Materials: The following are available online at http://www.mdpi.com/2072-6643/11/11/2804/s1, Table S1: Overview of previous EWAS of breastfeeding in humans, Table S2: EWAS Model Equations, Table S3: Associations of covariates with breastfeeding duration (six categories), Table S4: Discovery study subsample of children <10 years population characteristics, Table S5: Discovery study subsample of children >10 years population characteristics, Table S6: NTR replication sample population characteristics, Table S7: ALSPAC replication sample population characteristics, Table S8: Association of breastfeeding "never"/"ever" with CpG sites in basic Model 1 in total sample: top 100 and genomic annotation, Table S9: Association of breastfeeding duration (six categories) with CpG sites in basic Model 1 in total sample: top 100 and genomic annotation, Table S10: Association of breastfeeding "never"/"ever" with CpG sites in basic Model 1 in sub-sample <10 years: top 100 and genomic annotation, Table S11: Association of breastfeeding duration (six categories) with CpG sites in basic Model 1 in sub-sample <10 years: top 100 and genomic annotation, Table S12: Association of breastfeeding "never"/"ever" with CpG sites in adjusted Model 2 in total sample: top 100 and genomic annotation, Table S13: Association of breastfeeding duration (six categories) with CpG sites in adjusted Model 2 in total sample: top 100 and genomic annotation, Table S14: Association of breastfeeding "never"/"ever" with CpG sites in adjusted Model 2 in sub-sample <10 years: top 100 and genomic annotation, Table S15: Association of breastfeeding duration (six categories) with CpG sites in adjusted Model 2 in sub-sample <10 years: top 100 and genomic annotation, Table S16: Association of breastfeeding "never"/"ever" with CpG sites in adjusted Model 2 in sub-sample >10 years: top 100 and genomic annotation, Table S17: Follow-up of associations of breastfeeding with CpGs from literature in discovery total sample ($n = 1006$), Table S18: Follow-up of associations of breastfeeding with CpGs from literature in discovery sub-sample < 10 years old ($n = 517$), Table S19: Association of CpGs associated with breastfeeding in discovery study with mQTLs, Figure S1: QQ plots that shows epigenome-wide association study results of breastfeeding ever/never and breastfeeding duration (six categories) in total sample ($n = 1006$) and in sub-sample of children < 10 years ($n = 517$), Figure S2: Methylation level (beta) at significant CpGs in Model 1 in sub-sample of children <10 years ($n = 517$), Figure S3: Manhattan plot show the epigenome-wide association study results of breastfeeding 'never'/'ever' in discovery sample Model 2 ($n = 1006$), Figure S4: Methylation level (beta) at significant CpG in Model 2 in total sample ($n = 1006$), Figure S5: Methylation level (beta) at significant CpGs in Model 2 in sub-sample of children <10 years ($n = 517$) with outliers excluded from further replication, Figure S6: Methylation level (beta) at significant CpGs in Model 2 in sub-sample of children <10 years included in replication study, Figure S7: Methylation level (beta) at significant CpGs in Model 2 in NTR replication study ($n = 98$).

Author Contributions: Conceptualization, D.I.B., C.R., M.S., J.v.D.; Data curation, F.A.H., C.E.M.v.B. and J.v.D.; formal analysis, V.V.O. and J.v.D.; funding acquisition, D.I.B., M.B.,V.V.O.; investigation, F.A.H., C.E.M.v.B., E.A.E., N.A.K. and G.E.D.; methodology, D.I.B., M.S., J.v.D.; resources, E.A.E., G.E.D. and J.v.D.; software, V.V.O. and J.v.D.; supervision, D.I.B., C.R., M.B. and J.v.D.; visualization, V.V.O; validation, V.V.O., J.v.D., M.S., D.C. and C.R.; writing—original draft, V.V.O., M.S. and J.v.D.; writing—review & editing, F.A.H., C.E.M.v.B., D.C., G.T.S., V.F., D.I.B. and J.v.D.

Funding: V.V.O. is supported by Amsterdam Public Health research institute, and the Russian Foundation for Basic Research (20-015-00496). The Netherlands Twin Register acknowledges funding from multiple grants from the Netherlands Organization for Scientific research (NWO), including NWO-Grant 480-15-001/674: Netherlands Twin Registry Repository; the Biobanking and Biomolecular Resources Research Infrastructure (BBMRI—NL, 184.021.007 and 184.033.111); OCW_Gravity program—NWO-024.001.003 Consortium for Individual Development) and the Avera Institute for Human Genetics, Sioux Falls, South Dakota (USA). The methylation work, J.v.D. and F.H. were supported by funding from the European Union Seventh Framework Program (FP7/2007–2013) under grant agreement no 602768 (ACTION). J.v.D. is supported by the NWO-funded X-omics project (184.034.019) and D.I.B. the KNAW Academy Professor Award (PAH/6635). The UK Medical Research Council and Wellcome (Grant ref: 102215/2/13/2) and the University of Bristol provide core support for ALSPAC. A comprehensive list of grants funding is available on the ALSPAC website (http://www.bristol.ac.uk/alspac/external/documents/grant-acknowledgements.pdf). This research was specifically funded by the BBSRC (grant numbers BBI025751/1 and BB/I025263/1). C.R. and D.C. are funded by the MRC (grant numbers MC_UU_00011/5 and MC_UU_00011/1). This publication is the work of the authors and V.V.O. and J.v.D. will serve as guarantors for the contents of this paper.

Acknowledgments: We are very thankful to all NTR families for their participation in this study, which involved biosample collection and phenotyping. We are grateful to all the families who took part in the ALSPAC study, the midwives for their help in recruiting them, and the whole ALSPAC team, which includes interviewers, computer and laboratory technicians, clerical workers, research scientists, volunteers, managers, receptionists and nurses.

Conflicts of Interest: The authors declare no conflict of interest.

Abbreviations

CpG	cytosine-phosphate-guanine
EWAS	epigenome-wide association study
NTR	Netherlands Twin Register
ALSPAC	Avon Longitudinal Study of Parents and Children
SES	socio-economic status
GA	gestational age
BMI	body mass index
GEE	generalized estimating equations
mQTL	methylation quantitative trait locus
SNP	single nucleotide polymorphism
MZ	monozygotic
DZ	dizygotic

References

1. Godfrey, K.M.; Lillycrop, K.A.; Burdge, G.C.; Gluckman, P.D.; Hanson, M.A. Epigenetic Mechanisms and the Mismatch Concept of the Developmental Origins of Health and Disease. *Pediatr. Res.* **2007**, *61*, 5R–10R. [CrossRef]
2. Gluckman, P.D.; Hanson, M.A.; Low, F.M. The role of developmental plasticity and epigenetics in human health. *Birth Defects Res. Part C Embryo Today Rev.* **2011**, *93*, 12–18. [CrossRef]
3. Verduci, E.; Banderali, G.; Barberi, S.; Radaelli, G.; Lops, A.; Betti, F.; Riva, E.; Giovannini, M. Epigenetic Effects of Human Breast Milk. *Nutrients* **2014**, *6*, 1711–1724. [CrossRef]
4. Langley-Evans, S.C. Nutrition in Early Life and the Programming of Adult Disease: A Review. *J. Hum. Nutr. Diet.* **2015**, *28*, 1–14. [CrossRef]
5. Hochberg, Z.; Feil, R.; Constancia, M.; Fraga, M.; Junien, C.; Carel, J.-C.; Boileau, P.; Le Bouc, Y.; Deal, C.L.; Lillycrop, K.; et al. Child Health, Developmental Plasticity, and Epigenetic Programming. *Endocr. Rev.* **2011**, *32*, 159–224. [CrossRef]
6. Raghuraman, S.; Donkin, I.; Versteyhe, S.; Barrès, R.; Simar, D. The Emerging Role of Epigenetics in Inflammation and Immunometabolism. *Trends Endocrinol. Metab.* **2016**, *27*, 782–795. [CrossRef] [PubMed]
7. Paparo, L.; Di Costanzo, M.; Di Scala, C.; Cosenza, L.; Leone, L.; Nocerino, R.; Canani, R.B. The Influence of Early Life Nutrition on Epigenetic Regulatory Mechanisms of the Immune System. *Nutrients* **2014**, *6*, 4706–4719. [CrossRef] [PubMed]
8. Canani, R.B.; Di Costanzo, M.; Leone, L.; Bedogni, G.; Brambilla, P.; Cianfarani, S.; Nobili, V.; Pietrobelli, A.; Agostoni, C. Epigenetic mechanisms elicited by nutrition in early life. *Nutr. Res. Rev.* **2011**, *24*, 198–205. [CrossRef] [PubMed]
9. Wadhwa, P.D.; Buss, C.; Entringer, S.; Swanson, J.M. Developmental origins of health and disease: Brief history of the approach and current focus on epigenetic mechanisms. *Semin. Reprod. Med.* **2009**, *27*, 358–368. [CrossRef] [PubMed]
10. Gluckman, P.D.; Hanson, M.A.; Beedle, A.S. Early Life Events and Their Consequences for Later Disease: A Life History and Evolutionary Perspective. *Am. J. Hum. Biol.* **2007**, *19*, 1–19. [CrossRef]
11. Bateson, P.; Barker, D.; Clutton-Brock, T.; Deb, D.; D'Udine, B.; Foley, R.A.; Gluckman, P.; Godfrey, K.; Kirkwood, T.; Lahr, M.M.; et al. Developmental Plasticity and Human Health. *Nature* **2004**, *430*, 419–421. [CrossRef] [PubMed]
12. Choi, S.-W.; Friso, S. Epigenetics: A New Bridge between Nutrition and Health. *Adv. Nutr.* **2010**, *1*, 8–16. [CrossRef] [PubMed]

13. Peter, C.J.; Fischer, L.K.; Kundakovic, M.; Garg, P.; Jakovcevski, M.; Dincer, A.; Amaral, A.C.; Ginns, E.I.; Galdzicka, M.; Bryce, C.P.; et al. DNA Methylation Signatures of Early Childhood Malnutrition Associated With Impairments in Attention and Cognition. *Boil. Psychiatry* **2016**, *80*, 765–774. [CrossRef] [PubMed]

14. Waterland, R.A.; Michels, K.B. Epigenetic Epidemiology of the Developmental Origins Hypothesis. *Annu. Rev. Nutr.* **2007**, *27*, 363–388. [CrossRef]

15. Cutfield, W.S.; Hofman, P.L.; Mitchell, M.; Morison, I.M. Could Epigenetics Play a Role in the Developmental Origins of Health and Disease? *Pediatr. Res.* **2007**, *61*, 68R–75R. [CrossRef]

16. Tammen, S.A.; Friso, S.; Choi, S.-W. Epigenetics: The Link between Nature and Nurture. *Mol. Asp. Med.* **2013**, *34*, 753–764. [CrossRef]

17. Faa, G.; Manchia, M.; Pintus, R.; Gerosa, C.; Marcialis, M.A.; Fanos, V. Fetal programming of neuropsychiatric disorders. *Birth Defects Res. Part C Embryo Today Rev.* **2016**, *108*, 207–223. [CrossRef]

18. Heijmans, B.T.; Tobi, E.W.; Stein, A.D.; Putter, H.; Blauw, G.J.; Susser, E.S.; Slagboom, P.E.; Lumey, L.H. Persistent epigenetic differences associated with prenatal exposure to famine in humans. *Proc. Natl. Acad. Sci. USA* **2008**, *105*, 17046–17049. [CrossRef]

19. Tobi, E.W.; Goeman, J.J.; Monajemi, R.; Gu, H.; Putter, H.; Zhang, Y.; Slieker, R.C.; Stok, A.P.; Thijssen, P.E.; Müller, F.; et al. DNA methylation signatures link prenatal famine exposure to growth and metabolism. *Nat. Commun.* **2014**, *5*, 5592. [CrossRef]

20. Barker, D.J. The fetal and infant origins of adult disease. *BMJ* **1990**, *301*, 1111. [CrossRef]

21. Ruiz, L.; García-Carral, C.; Rodriguez, J.M. Unfolding the Human Milk Microbiome Landscape in the Omics Era. *Front. Microbiol.* **2019**, *10*, 1378. [CrossRef] [PubMed]

22. Fanos, V.; Reali, A.; Marcialis, M.A.; Bardanzellu, F. What You Have to Know about Human Milk Oligosaccharides. *J. Pediatr. Neonatal Individ. Med.* **2018**, *7*, e070137.

23. Hassiotou, F.; Beltran, A.; Chetwynd, E.; Stuebe, A.M.; Twigger, A.-J.; Metzger, P.; Trengove, N.; Lai, C.T.; Filgueira, L.; Blancafort, P.; et al. Breastmilk Is a Novel Source of Stem Cells with Multilineage Differentiation Potential. *Stem Cells* **2012**, *30*, 2164–2174. [CrossRef] [PubMed]

24. Bardanzellu, F.; Fanos, V.; Reali, A. "Omics" in Human Colostrum and Mature Milk: Looking to Old Data with New Eyes. *Nutrients* **2017**, *9*, 843. [CrossRef] [PubMed]

25. Dessì, A.; Briana, D.; Corbu, S.; Gavrili, S.; Marincola, F.C.; Georgantzi, S.; Pintus, R.; Fanos, V.; Malamitsi-Puchner, A. Metabolomics of Breast Milk: The Importance of Phenotypes. *Metabolites* **2018**, *8*, 79. [CrossRef]

26. Forsum, E.; Lonnerdal, B. Variation in the contents of nutrients of breast milk during one feeding. *Nutrients Rep. Int.* **1979**, *19*, 815–820.

27. Victora, C.G.; Bahl, R.; Barros, A.J.D.; França, G.V.A.; Horton, S.; Krasevec, J.; Murch, S.; Sankar, M.J.; Walker, N.; Rollins, N.C.; et al. Breastfeeding in the 21st Century: Epidemiology, Mechanisms, and Lifelong Effect. *Lancet* **2016**, *387*, 475–490. [CrossRef]

28. McInerny, T.K. Breastfeeding, Early Brain Development, and Epigenetics—Getting Children off to Their Best Start. *Breastfeed. Med.* **2014**, *9*, 333–334. [CrossRef]

29. Ip, S.; Chung, M.; Raman, G.; Chew, P.; Magula, N.; Devine, D.; Trikalinos, T.; Lau, J. Breastfeeding and maternal and infant health outcomes in developed countries. *Évid. Rep. Assess.* **2007**, *153*, 1–186.

30. Wisnieski, L.; Kerver, J.; Holzman, C.; Todem, D.; Margerison-Zilko, C. Breastfeeding and Risk of Metabolic Syndrome in Children and Adolescents: A Systematic Review. *J. Hum. Lact.* **2018**, *34*, 515–525. [CrossRef]

31. Chirico, G.; Marzollo, R.; Cortinovis, S.; Fonte, C.; Gasparoni, A. Antiinfective Properties of Human Milk. *J. Nutr.* **2008**, *138*, 1801S–1806S. [CrossRef] [PubMed]

32. Martin, C.R.; Ling, P.-R.; Blackburn, G.L. Review of Infant Feeding: Key Features of Breast Milk and Infant Formula. *Nutrients* **2016**, *8*, 279. [CrossRef] [PubMed]

33. Else-Quest, N.M.; Hyde, J.S.; Clark, R. *Breastfeeding, Bonding, and the Mother-Infant Relationship. Merrill-Palmer Quarterly*; Wayne State University Press: Detroit, MI, USA, 2003; pp. 495–517. [CrossRef]

34. WHO. *Protecting, Promoting, and Supporting Breastfeeding in Facilities Providing Maternity and Newborn Services: The Revised Baby-Friendly Hospital Initiative 2018*; WHO: Geneva, Switzerland, 2019.

35. Michaelsen, K.F.; Lauritzen, L.; Mortensen, E.L. Effects of Breast-Feeding on Cognitive Function. In *Breast-Feeding: Early Influences on Later Health*; Springer: Dordrecht, The Netherlands, 2009; Volume 639, pp. 199–215. [CrossRef]

36. Kramer, M.S.; Aboud, F.; Mironova, E.; Vanilovich, I.; Platt, R.W.; Matush, L.; Igumnov, S.; Fombonne, E.; Bogdanovich, N.; Ducruet, T.; et al. Breastfeeding and child cognitive development: New evidence from a large randomized trial. *Arch. Gen. Psychiatry* **2008**, *65*, 578–584. [CrossRef] [PubMed]

37. Horta, B.L.; De Mola, C.L.; Victora, C.G. Breastfeeding and intelligence: A systematic review and meta-analysis. *Acta Paediatr.* **2015**, *104*, 14–19. [CrossRef]

38. Blesa, M.; Sullivan, G.; Anblagan, D.; Telford, E.J.; Quigley, A.J.; Sparrow, S.A.; Serag, A.; Semple, S.I.; Bastin, M.E.; Boardman, J.P. Early breast milk exposure modifies brain connectivity in preterm infants. *NeuroImage* **2019**, *184*, 431–439. [CrossRef]

39. Bartels, M.; Van Beijsterveldt, C.E.M.; Boomsma, D.I. Breastfeeding, Maternal Education and Cognitive Function: A Prospective Study in Twins. *Behav. Genet.* **2009**, *39*, 616–622. [CrossRef]

40. Harder, T.; Bergmann, R.; Kallischnigg, G.; Plagemann, A. Duration of Breastfeeding and Risk of Overweight: A Meta-Analysis. *Am. J. Epidemiol.* **2005**, *162*, 397–403. [CrossRef]

41. Burke, V.; Beilin, L.J.; Simmer, K.; Oddy, W.H.; Blake, K.V.; Doherty, D.; Kendall, G.E.; Newnham, J.P.; Landau, L.I.; Stanley, F.J. Breastfeeding and Overweight: Longitudinal Analysis in an Australian Birth Cohort. *J. Pediatr.* **2005**, *147*, 56–61. [CrossRef]

42. Gillman, M.W.; Rifas-Shiman, S.L.; Camargo, J.C.A.; Berkey, C.S.; Frazier, A.L.; Rockett, H.R.H.; Field, A.E.; Colditz, G.A. Risk of Overweight Among Adolescents Who Were Breastfed as Infants. *JAMA* **2001**, *285*, 2461. [CrossRef]

43. Owen, C.G.; Whincup, P.H.; Odoki, K.; Gilg, J.A.; Cook, D.G. Infant feeding and blood cholesterol: A study in adolescents and a systematic review. *Pediatrics* **2002**, *110*, 597–608. [CrossRef]

44. Shoji, H.; Shimizu, T. Effect of Human Breast Milk on Biological Metabolism in Infants. *Pediatr. Int.* **2019**, *61*, 6–15. [CrossRef] [PubMed]

45. Klement, E.; Cohen, R.V.; Boxman, J.; Joseph, A.; Reif, S. Breastfeeding and risk of inflammatory bowel disease: A systematic review with meta-analysis. *Am. J. Clin. Nutr.* **2004**, *80*, 1342–1352. [CrossRef] [PubMed]

46. Amitay, E.L.; Keinan-Boker, L. Breastfeeding and childhood leukemia incidence: A meta-analysis and systematic review. *JAMA Pediatr.* **2015**, *169*, e151025. [CrossRef] [PubMed]

47. Horta, B.L.; Victora, C.G.; World Health Organization. *Short-Term Effects of Breastfeeding: A Systematic Review on the Benefits of Breastfeeding on Diarrhoea and Pneumonia Mortality*; World Health Organization: Geneva, Switzerland, 2013.

48. Horta, B.L.; De Mola, C.L.; Victora, C.G. Long-term consequences of breastfeeding on cholesterol, obesity, systolic blood pressure and type 2 diabetes: A systematic review and meta-analysis. *Acta Paediatr.* **2015**, *104*, 30–37. [CrossRef]

49. Lodge, C.; Tan, D.; Lau, M.; Dai, X.; Tham, R.; Lowe, A.; Bowatte, G.; Allen, K.; Dharmage, S.; Lodge, C. Breastfeeding and asthma and allergies: A systematic review and meta-analysis. *Acta Paediatr.* **2015**, *104*, 38–53. [CrossRef]

50. Fall, C.H.; Borja, J.B.; Osmond, C.; Richter, L.; Bhargava, S.K.; Martorell, R.; Stein, A.D.; Barros, F.C.; Victora, C.G.; COHORTS group. Infant-Feeding Patterns and Cardiovascular Risk Factors in Young Adulthood: Data from Five Cohorts in Low- and Middle-Income Countries. *Int. J. Epidemiol.* **2011**, *40*, 47–62. [CrossRef]

51. Hartwig, F.P.; Loret de Mola, C.; Davies, N.M.; Victora, C.G.; Relton, C.L. Breastfeeding Effects on DNA Methylation in the Offspring: A Systematic Literature Review. *PLoS ONE* **2017**, *12*, e0173070.

52. Obermann-Borst, S.A.; Eilers, P.H.; Tobi, E.W.; De Jong, F.H.; Slagboom, P.E.; Heijmans, B.T.; Steegers-Theunissen, R.P.; Slagboom, P. Duration of breastfeeding and gender are associated with methylation of the LEPTIN gene in very young children. *Pediatr. Res.* **2013**, *74*, 344–349. [CrossRef]

53. Naumova, O.Y.; Odintsova, V.V.; Arincina, I.A.; Rychkov, S.Y.; Muhamedrahimov, R.J.; Shneider, Y.V.; Grosheva, A.N.; Zhukova, O.V.; Grigorenko, E.L. A Study of the Association between Breastfeeding and DNA Methylation in Peripheral Blood Cells of Infants. *Russ. J. Genet.* **2019**, *55*, 749–755. [CrossRef]

54. Rossnerova, A.; Tulupova, E.; Tabashidze, N.; Schmuczerova, J.; Dostal, M.; Jr, P.R.; Gmuender, H.; Sram, R.J. Factors affecting the 27K DNA methylation pattern in asthmatic and healthy children from locations with various environments. *Mutat. Res. Mol. Mech. Mutagen.* **2013**, *741*, 18–26. [CrossRef]

55. Sherwood, W.B.; Bion, V.; Lockett, G.A.; Ziyab, A.H.; Soto-Ramírez, N.; Mukherjee, N.; Kurukulaaratchy, R.J.; Ewart, S.; Zhang, H.; Arshad, S.H.; et al. Duration of breastfeeding is associated with leptin (LEP) DNA methylation profiles and BMI in 10-year-old children. *Clin. Epigenet.* **2019**, *11*, 128. [CrossRef] [PubMed]

56. Joubert, B.R.; Felix, J.F.; Yousefi, P.; Bakulski, K.M.; Just, A.C.; Breton, C.; Reese, S.E.; Markunas, C.A.; Richmond, R.C.; Xu, C.-J.; et al. DNA Methylation in Newborns and Maternal Smoking in Pregnancy: Genome-wide Consortium Meta-analysis. *Am. J. Hum. Genet.* **2016**, *98*, 680–696. [CrossRef] [PubMed]

57. Küpers, L.K.; Monnereau, C.; Sharp, G.C.; Yousefi, P.; Salas, L.A.; Ghantous, A.; Page, C.M.; Reese, S.E.; Wilcox, A.J.; Czamara, D.; et al. Meta-analysis of epigenome-wide association studies in neonates reveals widespread differential DNA methylation associated with birthweight. *Nat. Commun.* **2019**, *10*, 1893. [CrossRef] [PubMed]

58. Soto-Ramírez, N.; Arshad, S.H.; Holloway, J.W.; Zhang, H.; Schauberger, E.; Ewart, S.; Patil, V.; Karmaus, W. The interaction of genetic variants and DNA methylation of the interleukin-4 receptor gene increase the risk of asthma at age 18 years. *Clin. Epigenet.* **2013**, *5*, 1. [CrossRef]

59. Tao, M.-H.; Marian, C.; Shields, P.G.; Potischman, N.; Nie, J.; Krishnan, S.S.; Berry, D.L.; Kallakury, B.V.; Ambrosone, C.; Edge, S.B.; et al. Exposures in early life: Associations with DNA promoter methylation in breast tumors. *J. Dev. Orig. Health Dis.* **2013**, *4*, 182–190. [CrossRef]

60. Hartwig, F.P.; Smith, G.D.; Simpkin, A.J.; Victora, C.G.; Relton, C.L.; Caramaschi, D. Association between Breastfeeding and DNA Methylation over the Life Course: Findings from the Avon Longitudinal Study of Parents and Children (ALSPAC). *bioRxiv* **2019**, 800722. [CrossRef]

61. Smith, A.K.; Kilaru, V.; Klengel, T.; Mercer, K.B.; Bradley, B.; Conneely, K.N.; Ressler, K.J.; Binder, E.B. DNA Extracted from Saliva for Methylation Studies of Psychiatric Traits: Evidence Tissue Specificity and Relatedness to Brain. *Am. J. Med. Genet. Part B Neuropsychiatr. Genet.* **2015**, *168*, 36–44. [CrossRef]

62. Papavassiliou, P.; York, T.P.; Gursoy, N.; Hill, G.; Nicely, L.V.; Sundaram, U.; McClain, A.; Aggen, S.H.; Eaves, L.; Riley, B.; et al. The phenotype of persons having mosaicism for trisomy 21/Down syndrome reflects the percentage of trisomic cells present in different tissues. *Am. J. Med. Genet. Part A* **2009**, *149*, 573–583. [CrossRef]

63. Van Dongen, J.; Ehli, E.A.; Jansen, R.; Van Beijsterveldt, C.E.M.; Willemsen, G.; Hottenga, J.J.; Kallsen, N.A.; Peyton, S.A.; Breeze, C.E.; Kluft, C.; et al. Genome-wide analysis of DNA methylation in buccal cells: A study of monozygotic twins and mQTLs. *Epigenet. Chromatin* **2018**, *11*, 54. [CrossRef]

64. Van Beijsterveldt, C.E.M.; Groen-Blokhuis, M.; Hottenga, J.J.; Franić, S.; Hudziak, J.J.; Lamb, D.; Huppertz, C.; de Zeeuw, E.; Nivard, M.; Schutte, N.; et al. The Young Netherlands Twin Register (YNTR): Longitudinal Twin and Family Studies in Over 70,000 Children. *Twin Res. Hum. Genet.* **2013**, *16*, 252–267. [CrossRef]

65. Boomsma, D.I.; Fanos Cagliari, V.; Mussap Genoa, M.; Del Vecchio Bari, A.; Sun Shanghai, B.; Faa Cagliari, G.; Giordano Philadelphia, A. Aggression in Children: Unravelling the Interplay of Genes and Environment through (Epi)Genetics and Metabolomics. *J. Pediatr. Neonatal Individ. Med.* **2015**, *4*. [CrossRef]

66. Bartels, M.; Hendriks, A.; Mauri, M.; Krapohl, E.; Whipp, A.; Bolhuis, K.; Conde, L.C.; Luningham, J.; Ip, H.F.; Hagenbeek, F.; et al. Childhood aggression and the co-occurrence of behavioural and emotional problems: Results across ages 3-16 years from multiple raters in six cohorts in the EU-ACTION project. *Eur. Child Adolesc. Psychiatry* **2018**, *27*, 1105–1121. [CrossRef] [PubMed]

67. Hagenbeek, F.A.; Roetman, P.J.; Pool, R.; Kluft, C.; Harms, A.C.; van Dongen, J.; Colins, O.F.; Talens, S.; Van Beijsterveldt, C.E.M.; de Zeeuw, E.L.; et al. Urinary Amine and Organic Acid Metabolites Evaluated as Markers for Childhood Aggression: The ACTION Biomarker Study. *Frontiers in Psychiatry* **2019**, submitted.

68. Van Dongen, J. Epigenome-Wide Association Study Meta-Analysis of Aggressive Behavior. In Proceedings of the ECS Workshop on Aggression, Nijmegen, The Netherlands, 31 October 2016.

69. Boyd, A.; Golding, J.; Macleod, J.; Lawlor, D.A.; Fraser, A.; Henderson, J.; Molloy, L.; Ness, A.; Ring, S.; Davey Smith, G. Cohort Profile: The 'Children of the 90s'—The Index Offspring of the Avon Longitudinal Study of Parents and Children. *Int. J. Epidemiol.* **2013**, *42*, 111–127. [CrossRef]

70. Fraser, A.; Macdonald-Wallis, C.; Tilling, K.; Boyd, A.; Golding, J.; Davey Smith, G.; Henderson, J.; Macleod, J.; Molloy, L.; Ness, A.; et al. Cohort Profile: The Avon Longitudinal Study of Parents and Children: ALSPAC Mothers Cohort. *Int. J. Epidemiol.* **2013**, *42*, 97–110. [CrossRef]

71. Relton, C.L.; Gaunt, T.; McArdle, W.; Ho, K.; Duggirala, A.; Shihab, H.; Woodward, G.; Lyttleton, O.; Evans, D.M.; Reik, W.; et al. Data Resource Profile: Accessible Resource for Integrated Epigenomic Studies (ARIES). *Int. J. Epidemiol.* **2015**, *44*, 1181–1190. [CrossRef]

72. CBS. *Standard Classification of Occupations*; Central Bureau of Statistics: Heerlen/Voorburg, The Netherlands, 2001.

73. Erikson, R.; Goldthorpe, J.H.; Portocarero, L. Intergenerational class mobility and the convergence thesis: England, France and Sweden. *Br. J. Sociol.* **2010**, *61*, 185–219. [CrossRef]

74. Dolan, C.V.; Geels, L.; Vink, J.M.; van Beijsterveldt, C.E.M.; Neale, M.C.; Bartels, M.; Boomsma, D.I. Testing Causal Effects of Maternal Smoking during Pregnancy on Offspring's Externalizing and Internalizing Behavior. *Behav. Genet.* **2016**, *46*, 378–388. [CrossRef]

75. Van Beijsterveldt, C.E.M.T.; Hoekstra, C.; Schats, R.; Montgomery, G.W.; Willemsen, G.; Boomsma, D.I. Mode of Conception of Twin Pregnancies: Willingness to Reply to Survey Items and Comparison of Survey Data to Hospital Records. *Twin Res. Hum. Genet.* **2008**, *11*, 349–351. [CrossRef]

76. Odintsova, V.V.; Dolan, C.V.; Van Beijsterveldt, C.E.M.; De Zeeuw, E.L.; Van Dongen, J.; Boomsma, D.I. Pre- and Perinatal Characteristics Associated with Apgar Scores in a Review and in a New Study of Dutch Twins. *Twin Res. Hum. Genet.* **2019**, *22*, 164–176. [CrossRef]

77. Meulenbelt, I.; Droog, S.; Trommelen, G.J.; Boomsma, D.I.; Slagboom, P.E. High-yield noninvasive human genomic DNA isolation method for genetic studies in geographically dispersed families and populations. *Am. J. Hum. Genet.* **1995**, *57*, 1252–1254. [PubMed]

78. Moran, S.; Arribas, C.; Esteller, M. Validation of a DNA methylation microarray for 850,000 CpG sites of the human genome enriched in enhancer sequences. *Epigenomics* **2016**, *8*, 389–399. [CrossRef] [PubMed]

79. Sinke, L.; van Iterson, M.; Cats, D.; Slieker, R.; Heijmans, B. DNAmArray: Streamlined Workflow for the Quality Control, Normalization, and Analysis of Illumina Methylation Array Data. *Zenodo* **2019**. [CrossRef]

80. Van Iterson, M.; Tobi, E.W.; Slieker, R.C.; Hollander, W.D.; Luijk, R.; Slagboom, P.; Heijmans, B.T. MethylAid: Visual and interactive quality control of large Illumina 450k datasets. *Bioinformatics* **2014**, *30*, 3435–3437. [CrossRef]

81. Van Iterson, M.; Cats, D.; Hop, P.; Heijmans, B.T.; BIOS Consortium. omicsPrint: Detection of data linkage errors in multiple omics studies. *Bioinformatics* **2018**, *34*, 2142–2143. [CrossRef]

82. Min, J.L.; Hemani, G.; Smith, G.D.; Relton, C.; Suderman, M. Meffil: Efficient normalization and analysis of very large DNA methylation datasets. *Bioinformatics* **2018**, *34*, 3983–3989. [CrossRef]

83. Pidsley, R.; Zotenko, E.; Peters, T.J.; Lawrence, M.G.; Risbridger, G.P.; Molloy, P.; Van Djik, S.; Muhlhausler, B.; Stirzaker, C.; Clark, S.J. Critical evaluation of the Illumina MethylationEPIC BeadChip microarray for whole-genome DNA methylation profiling. *Genome Boil.* **2016**, *17*, 208. [CrossRef]

84. Zheng, S.C.; Webster, A.P.; Dong, D.; Feber, A.; Graham, D.G.; Sullivan, R.; Jevons, S.; Lovat, L.B.; Beck, S.; Widschwendter, M.; et al. A novel cell-type deconvolution algorithm reveals substantial contamination by immune cells in saliva, buccal and cervix. *Epigenomics* **2018**, *10*, 925–940. [CrossRef]

85. Houseman, E.A.; Accomando, W.P.; Koestler, D.C.; Christensen, B.C.; Marsit, C.J.; Nelson, H.H.; Wiencke, J.K.; Kelsey, K.T. DNA methylation arrays as surrogate measures of cell mixture distribution. *BMC Bioinform.* **2012**, *13*, 86. [CrossRef]

86. Van Iterson, M.; Van Zwet, E.W.; BIOS Consortium; Heijmans, B.T. Controlling bias and inflation in epigenome- and transcriptome-wide association studies using the empirical null distribution. *Genome Boil.* **2017**, *18*, 19. [CrossRef]

87. Tukey, J.W. Exploratory Data Analysis. *Biom. J.* **1981**, *23*, 413–414.

88. Leek, J.T.; Storey, J.D. Capturing Heterogeneity in Gene Expression Studies by Surrogate Variable Analysis. *PLoS Genet.* **2007**, *3*, e161. [CrossRef] [PubMed]

89. Ritchie, M.E.; Phipson, B.; Wu, D.; Hu, Y.; Law, C.W.; Shi, W.; Smyth, G.K. limma powers differential expression analyses for RNA-sequencing and microarray studies. *Nucleic Acids Res.* **2015**, *43*, e47. [CrossRef] [PubMed]

90. Li, M.; Zou, D.; Li, Z.; Gao, R.; Sang, J.; Zhang, Y.; Li, R.; Xia, L.; Zhang, T.; Niu, G.; et al. EWAS Atlas: A Curated Knowledgebase of Epigenome-Wide Association Studies. *Nucleic Acids Res.* **2019**, *47*, D983–D988. [CrossRef]

91. Yousefi, P.; Huen, K.; Davé, V.; Barcellos, L.; Eskenazi, B.; Holland, N. Sex differences in DNA methylation assessed by 450 K BeadChip in newborns. *BMC Genom.* **2015**, *16*, 911. [CrossRef]

92. Husquin, L.T.; Rotival, M.; Fagny, M.; Quach, H.; Zidane, N.; McEwen, L.M.; MacIsaac, J.L.; Kobor, M.S.; Aschard, H.; Patin, E.; et al. Exploring the genetic basis of human population differences in DNA methylation and their causal impact on immune gene regulation. *Genome Boil.* **2018**, *19*, 222. [CrossRef]

93. Marioni, R.E.; Harris, S.E.; Zhang, Q.; McRae, A.F.; Hagenaars, S.P.; Hill, W.D.; Davies, G.; Ritchie, C.W.; Gale, C.R.; Starr, J.M.; et al. GWAS on family history of Alzheimer's disease. *Transl. Psychiatry* **2018**, *8*, 99. [CrossRef]

94. Xue, J.; Gharaibeh, R.Z.; Pietryk, E.W.; Brouwer, C.; Tarantino, L.M.; Valdar, W.; Ideraabdullah, F.Y. Impact of vitamin D depletion during development on mouse sperm DNA methylation. *Epigenetics* **2018**, *13*, 959–974. [CrossRef]

95. Sujit, K.M.; Sarkar, S.; Singh, V.; Pandey, R.; Agrawal, N.K.; Trivedi, S.; Singh, K.; Gupta, G.; Rajender, S. Genome-wide differential methylation analyses identifies methylation signatures of male infertility. *Hum. Reprod.* **2018**, *33*, 2256–2267. [CrossRef]

96. Kim, S.K. Identification of 613 new loci associated with heel bone mineral density and a polygenic risk score for bone mineral density, osteoporosis and fracture. *PLoS ONE* **2018**, *13*, e0200785. [CrossRef]

97. Morris, J.A.; Kemp, J.P.; Youlten, S.E.; Laurent, L.; Logan, J.G.; Chai, R.C.; Vulpescu, N.A.; Forgetta, V.; Kleinman, A.; Mohanty, S.T.; et al. An Atlas of Genetic Influences on Osteoporosis in Humans and Mice. *Nat. Genet.* **2019**, *51*, 258–266. [CrossRef] [PubMed]

98. Kemp, J.P.; Morris, J.A.; Medina-Gomez, C.; Forgetta, V.; Warrington, N.M.; Youlten, S.E.; Zheng, J.; Gregson, C.L.; Grundberg, E.; Trajanoska, K.; et al. Identification of 153 new loci associated with heel bone mineral density and functional involvement of GPC6 in osteoporosis. *Nat. Genet.* **2017**, *49*, 1468–1475. [CrossRef] [PubMed]

99. Estrada, K.; Styrkarsdottir, U.; Evangelou, E.; Hsu, Y.-H.; Duncan, E.L.; Ntzani, E.E.; Oei, L.; Albagha, O.M.E.; Amin, N.; Kemp, J.P.; et al. Genome-wide meta-analysis identifies 56 bone mineral density loci and reveals 14 loci associated with risk of fracture. *Nat. Genet.* **2012**, *44*, 491–501. [CrossRef] [PubMed]

100. Medina-Gómez, C.; Kemp, J.P.; Trajanoska, K.; Luan, J.; Chesi, A.; Ahluwalia, T.S.; Mook-Kanamori, D.O.; Ham, A.; Hartwig, F.P.; Evans, D.S.; et al. Life-Course Genome-wide Association Study Meta-analysis of Total Body BMD and Assessment of Age-Specific Effects. *Am. J. Hum. Genet.* **2018**, *102*, 88–102. [CrossRef] [PubMed]

101. Emilsson, V.; Ilkov, M.; Lamb, J.R.; Finkel, N.; Gudmundsson, E.F.; Pitts, R.; Hoover, H.; Gudmundsdottir, V.; Horman, S.R.; Aspelund, T.; et al. Co-regulatory networks of human serum proteins link genetics to disease. *Science* **2018**, *361*, 769–773. [CrossRef] [PubMed]

102. Sun, B.B.; Maranville, J.C.; Peters, J.E.; Stacey, D.; Staley, J.R.; Blackshaw, J.; Burgess, S.; Jiang, T.; Paige, E.; Surendran, P.; et al. Genomic atlas of the human plasma proteome. *Nature* **2018**, *558*, 73–79. [CrossRef] [PubMed]

103. De Vries, P.S.; Brown, M.R.; Bentley, A.R.; Sung, Y.J.; Winkler, T.W.; Ntalla, I.; Schwander, K.; Kraja, A.T.; Guo, X.; Franceschini, N.; et al. Multiancestry Genome-Wide Association Study of Lipid Levels Incorporating Gene-Alcohol Interactions. *Am. J. Epidemiol.* **2019**, *188*, 1033–1054. [CrossRef]

104. Imaizumi, A.; Adachi, Y.; Kawaguchi, T.; Higasa, K.; Tabara, Y.; Sonomura, K.; Sato, T.-A.; Takahashi, M.; Mizukoshi, T.; Yoshida, H.-O.; et al. Genetic basis for plasma amino acid concentrations based on absolute quantification: A genome-wide association study in the Japanese population. *Eur. J. Hum. Genet.* **2019**, *27*, 621–630. [CrossRef]

105. Shin, S.-Y.; The Multiple Tissue Human Expression Resource (MuTHER) Consortium; Fauman, E.B.; Petersen, A.-K.; Krumsiek, J.; Santos, R.; Huang, J.; Arnold, M.; Erte, I.; Forgetta, V.; et al. An atlas of genetic influences on human blood metabolites. *Nat. Genet.* **2014**, *46*, 543–550. [CrossRef]

106. Johnston, K.J.A.; Adams, M.J.; Nicholl, B.I.; Ward, J.; Strawbridge, R.J.; Ferguson, A.; McIntosh, A.M.; Bailey, M.E.S.; Smith, D.J. Genome-wide association study of multisite chronic pain in UK Biobank. *PLoS Genet.* **2019**, *15*, e1008164. [CrossRef]

107. Li, Y.; Sekula, P.; Wuttke, M.; Wahrheit, J.; Hausknecht, B.; Schultheiss, U.T.; Gronwald, W.; Schlosser, P.; Tucci, S.; Ekici, A.B.; et al. Genome-Wide Association Studies of Metabolites in Patients with CKD Identify Multiple Loci and Illuminate Tubular Transport Mechanisms. *J. Am. Soc. Nephrol.* **2018**, *29*, 1513–1524. [CrossRef] [PubMed]

108. Wellcome Trust Case Control Consortium. Genome-Wide Association Study of 14,000 Cases of Seven Common Diseases and 3000 Shared Controls. *Nature* **2007**, *447*, 661–678. [CrossRef] [PubMed]

109. Liu, M.; Jiang, Y.; Wedow, R.; Li, Y.; Brazel, D.M.; Chen, F.; Datta, G.; Davila-Velderrain, J.; McGuire, D.; Tian, C.; et al. Association studies of up to 1.2 million individuals yield new insights into the genetic etiology of tobacco and alcohol use. *Nat. Genet.* **2019**, *51*, 237–244. [CrossRef] [PubMed]

110. Linnér, R.K.; Biroli, P.; Kong, E.; Meddens, S.F.W.; Wedow, R.; Fontana, M.A.; Lebreton, M.; Tino, S.P.; Abdellaoui, A.; Hammerschlag, A.R.; et al. Genome-wide association analyses of risk tolerance and risky behaviors in over 1 million individuals identify hundreds of loci and shared genetic influences. *Nat. Genet.* **2019**, *51*, 245–257. [CrossRef]

111. Hosokawa, N.; Sasaki, T.; Iemura, S.; Natsume, T.; Hara, T.; Mizushima, N. Atg101, a Novel Mammalian Autophagy Protein Interacting with Atg13. *Autophagy* **2009**, *5*, 973–979. [CrossRef]

112. Ahola-Olli, A.V.; Würtz, P.; Havulinna, A.S.; Aalto, K.; Pitkänen, N.; Lehtimäki, T.; Kähönen, M.; Lyytikäinen, L.-P.; Raitoharju, E.; Seppälä, I.; et al. Genome-Wide Association Study Identifies 27 Loci Influencing Concentrations of Circulating Cytokines and Growth Factors. *Am. J. Hum. Genet.* **2017**, *100*, 40–50. [CrossRef]

113. Astle, W.J.; Elding, H.; Jiang, T.; Allen, D.; Ruklisa, D.; Mann, A.L.; Mead, D.; Bouman, H.; Riveros-McKay, F.; Kostadima, M.A.; et al. The Allelic Landscape of Human Blood Cell Trait Variation and Links to Common Complex Disease. *Cell* **2016**, *167*, 1415–1429. [CrossRef]

114. Rojas, D.; Rager, J.E.; Smeester, L.; Bailey, K.A.; Drobná, Z.; Rubio-Andrade, M.; Stýblo, M.; García-Vargas, G.; Fry, R.C. Prenatal Arsenic Exposure and the Epigenome: Identifying Sites of 5-Methylcytosine Alterations That Predict Functional Changes in Gene Expression in Newborn Cord Blood and Subsequent Birth Outcomes. *Toxicol. Sci.* **2015**, *143*, 97–106. [CrossRef]

115. McNally, K.E.; Faulkner, R.; Steinberg, F.; Gallon, M.; Ghai, R.; Pim, D.; Langton, P.; Pearson, N.; Danson, C.M.; Nägele, H.; et al. Retriever is a multiprotein complex for retromer-independent endosomal cargo recycling. *Nature* **2017**, *19*, 1214–1225. [CrossRef]

116. Johansson, Å.; Enroth, S.; Gyllensten, U. Continuous Aging of the Human DNA Methylome Throughout the Human Lifespan. *PLoS ONE* **2013**, *8*, e67378. [CrossRef]

117. Gervin, K.; Andreassen, B.K.; Hjorthaug, H.S.; Carlsen, K.C.L.; Carlsen, K.-H.; Undlien, D.E.; Lyle, R.; Munthe-Kaas, M.C. Intra-individual changes in DNA methylation not mediated by cell-type composition are correlated with aging during childhood. *Clin. Epigenet.* **2016**, *8*, 110. [CrossRef] [PubMed]

118. Dunn, E.C.; Soare, T.W.; Simpkin, A.J.; Suderman, M.J.; Zhu, Y.; Klengel, T.; Smith, A.D.A.C.; Ressler, K.; Relton, C.L. Sensitive Periods for the Effect of Childhood Adversity on DNA Methylation: Results from a Prospective, Longitudinal Study. *bioRxiv* **2018**. [CrossRef] [PubMed]

119. Simpkin, A.J.; Suderman, M.; Gaunt, T.R.; Lyttleton, O.; McArdle, W.L.; Ring, S.M.; Tilling, K.; Smith, G.D.; Relton, C.L. Longitudinal analysis of DNA methylation associated with birth weight and gestational age. *Hum. Mol. Genet.* **2015**, *24*, 3752–3763. [CrossRef] [PubMed]

120. Agha, G.; Hajj, H.; Rifas-Shiman, S.L.; Just, A.C.; Hivert, M.-F.; Burris, H.H.; Lin, X.; Litonjua, A.A.; Oken, E.; DeMeo, D.L.; et al. Birth weight-for-gestational age is associated with DNA methylation at birth and in childhood. *Clin. Epigenet.* **2016**, *8*, 118. [CrossRef] [PubMed]

121. Alfano, R.; Guida, F.; Galobardes, B.; Chadeau-Hyam, M.; Delpierre, C.; Ghantous, A.; Henderson, J.; Herceg, Z.; Jain, P.; Nawrot, T.S.; et al. Socioeconomic position during pregnancy and DNA methylation signatures at three stages across early life: Epigenome-wide association studies in the ALSPAC birth cohort. *Int. J. Epidemiol.* **2019**, *48*, 30–44. [CrossRef] [PubMed]

122. Richmond, R.C.; Simpkin, A.J.; Woodward, G.; Gaunt, T.R.; Lyttleton, O.; McArdle, W.L.; Ring, S.M.; Smith, A.D.A.C.; Timpson, N.J.; Tilling, K.; et al. Prenatal Exposure to Maternal Smoking and Offspring DNA Methylation across the Lifecourse: Findings from the Avon Longitudinal Study of Parents and Children (ALSPAC). *Hum. Mol. Genet.* **2015**, *24*, 2201–2217. [CrossRef]

123. Tobi, E.W.; Lumey, L.H.; Talens, R.P.; Kremer, D.; Putter, H.; Stein, A.D.; Slagboom, P.E.; Heijmans, B.T. DNA methylation differences after exposure to prenatal famine are common and timing- and sex-specific. *Hum. Mol. Genet.* **2009**, *18*, 4046–4053. [CrossRef]

124. Muniz, L.C.; Menezes, A.M.B.; Buffarini, R.; Wehrmeister, F.C.; Assunção, M.C.F. Effect of Breastfeeding on Bone Mass from Childhood to Adulthood: A Systematic Review of the Literature. *Int. Breastfeed. J.* **2015**, *10*, 31. [CrossRef]

125. Dessì, A.; Murgia, A.; Agostino, R.; Pattumelli, M.G.; Schirru, A.; Scano, P.; Fanos, V.; Caboni, P. Exploring the Role of Different Neonatal Nutrition Regimens during the First Week of Life by Urinary GC-MS Metabolomics. *Int. J. Mol. Sci.* **2016**, *17*, 265. [CrossRef]

126. San-Cristobal, R.; Navas-Carretero, S.; Milagro, F.I.; Riezu-Boj, J.I.; Guruceaga, E.; Celis-Morales, C.; Livingstone, K.M.; Brennan, L.; Lovegrove, J.A.; Daniel, H.; et al. Gene methylation parallelisms between peripheral blood cells and oral mucosa samples in relation to overweight. *J. Physiol. Biochem.* **2016**, *73*, 465–474. [CrossRef]

127. Van Dongen, J.; Nivard, M.G.; Willemsen, G.; Hottenga, J.-J.; Helmer, Q.; Dolan, C.V.; Ehli, E.A.; Davies, G.E.; Van Iterson, M.; Breeze, C.E.; et al. Genetic and environmental influences interact with age and sex in shaping the human methylome. *Nat. Commun.* **2016**, *7*, 11115. [CrossRef] [PubMed]

128. Colodro-Conde, L.; Sánchez-Romera, J.F.; Ordoñana, J.R. Heritability of Initiation and Duration of Breastfeeding Behavior. *Twin Res. Hum. Genet.* **2013**, *16*, 575–580. [CrossRef] [PubMed]

129. Merjonen, P.; Dolan, C.V.; Bartels, M.; Boomsma, D.I. Does Breastfeeding Behavior Run in Families? Evidence from Twins, Their Sisters and Their Mothers in the Netherlands. *Twin Res. Hum. Genet.* **2015**, *18*, 179–187. [CrossRef] [PubMed]

130. Eriksson, N.; Benton, G.M.; Do, C.B.; Kiefer, A.K.; Mountain, J.L.; Hinds, D.A.; Francke, U.; Tung, J.Y. Genetic variants associated with breast size also influence breast cancer risk. *BMC Med. Genet.* **2012**, *13*, 53. [CrossRef] [PubMed]

131. Cunningham, B.; Bass, S.; Fuh, G.; Wells, J. Zinc mediation of the binding of human growth hormone to the human prolactin receptor. *Science* **1990**, *250*, 1709–1712. [CrossRef]

132. Sukonina, V.; Ma, H.; Zhang, W.; Bartesaghi, S.; Subhash, S.; Heglind, M.; Foyn, H.; Betz, M.J.; Nilsson, D.; Lidell, M.E.; et al. FOXK1 and FOXK2 regulate aerobic glycolysis. *Nature* **2019**, *566*, 279–283. [CrossRef]

133. Hannenhalli, S.; Kaestner, K.H. The evolution of Fox genes and their role in development and disease. *Nat. Rev. Genet.* **2009**, *10*, 233–240. [CrossRef]

134. Tachmazidou, I.; Süveges, D.; Min, J.L.; Ritchie, G.R.; Steinberg, J.; Walter, K.; Iotchkova, V.; Schwartzentruber, J.; Huang, J.; Memari, Y.; et al. Whole-Genome Sequencing Coupled to Imputation Discovers Genetic Signals for Anthropometric Traits. *Am. J. Hum. Genet.* **2017**, *100*, 865–884. [CrossRef]

135. Crujeiras, A.B.; Pissios, P.; Moreno-Navarrete, J.M.; Díaz-Lagares, A.; Sandoval, J.; Gómez, A.; Ricart, W.; Esteller, M.; Casanueva, F.F.; Fernandez-Real, J.M. An Epigenetic Signature in Adipose Tissue Is Linked to Nicotinamide N-Methyltransferase Gene Expression. *Mol. Nutr. Food Res.* **2018**, *62*, 1700933. [CrossRef]

136. Bardanzellu, F.; Fanos, V.; Strigini, F.A.L.; Artini, P.G.; Peroni, D.G. Human Breast Milk: Exploring the Linking Ring Among Emerging Components. *Front. Pediatr.* **2018**, *6*, 215. [CrossRef]

137. Lehmann, G.M.; LaKind, J.S.; Davis, M.H.; Hines, E.P.; Marchitti, S.A.; Alcala, C.; Lorber, M. Environmental Chemicals in Breast Milk and Formula: Exposure and Risk Assessment Implications. *Environ. Health Perspect.* **2018**, *126*, 096001. [CrossRef] [PubMed]

138. Gómez-Gallego, C.; Morales, J.M.; Monleón, D.; Du Toit, E.; Kumar, H.; Linderborg, K.M.; Zhang, Y.; Yang, B.; Isolauri, E.; Salminen, S.; et al. Human Breast Milk NMR Metabolomic Profile across Specific Geographical Locations and Its Association with the Milk Microbiota. *Nutrients* **2018**, *10*, 1355. [CrossRef] [PubMed]

139. Hermansson, H.; Kumar, H.; Collado, M.C.; Salminen, S.; Isolauri, E.; Rautava, S. Breast Milk Microbiota Is Shaped by Mode of Delivery and Intrapartum Antibiotic Exposure. *Front. Nutr.* **2019**, *6*, 4. [CrossRef] [PubMed]

Article

Factors Associated with the Early Initiation of Breastfeeding in Economic Community of West African States (ECOWAS)

Osita Kingsley Ezeh [1], Felix Akpojene Ogbo [2], Garry John Stevens [3], Wadad Kathy Tannous [4], Osuagwu Levi Uchechukwu [5], Pramesh Raj Ghimire [1], Kingsley Emwinyore Agho [1,2,*] and Global Maternal and Child Health Research Collaboration (GloMACH) [†]

[1] School of Science and Health, University of Western Sydney, Locked Bag 1797, Penrith, NSW 1797, Australia; Ezehosita@yahoo.com (O.K.E.); Prameshraj@hotmail.com (P.R.G.)

[2] Translational Health Research Institute (THRI), School of Medicine, Western Sydney University, Campbelltown Campus, Locked Bag 1797, Penrith, NSW 2571, Australia; F.Ogbo@westernsydney.edu.au

[3] Humanitarian and Development Research Initiative (HADRI), School of Social Sciences and Psychology, Western Sydney University, Locked Bag1797, Penrith, NSW 2751, Australia; G.Stevens@westernsydney.edu.au

[4] School of Business, Western Sydney University, Locked Bag 1797, Penrith, NSW 2751, Australia; K.Tannous@westernsydney.edu.au

[5] Diabetes, Obesity and Metabolism Translational Research Unit, Western Sydney University, Campbelltown, NSW 2560, Australia; L.osuagwu@westernsydney.edu.au

* Correspondence: K.agho@westernsydney.edu.au; Tel.: +612-46203635

† Members are listed at the end of Acknowledgments.

Received: 30 September 2019; Accepted: 8 November 2019; Published: 14 November 2019

Abstract: The early initiation of breastfeeding (EIBF) within one hour after birth enhanced mother–newborn bonding and protection against infectious diseases. This paper aimed to examine factors associated with EIBF in 13 Economic Community of West African States (ECOWAS). A weighted sample of 76,934 children aged 0–23 months from the recent Demographic and Health Survey dataset in the ECOWAS for the period 2010 to 2018 was pooled. Survey logistic regression analyses, adjusting for country-specific cluster and population-level weights, were used to determine the factors associated with EIBF. The overall combined rate of EIBF in ECOWAS was 43%. After adjusting for potential confounding factors, EIBF was significantly lower in Burkina Faso, Cote d'Ivoire, Guinea, Niger, Nigeria, and Senegal. Mothers who perceived their babies to be average and large at birth were significantly more likely to initiate breastfeeding within one hour of birth than those mothers who perceived their babies to be small at birth. Mothers who had a caesarean delivery (AOR = 0.28, 95%CI = 0.22–0.36), who did not attend antenatal visits (ANC) during pregnancy, and delivered by non-health professionals were more likely to delay initiation of breastfeeding beyond one hour after birth. Male children and mothers from poorer households were more likely to delay introduction of breastfeeding. Infant and young child feeding nutrition programs aimed at improving EIBF in ECOWAS need to target mothers who underutilize healthcare services, especially mothers from lower socioeconomic groups.

Keywords: breastfeeding; infants; ECOWAS; antenatal care; pregnancy; infant mortality

---

1. Introduction

Globally, about 75% of all under-five deaths occur within the first year of life, and sub-Saharan African countries remain the hub of the world burden of all under-five deaths [1,2]. Early initiation of breastfeeding (EIBF) within one hour of birth remains a significant public health challenge, particularly in sub-Saharan African (SSA) countries, including the Economic Community of West African States (ECOWAS). In keeping with the Sustainable Development Goals, developing countries have committed to reducing under five mortality by 25 deaths per 1000 live births by 2030 [3]. Global strategies, including the Baby-Friendly Hospital Initiative (BFHI), Community Integrated Management of Childhood Illness (C-IMCI), and Infant and Young Child Feeding (IYCF) guidelines have been developed by United Nations Children's Fund (UNICEF) and the World Health Organization (WHO). These aim to substantially reduce child mortality, morbidity, and undernutrition [4,5]. They have succeeded in some parts of the world but remain refractory in Africa. In addition, there is a strong epidemiological evidence to suggest that the nutritional benefits of EIBF may include a reduced risk of hypoglycemia in at-risk infants (e.g., small for gestational age and macrosomia infants), which can cause significant morbidity and mortality [6]. This is due to EIBF containing antibodies for natural immunity, which provide protection against infectious diseases (such as diarrheal and pneumonia) and other potentially life-threatening ailments [7,8]. ECOWAS is a regional political and economic union of 15 countries in West Africa, founded in 1975 with members including Benin, Burkina Faso, Cabo Verde, Cote d'Ivoire, The Gambia, Ghana, Guinea, Guinea-Bissau, Liberia, Mali, Niger, Nigeria, Senegal, Sierra Leone, and Togo. The main aim of this alliance is to promote socioeconomic integration among member states to raise living standards and promote economic development, with wider implications for judicial and health service cooperation [9].

There are numerous recognized advantages of EIBF, however, the rates of EIBF in ECOWAS are typically low and still suboptimal, ranging from 17% to 62% [10,11]. These low rates mean a substantial proportion of newborn babies are deprived of mother–newborn bonding, resistance against infection, and the early breastfeeding needed to fight disease [7,8]. Studies highlight that infants initiated to breastfeeding within an hour of birth have a reduced incidence of life-threatening conditions, such as serious systemic bacterial infections, diarrhea, respiratory infections, as well as less severe infections of the middle ear and urinary tract, and allergic conditions [12–14]. A cohort study conducted in rural Ghana indicated that 22% of neonatal deaths could be prevented if breastfeeding started within the first hour of birth [15]. Similarly, a community-based randomized trial conducted in southern Nepal revealed that approximately 8–19% of newborn deaths could be averted annually in developing countries if all newborns were breastfed within an hour of delivery [16]. EIBF is not only beneficial to neonates but also to mothers, as it reduces the risk of postpartum hemorrhage [17].

Published studies conducted on EIBF in SSA, including ECOWAS countries, have indicated that EIBF is associated with demographic, socioeconomic, and obstetric characteristics. A nationwide cross-sectional study on trends and factors associated with EIBF in Nigeria found that mothers who had four or more antenatal visits (ANC) visits during pregnancy, mothers from wealthier households, and mothers who delivered their babies at the health facility are more likely to initiate breastfeeding within one hour of birth [18,19]. Another population-based study conducted in Nigeria that examined urban–rural differences in the rates of EIBF concluded that EIBF rates were significantly lower in rural areas than urban areas [20]. The perception of a lack of breastmilk, perception that the mother and the baby need rest after birth, performing post-birth activities (such as bathing), and the baby not crying for milk were observed as barriers to EIBF among mothers in Ghana [11]. In that study, delivery in a health facility was a facilitator to EIBF, as health practitioners encouraged early breastfeeding [11]. A nationwide cross-sectional study in Zimbabwe found that skilled delivery assistants, multiparity, and postpartum skin to skin contact between mother and newborn babies immediately after birth were significantly related to EIBF [21]. Similarly, in a community-based cross-sectional study on EIBF conducted in Ethiopia, maternal education, delivery at a health facility and visiting antenatal care services during pregnancy were also significantly associated with EIBF within one hour of delivery [22]. The low rate of EIBF practice in ECOWAS could have potentially contributed to nearly half of all under-five deaths, such as those due to prematurity or malnutrition in sub-Saharan Africa region in the year 2017 [1]. Prior to the current study, no studies have used combined country-specific local data across ECOWAS countries to examine factors associated with EIBF within one hour of birth. Evidence based on collective data will provide region-specific targeted interventions to scale up EIBF practices to reduce morbidity and mortality.

Hence, this study aims to investigate the possible characteristics influencing EIBF in 13 of the ECOWAS member countries using the most recent Demographic and Health Survey (DHS) survey datasets conducted between 2010 and 2018. Recognition of factors associated with EIBF in ECOWAS countries is essential for building effective nutrition education and behavior change communication interventions in targeting women at heightened risk of sub-optimal feeding behaviors. In addition, the identification of modifiable associated factors will help public health nutrition program managers and policymakers develop strategies to support these contributory factors. ECOWAS would gain enormous health and economic benefits by improving EIBF. Thus, the findings from this study could assist in the planning and monitoring of effective infant and breastfeeding promotion programs in each ECOWAS country, which may contribute to achieving the child survival sustainable development goal.

2. Materials and Methods

The most recent ECOWAS DHS datasets between 2010 and 2018 were used for this study. Cabo Verde and Guinea-Bissau were not included in the study because their last DHS surveys were not within the study period. As a result, the most recent DHS datasets from 13 ECOWAS countries were used for this study. These data sets were obtained from standardized population-based cross-sectional surveys with a high response rate and trained interviewers [10]. The DHS surveys are publicly available for download [10,23]. Each ECOWAS National Bureau of Statistics collects the DHS Data in collaboration with the United States Agency for International Development [10]. The DHS surveys use identical questionnaires in all countries to collect information on diverse topics, including fertility, reproductive health, maternal and child health, mortality, nutrition, and self-reported health behaviors among adults [5]. In this study and to reduce recall bias, we restricted the analysis to the last born child at 23 months and living with the respondent. This yielded a weighted total of 76,934 children.

2.1. Outcome and Confounding Factors

The outcome variable was EIBF. Women were asked how long after birth the baby was put to the breast for the first time. Responses were recorded in minutes and/or hours. EIBF within one hour of birth was estimated using the World Health Organization recommendation [24,25]. This indicator includes the EIBF rate (the proportion of children born aged 0–23 months who were put to the breast within one hour of birth). EIBF was categorized into a binary form of the outcome variables: "Yes" (1 = if a child was given EIBF) and "No" (0 = if a child was not given EIBF). The potential confounding factors were organised into four distinct groups: demographic factors (countries, place of residence, mother's age, marital status, combined birth rank and birth interval, sex of baby, child's age in category, perceived size of the baby); socio-economic factors (household wealth index, maternal work in the last 12 months, maternal education, and maternal literacy); access to media factors (frequency of reading newspapers/magazine, frequency of listening to radio, and frequency of watching television); and healthcare utilization factors (place of delivery, mode of delivery, type of delivery assistance, and antenatal clinic visits). For the pooled dataset, the household wealth index was constructed using the 'hv271' variable. The hv271 is a household's wealth index value generated by the product of standardized scores (z-scores) and factor coefficient scores (factor loadings) of wealth indicators [26]. In the household wealth index categories, the bottom 20% of households was arbitrarily referred to as the poorest households and the top 20% as richest households, and they were divided into poorest, poor, middle, rich, and richest.

2.2. Statistical Analysis

For the combined 13 ECOWAS datasets, a population-level weight, unique country-specific clustering, and strata were created. This was to manage large country weights being greater than small country weights and also because clusters were different in each country. Population-level weights were used for survey (SVY) tabulation that adjusted for a unique country-specific stratum, and clustering was used to determine the percentage, frequency count, and univariate analysis of all selected characteristics. Country-specific weights were used for the Taylor series linearization method [27] in the surveys when estimating the rates and 95% confidence intervals of EIBF in each country.

Population-level weights were used in the multivariate analyses. In the multivariate analyses, a four-stage hierarchical model was carried out. In the first stage model, demographic factors were entered into the model. In the second stage model, demographic factors were added to the socio-economic factors. A similar procedure was employed for the third stage model, which included the demographic and socio-economic level factors, as well as access to media factors. The fourth and final stage model, healthcare utilization factors, were added to demographic, socio-economic, and media factors. All statistical analyses were conducted using (statistics and data) STATA/MP Version.14.1 (StataCorp, College Station, TX, USA) and adjusted odds ratios (AORs) and their 95% confidence intervals (CIs) obtained from the adjusted multivariate logistic regression model were used to measure the factors associated with EIBF.

3. Results

3.1. Characteristics of the Sample and Univariate Analyses for EIBF

The pooled rate of EIBF in the 13 ECOWAS was 43% and EIBF lay between 52% and 62% in Togo, Niger, Mali, Benin, Sierra Leone, Liberia, Ghana, and Gambia, as indicated in Figure 1. The highest rates of EIBF were observed in Togo and Liberia, and the lowest was in Guinea (17%).

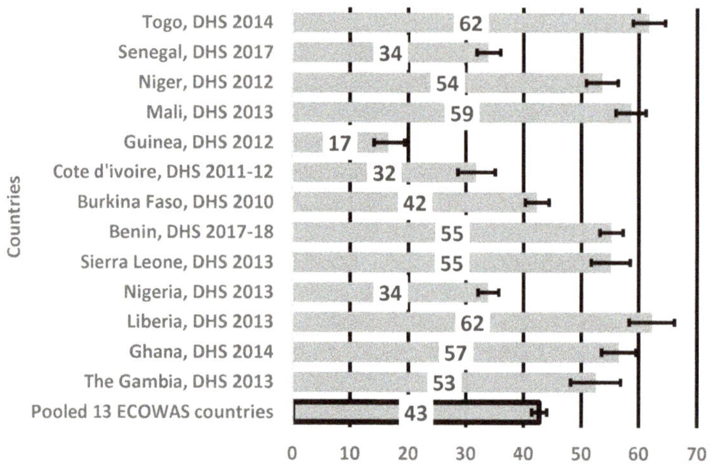

Figure 1. Rates of early initiation of breastfeeding in thirteen Economic Community of West African States. DHS = Demographic and Health Survey.

Table 1 illustrates the characteristics of the sample of children aged 0–23 months and the unadjusted odd ratios of EIBF in children aged 0–23 months. Overall, 21% of children aged 0–23 months were from Nigeria, and compared to Benin (7.7%), the odds of EIBF in Nigeria was reduced significantly, by 58% (OR = 0.42, 95% CI:0.36, 0.48). About 42% of EIBF babies were perceived by their mothers to be of average birth size compared to the 18% of EIBF babies perceived to be small at birth. The unadjusted odd ratios of EIBF were significant with babies perceived by their mothers to be the average birth size at birth, combined birth rank and birth interval, household wealth, frequency of listening to the radio, and watching television.

Table 1. Individual, household, and community level characteristics and unadjusted odds ratios (OR) (95% CI) of early initiation of breastfeeding (EIBF) among children aged 0–23 months in 13 Economic Community of West African States (ECOWAS).

Characteristic	n *	% *	%	EIBF		
				OR	95% CI		p-Value
Demographic factors							
Country							
Benin	5937	7.7	9.2	1.00			
Burkina Faso	6887	9.0	10.0	0.58	0.50	0.67	<0.001
Cote d'ivoire	4554	5.9	5.2	0.43	0.33	0.56	<0.001
The Gambia	5387	7.0	6.0	0.76	0.52	1.11	0.151
Ghana	3412	4.4	4.0	1.07	0.88	1.31	0.483
Guinea	3574	4.6	4.7	0.17	0.13	0.22	<0.001
Liberia	4241	5.5	5.1	1.30	0.96	1.77	0.093
Mail	4843	6.3	6.7	1.11	0.93	1.32	0.242
Niger	7044	9.2	8.1	0.90	0.75	1.08	0.262
Nigeria	15,993	20.8	20.7	0.42	0.36	0.48	<0.001
Senegal	5790	7.5	8.1	0.35	0.29	0.41	<0.001
Sierra Leone	6230	8.1	7.4	0.85	0.67	1.08	0.173
Togo	3042	4.0	4.8	1.49	1.23	1.79	<0.001
Residence							
Urban	27,769	36.1	31.0	1.00			
Rural	49,165	63.9	69.0	0.96	0.85	1.08	0.468
Mothers' age in years							
15–19	25,575	33.2	33.4	1.00			
20–34	36,855	47.9	46.9	1.09	1.02	1.17	0.012
35–49	14,503	18.9	19.8	1.08	0.99	1.17	0.084
Marital status							
Currently married	70,916	92.2	92.7	1.00			
Formerly married ^	1875	2.4	2.5	1.13	0.91	1.39	0.269
Never married	4142	5.4	4.8	0.98	0.83	1.17	0.863
Age of child (months)							
0–5	19,736	25.7	25.8	1.00			
6–11	20,308	26.4	26.9	1.02	0.93	1.12	0.657
12–17	20,402	26.5	26.0	1.07	0.98	1.16	0.138
18–23	16,488	21.4	21.3	1.10	1.00	1.21	0.047
Birth order or birth rank							
First-born	15,935	20.7	20.0	1.00			
Second–fourth	36,840	47.9	46.6	1.26	1.16	1.36	<0.001
Fifth or more	24,159	31.4	33.4	1.19	1.09	1.31	<0.001

Table 1. *Cont.*

Characteristic	n *	% *	%	EIBF		
				OR	95% CI		p-Value
Preceding birth interval (n = 76,748)						
No previous birth	15,935	20.7	20.0	1.00		
<24 months	8163	10.6	10.6	1.13	1.00	1.28	0.042
≥24 months	52,650	68.4	69.2	1.25	1.16	1.35	<0.001
Combined birth rank and birth interval						
Second/third birth rank, more than two years interval	30,717	39.9	39.0	1.00		
First birth rank	15,935	20.7	20.0	0.78	0.72	0.85	<0.001
Second/third birth rank, less than or equal to two years interval	6123	8.0	7.7	0.90	0.79	1.02	0.097
Fourth birth rank, more than two years interval	20,520	26.7	28.4	0.94	0.87	1.02	0.120
Fourth birth rank, less than or equal to two years interval	3638	4.7	5.0	0.89	0.77	1.02	0.103
Sex of baby						
Male	38,723	50.3	50.6	1.00		
Female	38,211	49.7	49.4	1.05	0.99	1.11	0.120
Size of baby (n = 76,255)						
Small	13,689	17.8	17.5	1.00		
Average	32,517	42.3	42.6	1.22	1.11	1.34	<0.001
Large	30,049	39.1	38.9	1.19	1.07	1.31	0.001
Socio-economic factors						
Household Wealth Index						
Poorest	12,753	16.6	20.1	1.00		
Poorer	13,520	17.6	19.7	1.35	1.19	1.54	<0.001
Middle	15,088	19.6	19.8	1.55	1.36	1.78	<0.001
Richer	14,854	19.3	20.1	1.33	1.16	1.52	<0.001
Richest	20,719	26.9	20.5	1.25	1.07	1.46	0.005
Work in the last 12 months (n = 76,918)						
Not working	30,637	39.8	38.2	1.00		
Working	46,281	60.2	61.8	0.94	0.86	1.02	0.150
Maternal education (n = 76,926)						
No education	45,527	59.2	61.6	1.00		
Primary	13,319	17.3	18.1	1.06	0.96	1.17	0.239
Secondary and above	18,080	23.5	20.4	1.03	0.93	1.14	0.549
Maternal Literacy (n = 76,549)						
Cannot read at all	57,292	74.5	77.4	1.00		
Able to read only part of sentences	19,257	25.0	22.1	0.96	0.88	1.04	0.341

Table 1. *Cont.*

Characteristic	n *	% *	%	EIBF		
				OR	95% CI		p-Value
Access to media						
Frequency of reading newspaper or magazine (n = 76,736)						
Not at all	68,490	89.0	91.5	1.00		
Less than once a week	4575	5.9	4.8	0.93	0.81	1.08	0.356
At least once a week	3644	4.7	3.7	1.10	0.92	1.32	0.281
Almost every day	27	0.0	0.0	2.13	0.80	5.69	0.132
Frequency of listening to radio (n = 76,808)						
Not at all	25,977	33.8	36.0	1.00		
Less than once a week	19,798	25.7	25.0	0.95	0.87	1.04	0.300
At least once a week	30,250	39.3	37.8	0.97	0.88	1.06	0.466
Almost every day	783	1.0	1.1	3.27	2.49	4.29	<0.001
Frequency of watching Television (n = 76,760)						
Not at all	42,468	55.2	59.8	1.00		
Less than once a week	12,287	16.0	15.1	1.01	0.90	1.13	0.861
At least once a week	21,294	27.7	24.1	0.84	0.76	0.93	0.001
Almost every day	711	0.9	0.8	2.60	1.97	3.44	<0.001
Healthcare utilization factors						
Place of delivery						
Home	31,183	40.5	41.3	1.00		
Health facility	45,751	59.5	58.7	1.34	1.23	1.47	<0.001
Mode of delivery (n = 76,729)						
Non-caesarean	73,763	95.9	96.3	1.00		
Caesarean section ‡	2966	3.9	3.5	0.39	0.32	0.49	<0.001
Type of delivery assistance (n = 63,553)						
Health professional	49,991	65.0	62.8	1.00		
Traditional birth attendant	1884	2.4	2.8	0.73	0.61	0.88	0.001
Other untrained	6959	9.0	10.7	0.76	0.64	0.90	0.001
No one	4699	6.1	6.0	0.61	0.51	0.72	<0.001
Antenatal clinic visits (n = 74,550)						
None	9478	12.3	12.6	1.00		
1–3	23,314	30.3	31.9	1.55	1.36	1.77	<0.001
4+	41,758	54.3	52.7	1.67	1.45	1.91	<0.001

^ divorce/separated/widowed; * Weighted sample sizes and percentages; ‡ Caesarean section is a combination of both general and regional anesthesia.

3.2. Factors Associated with Early Initiation of Breastfeeding

As shown in Table 2, EIBF within one hour of birth was significantly lower among children living in Nigeria, Cote d'Ivoire, Burkina Faso, Guinea, and Senegal. Mothers from poorest households, mothers' first birth, and delivery by non-health professionals were significantly associated with delayed initiation of breastfeeding. Mothers who underwent caesarean delivery were significantly more likely to delay the initiation of breastfeeding (adjusted OR = 0.28, 95% CI:0.22, 0.36; $p < 0.001$) than those who had non-caesarean delivery.

Mothers who perceived their babies to be average or larger at birth were more likely to initiate breastfeeding within one hour of delivery than mothers who perceived their babies to be small. EIBF was significantly higher among mothers who frequently listened to the radio, who delivered their babies at the health facilities, and those mothers who had four or more antenatal clinic visits. EIBF was significantly higher with female births (adjusted OR = 1.17, 95% CI:1.00, 1.37; $p = 0.046$) than male births.

Table 2. Adjusted odds ratio (AOR) (95% confidence intervals (CI)) of factors associated with the early initiation of breastfeeding (EIBF) among children aged 0–23 months in 13 Economic Community of West African States (ECOWAS).

Characteristic	Model 1				Model 2				Model 3				Model 4			
	AOR	95% CI		p-Value	AOR	95% CI		p-Value	AOR	95% CI		p-Value	AOR	95% CI		p-Value
Demographic factors																
Benin	1.00				1.00				1.00				1.00			
Burkina Faso	0.58	0.50	0.67	<0.001	0.56	0.48	0.65	<0.001	0.56	0.48	0.66	<0.001	0.58	0.48	0.69	<0.001
Cote d'ivoire	0.42	0.32	0.54	<0.001	0.43	0.33	0.56	<0.001	0.43	0.33	0.56	<0.001	0.43	0.33	0.55	<0.001
The Gambia	0.75	0.51	1.09	0.134	0.73	0.50	1.07	0.109	0.75	0.51	1.11	0.148	0.67	0.44	1.02	0.062
Ghana	1.08	0.88	1.34	0.460	1.11	0.89	1.38	0.351	1.15	0.92	1.43	0.212	1.33	1.03	1.72	0.028
Guinea	0.17	0.13	0.22	<0.001	0.17	0.13	0.23	<0.001	0.17	0.13	0.23	<0.001	0.19	0.14	0.25	<0.001
Liberia	1.30	0.95	1.79	0.106	1.34	0.96	1.87	0.081	1.34	0.96	1.87	0.085	1.44	1.04	1.98	0.028
Mali	1.10	0.93	1.32	0.271	1.30	1.07	1.58	0.009	1.34	1.09	1.63	0.005	1.36	1.09	1.71	0.006
Niger	0.94	0.78	1.13	0.520	0.86	0.71	1.04	0.132	0.87	0.72	1.05	0.144	0.98	0.80	1.22	0.882
Nigeria	0.41	0.36	0.48	<0.001	0.44	0.38	0.51	<0.001	0.44	0.38	0.52	<0.001	0.51	0.43	0.61	<0.001
Senegal	0.36	0.30	0.43	<0.001	0.37	0.31	0.45	<0.001	0.38	0.32	0.46	<0.001	0.40	0.33	0.48	<0.001
Sierra Leone	0.86	0.67	1.10	0.230	0.84	0.65	1.07	0.151	0.83	0.65	1.07	0.147	0.78	0.60	1.03	0.075
Togo	1.50	1.25	1.81	0.000	1.50	1.24	1.81	<0.001	1.29	1.05	1.59	0.015	1.56	1.25	1.94	<0.001
Residence																
Urban	1.00				1.00				1.00				1.00			
Rural	0.95	0.84	1.06	0.335	0.94	0.83	1.07	0.373	0.94	0.82	1.07	0.355	0.92	0.81	1.04	0.200
Mothers' age in years																
15–19	1.00				1.00				1.00				1.00			
20–34	0.98	0.90	1.07	0.732	0.99	0.91	1.08	0.777	0.98	0.90	1.07	0.605	0.98	0.90	1.08	0.731
35–49	0.98	0.86	1.11	0.725	0.98	0.86	1.11	0.727	0.96	0.85	1.09	0.538	0.94	0.82	1.08	0.372
Marital status																
Currently married	1.00				1.00				1.00				1.00			
Formerly married ^	0.99	0.79	1.26	0.957	0.99	0.78	1.26	0.940	0.99	0.78	1.26	0.952	0.98	0.76	1.26	0.894
Never married	0.96	0.78	1.18	0.692	0.93	0.76	1.14	0.501	0.95	0.77	1.17	0.610	0.98	0.79	1.21	0.843
Birth rank and birth interval																
Second/third birth rank, more than two years interval	1.00				1.00				1.00				1.00			
First birth rank	0.78	0.71	0.86	<0.001	0.78	0.70	0.86	<0.001	0.77	0.70	0.85	<0.001	0.78	0.69	0.87	<0.001
Second/third birth rank, less than or equal to two years interval	0.89	0.78	1.03	0.123	0.89	0.77	1.02	0.098	0.89	0.77	1.03	0.105	0.91	0.78	1.07	0.268
Fourth birth rank, more than two years interval	0.98	0.89	1.08	0.699	0.99	0.89	1.09	0.791	1.00	0.90	1.11	0.978	1.03	0.93	1.15	0.551
Fourth birth rank, less than or equal to two years interval	0.92	0.79	1.07	0.268	0.92	0.78	1.07	0.288	0.93	0.79	1.09	0.359	0.98	0.82	1.17	0.808
Sex of baby																
Male	1.00				1.00				1.00				1.00			
Female	1.07	1.01	1.13	0.027	1.07	1.01	1.13	0.024	1.07	1.01	1.13	0.025	1.08	1.01	1.16	0.021
Age of child (months)																
0–5	1.00				1.00				1.00				1.00			
6–11	1.03	0.94	1.13	0.546	1.03	0.94	1.13	0.538	1.03	0.93	1.13	0.570	1.02	0.92	1.14	0.664
12–17	1.09	1.00	1.20	0.056	1.10	1.00	1.20	0.045	1.09	0.99	1.19	0.071	1.10	0.98	1.22	0.096
18–23	1.10	1.00	1.21	0.048	1.10	1.00	1.21	0.047	1.10	1.00	1.21	0.052	1.10	0.99	1.23	0.075

Table 2. *Cont.*

Characteristic	Model 1				Model 2				Model 3				Model 4			
	AOR	95% CI		p-Value	AOR	95% CI		p-Value	AOR	95% CI		p-Value	AOR	95% CI		p-Value
Size of baby																
Small	1.00				1.00				1.00				1.00			
Average	1.21	1.09	1.33	<0.001	1.20	1.08	1.32	<0.001	1.20	1.09	1.32	<0.001	1.19	1.07	1.32	0.002
Large	1.23	1.11	1.37	<0.001	1.22	1.10	1.36	<0.001	1.22	1.10	1.36	<0.001	1.20	1.07	1.36	0.002
Socio-economic factors																
Household Wealth Index																
Poorest					1.00				1.00				1.00			
Poorer					1.29	1.13	1.48	<0.001	1.32	1.15	1.51	<0.001	1.19	1.02	1.38	0.028
Middle					1.38	1.20	1.59	<0.001	1.42	1.23	1.63	<0.001	1.22	1.03	1.43	0.018
Richer					1.34	1.17	1.55	<0.001	1.39	1.20	1.60	<0.001	1.17	0.98	1.39	0.072
Richest					1.16	0.98	1.36	0.083	1.19	1.00	1.43	0.052	1.03	0.84	1.26	0.782
Work in the last 12 months																
Not working					1.00				1.00				1.00			
Working					0.95	0.86	1.04	0.243	0.95	0.87	1.04	0.291	0.95	0.87	1.05	0.329
Maternal education																
No education					1.00				1.00				1.00			
Primary					1.02	0.92	1.13	0.732	1.03	0.92	1.14	0.627	1.00	0.90	1.12	0.992
Secondary and above					0.98	0.80	1.21	0.860	0.96	0.78	1.19	0.717	0.89	0.73	1.10	0.291
Maternal Literacy																
Cannot read at all					1.00				1.00				1.00			
Able to read only part of sentences					1.04	0.87	1.24	0.690	1.00	0.82	1.21	0.985	1.02	0.85	1.23	0.811
Access to media																
Frequency of reading newspaper or magazine																
Not at all									1.00				1.00			
Less than once a week									1.11	0.93	1.32	0.255	1.06	0.89	1.27	0.508
At least once a week									1.33	1.09	1.63	0.006	1.36	1.10	1.67	0.004
Almost every day									0.91	0.33	2.51	0.856	1.01	0.40	2.50	0.990
Frequency of listening to Radio																
Not at all									1.00				1.00			
Less than once a week									0.90	0.82	0.99	0.037	0.86	0.78	0.95	0.002
At least once a week									0.92	0.83	1.01	0.089	0.92	0.83	1.02	0.099
Almost every day									1.47	1.11	1.95	0.007	1.45	1.07	1.95	0.016
Frequency of watching Television																
Not at all									1.00				1.00			
Less than once a week									1.03	0.90	1.18	0.660	1.00	0.86	1.17	0.967
At least once a week									0.95	0.84	1.06	0.333	0.90	0.80	1.02	0.104
Almost every day									1.17	0.85	1.60	0.335	1.14	0.82	1.60	0.438
Healthcare utilization factors																
Place of delivery																
Home																
Health facility													1.00			
Mode of delivery													1.38	1.21	1.57	<0.001

Table 2. *Cont.*

Characteristic	Model 1			Model 2			Model 3			Model 4		
	AOR	95% CI	*p*-Value	AOR	95% CI	*p*-Value	AOR	95% CI	*p*-Value	AOR	95% CI	*p*-Value
Non-caesarean										1.00		
Caesarean section ‡										0.28	0.22 0.36	<0.001
Type of delivery assistance												
Health professional										1.00		
Traditional birth attendant.										0.77	0.60 0.98	0.037
Other untrained										0.96	0.79 1.16	0.663
No one										0.76	0.62 0.92	0.005
Antenatal Clinic visits												
None										1.00		
1–3										1.17	1.00 1.37	0.052
4+										1.20	1.01 1.41	0.034

^ divorce/separated/widowed.; ‡ Caesarean section is a combination of both general and regional anesthesia.

4. Discussion

Our results indicate that EIBF rates in the 13 ECOWAS are at a sub-optimum level, and need further improvement to fulfil the goal of optimal infant feeding in this subregion and the related Sustainable Development Goals (SDGs) of reducing neonatal mortality. During the study period (2010–2018), we noted the uneven prevalence of EIBF among infants aged between 0 and 23 months of age across the ECOWAS countries, ranging from a low of 17% (Guinea) to the highest of 61% (Togo). The reported rates of EIBF in this study are still well below the expected target of ≥70% recommended by the WHO [1] to further reduce infant mortality and morbidity in the sub-region by 25 deaths per 1000 live births by 2030. The wide variation in the rates of EIBF within the 13 ECOWAS countries may be due to socio-cultural, geographical location, and health inequalities, as well as economic issues among different populations. The study indicated that place of delivery (health institution), mode of delivery (vaginal delivery), ANC visits (≥4), and household wealth index (poorer or middle-class household) were significantly associated with EIBF within one hour of birth. Media exposure (newspaper/magazine or radio), baby size at birth (average and large), child's sex (female), and first birth order were also found to be positively related to EIBF. Delayed introduction to breastfeeding was significant in six ECOWAS member states (Gambia, Nigeria, Senegal, Guinea, Burkina Faso, and Cote d'Ivoire).

Child delivery at health facilities predicts EIBF practices within one hour of the birth. This finding is consistent with previous studies conducted in Nepal [28], Nigeria [29], Namibia [30], and Ethiopia [31]. However, this is not unexpected given that the majority of health institutions across ECOWAS countries have adopted BFHI, which is overseen by skilled health providers (midwife, nurse, or doctor) whose core aim is to encourage and assist mothers in achieving optimal breastfeeding practices, including EIBF. It has been suggested that the presence of trained breastfeeding and delivery assistants in health institutions improved mothers' EIBF practice [32]. This may have contributed to the substantial proportion (60%) of EIBF among mothers who gave birth at a hospital, as observed in the current study.

Similar to reports from earlier studies [22,29,30], the present study found that vaginal delivery was strongly correlated with EIBF practice when compared with caesarean birth. A plausible reason for this difference may be related to the adverse physical effects mothers and newborns often experience after CS surgery. Factors such as excruciating pain, anesthesia, prolonged labor, and related respiratory distress among newborns [33,34] may result in healthcare providers being preoccupied with helping both babies and mothers to stabilize rather than initiating breastfeeding [35,36].

Previously published studies indicated that pregnant women being unable to access or attend ANC services during pregnancy poses a significant barrier in initiating breastfeeding within one-hour post-birth [37,38]. However, in the present study, we found that mothers who had four or more ANC visits to a health institution prior to childbirth showed a greater likelihood of EIBF practice than those who had none or less than four ANC visits. This result is similar to that reported in past studies carried out in India [39] and Nepal [40]. Breastfeeding counseling and health messages received through the Baby-Friendly Hospital Initiative (BFHI) may have contributed to the increased likelihood of EIBF noted in the study. It has been suggested in previous work that breastfeeding counseling during ANC visits is positively related to the mother's adherence to the WHO optimal breastfeeding practices [41,42]. Therefore, promoting and supporting ANC uptake in ECOWAS countries could ensure remarkable improvement in EIBF practices.

Newborn babies whose mother perceived their size to be average or larger had an increased likelihood of EIBF within one hour of birth compared with those who were perceived to be small or smaller size. This finding is consistent with a cross-sectional study carried out in Nigeria in 2016, which indicated that newborn babies perceived by their mothers as average or larger size at birth were more likely to be EIBF [29]. Also, a cross-sectional study done in Nepal in 2014 reported that newborn babies whose size was large at birth were significantly more likely to receive EIBF within one hour of birth in comparison with babies whose size was small [28]. The significantly increased odds of EIBF practice among newborn babies perceived to be of average or larger size in ECOWAS countries may be attributed to their physical maturity and associated proper coordination of the suction–deglutination–respiratory cycle and breast-seeking reflex [28]. Conversely, the immaturity of small babies often hinders the abilities required to initiate breastfeeding within one hour of delivery due to poor breast-seeking reflex and sucking inability [43]. Additionally, illness often leads to newborn babies being kept away from their mothers [44], resulting in the early introduction of complementary drinks. Notably, female births were found to be associated with EIBF within one hour of delivery. This outcome is similar to earlier reports in Sri Lanka [45] and Ethiopia [46], which found that male babies were less likely to receive early breastfeeding than their female counterparts. This gender difference in EIBF practice may be influenced by socio-cultural beliefs and require further assessment. However, previous studies have suggested that the practice of prelacteal feeds (or the act of given foods to newborn babies before initiating breastfeeding) in male babies is common and acceptable in African [8,47] and Asian countries [47,48].

Newborn babies born to mothers from a poor household had significantly higher odds of initiating breastfeeding within one hour of birth compared with those from a wealthy household. This finding contradicts previous reports, which indicated that mothers from wealthy households were more likely to initiate breastfeeding in the first hour of delivery [28,35]. Nevertheless, our finding is similar to those reported from Nepal [49] and Namibia [30]. The significantly higher odds of EIBF practice among poor mothers noted in the current study may be attributed to lack of access to infant formula or inability to buy costly complementary substitutes to breast milk, which may result in poor mothers depending solely on breastfeeding. It has also been suggested that mothers from wealthy households were more likely have an elective caesarean section, resulting in delayed initiation of breastfeeding within one hour of birth [50]. It was not unexpected that mothers who frequently watch television or listen to radio were more likely to initiate breastfeeding within one hour of delivery. This finding is consistent with previous studies [51,52]. Apart from health facilities, exposure to other information sources regarding breastfeeding practices during antenatal and postnatal care can also play a key role in encouraging women to practice early initiation of breastfeeding.

Similar to previous reports [30,46], we found that mothers who were having their first child were less likely to initiate breastfeeding within one hour of birth compared with those with two or more deliveries. This may be related to inexperience regarding breastfeeding and relatively poor use of maternal health services by first time mothers. As shown in a past study, earlier breastfeeding experience is correlated with both intention and timely breastfeeding initiation [53].

5. Limitations and Strengths

The following limitations of this study should be considered when making specific conclusions about its findings: (a) the study was unable to establish a causal relationship between the study characteristics and EIBF, as cross-sectional data was used to identify the predictors of EIBF; (b) recall and measurement errors may have overestimated or underestimated the findings of this study, as data regarding some of the study factors was obtained from mothers up to five years after childbirth; (c) there were some unmeasured coexisting factors, such as instrumental vaginal birth, cultural beliefs, and health professionals' prior knowledge of EIBF or family dynamics, which may have also impacted the study findings. Additionally, detailed information concerning infant formula advertising via electronic or print media was not available at the time of the DHS survey. Despite the highlighted weaknesses, there are a number of strengths that should also be noted regarding this study. Firstly, the rate of EIBF was based on nationally representative data of 13 countries, and hence findings could be generalized across the subregion. Secondly, the results are comparable across the ECOWAS countries because all the variables used were similarly described across countries. Thirdly, the study used a combined nationally representative large sample size across the 13 ECOWAS, which meant that the potential impact of selection bias was minimal and unlikely to affect the estimates.

6. Conclusions

The findings of this study suggest that delivery in a hospital or health facility, vaginal delivery, frequent ANC visits (≥4), middle income household, access to electronic media (television or radio), average or larger baby size at birth, child's sex (female), and first birth order were significantly associated with EIBF within one hour of birth in the 13 ECOWAS countries. In the ECOWAS subregion, it is crucial that integrated nutritional health promotion campaigns be conducted both at the community and individual levels. At the community level, the role of midwives and community health officers (CHO) in promoting breastfeeding practices appeared evident, with CHO antenatal home visits contributing to the early initiation of breastfeeding and empowering the community with greater knowledge of the nutrition and health requirements of children under two years of age. At the individual level, interventions to promote the involvement of family members during pregnancy and child birth are needed in ECOWAS and should particularly target young mothers and families from lower socioeconomic groups with lower rates of EIBF.

Author Contributions: Conceptualization, K.E.A. and F.A.O.; methodology, K.E.A. and O.K.E.; software, K.E.A.; validation, O.L.U., F.A.O. and O.K.E.; formal analysis, K.E.A. and P.R.G.; investigation, O.K.E., P.R.G.; resources, W.K.T.; data curation, F.A.O., P.R.G. and O.L.U.; writing—original draft preparation, K.E.A. and O.K.E.; writing—review and editing, O.L.U., W.K.T., G.J.S.; visualization, G.J.S., O.L.U.; supervision, K.E.A., W.K.T. and G.J.S.; project administration, O.K.E., P.R.G. and K.E.A.

Funding: This research received no external funding.

Acknowledgments: The authors are grateful to Measure DHS, ICF International, Rockville, Marylands, USA for providing the data for this study. GloMACH: Members are Kingsley E Agho, Felix A Ogbo, Thierno Diallo, Osita E Ezeh, Osuagwu L Uchechukwu, Pramesh R Ghimire, Blessing Akombi, Paschal Ogeleka, Tanvir Abir, Abukari I Issaka, Kedir Yimam Ahmed, Rose Victor, Deborah Charwe, Abdon Gregory Rwabilimbo, Daarwin Subramanee, Mehak Mehak, Nilu Nagdev and Mansi Dhami.

Conflicts of Interest: The authors declare no conflict of interest. K.E.A is a Guest Editor for the Nutrients Special Issue on "Breastfeeding: Short and Long-Term Benefits to Baby and Mother" and did not have any role in the journal review and decision-making process for this manuscript.

References

1. Collective, G.B.; UNICEF; World Health Organization. *Global Breastfeeding Scorecard, 2017: Tracking Progress for Breastfeeding Policies and Programmes*; World Health Organization: Geneva, Switzerland, 2017.
2. UNICEF. *Levels and Trends in Child Mortality: Report 2011*; WHO Estimates Developed by the UN Inter-Agency Group for Child Mortality Estimation: New York, NY, USA, 2011.
3. United Nations. *Population Facts*; United Nations, Department of Economic and Social Affairs, Population Division: New York, NY, USA, 2017; Volume 6, pp. 1–2.
4. WHO; UNICEF. *Planning Guide for National Implementation of the Global Strategy for Infant and Young Child Feeding*; World Health Organization: Geneva, Switzerland, 2007.
5. WHO; UNICEF; Health WHODoCA. *Handbook IMCI: Integrated Management of Childhood Illness*; World Health Organization: Geneva, Switzerland, 2005.
6. Smith, E.R.; Hurt, L.; Chowdhury, R.; Sinha, B.; Fawzi, W.; Edmond, K.M.; Neovita Study Group. Delayed breastfeeding initiation and infant survival: A systematic review and meta-analysis. *PLoS ONE* **2017**, *12*, e0180722. [CrossRef]
7. Horta, B.L.; Victora, C.G. *Long-Term Effects of Breastfeeding*; World Health Organization: Geneva, Switzerland, 2013; Volume 74.
8. Agho, K.E.; Ogeleka, P.; Ogbo, F.A.; Ezeh, O.K.; Eastwood, J.; Page, A. Trends and Predictors of Prelacteal Feeding Practices in Nigeria (2003–2013). *Nutrients* **2016**, *8*, 462. [CrossRef]
9. Turner, B. Economic Community of West African States (ECOWAS). In *The Statesman's Yearbook*; Springer: Berlin, Germany, 2012; pp. 59–60.
10. DHS. *The Demographic and Health Survey Program*, 2016 ed.; United States Agency for International Development: Washington, DC, USA, 2016.
11. Tawiah-Agyemang, C.; Kirkwood, B.; Edmond, K.; Bazzano, A.; Hill, Z. Early initiation of breast-feeding in Ghana: Barriers and facilitators. *J. Perinatol.* **2008**, *28*, S46–S52. [CrossRef]

12. Leung, A.K.C.; Sauve, R.S. Breast is best for babies. *J. Nat. Med. Assoc.* **2005**, *97*, 1010–1019.

13. Hanieh, S.; Ha, T.T.; Simpson, J.A.; Thuy, T.T.; Khuong, N.C.; Thoang, D.D.; Tran, T.D.; Tuan, T.; Fisher, J.; Biggs, B.-A. Exclusive breast feeding in early infancy reduces the risk of inpatient admission for diarrhea and suspected pneumonia in rural Vietnam: A prospective cohort study. *BMC Public Health* **2015**, *15*, 1166. [CrossRef] [PubMed]

14. Arifeen, S.; Black, R.E.; Antelman, G.; Baqui, A.; Caulfield, L.; Becker, S. Exclusive breastfeeding reduces acute respiratory infection and diarrhea deaths among infants in Dhaka slums. *Pediatrics* **2001**, *108*, e67. [CrossRef] [PubMed]

15. Edmond, K.M.; Zandoh, C.; Quigley, M.A.; Amenga-Etego, S.; Owusu-Agyei, S.; Kirkwood, B.R. Delayed breastfeeding initiation increases risk of neonatal mortality. *Pediatrics* **2006**, *117*, e380–e386. [CrossRef] [PubMed]

16. Mullany, L.C.; Katz, J.; Li, Y.M.; Khatry, S.K.; LeClerq, S.C.; Darmstadt, G.L.; Tielsch, J.M. Breast-feeding patterns, time to initiation, and mortality risk among newborns in southern Nepal. *J. Nutr.* **2008**, *138*, 599–603. [CrossRef]

17. Exavery, A.; Kanté, A.M.; Hingora, A.; Phillips, J.F. Determinants of early initiation of breastfeeding in rural Tanzania. *Int. Breastfeed. J.* **2015**, *10*, 27. [CrossRef] [PubMed]

18. Ogbo, F.A.; Page, A.; Agho, K.E.; Claudio, F. Determinants of trends in breast-feeding indicators in Nigeria, 1999–2013. *Public Health Nutr.* **2015**, *18*, 3287–3299. [CrossRef]

19. Ogbo, F.A.; Agho, K.E.; Page, A. Determinants of suboptimal breastfeeding practices in Nigeria: Evidence from the 2008 demographic and health survey. *BMC Public Health* **2015**, *15*, 259. [CrossRef] [PubMed]

20. Adewuyi, E.O.; Zhao, Y.; Khanal, V.; Auta, A.; Bulndi, L.B. Rural-urban differences on the rates and factors associated with early initiation of breastfeeding in Nigeria: Further analysis of the Nigeria demographic and health survey, 2013. *Int. Breastfeed. J.* **2017**, *12*, 51. [CrossRef] [PubMed]

21. Mukora-Mutseyekwa, F.; Gunguwo, H.; Mandigo, R.G.; Mundagowa, P. Predictors of early initiation of breastfeeding among Zimbabwean women: Secondary analysis of ZDHS 2015. *Matern. Health Neonatol. Perinatol.* **2019**, *5*, 2. [CrossRef] [PubMed]

22. Bisrat, Z.; Kenzudine, A.; Bossena, T. Factors associated with early initiation and exclusive breastfeeding practices among mothers of infant's age less than 6 months. *J. Pediatr. Neonatal Care* **2017**, *7*, 00292.

23. Corsi, D.J.; Neuman, M.; Finlay, J.E.; Subramanian, S. Demographic and health surveys: A profile. *Int. J. Epidemiol.* **2012**, *41*, 1602–1613. [CrossRef]

24. UNICEF; World Health Organization. *Capture the Moment: Early Initiation of Breastfeeding: The Best Start for Every Newborn*; UNICEF: New York, NY, USA; World Health Organization: Geneva, Switzerland, 2018.

25. Bortz, K. WHO, UNICEF: 78 million newborns not breastfed within first hour after birth. *Infect. Dis. Child.* **2018**, *31*, 19.

26. Rutstein, S.O.; Johnson, K. The DHS Wealth Index. DHS Comparative Reports No. 6. (ORC Macro, Calverton, MD, 2004). Available online: https://dhsprogram.com/pubs/pdf/CR6/CR6.pdf (accessed on 20 August 2019).

27. Rutstein, S.O.; Rojas, G. *Guide to DHS Statistics*; ORC Macro: Calverton, MD, USA, 2006.

28. Adhikari, M.; Khanal, V.; Karkee, R.; Gavidia, T. Factors associated with early initiation of breastfeeding among Nepalese mothers: Further analysis of Nepal Demographic and Health Survey, 2011. *Int. Breastfeed. J.* **2014**, *9*, 21. [CrossRef]

29. Berde, A.S.; Yalcin, S.S. Determinants of early initiation of breastfeeding in Nigeria: A population-based study using the 2013 demograhic and health survey data. *BMC Pregnancy Childbirth* **2016**, *16*, 32. [CrossRef]

30. Ndirangu, M.; Gatimu, S.; Mwinyi, H.; Kibiwott, D. Trends and factors associated with early initiation of breastfeeding in Namibia: Analysis of the demographic and health surveys 2000–2013. *BMC Pregnancy Childbirth* **2018**, *18*, 171. [CrossRef]

31. Liben, M.L.; Gemechu, Y.B.; Adugnew, M.; Asrade, A.; Adamie, B.; Gebremedin, E.; Melak, Y. Factors associated with exclusive breastfeeding practices among mothers in dubti town, afar regional state, northeast Ethiopia: A community based cross-sectional study. *Int. Breastfeed. J.* **2016**, *11*, 4. [CrossRef]

32. Okafor, I.; Olatona, F.; Olufemi, O. Breastfeeding practices of mothers of young children in Lagos, Nigeria. *Niger. J. Paediatr.* **2014**, *41*, 43–47. [CrossRef]

33. Awi, D.; Alikor, E. Barriers to timely initiation of breastfeeding among mothers of healthy full-term babies who deliver at the University of Port Harcourt Teaching Hospital. *Niger. J. Clin. Pract.* **2006**, *9*, 57–64. [PubMed]

34. Liben, M.L.; Yesuf, E.M. Determinants of early initiation of breastfeeding in Amibara district, Northeastern Ethiopia: A community based cross-sectional study. *Int. Breastfeed. J.* **2016**, *11*, 7. [CrossRef] [PubMed]
35. Tilahun, G.; Degu, G.; Azale, T.; Tigabu, A. Prevalence and associated factors of timely initiation of breastfeeding among mothers at Debre Berhan town, Ethiopia: A cross-sectional study. *Int. Breastfeed. J.* **2016**, *11*, 27. [CrossRef] [PubMed]
36. Hobbs, A.J.; Mannion, C.A.; McDonald, S.W.; Brockway, M.; Tough, S.C. The impact of caesarean section on breastfeeding initiation, duration and difficulties in the first four months postpartum. *BMC Pregnancy Childbirth* **2016**, *16*, 90. [CrossRef] [PubMed]
37. Goldie, S.J.; Sweet, S.; Carvalho, N.; Natchu, U.C.M.; Hu, D. Alternative strategies to reduce maternal mortality in India: A cost-effectiveness analysis. *PLoS Med.* **2010**, *7*, e1000264. [CrossRef] [PubMed]
38. Darmstadt, G.L.; Bhutta, Z.A.; Cousens, S.; Adam, T.; Walker, N.; De Bernis, L. Evidence-based, cost-effective interventions: How many newborn babies can we save? Lancet Neonatal Survival Steering Team. *Lancet* **2005**, *365*, 977–988. [CrossRef]
39. Senanayake, P.; O'Connor, E.; Ogbo, F.A. National and rural-urban prevalence and determinants of early initiation of breastfeeding in India. *BMC Public Health* **2019**, *19*, 896. [CrossRef]
40. Subedi, N.; Paudel, S.; Rana, T.; Poudyal, A. Infant and young child feeding practices in Chepang communities. *J. Nepal Health Res. Counc.* **2012**, *10*, 141–146.
41. El Shafei, A.M.H.; Labib, J.R. Determinants of exclusive breastfeeding and introduction of complementary foods in rural Egyptian communities. *Glob. J. Health Sci.* **2014**, *6*, 236–244. [CrossRef]
42. Mekuria, G.; Edris, M. Exclusive breastfeeding and associated factors among mothers in Debre Markos, Northwest Ethiopia: A cross-sectional study. *Int. Breastfeed. J.* **2015**, *10*, 1. [CrossRef]
43. Walker, M. Breastfeeding the late preterm infant. *J. Obstet. Gynecol. Neonatal Nurs.* **2008**, *37*, 692–701. [CrossRef]
44. Mulready-Ward, C.; Sackoff, J. Outcomes and factors associated with breastfeeding for <8 weeks among preterm infants: Findings from 6 states and NYC, 2004–2007. *Matern. Child Health J.* **2013**, *17*, 1648–1657. [CrossRef]
45. Senarath, U.; Siriwardena, I.; Godakandage, S.S.; Jayawickrama, H.; Fernando, D.N.; Dibley, M.J. Determinants of breastfeeding practices: An analysis of the Sri Lanka Demographic and Health Survey 2006–2007. *Matern. Child Nutr.* **2012**, *8*, 315–329. [CrossRef]
46. John, J.R.; Mistry, S.K.; Kebede, G.; Manohar, N.; Arora, A. Determinants of early initiation of breastfeeding in Ethiopia: A population-based study using the 2016 demographic and health survey data. *BMC Pregnancy Childbirth* **2019**, *19*, 69. [CrossRef]
47. Nguyen, P.H.; Keithly, S.C.; Nguyen, N.T.; Nguyen, T.T.; Tran, L.M.; Hajeebhoy, N. Prelacteal feeding practices in Vietnam: Challenges and associated factors. *BMC Public Health* **2013**, *13*, 932. [CrossRef]
48. Khanal, V.; Adhikari, M.; Sauer, K.; Zhao, Y. Factors associated with the introduction of prelacteal feeds in Nepal: Findings from the Nepal demographic and health survey 2011. *Int. Breastfeed. J.* **2013**, *8*, 9. [CrossRef]
49. Khanal, V.; Scott, J.; Lee, A.; Karkee, R.; Binns, C. Factors associated with early initiation of breastfeeding in Western Nepal. *Int. J. Environ. Res. Public Health Nutr.* **2015**, *12*, 9562–9574. [CrossRef]
50. Rowe-Murray, H.J.; Fisher, J.R. Baby friendly hospital practices: Cesarean section is a persistent barrier to early initiation of breastfeeding. *Birth* **2002**, *29*, 124–131. [CrossRef]
51. Tamiru, D.; Mohammed, S. Maternal knowledge of optimal breastfeeding practices and associated factors in rural communities of Arba Minch Zuria. *Int. J. Nutr. Food Sci.* **2013**, *2*, 122–129. [CrossRef]
52. Hailemariam, T.W.; Adeba, E.; Sufa, A. Predictors of early breastfeeding initiation among mothers of children under 24 months of age in rural part of West Ethiopia. *BMC Public Health* **2015**, *15*, 1076. [CrossRef]
53. Lessen, R.; Crivelli-Kovach, A. Prediction of initiation and duration of breast-feeding for neonates admitted to the neonatal intensive care unit. *J. Perinat. Neonatal Nurs.* **2007**, *21*, 256–266. [CrossRef]

Article

Prevalence and Associated Factors of Caesarean Section and its Impact on Early Initiation of Breastfeeding in Abu Dhabi, United Arab Emirates

Zainab Taha [1],*, Ahmed Ali Hassan [2], Ludmilla Wikkeling-Scott [1] and Dimitrios Papandreou [1]

[1]   Department of Health Sciences, College of Natural and Health Sciences, Zayed University, Abu Dhabi, P.O. Box 144534, UAE; ludmilla.scott@gmail.com (L.W.-S.); Dimitrios.Papandreou@zu.ac.ae (D.P.)

[2]   Taami for Agricultural and Animal Production, Khartoum, Sudan; aa801181@gmail.com

*   Correspondence: Zainab.taha@zu.ac.ae; Tel.: +971-2-599-3111

Received: 18 October 2019; Accepted: 8 November 2019; Published: 10 November 2019

Abstract: The World Health Organization (WHO) recommends the early initiation of breastfeeding. Research shows that factors such as mode of delivery may interfere with the early initiation of breastfeeding. However, data in the United Arab Emirates (UAE) on these findings is limited. Thus, the aim of this study was to describe the prevalence of caesarean sections (CSs) and evaluate their effect on breastfeeding initiation among mothers of children under the age of two years in Abu Dhabi. Data were collected in clinical and non-clinical settings across various geographical areas in Abu Dhabi during 2017 using consent and structured questionnaires for interviews with mothers. Data analysis included both descriptive and inferential statistics. Among the 1624 participants, one-third (30.2%) reportedly delivered by CS, of which 71.1% were planned, while 28.9% were emergency CS. More than half of all mothers (62.5%) initiated early breastfeeding. Multivariable logistic regression indicated factors that were associated positively with CS included advanced maternal age, nationality, and obesity. However, gestational age (GA) was negatively associated with CS. This study shows that the prevalence of CS is high in Abu Dhabi, UAE. CS is associated with lower early initiation rates of breastfeeding. The early initiation rates of breastfeeding were 804 (79.2%) 95% confidence interval (CI) (76.4, 82.0), 162 (16.0%) 95% CI (10.4, 21.6), and 49 (4.8%) 95% CI (1.2, 10.8) among vaginal delivery, planned CS, and emergency CS, respectively. Regarding the mode of delivery, vaginal were 2.78 (Adjusted Odd Ratio (AOR)): CI (95%), (2.17–3.56, $p < 0.001$) times more likely related to an early initiation of breastfeeding. CS in general, and emergency CS, was the main risk factor for the delayed initiation of breastfeeding. The study provides valuable information to develop appropriate strategies to reduce the CS rate in UAE. Maternal literacy on CS choices, the importance of breastfeeding for child health, and additional guidance for mothers and their families are necessary to achieve better breastfeeding outcomes.

Keywords: caesarean section; initiation of breastfeeding; maternal age; gestational age; United Arab Emirates

---

1. Introduction

The rate of caesarean section (CS) in the United Arab Emirates (UAE) has increased from 10% in 1995 [1] to 24% in 2014 [2]. The World Health Organization (WHO) [3] suggests a CS rate of 10%–15% of all live births, which is significantly lower than those reported in the UAE. Researchers suggest that the mode of delivery influences breastfeeding initiation and duration [4,5], and may influence subsequent breastfeeding and duration [6].

CS is well documented to be associated with suboptimal consequences related to both the mother and her infant's health [7,8]. Among those consequences reported, endometritis, hemorrhage, cystitis, infant respiratory complications, and hypoglycemia [8,9] may have negative effects on breastfeeding. The effect of CS on the initiation of breastfeeding may be related to the adverse effects of anesthesia for both mothers and their newborns. Maternal distress, which often accompanies CS, may negatively affect the baby's feeding behaviors and breastfeeding outcomes.

An abundance of studies have shown that mothers who give birth via CS delivery may be less likely to breastfeed, and or more likely to delay breastfeeding initiation [10–12]. An early initiation of breastfeeding, i.e., within the first hour after delivery, has been recommended by the WHO as an important factor to extend breastfeeding duration [13–15]. An important practice recommended by the WHO, as part of the 10 steps of the Baby-Friendly Hospital Initiative (BFHI), is skin-to-skin contact [16]. To ensure this practice, the mother and her newborn infant must be conscious and fully awake. Therefore, babies born by CS may not benefit from skin-to-skin contact immediately after birth, and may be more susceptible to delayed breastfeeding [16]. Previous studies have found that delaying breastfeeding initiation that co-occurs with CS delivery is associated with factors such as mother–baby separation, impaired suckling skills, and insufficient milk production. This will ultimately impact the continuation of breastfeeding [10,17,18].

The health benefits of breastfeeding for mother and child have been well documented. Mothers who breastfeed their babies have a reduced incidence of type 2 diabetes mellitus, as well as breast and ovarian cancers [19]. Moreover, breastfed babies are less likely to develop childhood illnesses and obesity, and have higher levels of intelligence as an adult [20,21].

Following the WHO recommendations [14], the UAE has incorporated the Global Strategy for Infant and Young Child Feeding, and the Ministry of Health (MOH) has issued a national infant feeding policy [22]. This policy states that infants should be breast fed exclusively until six months of age and continue breastfeeding up to or beyond two years of age [22]. A previous study by Taha et al. revealed a breastfeeding initiation rate of 95.6% and early initiation of 62.5% [23]. However, only 7% of babies were still breastfeeding at six months of age, which prompted the authors to further investigate these findings.

Multiple studies from different countries on the effects of CS versus vaginal delivery on breastfeeding practices have shown contradicting results [24–27]. While some research has documented major barriers to breastfeeding after CS [28–30], other studies have indicated that CS had no effect on breastfeeding initiation [31,32] and duration [33]. Some studies reported that planned CS was associated with delayed breastfeeding [25,34] while others concluded the opposite [35].

Regardless of the discrepancies in previous results, CS will remain a significant and alarming concern, as it influences the initiation of breastfeeding. For a country such as the UAE, which follows the WHO recommendations of infants' and young children's feeding, but still encounters increased rates of CS deliveries, further investigation and more implementation is required to improve knowledge about predictive breastfeeding factors. Thus, the aim of this study was to describe the prevalence of CS and evaluate its effect on breastfeeding initiation among mothers of children under the age of two years in Abu Dhabi.

2. Materials and Methods

2.1. Participants and Data Collection

The sample for this study is based on secondary data from an original sample of clinical and non-clinical data obtained from mothers with at least one infant under the age of two years. Participants for the original study included UAE nationals and non-nationals in the Emirate of Abu Dhabi, which represents 87% of the geographical landmass of the UAE [36]. All data were collected between March and September 2017. The subjects were randomly selected from the community and seven out of a total of 11 maternal and child health centers serving children in Abu Dhabi Capital City (two rural, one suburban, four urban), which provided the possibility of data collection across various areas of Abu Dhabi. Among the 1578 mothers from the clinics who were invited to participate, 1555 mothers agreed and were included in the study. Another 267 mothers from the community also agreed to participate in the study, resulting in a total sample size of 1822 mothers. The geographical distribution of the sample was 54%, 10.8%, and 35.2% of mothers recruited from rural areas, suburban areas, and urban areas respectively.

Mothers with young children attending the centers were approached by the trained bilingual (Arabic and English) female research assistants, who provided oral and written information about the study. Consenting mothers who met the inclusion criteria of having at least one child under two years of age were interviewed by the research assistants using a structured questionnaire.

2.2. Study Instrument

A pre-tested questionnaire included family demographics (e.g., education, age, nationality, occupation), child's information (e.g., birth weight and height, delivery mode), and infant feeding practices (e.g., initiation of breastfeeding, and rooming-in). Details of the methodology of the primary data have been previously described [23].

2.3. Study Inclusion and Exclusion Criteria

From the 1822 mother–child pairs, there were 1624 with complete data on all the variables of interest that were included in the analysis.

2.4. Statistical Analysis

Data analysis was conducted using Statistical Package for the Social Science (IBM SPSS Statistics for Windows, Version 20.0. Armonk, NY: IBM Corp.). Both descriptive and inferential statistics were used to analyze the data. T-tests and chi-square tests were applied to analyze continuous and categorical data, respectively. Variables with significant p-value (<0.05) in univariate analysis were entered in multivariable logistic analysis with mode of delivery (vaginal delivery was coded as 0 and CS was coded as 1) as the dependent variable. In addition, body mass index (BMI) was categorized to underweight, normal, overweight, and obese in order to establish which category was associated with CS. Other variables such as sociodemographic characteristic (e.g., age, parent education, occupation, child gender, etc.) were considered as the independent variables. Odds ratio (OR) and 95% confidence intervals (CIs) were calculated with a significance level of p-value < 0.05. Furthermore, multinomial logistic regression analysis was used to investigate the impact of each mode of delivery (vaginal delivery, planned and emergency CS) on the early initiation of breastfeeding.

2.5. Ethics

The original study from which this data was extracted was approved by the Research Ethics Committee at Zayed University UAE (ZU17_006_F). Additional clearance was obtained from the Abu Dhabi Health Services Company. Informed consent was obtained from all participants prior to any data collection. Several measures were taken to ensure privacy and confidentiality throughout the study period by excluding personal identifiers during data collection.

2.6. Definitions

Early initiation of breastfeeding: when the infant initiated breastfeeding within one hour after birth [37].

Delayed initiation of breastfeeding: when the infant initiated breastfeeding within more than one hour after birth.

Rooming-in: the child stays with mother in the same room during hospital stay.

Exclusive breastfeeding: the infant being fed only breast milk without any other oral intake, except medications and vitamins, for the first six months of life; it was calculated based on the last 24 h.

Breastfeeding support: the support and encouragement from family (mother, husband, other relatives, and other non-relatives) on breastfeeding

Breastfeeding advice and or discussion: any received information, positive or negative things about breastfeeding before or after delivery

Gestational age: a measure of the age of a pregnancy that is taken from the beginning of the woman's last menstrual period

Preterm birth: the birth of a baby at <37 weeks GA

Arab nationality: all Emirati mothers and other Arab ones

Non-Arab nationality: Asian mothers and other nationalities

Cesarean section: a surgical procedure in which incisions are made through a woman's abdomen and uterus to deliver her baby

3. Results

Secondary data analysis included 1624 samples from the original study, of which 491 (30.2%) participants reportedly delivered by CS, of which 349 (71.1%) delivered by planned CS, and 142 (28.9%) delivered by emergency CS (Figure 1).

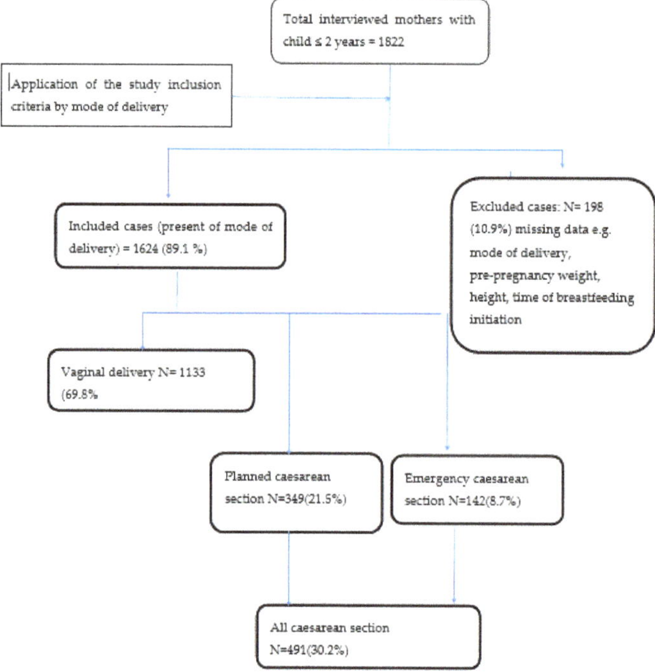

Figure 1. Study participants flow chart and main findings.

The mean and standard deviation (SD) of the mothers' age and children's age were 30.1 (5.2) years and 8.1 months (5.9), respectively. The mean (SD) gestational age at delivery was 39.1 (1.9) weeks, and 6.5% of infants were preterm (GA <37 weeks). More than half (62.5%) of the mothers initiated early breastfeeding. Bivariate analysis showed that child age, birth weight, pre-pregnancy BMI, maternal education, initiation of breastfeeding, and rooming-in were associated with the mode of delivery (Table 1). Exclusive breastfeeding among infants between 0–6 months was 46.5% (332/710).

Table 1. Bivariate analysis of factors associated with caesarean section among mothers with children two years old and younger in Abu Dhabi, United Arab Emirates (UAE).

Variables	Total ($n = 1624$)		Vaginal ($n = 1133$)		CS ($n = 491$)		p-Value
	Mean (SD), range		Mean (SD)		Mean (SD)		
Maternal age, years	30.1(5.2)		29.9(5.3)		31.0(4.8)		<0.001
Child age, months (median 6, interquartile range 9)	8.1(5.9)		8.3(6.0)		7.6(5.7)		0.042
Gestational age, weeks	39.1(1.9)		39.4(1.7)		38.6(2.1)		<0.001
Birth weight, grams	3079(518)		3110(463)		3008(621)		<0.001
Mother pre-pregnancy BMI	23.9(3.8), (15.2, 64.9)		23.7(3.7)		24.3(4.2)		0.002
	N	%	N	%	N	%	
Child Gender							
Male	799	49.2	560	49.4	239	48.7	0.781
Female	825	50.8	573	50.6	252	51.3	
Nationality by category							
Arab	1056	65.0	757	66.7	299	60.9	0.049
Non-Arab	568	35.0	376	33.3	192	39.1	
Marital status							
Married	1602	98.6	1116	98.5	486	99.0	0.440
Unmarried	22	1.4	17	1.5	5	1.0	
Initiation of breastfeeding							
Delayed initiated	609	37.5	329	29.0	280	57.0	< 0.001
Early initiated	1015	62.5	804	71.0	211	43.0	
Rooming-in							
Yes	1562	96.2	1101	97.2	461	93.9	0.002
No	62	3.8	32	2.8	30	6.1	
Mother's education							
Secondary level	65	4.0	53	4.7	12	2.4	0.035
≥secondary level	1559	96.0	1080	95.3	479	97.6	
Father's education							
<Secondary level	31	1.9	23	2.0	8	1.6	0.588
≥secondary level	1593	98.1	1110	98.0	483	98.4	
Mother occupation							
Housewives	1008	62.1	695	61.3	313	63.7	0.533
Employed	616	37.9	438	38.7	178	36.3	
Child order							
1st order	1038	63.9	399	35.2	187	38.1	0.269
>1st order	1038	63.9	304	61.9	304	61.9	

Multivariable logistic regression analysis indicated that several factors were positively associated with caesarean section (CS). These included: advanced maternal age (Adjusted Odd Ratio (AOR) = 1.04, 95% confidence interval (CI) = 1.02, 1.07), nationality (non-Arab) (AOR = 1.30, 95% CI = 1.02, 1.65), and body mass index (BMI) status (obesity) (AOR = 1.79, 95% CI = 1.15, 2.79). Gestational age (GA); however, (AOR = 0.80, 95% CI = 0.75, 0.86) was found to be negatively associated with CS (Table 2).

Table 2. Multivariable logistic regression analyses of factors associated with caesarean section among mothers with children two years old and younger in Abu Dhabi, UAE.

Variable		Adjusted Odds Ratio	95% Confidence Interval	*p*-Value
Advanced maternal age, years		1.04	1.02, 1.07	< 0.001
Gestational age, weeks		0.80	0.75, 0.86	<0.001
Child's birth weight, grams		1.00	1.00, 1.00	0.283
Nationality	Non-Arab	1.36	1.08, 1.71	0.009
	Arab	Reference		
Maternal education	≥Secondary	1.88	0.94, 3.75	0.072
	<Secondary	Reference		
Mother pre-pregnancy BMI	Underweight (<18.5)	1.12	0.56, 2.26	0.746
	Normal (18.5–24.9)	Reference		
	Overweight (25–29.9)	1.10	0.85, 1.42	0.477
	Obese (≥30)	1.79	1.15, 2.79	0.010

From the total (1624), 1133 (69.8%), 349 (21.5%), and 142 (8.7%) delivered vaginally, by planned CS and by emergency CS, respectively. Among mothers who initiated breastfeeding early (1015), the rates of an early initiation of breastfeeding were 804 (79.2%) 95% CI (76.4, 82.0), 162 (16.0%) 95% CI (10.4, 21.6), and 49 (4.8%) 95% CI (1.2, 10.8) for vaginal deliveries, planned CS, and emergency CS, respectively.

By using logistic regression analysis, the different impacts of each mode of delivery (vaginal, planned CS, and emergency CS), and their impact on early initiation of breastfeeding, the study findings showed that vaginal delivery mothers were most likely to initiate early breastfeeding, followed by planned CS and emergency CS delivery (Table 3).

Table 3. Impact of mode of delivery (vaginal delivery, planned CS, and emergency CS) in comparison to each other on an early initiation of breastfeeding using logistic regression analysis.

Mode of Delivery	95% Confidence Interval (CI)	*p*-Value	AOR [1] (95% CI)	*p*-Value	AOR (95% CI)	*p*-Value
Vaginal delivery	4.5 (3.16, 6.61)	< 0.001	2.78 (2.17, 3.56)	< 0.001	Reference	
Planned CS	1.64 (1.09, 2.46)	0.017	Reference		0.36 (0.28, 046)	< 0.001
Emergency CS	Reference		0.61 (0.41, 0.91)	0.017	0.22 (0.15, 0.32)	< 0.001

[1] AOR: Adjusted for advance maternal age, gestational age, and mother's BMI.

4. Discussion

One of the main findings was the high rate of CS (30.2%). This rate is more than double in comparison to the WHO set upper limit of 15% [3], and three times that of rates previously reported in the UAE (10% in 1995%) [1]. These data support the WHO reports that the rate of CS is increasing [2]. The results from this study exceed those for other Arab nations such as Saudi Arabia where a low rate of CS was reported (13.7%) [38] and Sudan (17.8%) [39]. Researchers attributed the varying trend of CS rate among nations to multiple factors including affluence [40], urbanization [40], and increases in preterm births [41]. For example, Boatin et al. [42] reported a low CS rate in South Sudan of 0.6% in comparison to the Dominican Republic, which reported 58.9%. CS rates also vary between public and private hospitals [43,44], which may be attributed to access to care.

This study indicated that non-Arab mothers were 1.3 times more likely to deliver by CS compared to Arab mothers. In multinational countries such as the UAE, the interpretation of such differences requires additional research. It should be noted that maternal age among non-Arab mothers was higher than Arab mothers, which may have contributed to the increased rates of CS in this group.

Advanced GA was found to be a protective factor for CS, (AOR 0.80, 95% CI = 0.75, 0.86) i.e., full-term babies were less likely to have been delivered by CS in comparison to preterm babies, which is a finding that is supported by previous research [41,45]. In addition, advanced maternal age was found to be associated with preterm birth [46–48]. Thus, we can conclude that factors such as advanced maternal age, preterm birth, and nationality are correlated.

In contrast to our results, previous studies have found birth weight [43,49] and maternal education [40,49] to be associated with mode of delivery. Differences in the scope and approach to data collection may limit the ability to adequately compare current results with previous studies and warrant additional research.

As with previous findings [50–52], the current study found an increased risk of CS in obese women. However, overweight was not associated with CS, and the findings of the current study were confined to the obese mothers. Regardless of the association between obesity and CS, it is well documented that maternal obesity is associated with a decreased intention and initiation of breastfeeding, a shortened duration of breastfeeding, a less adequate milk supply, and a delayed onset of lactogenesis II, and can thus be considered as a risk factor for adverse breastfeeding outcomes [53–55].

In addition to the CS prevalence and its associated factors, including the breastfeeding ones, the study also documented the impact of each mode of delivery on breastfeeding initiation. In line with others, the present results documented CS as a key risk factor for the delayed initiation of breastfeeding [24,28,56,57]. The researchers attributed the delay to many reasons, including pain [28], delayed lactogenesis [56], and feeding the newborn with formula [28,57].

The current study found that women who had a vaginal delivery 4.57 (3.16, 6.61) versus planned CS delivery 1.64 (1.09, 2.46) were four times more likely to initiate breastfeeding early. This finding is in line with several studies that documented vaginal delivery to be associated with an early initiation of breastfeeding [26,28–30]. This finding supports previous research that suggests rooming-in as a factor for the early initiation of breastfeeding, and CS decreases that opportunity.

However, controversy exists between planned CS and emergency CS. Some researchers propose that planned CS is associated with a delayed initiation of breastfeeding [34] while others report that emergency CS is associated with a delayed initiation of breastfeeding [35], similar to our findings. The delayed initiation of breastfeeding after emergency CS could be attributed to the stress accompanying the labor and delivery, which is associated with a delayed onset of lactation [35].

Our study is the first of this kind in the UAE, using clinical and non-clinical data to describe CS and the initiation of breastfeeding. Several limitations should be noted. First, maternal indications and/or fetal indications of CS such as fetal distress, the type of anesthesia (general or regional) used during CS, and maternal and child condition such as morbidity and history of hospitalization after delivery by CS were not reported. These issues can affect the mode of delivery and its effect on breastfeeding practices. Future studies are needed to better describe the factors associated with CS in UAE populations.

5. Conclusions

This study suggests that the prevalence of CS is high in Abu Dhabi, UAE and that CS is associated with advanced maternal age, GA, nationality, and obesity. CS, and emergency CS in particular, was the main risk factor for a delayed initiation of breastfeeding. The study provides valuable information to aid with the development of appropriate strategies to reduce the CS rate in the UAE. Maternal literacy on CS choices, the importance of breastfeeding for child health, and additional guidance so that mothers and their families can achieve better breastfeeding outcomes are recommended to ensure successful breastfeeding practices.

Author Contributions: Z.T. designed the study and recruited the participants. Z.T. and A.A.H. analyzed the data and wrote the manuscript. L.S. and D.P. contributed to the design of the study, data collection, and manuscript writing. All contributing authors of this original manuscript authorized the final version of the manuscript. All authors read and approved the final manuscript.

Funding: The study was funded (R17042) by the Research Office at Zayed University.

Acknowledgments: The authors are grateful to the Abu Dhabi Health Services Company (SEHA) for granting access and approval to seven public ambulatory health care centers across the capital city of Abu Dhabi. We would like to express our gratitude to the mothers for their sincere cooperation and the provision of valuable information. Furthermore, we would like to thank the research assistants Amira Badr Eldin, Razan Abdelrahman, Nahed Yaghi, Nour Mohammed, Dhuha Abdulla Naser, Ayesha Rashed, and Jawaher Saeed for their time and commitment. We appreciate the work done by Teresa Arora on improving the English language. Special thanks go to Malin Garemo and Joy Nanda for their technical assistance in the study design.

Conflicts of Interest: The authors declare no competing interests.

References

1. World Health Organization. *The World Health Report 2005 Statistical Annex*; World Health Organization: Geneva, Switzerland, 2005.

2. World Health Organization. *World health Statistics 2015*; World Health Organization: Geneva, Switzerland, 2015.

3. World Health Organization. *WHO Statement on Caesarean Section Rates*; World Health Organization: Geneva, Switzerland, 2015.

4. Smith, L.J. Impact of birthing practices on the breastfeeding dyad. *J. Midwifery Womens Health* **2007**, *52*, 621–630. [CrossRef]

5. Procianoy, R.S.; Fernandes-Filho, P.H.; Lazaro, L.; Sartori, N.C. Factors affecting breastfeeding: The influence of caesarean section. *J. Trop. Pediatr.* **1984**, *30*, 39–42. [CrossRef]

6. Jaafar, S.H.; Ho, J.J.; Lee, K.S. Rooming-in for new mother and infant versus separate care for increasing the duration of breastfeeding. *Cochrane Database Syst. Rev.* **2016**, Cd006641. [CrossRef]

7. Bodner, K.; Wierrani, F.; Grunberger, W.; Bodner-Adler, B. Influence of the mode of delivery on maternal and neonatal outcomes: A comparison between elective cesarean section and planned vaginal delivery in a low-risk obstetric population. *Arch. Gynecol. Obstet.* **2011**, *283*, 1193–1198. [CrossRef]

8. Wax, J.R. Maternal request cesarean versus planned spontaneous vaginal delivery: Maternal morbidity and short term outcomes. *Semin. Perinatol.* **2006**, *30*, 247–252. [CrossRef]

9. Karlstrom, A.; Lindgren, H.; Hildingsson, I. Maternal and infant outcome after caesarean section without recorded medical indication: Findings from a Swedish case-control study. *BJOG Int. J. Obstet. Gynaecol.* **2013**, *120*, 479–486; discussion 486. [CrossRef]

10. Rowe-Murray, H.J.; Fisher, J.R. Baby friendly hospital practices: Cesarean section is a persistent barrier to early initiation of breastfeeding. *Birth* **2002**, *29*, 124–131. [CrossRef]

11. Watt, S.; Sword, W.; Sheehan, D.; Foster, G.; Thabane, L.; Krueger, P.; Landy, C.K. The effect of delivery method on breastfeeding initiation from the The Ontario Mother and Infant Study (TOMIS) III. *J. Obstet. Gynecol. Neonatal Nurs.* **2012**, *41*, 728–737. [CrossRef]

12. Esteves, T.M.; Daumas, R.P.; Oliveira, M.I.; Andrade, C.A.; Leite, I.C. Factors associated to breastfeeding in the first hour of life: Systematic review. *Rev. Saude Publica* **2014**, *48*, 697–708. [CrossRef]

13. Meedya, S.; Fahy, K.; Kable, A. Factors that positively influence breastfeeding duration to 6 months: A literature review. *Women Birth* **2010**, *23*, 135–145. [CrossRef]

14. World Health Organization and United Nations Children's Fund (UNICEF). *Baby-Friendly Hospital Initiative Revised, Updated and Expanded for Integrated Care*; World Health Organization and United Nations Children's Fund: Geneva, Switzerland, 2009.

15. World Health Organization. *Global Strategy for Infant and Young Child Feeding*; World Health Organization: Geneva, Switzerland, 2003; pp. 1–30.

16. Moore, E.R.; Bergman, N.; Anderson, G.C.; Medley, N. Early skin-to-skin contact for mothers and their healthy newborn infants. *Cochrane Database Syst. Rev.* **2016**, *11*, Cd003519. [CrossRef]

17. Patel, R.R.; Liebling, R.E.; Murphy, D.J. Effect of operative delivery in the second stage of labor on breastfeeding success. *Birth* **2003**, *30*, 255–260. [CrossRef]

18. Zanardo, V.; Pigozzo, A.; Wainer, G.; Marchesoni, D.; Gasparoni, A.; Di Fabio, S.; Cavallin, F.; Giustardi, A.; Trevisanuto, D. Early lactation failure and formula adoption after elective caesarean delivery: Cohort study. *Arch. Dis. Child. Fetal Neonatal Ed.* **2013**, *98*, F37–F41. [CrossRef]

19. Ip, S.; Chung, M.; Raman, G.; Chew, P.; Magula, N.; DeVine, D.; Trikalinos, T.; Lau, J. Breastfeeding and maternal and infant health outcomes in developed countries. *Evid. Rep. Technol. Assess.* **2007**, *153*, 1–186.

20. Yan, J.; Liu, L.; Zhu, Y.; Huang, G.; Wang, P.P. The association between breastfeeding and childhood obesity: A meta-analysis. *BMC Public Health* **2014**, *14*, 1267. [CrossRef]

21. Victora, C.G.; Horta, B.L.; Loret de Mola, C.; Quevedo, L.; Pinheiro, R.T.; Gigante, D.P.; Goncalves, H.; Barros, F.C. Association between breastfeeding and intelligence, educational attainment, and income at 30 years of age: A prospective birth cohort study from Brazil. *Lancet Glob. Health* **2015**, *3*, e199–e205. [CrossRef]

22. Taha, Z. Trends of breastfeeding in the United Arab Emirates (UAE). *Arab. J. Nutr. Exerc.* **2017**, *2*, 152–159. [CrossRef]

23. Taha, Z.; Garemo, M.; Nanda, J. Patterns of breastfeeding practices among infants and young children in Abu Dhabi, United Arab Emirates. *Int Breastfeed. J.* **2018**, *13*, 48. [CrossRef]

24. Perez-Rios, N.; Ramos-Valencia, G.; Ortiz, A.P. Cesarean delivery as a barrier for breastfeeding initiation: The Puerto Rican experience. *J. Hum. Lact.* **2008**, *24*, 293–302. [CrossRef]

25. Prior, E.; Santhakumaran, S.; Gale, C.; Philipps, L.H.; Modi, N.; Hyde, M.J. Breastfeeding after cesarean delivery: A systematic review and meta-analysis of world literature. *Am. J. Clin. Nutr.* **2012**, *95*, 1113–1135. [CrossRef]

26. Perez-Escamilla, R.; Maulen-Radovan, I.; Dewey, K.G. The association between cesarean delivery and breast-feeding outcomes among Mexican women. *Am. J. Public Health* **1996**, *86*, 832–836. [CrossRef] [PubMed]

27. Brown, A.; Jordan, S. Impact of birth complications on breastfeeding duration: An internet survey. *J. Adv. Nurs.* **2013**, *69*, 828–839. [CrossRef] [PubMed]

28. Albokhary, A.A.; James, J.P. Does cesarean section have an impact on the successful initiation of breastfeeding in Saudi Arabia? *Saudi. Med. J.* **2014**, *35*, 1400–1403. [PubMed]

29. Badaya, N.; Jain, S.; Kumar, N. Time of initiation of breastfeeding in various modes of delivery and to observe the effect of low birth weight and period of gestation on initiation of breastfeeding. *Int. J. Contemp. Pediatr.* **2018**, *5*, 1509–1517. [CrossRef]

30. Regan, J.; Thompson, A.; DeFranco, E. The influence of mode of delivery on breastfeeding initiation in women with a prior cesarean delivery: A population-based study. *Breastfeed. Med.* **2013**, *8*, 181–186. [CrossRef] [PubMed]

31. Kiani, S.N.; Rich, K.M.; Herkert, D.; Safon, C.; Perez-Escamilla, R. Delivery mode and breastfeeding outcomes among new mothers in Nicaragua. *Matern. Child. Nutr.* **2018**, *14*, e12474. [CrossRef] [PubMed]

32. Hassan, A.A.; Taha, Z.; Ahmed, M.A.A.; Ali, A.A.A.; Adam, I. Assessment of initiation of breastfeeding practice in Kassala, Eastern Sudan: A community-based study. *Int. Breastfeed. J.* **2018**, *13*, 34. [CrossRef]

33. Rabiepoor, S.; Sadeghi, E.; Hamidiazar, P. The relationship between type of delivery and successful breastfeeding. *Int. J. Paediatr.* **2017**, *5*, 4899–4907.

34. Palla, H.; Kitsantas, P. Mode of delivery and breastfeeding practices. *Int. J. Pregnancy Child Birth* **2017**, *2*, 167–172.

35. Grajeda, R.; Perez-Escamilla, R. Stress during labor and delivery is associated with delayed onset of lactation among urban Guatemalan women. *J. Nutr.* **2002**, *132*, 3055–3060. [CrossRef]

36. Abu Dhabi Government. *Abu Dhabi Emirate: Facts and Figures*; Abu Dhabi Government: Abu Dhabi, UAE, 2018.

37. World Health Organization. *Protecting, Promoting and Supporting Breastfeeding in Facilities Providing Maternity and Newborn Services*; World Health Organization: Geneva, Switzerland, 2017.

38. Ghadeer, Z.; Abdullah, A.; Afnan, B.; Abdullah, A.; Abrar, H.; Hanoof, A.; Aeshah, A.; Fatimah, A.; Duaa, F.; Amair, A. Prevalence of caesarean section and its indicating factors among pregnant women attending delivery at King Abdulaziz University Hospital, Jeddah city during 2016. *EC Gynaecol.* **2018**, *7*, 43–51.

39. Elmugabil, A.; Rayis, D.A.; Hassan, A.A.; Ali, A.A.A.; Adam, I. Epidemiology of cesarean delivery in Kassala, Eastern Sudan: A community-based study 2014–2015. *Sudan JMS* **2016**, *11*, 49–54.

40. Khan, M.N.; Islam, M.M.; Shariff, A.A.; Alam, M.M.; Rahman, M.M. Socio-demographic predictors and average annual rates of caesarean section in Bangladesh between 2004 and 2014. *PLoS ONE* **2017**, *12*, e0177579. [CrossRef] [PubMed]

41. Barros, F.C.; Rabello Neto, D.L.; Villar, J.; Kennedy, S.H.; Silveira, M.F.; Diaz-Rossello, J.L.; Victora, C.G. Caesarean sections and the prevalence of preterm and early-term births in Brazil: Secondary analyses of national birth registration. *BMJ Open* **2018**, *8*, e021538. [CrossRef]

42. Boatin, A.A.; Schlotheuber, A.; Betran, A.P.; Moller, A.B.; Barros, A.J.D.; Boerma, T.; Torloni, M.R.; Victora, C.G.; Hosseinpoor, A.R. Within country inequalities in caesarean section rates: Observational study of 72 low and middle income countries. *BMJ* **2010**, *360*, k55. [CrossRef]

43. Al Rifai, R. Rising cesarean deliveries among apparently low-risk mothers at university teaching hospitals in Jordan: Analysis of population survey data, 2002–2012. *Glob. Health Sci. Pract.* **2014**, *2*, 195–209. [CrossRef]

44. Einarsdottir, K.; Haggar, F.; Pereira, G.; Leonard, H.; de Klerk, N.; Stanley, F.J.; Stock, S. Role of public and private funding in the rising caesarean section rate: A cohort study. *BMJ Open* **2013**, *3*, e002789. [CrossRef]

45. Delnord, M.; Blondel, B.; Drewniak, N.; Klungsoyr, K.; Bolumar, F.; Mohangoo, A.; Gissler, M.; Szamotulska, K.; Lack, N.; Nijhuis, J.; et al. Varying gestational age patterns in cesarean delivery: An international comparison. *BMC Pregnancy Childbirth* **2014**, *14*, 321. [CrossRef]

46. Cavazos-Rehg, P.A.; Krauss, M.J.; Spitznagel, E.L.; Bommarito, K.; Madden, T.; Olsen, M.A.; Subramaniam, H.; Peipert, J.F.; Bierut, L.J. Maternal age and risk of labor and delivery complications. *Matern. Child. Health J.* **2015**, *19*, 1202–1211. [CrossRef]

47. Fuchs, F.; Monet, B.; Ducruet, T.; Chaillet, N.; Audibert, F. Effect of maternal age on the risk of preterm birth: A large cohort study. *PLoS ONE* **2018**, *13*, e0191002. [CrossRef]

48. Waldenstrom, U.; Cnattingius, S.; Vixner, L.; Norman, M. Advanced maternal age increases the risk of very preterm birth, irrespective of parity: A population-based register study. *BJOG Int. J. Obstet. Gynaecol.* **2017**, *124*, 1235–1244. [CrossRef] [PubMed]

49. Apanga, P.A.; Awoonor-Williams, J.K. Predictors of caesarean section in northern Ghana: A case-control study. *Pan Afr. Med. J.* **2018**, *29*, 20. [CrossRef] [PubMed]

50. Cresswell, J.A.; Campbell, O.M.; De Silva, M.J.; Slaymaker, E.; Filippi, V. Maternal obesity and Caesarean delivery in sub-Saharan Africa. *Trop. Med. Int. Health* **2016**, *21*, 879–885. [CrossRef] [PubMed]

51. Ovesen, P.; Rasmussen, S.; Kesmodel, U. Effect of prepregnancy maternal overweight and obesity on pregnancy outcome. *Obstet. Gynecol.* **2011**, *118*, 305–312. [CrossRef]

52. Tsvieli, O.; Sergienko, R.; Sheiner, E. Risk factors and perinatal outcome of pregnancies complicated with cephalopelvic disproportion: A population-based study. *Arch. Gynecol. Obstet.* **2012**, *285*, 931–936. [CrossRef]

53. Donath, S.M.; Amir, L.H. Maternal obesity and initiation and duration of breastfeeding: Data from the longitudinal study of Australian children. *Matern. Child. Nutr.* **2008**, *4*, 163–170. [CrossRef]

54. Ramji, N.; Quinlan, J.; Murphy, P.; Crane, J.M. The impact of maternal obesity on breastfeeding. *J. Obstet. Gynaecol. Can.* **2016**, *38*, 703–711. [CrossRef]

55. Verret-Chalifour, J.; Giguere, Y.; Forest, J.-C.; Croteau, J.; Zhang, P.; Marc, I. Breastfeeding initiation: Impact of obesity in a large Canadian perinatal cohort study. *PLoS ONE* **2015**, *10*, e0117512. [CrossRef]

56. Isik, Y.; Dag, Z.O.; Tulmac, O.B.; Pek, E. Early postpartum lactation effects of cesarean and vaginal birth. *Ginekol. Pol.* **2016**, *87*, 426–430. [CrossRef]

57. Chen, C.; Yan, Y.; Gao, X.; Xiang, S.; He, Q.; Zeng, G.; Liu, S.; Sha, T.; Li, L. Influences of Cesarean Delivery on Breastfeeding Practices and Duration: A Prospective Cohort Study. *J. Hum. Lact.* **2018**, *34*, 526–534. [CrossRef]

 *nutrients* 

Article

Impact of Fenugreek on Milk Production in Rodent Models of Lactation Challenge

Thomas Sevrin [1,2], Marie-Cécile Alexandre-Gouabau [2], Blandine Castellano [2], Audrey Aguesse [2], Khadija Ouguerram [2], Patrick Ngyuen [3], Dominique Darmaun [2,4] and Clair-Yves Boquien [2,*]

[1] Laboratoire FRANCE Bébé Nutrition, 53000 Laval, France; thomas.sevrin@etu.univ-nantes.fr
[2] Nantes Université, INRA, UMR 1280 PhAN, CRNH-Ouest, IMAD, F-44000 Nantes, France;
     Marie-Cecile.Alexandre-Gouabau@univ-nantes.fr (M.-C.A.-G.); Blandine.Castellano@univ-nantes.fr (B.C.);
     Audrey.Aguesse@univ-nantes.fr (A.A.); Khadija.Ouguerram@univ-nantes.fr (K.O.);
     Dominique.Darmaun@chu-nantes.fr (D.D.)
[3] ONIRIS, NP3, CRNH-Ouest, F-44000 Nantes, France; patrick.nguyen@oniris-nantes.fr
[4] CHU Nantes, F-44000 Nantes, France
*   Correspondence: clair-yves.boquien@univ-nantes.fr

Received: 30 September 2019; Accepted: 21 October 2019; Published: 24 October 2019

Abstract: Fenugreek, a herbal remedy, has long been used as galactologue to help mothers likely to stop breastfeeding because of perceived insufficient milk production. However, few studies highlight the efficacy of fenugreek in enhancing milk production. The aims of our study were to determine whether fenugreek increased milk yield in rodent models of lactation challenge and if so, to verify the lack of adverse effects on dam and offspring metabolism. Two lactation challenges were tested: increased litter size to 12 pups in dams fed a 20% protein diet and perinatal restriction to an 8% protein diet with eight pups' litter, with or without 1 g.kg^{-1}.day^{-1} dietary supplementation of fenugreek, compared to control dams fed 20% protein diet with eight pups' litters. Milk flow was measured by the deuterium oxide enrichment method, and milk composition was assessed. Lipid and glucose metabolism parameters were assessed in dam and offspring plasmas. Fenugreek increased milk production by 16% in the litter size increase challenge, resulting in an 11% increase in pup growth without deleterious effect on dam-litter metabolism. Fenugreek had no effect in the maternal protein restriction challenge. These results suggest a galactologue effect of fenugreek when mothers have no physiological difficulties in producing milk.

Keywords: fenugreek; milk flow; milk composition; litter size; maternal protein restriction; plasma metabolic parameters

---

1. Introduction

The World Health Organisation recommends exclusive breastfeeding for infants up to six months of age, based on the clear health benefits of breastfeeding on mother-infant dyad [1]. Indeed, there is a consensus regarding the association of breastfeeding with a reduced risk of respiratory and gastro-intestinal infections during the first year of life. Infants, who were breastfed for longer periods, may also have a lower risk of developing obesity and type II diabetes at adulthood [2,3]. For mothers, breastfeeding could limit the risk of developing ovarian cancer and type II diabetes [4,5]. Despite these benefits, breastfeeding prevalence remains relatively low, particularly in several high-income countries in North America and Europe, where only 40% of mothers breastfeed six months after delivery [6]. Exclusive breastfeeding rate is about 60% at four months in Scandinavian countries, 35% in the Netherlands, 16% in the UK [7], and 10% in France [7,8].

Early cessation of breastfeeding clearly is multifactorial [9]. One of the main factors is the maternal perception of insufficient milk secretion to quell infant's hunger or support infant growth and leads to early cessation of breastfeeding in 35% of cases [9–11]. Perception of insufficient milk production is a complex, multifactorial issue that can have biological, social, or psychological determinants, and it often remains unclear whether the low milk secretion is real or only perceived [9]. Indeed, as a result of the lack of an objective marker of insufficient milk production and the importance of maternal psychology in breastfeeding duration, perceived milk insufficiency is probably much more common than true insufficient production [9–11]. True insufficient milk secretion can result from many causes, ranging from inability to lactate due to breast abnormalities or endocrine disorder (5% of women) to difficulties in breastfeeding management, maternal stress and anxiety, or early food diversification in the infant. Although maternal milk production can be often increased through psychological support or maternal breastfeeding education [9,10,12], many healthy mothers are eager to enhance their milk production through various nutritional supplements.

Several drugs and herbal preparations have traditionally been prescribed as galactologues: i.e., substances that promote initiation or increase of lactation [13]. Drugs like domperidone®or metoclopramide®carry the risk of adverse side effects such as arrhythmia or hypothyroidism in mother-children dyad [12]. That is why herbal galactologues like fennel, anise, barley, milk thistle, or garlic are becoming more and more popular for increasing lactation. Among these herbal compounds, fenugreek probably is the most widely consumed [12–14]. Fenugreek has been used since antiquity in traditional Persian, Chinese, and Egyptian medicine for its range of therapeutic effects. It is now increasingly consumed in Western countries for its presumed protective effects against diabetes, atherosclerosis, inflammation, and hypertriglyceridemia, as well as its putative role as a galactologue [12,15] due, in part, to trigonelline, one of its main active ingredients [15]. There is, however, little evidence for its effectiveness on milk yield [13]. Whereas a positive effect of fenugreek on milk production was observed in various mammals, such as rabbit [16], buffalo [17], goat [18], or ewe [19]; the wide range of doses tested (from 180 mg.kg-1.d-1 [17] to 2.1 g.kg-1.d-1 [18]) led to large discrepancies in the reported effect on milk production (ranging from a 10% increase [16] to a 110% [18] increase), which makes it difficult to determine the effective dose of fenugreek. Moreover, other studies failed to demonstrate an effect in rabbits [20] and goats [21]. In these studies, milk production was measured either by the weight–suckle–weight method (in rabbit), or by milking (in larger animals). The latter methods do not, however, directly assess milk production in response to physiological suckling by the pups. Moreover, these studies evaluated the effectiveness of fenugreek to increase milk production with the aim of productivity, under otherwise optimal conditions of lactation, as opposed to mothers challenged by their perception of insufficient milk production. Finally, none of these studies evaluated the metabolic status of mother-infant dyad following fenugreek supplementation. Thus, the true efficacy of fenugreek on milk production in an animal model submitted to a breastfeeding challenge remains to be ascertained.

To that purpose, the current study used the deuterium oxide enrichment method [22], which meets all criteria to adequately assess breast milk supply even in small animals and, in turn, evaluate the putative galactologue effect of substances such as fenugreek. Indeed, it has been shown to be precise and to provide a smoothed value of several days milk production obtained under physiological conditions of lactation [22]. Moreover, in order to simulate conditions where mothers are unable to feed their own pups optimally, two classical models of lactation challenges were used. The first one mimics conditions in which mothers can adapt their milk production but still fail to adequately cover the litters' needs, leading to suboptimal growth of the offspring. This model, obtained by increasing litter size through pup adoption, is known to induce extra-uterine growth restriction (EUGR) in the offspring due to a reduced milk intake in individual pups due to their larger number [23,24]. The second model mimics conditions in which mothers are truly unable to produce sufficient milk due to a perinatal restriction in dietary protein intake when fed a diet containing 8% protein instead of 20% in the standard diet. This model is also notably known to induce EUGR [25] due to a 34% drop in milk production [22].

The specific aims of this study were: (a) to verify the ability of our stable isotope method to detect changes in milk production in two rodent models of lactation challenge; (b) to test the galactologue effect of fenugreek on milk production and composition in these models of lactation challenge; and (c) to explore the effect of fenugreek on maternal metabolism during the lactation period and the short and long-term metabolic outcome in the offspring.

2. Materials and Methods

2.1. Animal Experiment

Housing and Diets

The experimental protocol was approved by the Animal Ethics committee and the French Ministry of Research (protocol APAFIS 2018121018129789). Pregnant Sprague-Dawley rats were purchased from Janvier Labs (Le Genest-Saint-Isle, France) at gestational day one (G1). They were housed individually in cages with wood chips located on ventilated racks kept at a constant temperature of 22 ± 1 °C and at a relative humidity of 50% ± 3%. Cages were placed in a room with a fixed 12 h light–dark cycle (light from 7:00 a.m. to 7:00 p.m.). Pregnant rats had access to water and food ad libitum.

During gestation, dams received a standard normal protein diet based on AIN-93G diet [26] with 20 g protein per 100 g of food (NP diet) or an isoenergetic, low-protein diet with 8 g protein per 100 g of food (LP diet).

During lactation, dams received experimental diets based on NP and LP diets and supplemented with a dry water extract of fenugreek seeds (Plantex, Sainte-Geneviève-des-Bois, France) named NPF and LPF diets, respectively, in the following. Fenugreek, as an oxytocic substance [13], was provided to dams only during the lactation period. The amount of fenugreek in NPF and LPF diets was calculated to reach a consumption of 1 g.kg body weight^{-1}.day^{-1} in rat as an equivalent as the traditionally recommended [12,13] 6 g daily dose in women weighing 60 kg and assuming that metabolic rate per unit of body weight is 10 fold higher in rats than in humans [27]. The four diets (i.e., NP, LP, NPF, and LPF) were manufactured by the "Unité de Préparation des Aliments Expérimentaux" (INRA-UPAE, Jouy-en-Josas, France). The composition and energy of each diet is provided in Supplementary Table S1.

After weaning, offspring were fed ad libitum with a standard growth diet A03 (SAFE, Augy, France).

2.2. Experimental Design

On the first day of gestation (G1), seventy-two female rats (i.e., 4 consecutive series of 18 animals), were randomly assigned to be fed experimental diets during gestation: 52 females received the NP diet and 20 females received the LP diet. Delivery occurred at the 21st day of gestation that was considered as day 0 of lactation (L0). At birth, pups born from protein-restricted dams (LP dams) were discarded and killed to avoid the bias due to intrauterine growth restriction. Only pups born from control dams (NP dams) were randomly adopted by NP or LP dams, taking care to balance pup's birth weights between groups. The litter size was adjusted to either 8 or 12 pups per dam with a female/male ratio of 1/1, as described in Figure 1.

Five groups were defined (47 dams). In the NP:8 group (our reference group, *n* = 8), dams suckled 8 pups per litter and were fed the NP diet. The NP:12 group (*n* = 11) corresponded to the first model of lactation challenge by increasing litter size (12 pups per litter) with dams under the NP diet. In the LP:8 group (*n* = 8), dams suckled 8 pups per litter and were fed the LP diet. The LP:8 group corresponded to the second model of lactation challenge with a known decreased milk flow due to perinatal protein restriction [22,25]. In the two other groups NPF:12 (12 suckled pups per litter, *n* = 11) and LPF:8 (8 suckled pups per litter, *n* = 9), dams were respectively fed an NP and LP diet supplemented with fenugreek.

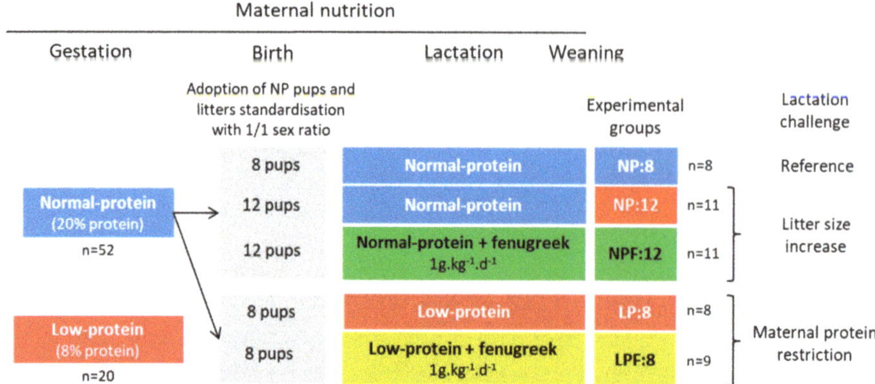

Figure 1. Representation of the 5 experimental groups of dams.

At weaning (L20), 4 pups (2 males and 2 female) per litter were killed and perirenal, subcutaneous and brown adipose tissues were collected and weighed. Two other males and two females per litter were weaned and placed with 3 animals per cage with ad libitum access to food and water until post-natal day 75 (PND75), and then killed. Liver, gastrocnemius muscle, perirenal, subcutaneous, and visceral adipose tissues were removed from PND75 offspring and weighed.

At L21, mothers were fasted for 4h and then sacrificed. Liver, left inguinal mammary gland, perirenal, subcutaneous, and visceral adipose tissues were removed, weighed, sampled, and immediately frozen in liquid nitrogen before storage at −80 °C.

Sacrifice was performed by intracardiac injection of 0.5 mL Exagon®(Richter pharma, Wels, Austria) when animal tissues were collected, whereas supernumerary dams or supernumerary weaned pups were killed by carbon dioxide anaesthesia or by decapitation for pups at birth.

During lactation, dams' weight, food and water consumption, and both male and female litters' weight were recorded every two days from L0 to L11, and every day from L11 to L21. For dams, weight loss (in g) during the lactation period was calculated by subtracting weight at L0 (delivery) from the daily weight. For male and female offspring, pups' mean weight (expressed as g) was obtained by dividing litter weight by the number of pups. Weight gain (in g) was calculated by subtracting birth weight (L0) from the daily weight. Daily growth rate (in g.day^{-1}) was calculated by subtracting weight at day d-1 from the weight at day d. Food and water relative intakes (in g.kg^{-1}.day^{-1}) were obtained by dividing daily intakes by daily weight.

2.3. Biological Samples Collection

Milk samples were collected from lactating dams at L18, as previously described [28]. Briefly, dams were separated from their pups and received an intraperitoneal injection of oxytocin (1 unit of Syntocinon®; Sigma-Tau, Ivry-sur-Seine, France) to stimulate milk ejection. After 20 min, dams were anaesthetized with 4% of isoflurane, and by applying manual pressure to nipples, about 200 to 400 μL of milk were collected before storage at −20 °C.

Blood samples were collected before animal sacrifice by an intra-cardiac puncture in EDTA-tubes (Pfizer-Centravet, Plancoët, France). Otherwise, blood samples were collected on alert dams by a tail snip in EDTA-tubes. Blood samples were centrifuged at 1132 g for 15 min at 4 °C, and plasma was collected in Eppendorf before storage at −20 °C until analysis.

For urine, pups were first separated from the mother for 30 min to avoid urine loss by maternal stimulation. Urine samples were then collected from pups by stimulation of lower bellies with an iced cotton bud and pooled for the male and female litter.

2.4. Milk Flow Measurement by Water Turnover Method

The water turnover method was used as previously described [22]. Briefly, at L8, plasma and urine samples were collected prior to maternal deuterated water (D_2O) injection to determine baseline body D_2O abundance in dams and litter, respectively. At L11, when lactation was well established, following 4% isoflurane anaesthesia, mothers received an intravenous tail injection of 4.95 ± 0.13 g.kg^{-1} D_2O (99.9 mole % D_2-enrichment) (Sigma-Aldrich, Lyon, France).

At 2 h, 24 h, 48 h, 72 h, and 96 h following D_2O injection, dams' blood samples (300 µL) were collected by a tail snip. At 24 h, 48 h, 72 h, 96 h, and 168 h following D_2O-injection, a pool of urine (about 300 µL) was collected from both male and female pups of each litter. The D_2O enrichment of both plasma and urine samples was measured using the Fourier Transform infrared spectrophotometer Alpha II®(Brucker, Rheinstetten, Germany).

Milk flow calculation, using water turnover method, has been previously described [22] and was refined in order to take into account both intra-litter sexual dimorphism and litter size differences between experimental groups. Regarding sexual dimorphism, instead of a bi-compartmental model (dam-litter), a four-compartment model (Figure 2) was used, in which 2 single-compartments corresponded to the turnover of total body water (TBW) of both male (3) and female (4) pups in the same litter, and the other 2 single-compartments corresponded to the turnover of TBW of their own dam (1 and 2, with 1 = 2). Each mother compartment was related by water flow to one pup compartment (male or female). Absolute production rates are represented in Figure 2: R10, R20 (R20 = R10), R30, and R40 are the inputs into the body of dam and its male and female litter, respectively, arising from water drinking and non-dietary water as metabolic water production; R01, R02, R03, and R04 are the outputs of water by transpiration, urine or faeces of dam and its male and female litter, respectively; R31 and R42 are water flows from the dam to male litter and from the dam to female litter, respectively. The model has five unknown parameters: i) K01 and K02 are equal and represent the output flow constants of the dam; ii) K03 and K04 are the output flow constants from male and female litter, respectively and iii) K31 and K42 are the output flow constants from the dam to its male and female litter, respectively.

Using the isotope dilution method, the mass of dam's TBW (TBWd, in g) was calculated by dividing the amount of D_2O injected by the D_2O value extrapolated from the D_2O concentration curve to the intercept with the y-axis at time 0 (Supplementary Table S2). TBWd (in %) was calculated by dividing the mass of TBWd (in g) by dams' mean mass from L11 to L15.

Flow constants of the model (K01, K03, K04, K31, and K42) were then determined using the SAAM II®software. As litter size (8 or 12 pups) presumably impacts D_2O dilution in the litter, we used values of D_2O mass instead of D_2O concentrations for calculations in our model. To this end, D_2O concentration, obtained after baseline concentration deduction, was multiplied by animal mass at each day of sampling, following Equation (1):

$$D_2O \text{ concentration } \left(\mu g.g^{-1}\right) * animal\ mass\ (g) = D_2O\ mass\ (\mu g) \tag{1}$$

This calculation implicitly assumes that dams' and pups' TBW (in %) are equivalent, which is probably the case considering that suckled pups have about 75% of TBW [29], which is similar to suckling dams' TBW [22]. Moreover, in order to indicate that compartment (1) and (2) represent the same individual (dam), the same data were incremented for both compartments, and K02 was forced to be equal to K01. Flow constants of the model were obtained directly with SAAM II®from a fit of plasma and urine D_2O mass–time curves. The water flows, from mother to male litter R31 (g.h^{-1}) and from the dam to female litter R42 (g.h^{-1}), were calculated as the product of TBWd (g) and K31 or K42, respectively. These values were then multiplied by 24 to obtain the daily milk production (g.day^{-1}).

In this model, R31 and R42 were associated to milk flows between the dam and its male litter or between the dam and its female litter, respectively, assuming milk was the only external source of water for the pups. Thus, milk flow corresponds to the milk produced by the dam, as determined from the D2O transfer from the dam to litter, after the dam received a D2O injection. As the water content of each individual milk was not accurately measured, milk flow was directly calculated from water flow without correction. We assumed that this does not alter conclusions concerning group comparisons. Total milk production of the dam was calculated by summing R31 and R42 and represented the mean milk production throughout the sampling period (L11 to L18). Milk consumption by male and female litters was obtained by dividing R31 and R42, respectively, by the number of males or females pups (4 in NP:8, LP:8 and LPF:8 groups or 6, in NP:12 and NPF:12 groups).

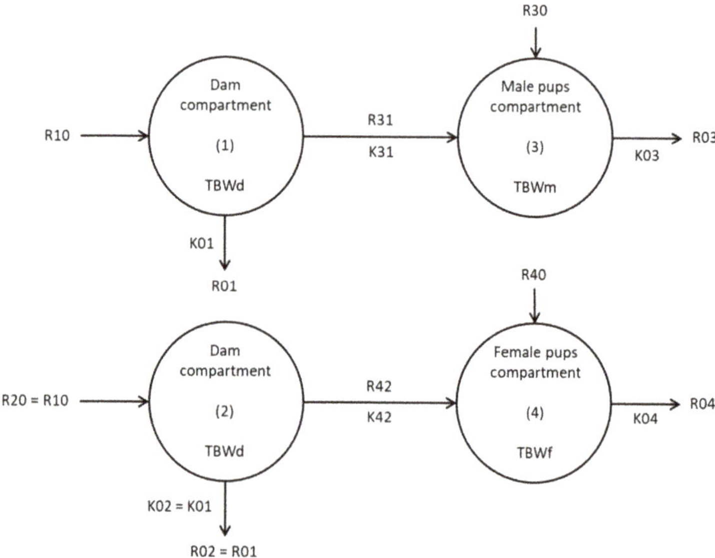

Figure 2. Four-compartment model of water turnover. TBWd, TBWm, TBWf: total body water (expressed in g) of dam, male, and female litter respectively; K01 and K02 (in h^{-1}) are equal and represent the output flow constants from the dam; K03 and k04 (in h^{-1}) are the output flow constants from the male and female litters, respectively; K31 and K42: (in h^{-1}) are the output flow constants from the dam to its male litter and from the dam to its female litter, respectively; R10, R20, R30, and R40 (in $g.h^{-1}$) are input water flows into the body of dam (R10 is equal to R20) and its male and female litter respectively; R01 and R02 (in $g.h^{-1}$) are equal and represent output water flows from the dam; R03 and R04 (in $g.h^{-1}$) are output water flows from male and female litters respectively; R31 and R42 (in $g.h^{-1}$) are water flows from the dam to its male litter and from the dam to its female litter and are associated to milk flow.

2.5. Milk Protein, Lactose, and Fatty Acid Analysis

Milk protein concentration was determined using a colourimetric Pierce BCA Protein Assay Kit (ThermoFisher Scientific, Waltham, MA, USA) with milk diluted at 1/40 in osmosed water and bovine serum albumin (fraction V) as standard. Milk lactose concentration was estimated using an enzymatic Lactose/D-Galactose Assay Kit K-LACGAR®(Megazyme, Bray, Ireland) with milk diluted at 1/20 in osmosed water and α-lactose monohydrate as standard. Milk total lipids were not fractionated. Fatty acids (FAs) were extracted using the modified liquid–liquid extraction method of Bligh-Dyer, as previously described [28]. Briefly, FAs were extracted from 30 μL milk in methanol–chloroform mix (1:1, *v/v*). Heptadecanoic acid was used as the internal standard. Total FAs were transesterified using boron trifluoride in methanol, and fatty acid methyl esters were analysed by gas chromatography using an Agilent Technologies 7890A®instrument. Each FA was expressed as a percentage of the total identified FAs, and only FAs whose percentage was above 0.5% were taken into account. The sum of FAs (in g.L^{-1}) was assumed to represent total milk lipid (free fatty acids, triacylglycerols and phospholipids) content.

The energy content of milk was calculated by multiplying lactose, proteins, and lipid content by their energy content, assuming 4 kcal.g^{-1} for both carbohydrate (lactose) and protein and 9 kcal.g^{-1} for fat. Milk macronutrient production and milk energy production by the dam were then calculated by multiplying total milk production by macronutrient concentrations or energy content of milk respectively.

2.6. Dams and Offspring Metabolic Markers and Offspring Glucose Tolerance Test

At PND60, blood samples (300 µL) were collected from 6 h fasted offspring. An oral glucose tolerance test (OGTT) was performed in offspring at PND61 and PND62. Following 6 hours of fasting, all rats received a 2 g/kg BW dose of glucose by gavage. Blood samples (100 µL) were collected to determine plasma insulin concentration before (T0) and after 15 (T15) and 30 min (T30) gavage with glucose. Blood glucose was measured at T0, T15, T30, T45, T60, T90, and T120 after glucose intake, using a Performa Accu-Chek®glucometer (Roche Diabetes Care France, Meylan, France).

Insulin (Rat Insulin ELISA kit®, ALPCO, Salem, USA), glucose (Glucose GOD FS®, DiaSys, Holzheim, Germany), triglycerides (Triglycerides FS®, DiaSys, Holzheim, Germany), and cholesterol (Cholesterol FS®, DiaSys, Holzheim, Germany) were measured in plasma, following manufacturers' instructions. Optical density was read with a microplate reader Varioskan Lux®(ThermoFisher Scientific, Waltham, USA).

2.7. Trigonelline Quantification in Experimental Diet, Dams' Plasma, and Milk

Ten µL of the labelled internal standard: trigonelline-D3 (CIL, Sainte Foy la Grande, France) at 50 µM, and 100 µL of acetonitrile were added to 10µL of diet solution (mixed diet in water, 100 g.L^{-1}), dams' plasma or milk. Samples were centrifuged for 10 min at 11,000 g, and the supernatant was injected into the LC–MS/MS using a hydrophilic interaction liquid chromatography system (Acquity H-Class®UPLC^{TM} device, Waters) coupled with a triple quadrupole mass spectrometry detector (Xevo®TQD) with an electrospray interface. Data acquisition and analyses were performed with MassLynx®and TargetLynx®software, both versions 4.1 (Waters). Trigonelline was separated over 6 min on a HILIC column (2.1 × 100 mm; 1.7 µM), (Waters) held at 45 °C with a linear gradient of mobile phase A (10 mM ammonium acetate in water) in mobile phase B (98% acetonitrile in water), each containing 0.1% formic acid, at a flow rate of 400 µL/min. Trigonelline was detected by the Xevo®TQD that allowed the multiple reaction monitoring (MRM) mode to be performed, with the electrospray interface operating in the positive ion mode (capillary voltage, 1.5 kV; desolvation gas (N$_2$) flow and temperature, 650 L/h and 150 °C; source temperature, 150 °C). Based on specific collision-induced fragmentation of precursor ions, the MRM precursor/fragment pairs were based on the following transitions *m/z*: 138.06→93.98 amu for trigonelline and 141.03→97.00 amu for the internal standard. Chromatographic peak area ratios between trigonelline and its internal standard constituted the detector response. Trigonelline standard solutions (concentrations from 1 nM to 10 µM) were used for calibration.

2.8. Statistical Analysis

To maximise the power of analysis, parametric tests were favoured using one-way ANOVA or two-way ANOVA with an experimental group factor and a day or a sex factor. The validity of the parametric tests was checked by assessing normality of residuals with a Shapiro–Wilk test. In the case of absence of residual normality non-parametric, the Kruskal–Wallis test was used instead of one-way ANOVA. Multiple comparison tests used after ANOVA were: Dunnett's post-hoc test to compare several levels to one level of interest (Table 1), Tukey's post-hoc test to compare levels altogether (Table 2) and Sidak's post-hoc test when there were only two levels in a factor (Table 4). After the Kruskal–Wallis test, the non-parametric Dunn's post-hoc test was used. To determine the strength of the link between two variables, Pearson's correlation tests were used, and simple linear regression was performed to determine the relation between two variables. All tests were performed with the GraphPad prism®software, version 6.

3. Results

3.1. Dams and Litter Characteristics in the Reference Group NP:8

For all groups, we took care to balance pups' birth weight, following the adoption of NP pups and litter standardization. Mean pup birth weight was 7.1 ± 0.0 g, with no difference between groups ($p = 0.37$), although males were significantly bigger than females (7.3 ± 0.0 g versus 6.9 ± 0.0 g, $p < 0.001$). NP:8 dams weighed 344.3 ± 8.7 g at delivery. Maternal weight loss during the lactation period and food and water intakes are given in Table 1. Litter growth rate was 22.9 ± 0.7 g·day^{-1}, which represented a pup growth rate of 2.86 ± 0.09 g·day^{-1} and did not differ between genders ($p = 0.23$). In the water turnover models, output flow constants ($K3,1$ and $K4,2$), as reported in Supplementary Table S3, were determined with good accuracy due to a forecast standard deviation (FSD) of each group lower than 5%, allowing the milk flows to be determined with confidence. For the NP:8 group, milk flow was 46.0 ± 1.6 g·day^{-1}, which represented a mean milk consumption of 5.74 ± 0.20 g·day^{-1} per pup, with no significant sexual dimorphism ($p = 0.36$).

Table 1. Effect of fenugreek on dams' lactation follow up, pups growth, and milk flow in two models of lactation challenges. Results are expressed as mean $\pm$ SEM.

Lactation Challenges	*Control*	*Litter Size Increase*		*Maternal Protein Restriction*	
Experimental Groups	**NP:8**	**NP:12**	**NPF:12**	**LP:8**	**LPF:8**
n	8	11	10	7	9
Dams (L0–L21)					
Weight loss, g	-6.2 ± 2.4	-13.5 ± 4.0	-7.7 ± 2.9	-8.0 ± 3.7 **	-7.9 ± 2.2 $^{\$\$\$}$
Food intake, g.d^{-1}	43.3 ± 1.5	50.7 ± 1.0 ***	57.4 ± 1.2 $^{\$\$\$ £££}$	37.2 ± 1.3 **	37.7 ± 0.7 $^{\$\$}$
Water intake, g.d^{-1}	49.6 ± 1.5	58.0 ± 1.4 ***	58.5 ± 2.0 $^{\$\$}$	30.3 ± 1.5 ***	29.1 ± 1.1 $^{\$\$\$}$
Pup growth (L11–L18)					
Litter growth rate, g.d^{-1}	22.9 ± 0.7	27.6 ± 0.7 ***	30.1 ± 0.8 $^{\$\$\$ £}$	10.3 ± 0.6 ***	10.0 ± 0.3 $^{\$\$\$}$
Pups growth rate, g.d^{-1}	2.86 ± 0.09	2.30 ± 0.06 ***	2.51 ± 0.06 $^{\$\$ \cdot}$	1.31 ± 0.08 ***	1.25 ± 0.04 $^{\$\$\$}$
Milk flow (L11–L18)					
Total milk production, g.d^{-1}	46.0 ± 1.6	54.3 ± 2.8 *	63.0 ± 3.1 $^{\$\$\$ £}$	25.9 ± 1.3 ***	26.6 ± 1.2 $^{\$\$\$}$
Pups milk consumption, g.d^{-1}	5.74 ± 0.20	4.52 ± 0.23 ***	5.25 ± 0.25 $^{\cdot}$	3.24 ± 0.16 ***	3.32 ± 0.16 $^{\$\$\$}$

NP:8, 20% protein diet with 8 pups per litter; NP:12, 20% protein diet with 12 pups per litter, NPF:12, 20% protein diet with fenugreek (1 g.kg BW^{-1}.d^{-1}) with 12 pups per litter, LP:8, 8% protein diet with 8 pups per litter and LPF:8, 8% protein diet with fenugreek (1g.kg BW^{-1}·day^{-1}) with 8 pups per litter. Dams and pup growth results for each individual are mean value through the indicated period (L0 to L21 or L11 to L18). The two lactation challenges (NP:12 and LP:8) were compared to NP:8 with one-way ANOVAs followed by Dunnett's post-hoc tests. * $p<0.05$, ** $p<0.01$ and *** $p<0.001$ compared to NP:8 group. Each fenugreek group (NPF:12 and LPF:8) were compared to its lactation challenge control (NP:12 or LP:8) and to NP:8 with one-way ANOVA followed by Dunnett's post-hoc test. \$ $p < 0.05$, \$\$ $p < 0.01$, \$\$\$ $p < 0.001$ compared to NP:8 and $\cdot$ $p < 0.10$, £ $p < 0.05$, ££ $p < 0.01$, £££ $p < 0.001$ compared to challenge model control.

3.2. Lactation Challenges Impact both Physiological Characteristics of Dam and Litter Compared to the Control Group NP:8

As no global sex effect was observed for pups' growth and milk flow variables, results from the female and male offspring were pooled. Except for weight at delivery (similar for all groups, mean of 348.4 ± 5.2 g), every variable characterizing lactating dams, pups' growth and milk flow was significantly affected by both lactation challenges (Table 1).

In the litter size increase challenge, dam's weight loss during the entire lactation period was similar between NP:8 and NP:12 dams. Food and water intakes of lactating dams increased significantly (+17%) in NP:12, compared with the NP:8 group. Similarly, the litter growth rate between L11 and L18 was significantly increased (+21%). The increase of litter size led to extra-uterine growth restriction

(EUGR) in the offspring during the lactation period since pup growth rate was 20% lower in NP:12, compared to the NP:8 litters. The enhancement in milk production with larger litter size remained insufficient to cope with increased demand: although total milk production of NP:12 dams was 18% greater than for NP:8 dams ($p = 0.029$), NP:12 pups' milk consumption was 21% lower than in NP:8 pups ($p < 0.001$).

In the maternal protein restriction challenge, dam's weight loss during the entire lactation period was 4.6-fold greater for LP:8 than for the NP:8 ($p = 0.002$) group. Food and water intakes of lactating dams were significantly decreased for LP:8 dams compared to NP:8 dams (−14% and −40%, respectively). Similarly, litter and pup growth rates were significantly decreased (−53%) in LP:8 pups, confirming EUGR occurred in response to maternal protein restriction. The maternal dietary protein restriction challenge was very effective since both values of total milk production, and milk consumption by pups was much lower for the LP:8 group than for the NP:8 group (−44%).

3.3. Correlation between Milk Flow Variables, Pups' Growth Variables, and Lactating Dams' Intakes Variables

Pearson's correlations were calculated between milk flow variables and lactating dam's intakes or pup's growth variables. Total milk production was strongly correlated with litter growth rate (Figure 3a), dams' food ($r = 0.86$, $p < 0.001$) and water ($r = 0.93$, $p < 0.001$) intakes. Similarly, pups milk consumption was strongly correlated with pup growth rate (Figure 3b).

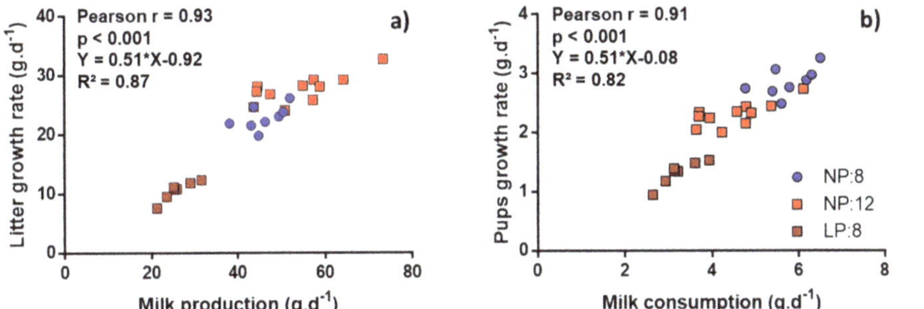

Figure 3. Correlation between pup growth and milk flow variables in three different conditions of lactation physiology (NP:8, NP:12, and LP:8). (**a**) Correlation between total milk production and litter growth rate from L11 to L18. (**b**) Correlation between pup milk consumption and pup growth rate from L11 toL18. NP:8, 20% protein diet with 8 pups per litter; NP:12, 20% protein diet with 12 pups per litter, and LP:8 groups, 8% protein diet with 8 pups per litter.

3.4. Determination of Galactologue Effect of Fenugreek in Two Models of Lactation Challenges

Fenugreek intake was close to 1 g.kg BW^{-1}.d^{-1} for both supplemented groups (i.e., 1.01 ± 0.02 g.kg BW^{-1}.d^{-1} for NPF:12 dams and 0.93 ± 0.02 g.kg BW^{-1}.day^{-1} for LPF:8 dams), but consumption was 7.8% higher in the NPF:12 group than in LPF:8 group ($p = 0.018$). Trigonelline was measured in the diet as a marker of fenugreek content, and was 12.5 µg.kg^{-1} and 14.9 µg.kg^{-1} for the NPF:12 and LPF:8 diet, respectively (SD1), leading to a slightly higher trigonelline intake for NPF:12 (1.99 ± 0.05 mg.kg BW^{-1}.day^{-1}) than LPF:8 dams (1.73 ± 0.04 mg.kg BW^{-1}.day^{-1}).

In the litter size increase challenge, no significant difference was observed for NPF:12 dams on delivery weight (367.0 ± 11.1 g, $p = 0.28$), weight loss, or water intake during lactation when compared to NP:12 dams (Table 1), although food intake was 13.1% higher in NPF:12 than NP:12 dams $p = 0.001$. However, when intake was reported to dam's body weight, the difference in food intake was no longer significant (153.8 ± 2.1 and 161.0 ± 3.6 g.kg^{-1}.day^{-1}, respectively). Moreover, water intake of NPF:12 dams (163.9 ± 3.7 g.kg^{-1}.day^{-1}) became significantly lower compared to NP:12

dams (175.8 ± 4.2 g.kg^{-1}.d^{-1}, p = 0.045), but was still significantly higher compared to NP:8 dams (146.8 ± 2.2 g.kg $^{-1}$.day^{-1}, p = 0.008).

In this challenge, fenugreek promoted pup growth when comparing NPF:12 and NP:12 groups (Figure 4a). This significant difference was observed from L12 (p = 0.022) to L18 (p = 0.002). However, NPF:12 pups failed to reach the growth observed in NP:8 pups. The final weight gain of NPF:12 pups was increased by 10.6% compared to NP:12 pups but was 12.2% lower compared to NP:8 pups (Figure 4b).

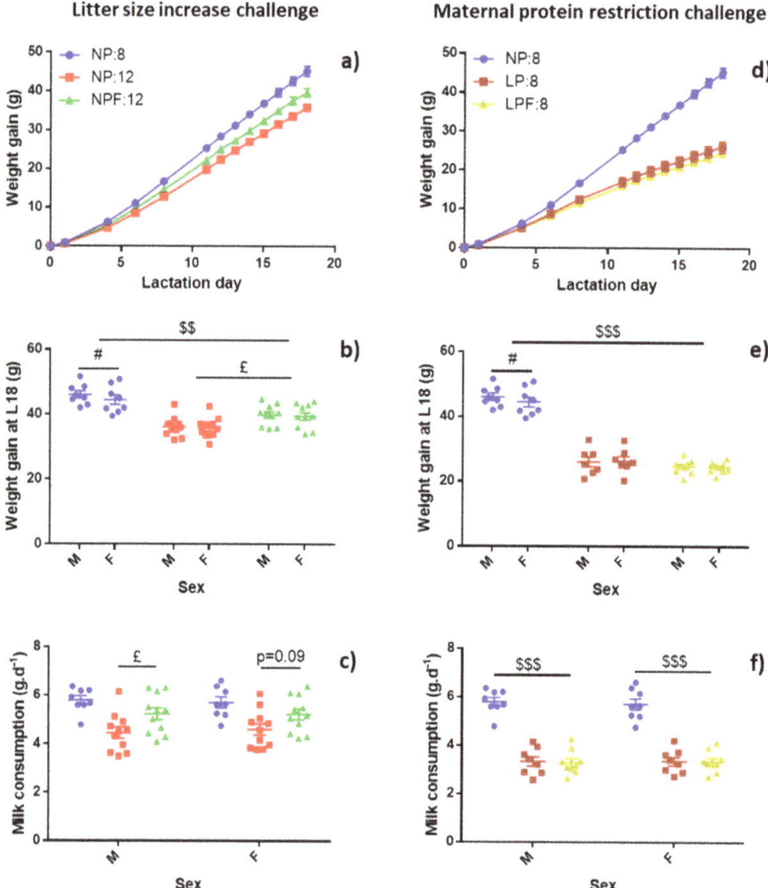

Figure 4. Effect of fenugreek on pup growth and pup milk consumption in 2 models of lactation challenges by litter size increase (**a**,**b**,**c**) or by maternal protein restriction (**d**,**e**,**f**). Weight gain during lactation was represented in graphs (**a**) and (**d**). Values were mean of male and female pups' weight gain. Final weight gain at L18 was represented in graphs (**b**) and (**e**) and pup milk consumption was represented in graphs (**c**) and (**f**) for both males (M) and females (F). Results were analysed with two-way ANOVAs with group and day factors for graphs (**a**) and (**d**) and with group and sex factors for other graphs. Pairwise comparisons were realised with Dunnett's post-hoc test to compare fenugreek supplemented groups to their challenge control and NP:8 and with Sidak's *post-hoc* tests for sex factors. £ p < 0.05 represented the significant difference between NPF:12 or LPF:8 and their own lactation challenge model control. $$ p < 0.01 and $$$ p < 0.001 represented the significant difference with NPF:12 or LPF:8 and NP:8. # p < 0.05 represented the significant difference between male and female pups.

Fenugreek promoted milk flow: total milk production was 16.1% higher in NPF:12 dams (63.0 ± 3.1 g.day^{-1}), compared with NP:12 dams ($p = 0.048$) (Table 1). Milk consumption per pup also tended to be higher ($p = 0.059$) with a greater increase for males ($+17.7\%$, $p = 0.028$) than females ($+13.8\%$, $p = 0.088$), although no overall sex effect was observed (Figure 4c).

In the maternal protein restriction challenge, fenugreek had no effect on dam's weight loss during lactation, nor on water and food intakes when comparing with the non-supplemented group LP:8. No significant difference was also observed for pup growth between LPF:8 and LP:8 groups neither for growth rate (Table 1), nor for weight gain during overall lactation or at a specific day (Figure 4d,e). Fenugreek had no effect on milk production (Table 1) and pups milk consumption (Figure 4f).

3.5. Fenugreek Enhances Milk Lactose and Trigonelline Content in the Litter Size Increase Challenge

In the litter size increase challenge, milk composition was similar between the NP:8 and NP:12 groups. Fenugreek had no effect on milk fatty acids (FAs), protein, and energy content, but led to a 27% increase in lactose concentration compared to NP:12 (Table 2). Every macronutrient flow was significantly increased in the NPF:12 group compared to the NP:12 group ($+29.6\%$, $+49.1\%$, and $+28.9\%$ for protein, lactose and FAs, respectively), leading to a 30.5% increase in energy flow for the NPF:12 group, while no difference was observed between the NP:12 and NP:8 groups.

The maternal protein restriction challenge had no significant effect on lactose and FAs concentrations and energy of milk but resulted in a significant 16% decrease in protein concentration (LP:8 vs. NP:8). Flow was significantly decreased for every macronutrient and for energy in the LP:8 group compared to NP:8 (-52.8%, -51.1%, -34.0%, and -38.9% for protein, lactose, FAs, and energy, respectively). Fenugreek failed to induce a modification in macronutrient composition or energy (in concentration and in flow) between LPF:8 and LP:8 milk.

Table 2. Effect of fenugreek on energy and macronutrient composition of milk and on energy and macronutrient flows.

Lactation Challenges	Control	Litter Size Increase		Maternal Protein Restriction	
Experimental Groups	NP:8	NP:12	NPF:12	LP:8	LPF:8
n	7	8	9	7	9
Macronutrient concentration, g.L^{-1}					
Protein	97.1 ± 4.9	91.7 ± 1.5	98.3 ± 3.3	81.5 ± 2.2 **	77.4 ± 2.1 $^{\$\$\$}$
Lactose	30.1 ± 1.9	29.4 ± 1.2	37.4 ± 0.9 $^{\$\$ \text{£££}}$	27.4 ± 1.3	25.3 ± 1.1
Fatty acids	147.1 ± 12.8	129.4 ± 6.6	137.8 ± 7.2	174.3 ± 13.6	176.5 ± 9.2
Energy, kcal.dL^{-1}	183.2 ± 12.0	165.1 ± 5.5	178.3 ± 6.9	200.4 ± 11.7	199.9 ± 8.8
Macronutrient flow, g.day^{-1}					
Protein	4.49 ± 0.32	4.87 ± 0.26	6.31 ± 0.41 $^{\$\$ \text{£}}$	2.12 ± 0.12 ***	2.06 ± 0.12 $^{\$\$\$}$
Lactose	1.38 ± 0.09	1.60 ± 0.12	2.38 ± 0.10 $^{\$\$\$ \text{£££}}$	0.71 ± 0.04 ***	0.68 ± 0.06 $^{\$\$\$}$
Fatty acids	6.79 ± 0.66	6.81 ± 0.31	8.78 ± 0.53 $^{\$ \text{£}}$	4.48 ± 0.34 *	4.72 ± 0.42 $^{\$}$
Energy flow, kcal.day^{-1}	84.6 ± 6.6	87.2 ± 4.0	113.8 ± 6.2 $^{\$\$ \text{££}}$	51.6 ± 3.0 ***	± 4.4 $^{\$\$\$}$

Values were mean $\pm$ SEM and were analysed by one-way ANOVA followed by a Tukey post-hoc test to compare all groups of each lactation challenge. * $p < 0.05$, ** $p < 0.01$, *** $p < 0.001$ represented significant differences between NP:12 or LP:8 and NP:8. £ $p < 0.05$, ££ $p < 0.01$, £££ $p < 0.001$ represented significant differences between NPF:12 or LPF:8 and their own lactation challenge control. \$ $p < 0.05$, \$\$ $p < 0.01$, \$\$\$ $p < 0.001$ represented significant differences between NPF:12 or LPF:8 and NP:8.

Trigonelline concentration was measured in milk at L18 as a marker of fenugreek content. NPF:12 milk had a trigonelline concentration 17.1-fold higher than NP:12 milk (2575.6 ± 132.9 nM and 150.3 ± 22.9 nM, respectively, $p < 0.001$). Similarly, LPF:8 milk had a significant 5-fold increased trigonelline concentration compared to LP:8 milk (2683.2 ± 141.9 nM and 533.7 ± 118.2 nM respectively, $p < 0.001$).

3.6. Effect of Fenugreek on Metabolic Status of Dams and Offspring at Short and Long Term for the Litter Size Increase Challenge

As fenugreek had a positive effect on milk production and quality only in the litter size increase lactation challenge, the metabolic status of dams and offspring was assessed exclusively in the NP:8, NP:12, and NPF:12 groups.

Dams' metabolic parameters were determined in plasma sampled at L12 without a fasting period and at L21 after a 4 h-fasting period (Table 3). Milk cholesterol, triglycerides (TGs), and glucose concentrations rose significantly between L12 (mid lactation) and L21 (end of lactation). A larger rise in cholesterol was observed in the NPF:12 group (+20,7%, $p < 0.001$) compared to the NP:12 group (+5%), but no significant difference was observed between the NPF:12 and NP:12 groups, at both days of lactation. Insulin concentration was significantly lower in the NPF:12 group than in NP:12 group (-42.4%, $p = 0{,}018$) transitory at L12, but no significant difference was observed between the NPF:12 group and NP:8 group.

Table 3. Effect of fenugreek on metabolic parameters in dams' plasma at L12 and L21 in the litter size increase lactation challenges.

Parameter	Groups			Two-Way ANOVA		
	NP:8	NP:12	NPF:12	Global Effects		
n	8	11	11	Inter	Group	Day
Cholesterol, mg.dL^{-1}						
L12	101.3 ± 6.0 [a,1]	94.6 ± 3.9 [a,1]	92.2 ± 3.8 [a,1]	0.050	0.57	0.002
L21	105.8 ± 5.6 [a,1]	99.2 ± 4.1 [a,1]	111.3 ± 6.9 [a,2]			
Triglycerides, mg.dL^{-1}						
L12	54.6 ± 7.0 [a,1]	43.5 ± 2.8 [a,1]	52.8 ± 5.2 [a,1]	0.027	0.062	<0.001
L21	132.0 ± 16.6 [a,2]	189.9 ± 25.1 [ab,2]	224.3 ± 23.7 [b,2]			
Glucose, mg.dL^{-1}						
L12	125.1 ± 6.4 [a,1]	120.6 ± 3.7 [a,1]	117.4 ± 3.7 [a,1]	0.72	0.62	<0.001
L21	148.0 ± 3.1 [a,2]	150.7 ± 6.5 [a,2]	144.7 ± 6.2 [a,2]			
Insulin, ng.mL^{-1}						
L12	0.60 ± 0.09 [a,1]	1.65 ± 0.25 [b,1]	0.95 ± 0.26 [a,1]	0.12	0.004	0.22
L21	0.68 ± 0.06 [a,1]	0.98 ± 0.09 [a,2]	0.94 ± 0.15 [a,1]			

Values were mean ± SE and were analysed with two-way ANOVA with group and day factors and with repeated values for day factor. ANOVA was followed by Tuckey's post-hoc test for comparisons between groups and by Sidak's post-hoc test for comparisons between days. For each biomarker, different letters represented significant differences ($p < 0.05$) between groups at each day and the difference between numbers represented significant difference ($p < 0.05$) between days for each group.

As expected, trigonelline content was largely increased at L12 in NPF:12 dams' plasma compared to NP:12 (3258.7 ± 200.4 nM and 464.3 ± 78.5 nM respectively, $p < 0.001$).

Offspring's metabolism was determined in the short (PND20) and long-term (PND60) when pups reached early adulthood. Selected plasma parameters of lipid and glucose metabolism are presented in Table 4. At PND 20, lactation challenge by litter size increase did not affect cholesterol, TGs, or insulin concentration in offspring's plasma, whereas it tended to increase glucose concentration (+7.5%, $p = 0.052$) in the NP:12 group vs. the NP:8 group. At PND 60, it tended to a decrease TGs concentration (-17.8%, $p = 0.10$) mainly for males (-21.7%, $p = 0.047$). Fenugreek had no effect on glucose, insulin, and TGs concentration although, at PND 20, a significant decrease in plasma cholesterol concentration was observed in the NPF:12 group compared to the NP:12 group (-10.0%, $p = 0.024$), greater for female (-11.8%) than for male (-8.1%). Fenugreek no longer had any effect on cholesterol concentration at PND 60.

Table 4. Effect of fenugreek on offspring's metabolism in the short and long term.

Parameter	Groups			Two-way ANOVA		
	NP:8	NP:12	NPF:12	Global Effects		
		PND 20		Inter	Group	Sex
n for each sex	8	22	11			
Cholesterol, mg.dL^{-1}						
Male	143.2 ± 10.1 [a,1]	144.5 ± 3.9 [a,1]	132.8 ± 3.0 [a,1]	0.225	0.005	0.039
Female	167.5 ± 12.4 [a,1]	151.5 ± 4.1 [a,b,1]	133.5 ± 6.1 [b,1]			
Triglycerides, mg.dL^{-1}						
Male	260.2 ± 39.2 [a,1]	272.0 ± 21.1 [a,1]	258.8 ± 56.3 [a,1]	0.853	0.763	0.969
Female	284.5 ± 46.1 [a,1]	267.8 ± 23.3 [a,1]	235.0 ± 35.9 [a,1]			
Glucose, mg.dL^{-1}						
Male	157.1 ± 7.4 [a,1]	164.4 ± 3.8 [a,1]	164.5 ± 3.5 [a,1]	0.446	0.055	0.655
Female	152.7 ± 3.7 [a,1]	168 ± 3.0 [a,1]	159.7 ± 3.5 [a,1]			
Insulin, ng.mL^{-1}						
Male	0.19 ± 0.04 [a,1]	0.35 ± 0.05 [a,1]	0.41 ± 0.09 [a,1]	0.958	0.037	0.282
Female	0.28 ± 0.04 [a,1]	0.41 ± 0.05 [a,1]	0.46 ± 0.07 [a,1]			
		PND 60				
n for each sex	10	22	16			
Cholesterol, mg.dL^{-1}						
Male	76.8 ± 2.5 [a,1]	85.0 ± 2.5 [a,1]	81.5 ± 2.1 [a,1]	0.433	0.208	0.170
Female	84.8 ± 3.1 [a,1]	85.7 ± 2.6 [a,1]	82.3 ± 3.0 [a,1]			
Triglycerides, mg.dL^{-1}						
Male	135.8 ± 11.7 [a,1]	106.4 ± 7.0 [b,1]	123.2 ± 12.0 [ab,1]	0.390	0.095	<0.001
Female	66.0 ± 3.7 [a,2]	59.5 ± 5.2 [a,2]	64.1 ± 6.3 [a,2]			

Values were mean ± SE and were analysed with two-way ANOVA with group and sex factors followed by Tukey's post-hoc test for comparisons between groups and by Sidak's *post-hoc* test for comparisons between sexes. For each biomarker, different letters represented significant ($p < 0.05$) differences between groups for each sex and different numbers represented significant differences ($p < 0.05$) between sexes for each group.

Long-term glucose metabolism was assessed in offspring by OGTT only for NP:12 and NPF:12 groups (Figure 5). Before oral glucose gavage (T0), no significant difference was observed between NPF:12 and NP:12 groups for plasma glucose and insulin concentrations, but females had significantly lower concentrations than males (-25.0%, $p = 0.014$), greater in the NPF:12 group (-37.1%, $p = 0.064$) than in the NP:12 group (-17.4%, $p = 0.33$). After oral glucose gavage, no difference was observed in glycaemia neither between sexes nor between groups (Figure 5a,b). Fenugreek decreased the insulin area under curve (AUC) in NPF:12 compared to the NP:12 group (-38.4%, $p = 0.004$) (Figure 5d) and this difference was greater for females (-60%, $p = 0.044$) than for males (-29.7%, $p = 0.094$). Insulin peak was delayed in the NPF:12 (30 min) group compared to the NP:12 group (15 min) (Figure 5c).

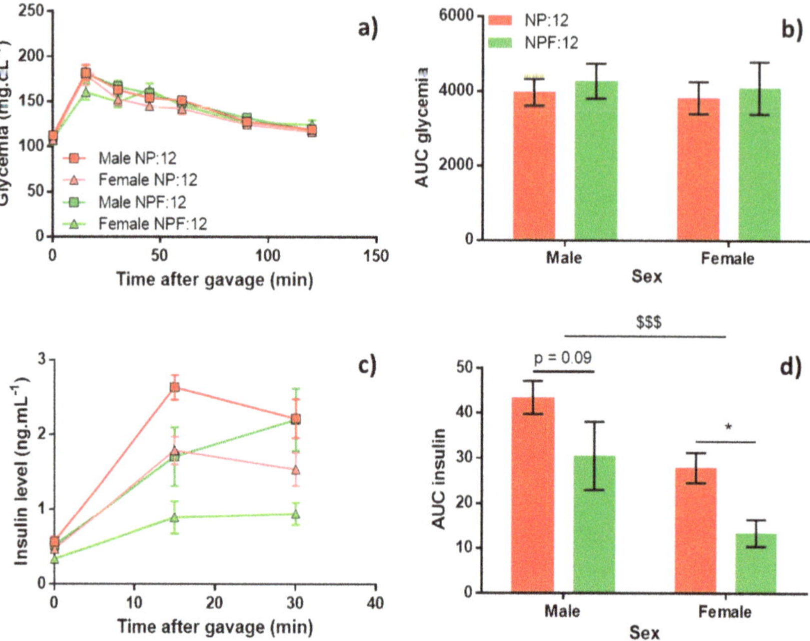

Figure 5. Effect of fenugreek on long-term glucose metabolism assessed by oral glucose tolerance test for NP:12 and NPF:12 groups. At PND 60, after 6 h fasting, 2 g.kg^{-1} of glucose was injected to rats by gavage. (**a**) Time course of glycaemia after glucose gavage; (**b**) area under curve (AUC) of glycaemia; (**c**) time course of insulin concentration after glucose gavage; (**d**) AUC of insulin concentration. For NP:12, n = 12 per sex and for NPF:12, n = 8 per sex. Values were mean ± SEM and were analysed with two-way ANOVA with group and sex factors followed by Sidak's post-hoc test. \$\$\$ $p < 0.001$ represents significant difference between sexes and * $p < 0.05$ represents significant difference between groups.

4. Discussion

To the best of our knowledge, this study is the first to explore the effect of fenugreek on milk production in a rat model. The effect on milk production, measured using stable isotope labelled water, was assessed in two separate models of lactation challenge: (a) litter size increase from 8 to 12 pups, and (b) maternal, perinatal dietary protein restriction (from 20% to 8%). We found that when dams were under appropriate physiological conditions for lactation and confronted with a litter size increase, fenugreek produced an increase in milk flow. In contrast, when dams were placed under inappropriate physiological conditions of lactation following dietary protein restriction, fenugreek was ineffective.

4.1. Effect of Lactation Challenges on Pup Growth and Milk Production

First, we verified that litter size increase and maternal, perinatal protein restriction both led to a decrease in milk consumption by the pups, resulting in Extra-Uterine Growth restriction (EUGR), confirming that dams had difficulties producing milk sufficiently to meet the demand.

When litter size was increased from 8 to 12 pups, dams adjusted their milk production, suggesting dams' adaptations to the increased demand from pups. Indeed, the litter size increase led to a 21% increase in litter growth rate, along with an 18% increase in milk flow. These results are in accordance with those from Morag et al. [23] and Kumaresan et al. [30] who reported a 43% and 22% increase of milk yield, measured by weight-suckle weight method, for a litter size increase from 9 or 8 pups to 12 pups, respectively. However, this rise of milk production was not sufficient to compensate the litter size increase, resulting in EUGR with a 26% lower overall pup weight gain, presumably due to a 21% decrease in milk consumption per pup. Kumaresan et al. [30] found similar results, with a 20% decrease in milk availability per pup at L14 when litter size was increased from 8 to 12 pups, and an 11% decreased in pup weight at L18 (31.1 g for 8 pups to 27.6 g for 12 pups) compared to the 18% decrease in our study (52.3 g for NP:8 and 42.9 g for NP:12 at L18). Our values of pup growth and milk consumption are probably closer to physiological values than those of Morag et al. [23] and Kumaresan et al. [30] as we considerably reduced the stress of dam/pup separation (from 8 h to 10 h in these studies compared to 30 min in ours). These conditions in which the dams' capacity to increase milk production is preserved but insufficiently to produce optimal pup growth, as observed in a litter of 12 or more pups, are likely the most suitable to test the effect of a galactologue compound [23,30].

In the other lactation challenge (maternal dietary protein restriction), we confirmed that 8% perinatal protein restriction led to an EUGR due to an impaired ability of dams to produce milk. This challenge led to a 44% reduced milk flow, confirming data from our earlier studies [22], and resulted in impaired pup growth with a 42% decrease in pup weight gain at L18 (as already observed by Bautista et al. [25] and Martin-Agnoux et al. [31]). The impaired ability of mothers to produce milk suggests the dams' physiological status was impaired due to undernutrition during gestation and lactation. Indeed, during lactation, LP:8 dams exhibited a 4.5-fold greater weight loss than NP:8 dams, which is a well-described consequence of nitrogen store depletion on milk-protein synthesis in states of malnutrition [32,33]. We also observed a significant 36% reduction in mammary gland weight at the end of lactation in the LP:8 group compared to the NP:8 group (5.73 ± 0.67 g for NP:8 vs. 3.67 ± 0.75g for LP:8 dams), as already reported in protein-restricted dams [32]. Finally, secretion of prolactin, the principal hormone promoting milk synthesis, has been shown to decrease by 70% in the serum of dams fed a low protein diet [32]. Altogether, these physiological changes contribute to a decrease in dams' capacity to produce sufficient milk to meet the offspring requirement.

4.2. Correlation between Milk Flow, Pups' Growth, and Lactating Dams' Intakes of Food and Water

Secondly, we confirmed that the use of the water turnover method, with D_2O values as mass, reliably measures milk flow. When considering the three lactation models altogether, the changes in total milk production were strongly correlated with litter growth rate ($r = 0.93$), and changes in milk production and milk consumption accounted for at 87% and 82% of the change in growth rate in the two models tested (Figure 3). As litter or pup growth rate and milk production or consumption are expressed in the same unit (g.day^{-1}), the slopes of 0.51 with intercept close to 0 in both regression lines suggest that the consumption of 1g of milk produced a weight gain of 0.51 g per day between L11 and L18. These results are also supported by the strong correlations between milk production and the dams food and water intake ($r = 0.86$ and $r = 0.91$, respectively), representing the increasing needs of the lactating mother for milk production with large litters [33]. Altogether these results suggest that the water turnover method, with values of D_2O mass, accurately measures milk production regardless of the dam conditions.

4.3. Galactologue Effect of Fenugreek in Two Models of Lactation Challenge

In the litter size increase challenge, dietary supplementation of fenugreek at a dose of 1 g.kg^{-1}.d^{-1} increased milk flow by 16% and increased pup growth rate and final weight gain, by 9% and 11%, respectively. The galactologue effect of fenugreek is consistent with the rise in milk production observed in other mammals (+13% at a fenugreek dose of 2 g.kg^{-1}.day^{-1}in goat [18] and +18% at a fenugreek dose of 270 mg.kg^{-1}.d^{-1} in buffalo [17]). Other authors found a stronger effect of fenugreek with a 42% greater pup growth at the end of lactation in rabbit (fenugreek dose of 0.5 g.kg^{-1}.day^{-1}) [16] and a 110% increase in milk yield in ewe (fenugreek dose of 1.2 g.kg^{-1}.day^{-1}) [19]. The stronger galactologue effects in the rabbit could be explained by the fact that they only nurse their litter once a day [24], suggesting that pups have important growth with only a small milk intake. Otherwise, in ewes, fenugreek supplementation is associated with greater crud protein and energy content in the ration compared to the control group, which could also affect milk production.

In contrast, in the maternal protein restriction challenge, fenugreek failed to affect pup growth or milk production. Undernutrition likely produced profound alterations in global metabolism and mammary gland development and function that could not be reversed by a galactologue

The mechanisms underlying the impact of fenugreek in the litter size challenge raise many questions. The reported increase in plasma prolactin and growth hormone may play an important role [17,18]. Fenugreek might also act by allowing dams to maintain their weight during lactation. Indeed, a significant correlation was found between dam mass and pup growth in rats [24]. Yet, NPF:12 dams' weight at L11 was significantly higher than NP:12 dams ($p = 0.021$) although no significant difference was observed at L0. Increased food intake in the fenugreek-supplemented group likely played a minor role in maintaining dam weight since the increased energy intake barely compensated for the increasing energy output that accounted for increased milk production. Thus, fenugreek probably promotes energy storage, which in turn could positively affect the mother's lactation performance [24]. Finally, the effect of fenugreek may be mediated by trigonelline [15]. Indeed, a 7-fold rise in dam plasma trigonelline was observed following fenugreek supplementation at 1 g.kg^{-1}.day^{-1}. Trigonelline is a precursor of niacin or nicotinic acid (B3 vitamin) involved in the formation of nicotinamide adenine dinucleotide (NAD$^+$) [34]. This coenzyme factor may play a key role in various metabolic pathways such as i) ATP formation, via its reduction in NADH in glycolysis, beta-oxidation, and citric acid cycle [35], ii) cell survival, as NAD$^+$ is the sole precursor of PARP, a DNA repair enzyme, and iii) transcriptional regulation as it is a main cofactor of sirtuins [36]. Thus, in the mammary gland, an increase in trigonelline could increase NAD$^+$ content, and thus, enhance lactation by promoting energy supply for milk synthesis, as well as mammary cell longevity and function.

4.4. Effect of Fenugreek on Milk Composition

A separate collection of milk secreted at the beginning and at the end of each suckling is not feasible. Milk composition was assessed at L18, which corresponds to the end of the lactation period in the rat. As rat milk composition changes during lactation [37], macronutrient composition reported at L18 may not reflect day to day variations in milk composition during the entire period during which the milk production was determined (L11–L18). As observed by earlier studies [38], the increase in litter size did not alter milk macronutrient concentration. Fenugreek increased milk lactose by 27%, whereas lipids and proteins remained unchanged. Discrepant results of fenugreek supplementation on milk macronutrient content have been reported in the literature in several animal models (rabbit, buffalo, goat, and ewe) [16–19]. However, lactose was the most affected by fenugreek, especially in our model, and its concentration is increased in most cases [16,17]. The key osmotic regulatory role of lactose on milk secretion could explain the positive effect of fenugreek on total milk production through an increase in water flow from the mammary epithelial cells into the mammary secretory vesicles and subsequently into the alveolar lumen [39].

Increasing litter size did not increase macronutrient and energy flow although milk flow was significantly increased, suggesting that the increase in milk production was offset by a dilution of macronutrients, likely due to the osmotic role of lactose. Indeed, NP:12 milk had the lower mean value of protein, fat, and energy, leading to even lower intakes for the pups. In contrast, macronutrient and energy flows were all increased when dams were supplemented with fenugreek suggesting that, with the increase in total milk production, the activity of the synthesis pathways of the three milk macronutrients was enhanced to maintain baseline concentrations. Further molecular investigations would be needed at the mammary gland level to confirm this hypothesis. In terms of pups' intake, the increase in macronutrient and energy flow led to a similar intake of the three macronutrients by the NPF:12 pups and NP:8 pups (Figure S2). As lactose concentration was increased in NPF:12 milk, the mean lactose intake by NPF:12 pups was slightly, but not significantly larger than the lactose intake in NP:8 pups. Larger lactose intake, however, did not allow NPF:12 pups to achieve the same growth rate as NP:8 pups. This suggests that lactose is not the major nutrient that promotes pup growth. Moreover, as the three macronutrients were consumed in similar amounts in both groups despite different growth trajectories, this suggests that micronutrients and/or other bioactive compounds of milk likely impact pup growth. In the maternal protein restriction model, fenugreek had no effect on milk composition. Once again, important modifications of milk composition due to maternal undernutrition [28] cannot be overcome by fenugreek.

4.5. Effect of Fenugreek on Dams Metabolism during Lactation and on Short- and Long-Term Offspring Metabolism

At mid-lactation, fenugreek did not modify dams' plasma cholesterol, triglycerides, and glucose concentrations as previously found in ewe [19] or in goat [21], suggesting that fenugreek did not alter maternal blood metabolic biomarkers during lactation. However, plasma insulin concentration was significantly increased in the NP:12 group and returned near to the baseline level in the NPF:12 group. Among metabolic adaptations reported during lactation, mammary gland displays enhanced insulin sensitivity while other tissues developed insulin resistance, leading to a redirection of energy substrates toward the mammary gland [40]. These changes result in a decrease in plasma insulin concentration, which is inversely related to milk production [41] and is consistent with the increased milk production and lower insulin concentration observed in NPF:12 dams at L12.

Contrary to the overall lactation period, the suckling phase leads to a rise in plasma insulin concentration [41,42]. Yet, plasma was sampled just after dam/pup separation, and pups were allowed to suckle until separation. The greater number of pups in NP:12 litters compared to NP:8 litters probably led to greater suckling, explaining the greater dams' insulin level observed. All these results in insulin must be relativized, knowing that at L12, dams' plasma was sampled without a fasting period.

At the end of the lactation period, both cholesterol and triglycerides were greatly increased in NPF:12 than in NP:12 dams' plasma compared to mid-lactation, which could be explained by the higher milk production of NPF:12 dams. Indeed, during lactation, the mammary gland becomes the main site of lipogenesis with a rate 5-fold higher than in liver [33], which in contrast, increases its hepatic cholesterol synthetic activity [43] and in detriment of adipose tissue, which exhibits an increased lipolytic activity [33]. These changes are associated with a large rise in triglyceride and cholesterol uptake from circulating chylomicrons in the mammary gland thanks to the increased activity of prolactin-mediated lipoprotein lipase (LPL) [44]. At the end of lactation, the prolactin-mediated LPL activity in the mammary gland drops rapidly [43], although lipolytic activity in adipose tissues and cholesterol synthesis in liver persist for a while. This leads to higher plasma triglyceride [43] and cholesterol concentration, which could presumably be further increased when milk production is enhanced. This would be consistent with the differences observed in NP:12 and NPF:12 plasma and with the larger rise in plasma triglyceride concentration in NP:12 dams compared to NP:8 dams.

Finally, the current study was the first to demonstrate the absence of adverse effects of dams' fenugreek supplementation on offspring metabolism. In the short-term, only cholesterol plasma concentration was decreased by fenugreek supplementation. Fenugreek is known to have a hypocholesterolemic effect that has been attributed to the large amount of fibre in fenugreek seed [45], or to steroid saponin such as diosgenin [15]. The effect of fenugreek on plasma cholesterol concentration of the offspring in the short-term may be mediated by the consumption of these components in milk. Indeed, milk trigonelline concentration increased 17-fold upon maternal supplementation, implying many components of fenugreek appear in mothers' milk. Alternatively, as pups begin to eat dams' pellet food at the end of lactation, the hypocholesterolemic effect of fenugreek might be directly mediated by solid food consumption. When fenugreek consumption stopped at L20, cholesterol concentration became indistinguishable from those of control offspring at PND 60, suggesting only a short-term hypocholesterolemic effect of fenugreek.

In the long run, no difference was observed between NPF:12 and NP:12 offspring in glucose concentration after glucose gavage, although NPF:12 offspring had lower concentrations of insulin with a delayed peak. Although a delayed peak of insulin is generally associated with insulin resistance [46], it is often related to an increased glucose concentration and area-under-curve, which is not the case in our study. Conversely, the decrease in insulin area under curve could reflect greater insulin sensitivity, and fenugreek is known to have antidiabetic properties [15], notably via the action of trigonelline [47]. The increased consumption of milk trigonelline by pups during lactation may help them to develop higher insulin sensitivity in the short-term and thus favour long-term glucose metabolism. However, despite the apparent absence of metabolic alterations, fenugreek should not be consumed by women with asthma, digestive disorders, hypertension, heart cardiovascular disease, or hypothyroidism because of its possible side effects [15,48].

5. Conclusions

The current study confirms the galactologue effect of fenugreek in another mammal model, suggesting that this dietary supplement may be helpful in humans. We tested the capacity of dams' fenugreek to increase milk production and subsequent pup growth in two models of the lactation challenges. The galactologue effect of fenugreek was confirmed, but only in the challenge by litter size increase when dams had no physiological impairment in their ability to produce milk. In contrast, the lack of effect of fenugreek under maternal dietary protein restriction suggests that fenugreek cannot overcome the lactation impairment due to undernutrition. Thus, fenugreek supplementation might enhance milk production in the case of insufficient maternal milk production, due to maternal stress, difficulties in breastfeeding management, first parity, or when mothers are breastfeeding twins, but fenugreek is unlikely to be effective in situations that affect lactation physiology, such as undernutrition deficiency, mammary hypoplasia, and hormonal deregulation. Finally, we observed no adverse metabolic effect neither on the dam at mid- and end-lactation nor on offspring, with preliminary evidence that fenugreek might even enhance insulin sensitivity in the long run. The 16% increase in milk production, thanks to fenugreek, is of significance and clearly warrants the design of clinical trials in breastfeeding women

Supplementary Materials: The following are available online at http://www.mdpi.com/2072-6643/11/11/2571/s1, Figure S1: Time course of D2O concentrations in dam's plasma, Table S1: Composition and characteristics of the experiment based on AIN-93G diets., Table S2: Model coefficients for measurement of milk flow by the water turnover method

Author Contributions: Conceptualization, T.S., M.-C.A.-G., D.D. and C.-Y.B.; data curation, T.S. and C.-Y.B.; formal analysis, T.S.; funding acquisition, M.-C.A.-G., D.D. and C.-Y.B.; investigation, T.S. and B.C.; methodology, A.A. and K.O.; project administration, C.-Y.B.; resources, A.A. and P.N.; software, K.O.; supervision, M.-C.A.-G. and C.-Y.B.; writing—original draft, T.S.; writing—review and editing, T.S., M.-C.A.-G., D.D. and C.-Y.B.

Funding: This research was funded by Laboratoire FRANCE Bébé Nutrition (Laval, France), grant number Cifre 2017.0004.

Acknowledgments: The authors want to thank specifically thank Alexis Gandon (PhAN, Nantes, France), Sandrine Suzanne and Amandine Lefebvre (UTE, Nantes, France) for their precious help during all the animal experiments. We thank Isabelle Grit (PhAN, Nantes, France) for help with every metabolic analysis. Finally, we thank the Mass Spectrometry Core Facility of CNRH-O (Centre de Recherche en Nutrition Humaine Ouest) and Biogenouest, particularly the Corsaire network of analytical platforms, for their support of this project.

Conflicts of Interest: Thomas Sevrin is employed by France Bébé Nutrition, which funded this study.

References

1. WHO. *Indicators for Assessing Infant and Young Child Feeding Practices: Part 1: Definitions: Conclusions of a Consensus Meeting Held 6–8 November 2007 in Washington, DC, USA*; WHO: Geneva, Switzerland, 2008.

2. Rito, A.I.; Buoncristiano, M.; Spinelli, A.; Salanave, B.; Kunešová, M.; Hejgaard, T.; García Solano, M.; Fijałkowska, A.; Sturua, L.; Hyska, J.; et al. Association between Characteristics at Birth, Breastfeeding and Obesity in 22 Countries: The WHO European Childhood Obesity Surveillance Initiative—COSI 2015/2017. *Obes. Facts* **2019**, *12*, 226–243. [CrossRef]

3. Eidelman, A.I.; Schanler, R.J. Breastfeeding and the Use of Human Milk. *Pediatrics* **2012**, *129*, e827–e841. [CrossRef]

4.   Danforth, K.N.; Tworoger, S.S.; Hecht, J.L.; Rosner, B.A.; Colditz, G.A.; Hankinson, S.E. Breastfeeding and risk of ovarian cancer in two prospective cohorts. *Cancer Causes Control* **2007**, *18*, 517–523. [CrossRef] [PubMed]

5.   Gunderson, E.P.; Lewis, C.E.; Lin, Y.; Sorel, M.; Gross, M.; Sidney, S.; Jacobs, D.R., Jr.; Shikany, J.M.; Quesenberry, C.P., Jr. Lactation Duration and Progression to Diabetes in Women across the Childbearing Years: The 30-Year CARDIA Study. *JAMA Intern. Med.* **2018**, *178*, 328–337. [CrossRef] [PubMed]

6.   Victora, C.G.; Bahl, R.; Barros, A.J.; Franca, G.V.; Horton, S.; Krasevec, J.; Murch, S.; Sankar, M.J.; Walker, N.; Rollins, N.C.; et al. Breastfeeding in the 21st century: Epidemiology, mechanisms, and lifelong effect. *Lancet* **2016**, *387*, 475–490. [CrossRef]

7.   Bonet, M.; Marchand, L.; Kaminski, M.; Fohran, A.; Betoko, A.; Charles, M.A.; Blondel, B. Breastfeeding duration, social and occupational characteristics of mothers in the French 'EDEN mother-child' cohort. *Matern. Child Health J.* **2013**, *17*, 714–722. [CrossRef] [PubMed]

8.   Salanave, B.d.L.C.; Boudet-Berquier, J.; Castetbon, K. Durée de l'allaitement maternel en France (Épifane 2012–2013). *Bull. Epidémiol. Hebd.* **2014**, *27*, 450–457. [CrossRef]

9.   Thulier, D.; Mercer, J. Variables associated with breastfeeding duration. *J. Obstet. Gynecol. Neonatal Nurs.* **2009**, *38*, 259–268. [CrossRef]

10. Gatti, L. Maternal perceptions of insufficient milk supply in breastfeeding. *J. Nurs. Scholarsh.* **2008**, *40*, 355–363. [CrossRef]

11. Murase, M.; Wagner, E.A.; Chantry, C.J.; Dewey, K.G.; Nommsen-Rivers, L.A. The Relation between Breast Milk Sodium to Potassium Ratio and Maternal Report of a Milk Supply Concern. *J. Pediatr.* **2017**, *181*, 294–297. [CrossRef]

12. Javan, R.; Javadi, B.; Feyzabadi, Z. Breastfeeding: A Review of Its Physiology and Galactogogue Plants in View of Traditional Persian Medicine. *Breastfeed. Med.* **2017**, *12*, 401–409. [CrossRef] [PubMed]

13. Forinash, A.B.; Yancey, A.M.; Barnes, K.N.; Myles, T.D. The use of galactogogues in the breastfeeding mother. *Ann. Pharmacother.* **2012**, *46*, 1392–1404. [CrossRef] [PubMed]

14. Abascal, K.; Yarnell, E. Botanical Galactagogues. *Altern. Complement. Ther.* **2008**, *14*, 288–294. [CrossRef]

15. Bahmani, M.; Shirzad, H.; Mirhosseini, M.; Mesripour, A.; Rafieian-Kopaei, M. A Review on Ethnobotanical and Therapeutic Uses of Fenugreek (*Trigonella foenum-graceum* L.). *J. Evid. Based Complement. Altern. Med.* **2016**, *21*, 53–62. [CrossRef] [PubMed]

16. Abdel-Rahman, H.; Fathalla, S.I.; Ezzat Assayed, M.; Ramadan Masoad, S.; Abdelaleem Nafeaa, A. Physiological Studies on the Effect of Fenugreek on Productive Performance of White New-Zealand Rabbit Does. *Food Nutr. Sci.* **2016**, *7*, 1276–1289. [CrossRef]

17. Mahgoub, A.A.S.; Sallam, M.T. Effect of Extract Crushed Fenugreek Seeds as Feed Additive on some Blood Parameters, Milk Yield and Its Composition of Lactating Egyptian Buffaloes. *J. Anim. Poult. Sci.* **2016**, *7*, 269–273. [CrossRef]

18. Alamer, M.A.; Basiouni, G.F. Feeding effect of Fenugreek seed (*Trigonella foenum-graecum* L.) on lactation performance, some plasma constituent and growth hormone level in goats. *Pak. J. Biol. Sci.* **2005**, *8*, 1553–1556.

19. Al-Sherwany, D.A.O. Feeding effects of fenugreek seeds on intake, milk yield, chemical composition of milk and some biochemical parameters in Hamdani ewes. *Al-Anbar J. Vet. Sci.* **2015**, *8*, 49–54.

20. Eiben, C.; Rashwan, A.; Kustos, K.; Gódor-Surmann, K.; Szendrő, Z. Effect of Anise and Fenugreek supplementation on performance of rabbit does. In Proceedings of the 8th World Rabbit Congress, Puebla, Mexico, 7–10 September 2004.

21. Al-Shaikh, M.A.; Al-Mufarrej, S.I.; Mogawer, H.H. Effect of Fenugreek Seeds (*Trigonella foenumgraecum* L.) on Lactational Performance of Dairy Goat. *J. Appl. Anim. Res.* **1999**, *16*, 177–183. [CrossRef]

22. Sevrin, T.; Alexandre-Gouabau, M.C.; Darmaun, D.; Palvadeau, A.; Andre, A.; Nguyen, P.; Ouguerram, K.; Boquien, C.Y. Use of water turnover method to measure mother's milk flow in a rat model: Application to dams receiving a low protein diet during gestation and lactation. *PLoS ONE* **2017**, *12*, e0180550. [CrossRef]

23. Morag, M.; Popliker, F.; Yagil, R. Effect of litter size on milk yield in the rat. *Lab. Anim.* **1975**, *9*, 43–47. [CrossRef] [PubMed]

24. Rodel, H.G.; Prager, G.; Stefanski, V.; von Holst, D.; Hudson, R. Separating maternal and litter-size effects on early postnatal growth in two species of altricial small mammals. *Physiol. Behav.* **2008**, *93*, 826–834. [CrossRef] [PubMed]

25. Bautista, C.J.; Boeck, L.; Larrea, F.; Nathanielsz, P.W.; Zambrano, E. Effects of a maternal low protein isocaloric diet on milk leptin and progeny serum leptin concentration and appetitive behavior in the first 21 days of neonatal life in the rat. *Pediatr. Res.* **2008**, *63*, 358–363. [CrossRef] [PubMed]

26. Reeves, P.G.; Nielsen, F.H.; Fahey, G.C., Jr. AIN-93 purified diets for laboratory rodents: Final report of the American Institute of Nutrition ad hoc writing committee on the reformulation of the AIN-76A rodent diet. *J. Nutr.* **1993**, *123*, 1939–1951. [CrossRef] [PubMed]

27. Osowska, S.; Duchemann, T.; Walrand, S.; Paillard, A.; Boirie, Y.; Cynober, L.; Moinard, C. Citrulline modulates muscle protein metabolism in old malnourished rats. *Am. J. Physiol. Endocrinol. Metab.* **2006**, *291*, E582–E586. [CrossRef] [PubMed]

28. Martin Agnoux, A.; Antignac, J.P.; Boquien, C.Y.; David, A.; Desnots, E.; Ferchaud-Roucher, V.; Darmaun, D.; Parnet, P.; Alexandre-Gouabau, M.C. Perinatal protein restriction affects milk free amino acid and fatty acid profile in lactating rats: Potential role on pup growth and metabolic status. *J. Nutr. Biochem.* **2015**, *26*, 784–795. [CrossRef]

29. Von Dreele, M.M. Age-related changes in body fluid volumes in young spontaneously hypertensive rats. *Am. J. Physiol. Ren. Physiol.* **1988**, *255*, F953–F956. [CrossRef]

30. Kumaresan, P.; Anderson, R.R.; Turner, C.W. Effect of litter size upon milk yield and litter weight gains in rats. *Rev. Tuberc. Pneumol.* **1966**, *30*, 41–45. [CrossRef]

31. Martin Agnoux, A.; Alexandre-Gouabau, M.C.; Le Drean, G.; Antignac, J.P.; Parnet, P. Relative contribution of foetal and post-natal nutritional periods on feeding regulation in adult rats. *Acta Physiol.* **2014**, *210*, 188–201. [CrossRef]

32. Moretto, V.L.; Ballen, M.O.; Gonçalves, T.S.S.; Kawashita, N.H.; Stoppiglia, L.F.; Veloso, R.V.; Latorraca, M.Q.; Martins, M.S.F.; Gomes-da-Silva, M.H.G. Low-Protein Diet during Lactation and Maternal Metabolism in Rats. *ISRN Obstet. Gynecol.* **2011**, *2011*, 876502. [CrossRef]

33. Williamson, D.H. Integration of metabolism in tissues of the lactating rat. *FEBS Lett.* **1980**, *117*, K93–K105. [CrossRef]

34. Mehrafarin, A.; Qaderi, A.; Rezazadeh, S.; Naghdi Badi, H.; Noormohammadi, G.; Zand, E. Bioengineering of important secondary metabolites and metabolic pathways in fenugreek (*Trigonella foenum-graecum* L.). *J. Med. Plants* **2010**, *3*, 1–18.

35. Stein, L.R.; Imai, S.-i. The dynamic regulation of NAD metabolism in mitochondria. *Trends Endocrinol. Metab.* **2012**, *23*, 420–428. [CrossRef] [PubMed]

36. Braidy, N.; Poljak, A.; Grant, R.; Jayasena, T.; Mansour, H.; Chan-Ling, T.; Guillemin, G.; Smythe, G.; Sachdev, P. Mapping NAD+ metabolism in the brain of ageing Wistar rats: Potential targets for influencing brain senescence. *Biogerontology* **2013**, *15*, 177–198. [CrossRef] [PubMed]

37. Nicholas, K.R.; Hartmann, P.E. Milk secretion in the rat: Progressive changes in milk composition during lactation and weaning and the effect of diet. *Comp. Biochem. Physiol. Part A Physiol.* **1991**, *98*, 535–542. [CrossRef]

38. Fiorotto, M.L.; Burrin, D.G.; Perez, M.; Reeds, P.J. Intake and use of milk nutrients by rat pups suckled in small, medium, or large litters. *Am. J. Physiol.* **1991**, *260*, R1104–R1113. [CrossRef]

39. Bleck, G.T.; White, B.R.; Miller, D.J.; Wheeler, M.B. Production of bovine α-lactalbumin in the milk of transgenic pigs1. *J. Anim. Sci.* **1998**, *76*, 3072–3078. [CrossRef]

40. Jones, R.G.; Ilic, V.; Williamson, D.H. Physiological significance of altered insulin metabolism in the conscious rat during lactation. *Biochem. J.* **1984**, *220*, 455–460. [CrossRef]

41. Widström, A.M.; Winberg, J.; Werner, S.; Hamberger, B.; Eneroth, P.; Uvnäs-Moberg, K. Suckling in lactating women stimulates the secretion of insulin and prolactin without concomitant effects on gastrin, growth hormone, calcitonin, vasopressin or catecholamines. *Early Hum. Dev.* **1984**, *10*, 115–122. [CrossRef]

42. Eriksson, M.; Lindén, A.; Uvnäs-Moberg, K. Suckling increases insulin and glucagon levels in peripheral venous blood of lactating dogs. *Acta Physiol. Scand.* **1987**, *131*, 391–396. [CrossRef]

43. Smith, J.L.; Lear, S.R.; Forte, T.M.; Ko, W.; Massimi, M.; Erickson, S.K. Effect of pregnancy and lactation on lipoprotein and cholesterol metabolism in the rat. *J. Lipid Res.* **1998**, *39*, 2237–2249.

44. Scow, R.O.; Chernick, S.S.; Fleck, T.R. Lipoprotein lipase and uptake of triacylglycerol, cholesterol and phosphatidylcholine from chylomicrons by mammary and adipose tissue of lactating rats in vivo. *Biochim. Biophys. Acta (BBA) Lipids Lipid Metab.* **1977**, *487*, 297–306. [CrossRef]

45. Wani, S.A.; Kumar, P. Fenugreek: A review on its nutraceutical properties and utilization in various food products. *J. Saudi Soc. Agric. Sci.* **2018**, *17*, 97–106. [CrossRef]
46. Hayashi, T.; Boyko, E.J.; Sato, K.K.; McNeely, M.J.; Leonetti, D.L.; Kahn, S.E.; Fujimoto, W.Y. Patterns of Insulin Concentration During the OGTT Predict the Risk of Type 2 Diabetes in Japanese Americans. *Diabetes Care* **2013**, *36*, 1229–1235. [CrossRef] [PubMed]
47. Zhou, J.; Zhou, S.; Zeng, S. Experimental diabetes treated with trigonelline: Effect on β cell and pancreatic oxidative parameters. *Fundam. Clin. Pharmacol.* **2013**, *27*, 279–287. [CrossRef] [PubMed]
48. Panda, S.; Tahiliani, P.; Kar, A. Inhibition of triiodothyronine production by fenugreek seed extract in mice and rats. *Pharmacol. Res.* **1999**, *40*, 405–409. [CrossRef]

Article

No Association between Glucocorticoid Diurnal Rhythm in Breastmilk and Infant Body Composition at 3 Months

Jonneke Hollanders [1,†,*], Lisette R. Dijkstra [1,†], Bibian van der Voorn [2], Stefanie M.P. Kouwenhoven [3], Alyssa A. Toorop [1], Johannes B. van Goudoever [3], Joost Rotteveel [1] and Martijn J.J. Finken [1]

[1] Emma Children's Hospital, Amsterdam UMC, Pediatric Endocrinology, Vrije Universiteit Amsterdam, 1000-1183 Amsterdam, The Netherlands; l.dijkstra@amsterdamumc.nl (L.R.D.); a.toorop@amsterdamumc.nl (A.A.T.); j.rotteveel@amsterdamumc.nl (J.R.); m.finken@amsterdamumc.nl (M.J.J.F.)

[2] Department of Paediatric Endocrinology, Obesity Center CGG, Sophia Children's Hospital, 3000-3099 Rotterdam, The Netherlands; b.vandervoorn@erasmusmc.nl

[3] Emma Children's Hospital, Amsterdam UMC, Department of Pediatrics, Vrije Universiteit Amsterdam, 1000-1183 Amsterdam, The Netherlands; s.kouwenhoven@amsterdamumc.nl (S.M.P.K.); h.vangoudoever@amsterdamumc.nl (J.B.v.G.)

*   Correspondence: j.hollanders@amsterdamumc.nl; Tel.: +31-(0)20-4443137

†   Both authors contributed equally to the article.

Received: 13 August 2019; Accepted: 24 September 2019; Published: 2 October 2019

Abstract: Objective: Glucocorticoids (GCs) in breastmilk have previously been associated with infant body growth and body composition. However, the diurnal rhythm of breastmilk GCs was not taken into account, and we therefore aimed to assess the associations between breastmilk GC rhythmicity at 1 month and growth and body composition at 3 months in infants. Methods: At 1 month postpartum, breastmilk GCs were collected over a 24-h period and analyzed by LC-MS/MS. Body composition was measured using air-displacement plethysmography at 3 months. Length and weight were collected at 1, 2, and 3 months. Results: In total, 42 healthy mother–infant pairs were included. No associations were found between breastmilk GC rhythmicity (area-under-the-curve increase and ground, maximum, and delta) and infant growth trajectories or body composition (fat and fat free mass index, fat%) at 3 months. Conclusions: This study did not find an association between breastmilk GC rhythmicity at 1 month and infant's growth or body composition at 3 months. Therefore, this study suggests that previous observations linking breastmilk cortisol to changes in infant weight might be flawed by the lack of serial cortisol measurements and detailed information on body composition.

Keywords: cortisol; cortisone; growth; circadian rhythm; human milk

---

1. Introduction

Growing attention is focused on the etiology of obesity, and it has been hypothesized that part of its origin can be traced back to events occurring in early life (i.e., the Developmental Origins of Health and Disease (DOHaD) hypothesis) [1].

Given its effects on fat disposition and metabolism, the hypothalamus-pituitary-adrenal (HPA) axis has been implicated to play a role in the pathway leading to obesity [2,3]. Not only endogenous but also maternal glucocorticoids (GCs) appear to be involved. Evidence from animal experiments indicates that increased transplacental supply of maternal GCs may be associated with a lower birth weight and cardiovascular correlates, such as hypertension and hyperglycemia [4]. In humans, fetal exposure to excess maternal cortisol, e.g., due to maternal anxiety or depression, has been associated with a higher risk of childhood adiposity [5].

After birth, small amounts of maternal GCs appear to be transferred to the developing infant through breastmilk. Maternal GCs in breastmilk have been shown to cross the intestinal barrier in animals [6], and have been associated with growth and body composition. Hinde et al. (2015) [7] found that cortisol in the breastmilk of rhesus macaques was positively associated with weight gain in offspring. In humans, Hahn-Holbrook et al. (2016) [8] showed that cortisol in breastmilk at the age of 3 months was inversely associated with body mass index (BMI) percentile gains in the first 2 years of life. Whether the findings from these studies are contradictory is unclear, since length was not taken into account by Hinde et al. [7]. Moreover, the effect of GCs on growth might change between the ages of 3 months and 2 years.

Our group has previously shown that GCs in breastmilk follow maternal HPA-axis activity, with a peak in the morning and a nadir at night [9]. Although previous studies have found associations between cortisol in breastmilk and growth of offspring, none of them took GC rhythmicity into account. However, obesity has previously been associated with a flatter diurnal cortisol slope in adults [10], and there is also some evidence that a blunted GC rhythm is associated with obesity in children [11,12]. Both Hinde et al. (2015) [7] and Hahn-Holbrook et al. (2016) [8] did not collect samples around peak GC levels, while Hahn-Holbrook et al. (2016) also had a wide time window during which samples could be collected (11:30–16:00).

We therefore aimed to assess the associations between breastmilk GC rhythmicity and infant growth and body composition. We measured cortisol and cortisone in breastmilk at 1 month of age over a 24-hour period, measured body composition using air-displacement plethysmography at 3 months of age, and collected length and weight data monthly up to that age. Due to associations found between a blunted endogenous GC rhythm and obesity in both children and adults [10–12], we hypothesized that less GC variability in breastmilk could be associated with a higher fat mass in the infants.

2. Materials and Methods

2.1. Population

Healthy mother–infant pairs were recruited at the maternity ward of the Amsterdam UMC, location VUMC (a tertiary hospital) in the Netherlands between March 2016 and July 2017. Subjects were eligible for inclusion when infants were born at term age (37–42 weeks) with a normal birth weight (−2 to +2 SDS), and when mothers had the intention to breastfeed for a minimum of three months. Exclusion criteria were: (1) Major congenital anomalies, (2) multiple pregnancy, (3) pre-eclampsia or HELLP, (4) medication use other than "over the counter" drugs, (5) maternal alcohol consumption of >7 IU/week, and/or 6) a maternal temperature of >38.5 °C at the time of sampling. Approval of the Medical Ethics Committee of the VUMC was obtained (protocol number 2015.524), and written informed consent was obtained from all participating mothers.

2.2. Data Collection

2.2.1. Peripartum

Shortly after inclusion, within the first days postpartum, mothers filled in a questionnaire pertaining to their pregnancy and birth, as well as maternal and infant anthropometric and demographic data.

2.2.2. One Month Postpartum

At 30 days postpartum (±5 days), mothers collected a portion of breastmilk (1–2 mL) prior to each feeding moment, over a 24-h period (i.e., five to eight times). Although only foremilk was collected through this method, previous research has shown that GC concentrations are similar in fore- and hindmilk [13]. Mothers could follow their own feeding schedule and were therefore asked to report the exact time of sampling. Milk was collected manually or with a breast pump; we requested that mothers used the same method for all samples. Milk was stored in the mother's freezer, and subsequently in the laboratory at −20 °C for less than 3 months prior to analysis.

At the time of sampling, maternal distress was quantified with the Hospital Anxiety and Depression Scale (HADS) [14]. This questionnaire contains 14 questions scored from 0 to 3, which assess self-reported levels of depression and anxiety symptoms. Seven questions concern depressive symptoms (HDS) and seven questions assess anxiety symptoms (HAS). A score of ≥8 on a subscale is indicative of clinically relevant depression and/or anxiety symptoms.

2.2.3. Three Months Postpartum

At 3 months postpartum (±2 weeks), the body composition of the infants was assessed with the Pea Pod, an air-displacement plethysmography (ADP) system (COSMED USA, Inc., Concord, CA, USA) [15]. It is based on a bi-compartmental model, which uses pressure and volume changes in the chamber through which body density was determined. Age- and sex-specific fat and fat free mass density values were subsequently used to calculate fat mass (FM) and fat free mass (FFM) [15].

As part of the national standard care, weight and length at 1, 2, and 3 months of age were measured by the staff of the child health clinic and were obtained through a questionnaire. Weight was measured fully undressed on a balance scale with an accuracy of 1 g. Length was measured in the supine position to the nearest 0.1 cm. Additionally, all mothers were asked if their infants were still breastfed for >80% at the age of 3 months.

2.3. Laboratory Analysis

Cortisol and cortisone concentrations in breastmilk were determined by isotope dilution liquid chromatography–tandem mass spectrometry (LC–MS/MS), as previously described [16]. In brief, internal standards (13C3-labeled cortisol and 13C3-labeled cortisone) were added to 200 µL of the samples. Then, breastmilk was washed 3 times with 2 mL of hexane to remove lipids. Finally, samples were extracted and analyzed using Isolute plates (Biotage, Uppsala, Sweden) and analyzed by LC-MS/MS (Acquity with Quattro Premier XE, Milford MA, USA, Waters Corporation). The intra-assay coefficients of variation (CV%) were 4 and 5% for cortisol levels of 7 and 23 nmol/L, and 5% for cortisone levels of 8 and 33 nmol/L for LC-MS/MS measurements. The inter-assay CV% was <9% for both cortisol and cortisone. The lower limit of quantitation (LLOQ) was 0.5 nmol/L for both cortisol and cortisone. All samples were measured in duplicate.

2.4. Statistics

First, data of GC concentrations in breastmilk were converted into the following rhythm parameters, in order to provide a full overview of GC rhythmicity:

- The maximum GC concentration, as a proxy for peak concentrations;
- The delta between maximum and minimum GC concentrations, as a measure of rhythm variability; and
- Area under the curve (AUC) ground (g) and increase (i), using the trapezoid rule [17].Calculations were corrected for total sampling time, since this differed between mothers. AUCg is a measure of total GC exposure, while AUCi provides information on GC variability.

Mother–infant pairs were excluded from analyses when no valid GC data was available around the time of the expected morning peak (5:00–10:00) and/or when the total sample collection was <8 h.

Fat% was determined from the FM and FFM values. The fat mass index (FMI) and fat free mass index (FFMI) were calculated by dividing FM and FFM values (in kg), respectively, by infant length squared (m^2), since fat mass and fat free mass are known to change with length [18]. Length and weight data were converted to SDS [19,20]. Body mass index (BMI) was only calculated at 3 months of age, and converted to SDS [19].

Linear regressions were used to assess the associations between GC rhythm parameters at 1 month of age and length SDS, weight SDS, BMI SDS, FMI, and FFMI at 3 months of age. First, unadjusted regression analyses were performed. Next, the following potential confounders were tested. Sex, HADS-score (HAS and/or HDS ≥ 8), pre-pregnancy BMI, ethnicity (Caucasian vs. non-Caucasian), socio-economic status, birth weight SDS, gestational age, weight gain during pregnancy, parity (1 vs. >1), mode of delivery (vaginal vs. caesarian section), and % breastmilk at 3 months of age (< or >80%). Due to our sample size, the three variables with the largest confounding effect (i.e., largest change in β of the independent variable) were used for the multiple linear regression analyses. Thus, weight gain during pregnancy, % breastmilk at 3 months of age, and ethnicity were included in the final model assessing the association between GC rhythm parameters and body composition outcomes. No effect modification was found for infant sex, and analyses were therefore not stratified.

Lastly, length and weight SDS growth trajectories between 1 and 3 months of age were plotted against AUCi and AUCg values by using generalized estimating equations (GEEs), and 95% confidence intervals were calculated according to the method described by Figueiras et al. [21]. AUCi and AUCg outcomes for cortisol and cortisone were categorized as ≤p25, p25–75, and ≥p75.3.

3. Results

3.1. Population

Forty-four mother–infant pairs were included in the study. One mother–infant pair was lost to follow-up, three mother–infant pairs returned the growth questionnaires but did not consent to the Pea Pod measurement, and one pair was excluded because no samples were collected between 5:00 and 10:00 and/or because the total sampling time was <8 h. Therefore, a total of 42 mother–infant pairs were included in the growth trajectory analyses, whereas 39 mother-infant pairs were included in the body composition analyses at 3 months of age. Of the included mother–infant pairs, 59.5% were mother–son pairs. Table 1 shows the characteristics of the population. Supplementary Table S1 shows the cortisol and cortisone concentrations in breastmilk in 4-h intervals.

Table 1. Characteristics of the study population ($n = 42$).

	Unit	Mean ± SD or n (%)
Gestational Age	Weeks	39.9 ± 1.3
Birth weight	grams	3561 ± 498
	SDS	0.2 ± 1.0
Birth length *	cm	52.0 ± 2.6
	SDS	1.0 ± 1.6
Male sex		25 (59.5)
Primiparity		23 (54.8)
Caesarian section		21 (51.2)
HAS and/or HDS ≥ 8 at 1-month pp		6 (14.6)
Pre-pregnancy maternal BMI	kg/m²	22.3 ± 2.8
Weight gain during pregnancy	kg	13.1 ± 3.2
Maternal age	years	36.0 ± 4.7
Non-Caucasian ethnicity		8 (20.0)
Socioeconomic status	SDS	0.6 ± 1.2
>80% breastfed at 3 months of age		35 (87.5)
Age at breastmilk sampling	days	30.8 ± 2.6
Age at Pea Pod measurement **	days	90.5 ± 7.0

Values represent Mean ± SD or n (%); pp = postpartum. HAS, Hospital Anxiety Score; HDS, Hospital Depression Score. * $n = 31$, ** $n = 39$.

3.2. Linear Regression Analyses

No associations were found between the GC rhythm parameters (AUCi, AUCg, maximum, and delta) and body composition in the unadjusted analyses. Adjusting the analyses for weight gain during pregnancy, % breastmilk at 3 months of age, and ethnicity did not change the results (Table 2).

Table 2. Adjusted associations between breastmilk glucocorticoid rhythmicity at 1 month of age and infant body composition at 3 months of age ($n = 39$).

		Length		Weight		BMI		FMI		FFMI		Fat %	
		B	95% CI	B	95% CI	B	95% CI	B	95% CI	B	95% CI	B	95% CI
Cortisol	Maximum	0.006	(−0.04 to 0.05)	0.022	(−0.02 to 0.07)	0.022	(−0.02 to 0.06)	0.003	(−0.04 to 0.05)	−0.006	(−0.04 to 0.03)	0.048	(−0.17 to 0.27)
	Delta	0.006	(−0.04 to 0.05)	0.024	(−0.02 to 0.07)	0.025	(−0.02 to 0.07)	0.003	(−0.04 to 0.05)	−0.004	(−0.04 to 0.03)	0.043	(−0.18 to 0.26)
	AUCi	0.025	(−0.15 to 0.20	0.101	(−0.07 to 0.28)	0.1	(−0.06 to 0.26)	0.06	(−0.10 to 0.23)	−0.095	(−0.23 to 0.04)	0.53	(−0.34 to 1.39)
	AUCg	0.029	(−0.13 to 0.19)	0.06	(−0.11 to 0.23)	0.046	(−0.11 to 0.20)	0.06	(−0.10 to 0.22)	−0.107	(−0.24 to 0.02)	0.53	(−0.28 to 1.33)
Cortison	Maximum	−0.002	(−0.04 to 0.03)	0.01	(−0.03 to 0.05)	0.014	(−0.02 to 0.05)	−0.006	(−0.04 to 0.03)	−0.006	(−0.04 to 0.02)	−0.007	(−0.19 to 0.18)
	Delta	−0.002	(−0.04 to 0.04)	0.018	(−0.02 to 0.06)	0.024	(−0.01 to 0.06)	−0.003	(−0.04 to 0.03)	−0.001	(−0.03 to 0.03)	0.005	(−0.19 to 0.20)
	AUCi	−0.005	(−0.09 to 0.08)	0.042	(−0.05 to 0.13)	0.055	(−0.02 to 0.14)	0.002	(−0.08 to 0.09)	−0.008	(−0.08 to 0.06)	0.034	(−0.41 to 0.48)
	AUCg	−0.001	(−0.07 to 0.07)	0.003	(−0.07 to 0.07)	0.004	(−0.06 to 0.07)	−0.013	(−0.08 to 0.05)	−0.02	(−0.08 to 0.04)	−0.02	(−0.37 to 0.33)

Values represent β (95% CI) as analyzed with linear regression. Analyses were adjusted for weight gain during pregnancy, % breastmilk at 3 months of age, and ethnicity. AUCi or g, area under the curve increase or ground; FMI, fat mass index; FFMI, fat free mass index.

3.3. Growth Trajectories

Figure 1 shows the growth trajectories for length and weight SDS according to breastmilk cortisone AUCi and AUCg outcomes. No differences were found between the categories ≤p25, p25–75, and ≥p75. Results for breastmilk cortisol AUCi and AUCg were similar (data not shown).

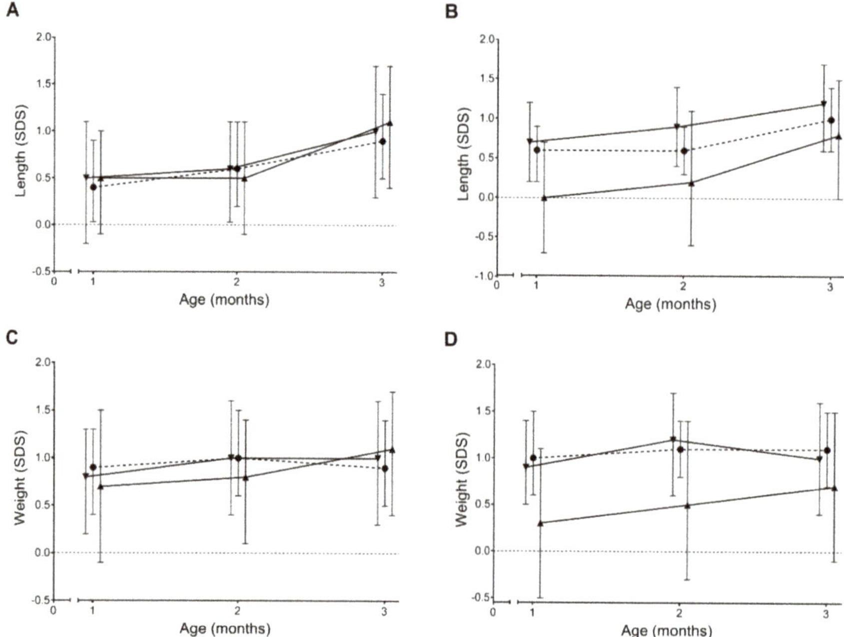

Figure 1. Growth trajectories between 1 and 3 months of age for length and weight, according to breastmilk cortisone AUC outcomes (*n* = 42). Results for breastmilk cortisol AUCi and AUCg were similar (data not shown). A = length for AUCi, B = weight for AUCg, C = weight for AUCi, D = weight for AUCg; ▼ = AUCi or g < p25, ● = AUCi or g = p25–75, ▲ = AUCi or g > p75, AUCi or g, area under the curve increase or ground.

4. Discussion

In this study, despite increased evidence for associations between blunted endogenous GC rhythms and obesity in both children and adults [10–12], no associations were found between GC rhythmicity in breastmilk sampled at 1 month and infant body composition or growth at 3 months. Therefore, our study could not confirm previous observations in animals and humans. Hinde et al. (2015) [7] measured cortisol in breastmilk of 108 rhesus macaques at 1 month of age, and analyzed growth outcomes at 3.5 months of age. They found that higher cortisol concentrations were associated with greater weight gain over time. Hahn-Holbrook et al. (2016) [8] studied associations between breast-milk cortisol and BMI gains up until the age of 2 years in 51 mother-infant pairs. They found that higher milk cortisol concentrations were associated with smaller BMI gains in offspring.

The different results between this study and previous studies could have several explanations. First, cortisol sampling in the previous studies did not take the diurnal rhythm of breastmilk GCs into account. Hinde et al. [7] sampled between 11:30 and 13:00, which did not capture the peak GC concentration, since in Rhesus macaques, similar to humans, this occurs at around 8:00. Hahn-Holbrook et al. [8] collected a single breastmilk sample within a wide time window (11:30–16:00), which also did not capture peak concentrations. Analyses were corrected for time of collection, but it has previously been shown that correcting for the time of sampling cannot account for all the variability observed in cortisol levels [9,22]. Second, in our study, GC concentrations were determined by LC-MS/MS, which has been shown to be more sensitive and reliable than radioimmunoassay and chemiluminescent immunoassay [23], which were used by Hinde et al. and Hahn-Holbrook et al., respectively. Lastly, it has previously been shown that increases in fat mass are specifically associated with mid-childhood overweight and obesity [24]. Therefore, in this study, body composition was measured by ADP, which is able to differentiate between fat mass and fat free mass. In contrast, weight gain and changes in BMI were used as outcomes measured by Hinde et al. [7] and Hahn-Holbrook et al. [8], respectively, both of which are less precise methods to determine body composition. Our more detailed methods when measuring GC concentrations in breastmilk as well as when determining body composition might therefore have led to more accurate conclusions.

Alternatively, the absence of associations might be due to the small sample size in this study, especially compared to Hinde et al. [7], who included 108 mother–infant pairs, resulting in more power to detect small differences. However, this should be balanced against the use of air-displacement plethysmography in this study, which is superior to weight gain for the assessment of body composition. Additionally, our follow-up until the age of 3 months was rather short. In contrast, follow-up took place up to 2 years of age in Hahn-Holbrook et al.'s study [8]. It is therefore possible that effects of GCs in breastmilk might only be noticeable at a later age. On the other hand, an increasing number of nutritional, lifestyle, and family factors determine body composition with advancing age, and it is therefore progressively more difficult to determine to what extent breastmilk cortisol explains BMI gains.

This study has several strengths and limitations. This was the first study to assess the association between GC rhythmicity in breastmilk and body composition in offspring. Body composition and GC rhythmicity were analyzed in detail, respectively, by the use of ADP and by measuring both cortisol and cortisone in breastmilk using samples that were collected over a 24-h period. Cortisone concentrations have been shown to be more reliable than cortisol measurements, at least in saliva and hair [25,26]. This is possibly due to the local conversion of cortisol by 11β-hydroxysteroid dehydrogenase type 2, which leads to higher concentrations of cortisone [27]. However, this study also has its limitations. The sample size of this study was relatively small, and it is therefore possible that modest effects could not be detected. It was also not possible to correct for all potential confounders. However, many confounders were considered, and the three variables with the largest confounding effect were included in the final model, which did not change the results compared to unadjusted analyses. It is therefore unlikely that adjusting for more variables would have altered the results. Second, the follow-up in this study was relatively short, and it is therefore possible that breastmilk GC rhythmicity has an effect that is only noticeable at a later age. Additionally, selection bias cannot be ruled out, since we did not collect data on mothers who were eligible for inclusion but decided against participation and since we included mother–infant pairs at a (tertiary) hospital. The study population might therefore not reflect the general population; for example, 51% of the mothers gave birth via Caesarian section, compared to approximately 17% in the general population [28]. Lastly, the interplay between GCs and infant body composition is complex, and could be moderated by, for example, exposure to GCs and other conditions in utero, the number of feeds per day, and the extrauterine environment of the infants, including synchrony in mother–infant interactions as well as stressful events. Unfortunately, we were not able to take these factors into account.

5. Conclusions

This study did not find an association between breastmilk GC rhythmicity at 1 month of age and growth trajectories as well as body composition of the offspring at 3 months of age. Therefore, this study suggests that previous observations linking breastmilk cortisol to changes in infant weight might be flawed by the lack of serial cortisol measurements and detailed information on body composition.

Supplementary Materials: The following are available online at http://www.mdpi.com/2072-6643/11/10/2351/s1, Table S1. Cortisol and cortisone concentrations in breastmilk in 4-h intervals.

Author Contributions: J.H. conceptualized the study, collected the data, performed the analyses and drafted and corrected the manuscript. L.R.D. conceptualized the study, collected the data, performed the analyses and drafted the manuscript. B.v.d.V. conceptualized the study, collected the data and reviewed the manuscript. O.M.P.K. helped with the study design and the body composition analyses, and reviewed the manuscript. A.A.T. conceptualized the study, collected the data and reviewed the manuscript. J.B.v.G. helped with the study design and reviewed the manuscript. J.R. conceptualized the study, helped with the data analyses and reviewed the manuscript. M.J.J.F. conceptualized the study, helped with the data analyses and reviewed the manuscript.

Funding: No funding was received for this research.

Acknowledgments: The study participants, for giving their time and energy to this study. D.L. for including mother-infant pairs and analyzing breastmilk and saliva samples. A.C.H., F.M., A.F. and other colleagues at the endocrinology laboratory for their contribution to the GC analyses.

Conflicts of Interest: The authors declared not conflict of interest.

References

1. Barker, D.J.; Osmond, C. Infant mortality, childhood nutrition, and ischaemic heart disease in England and Wales. *Lancet* **1986**, *1*, 1077–1081. [CrossRef]

2. Finken, M.J.; van der Voorn, B.; Heijboer, A.C.; de Waard, M.; van Goudoever, J.B.; Rotteveel, J. Glucocorticoid Programming in Very Preterm Birth. *Horm. Res. Paediatr.* **2016**, *85*, 221–231. [CrossRef] [PubMed]

3. Rosmond, R.; Bjorntorp, P. The hypothalamic-pituitary-adrenal axis activity as a predictor of cardiovascular disease, type 2 diabetes and stroke. *J. Intern. Med.* **2000**, *247*, 188–197. [CrossRef] [PubMed]

4. Drake, A.J.; Tang, J.I.; Nyirenda, M.J. Mechanisms underlying the role of glucocorticoids in the early life programming of adult disease. *Clin. Sci.* **2007**, *113*, 219–232. [CrossRef] [PubMed]

5. Entringer, S. Impact of stress and stress physiology during pregnancy on child metabolic function and obesity risk. *Curr. Opin. Clin. Nutr. Metab. Care* **2013**, *16*, 320–327. [CrossRef] [PubMed]

6. Angelucci, L.; Patacchioli, F.R.; Scaccianoce, S.; Di Sciullo, A.; Cardillo, A.; Maccari, S. A model for later-life effects of perinatal drug exposure: Maternal hormone mediation. *Neurobehav. Toxicol. Teratol.* **1985**, *7*, 511–517.

7. Hinde, K.; Skibiel, A.L.; Foster, A.B.; Rosso, L.D.; Mendoza, S.P.; Capitanio, J.P. Cortisol in mother's milk across lactation reflects maternal life history and predicts infant temperament. *Behav. Ecol.* **2015**, *26*, 269–281. [CrossRef]

8. Hahn-Holbrook, J.; Le, T.B.; Chung, A.; Davis, E.P.; Glynn, L.M. Cortisol in human milk predicts child BMI. *Obesity (Silver Spring)* **2016**, *24*, 2471–2474. [CrossRef]

9. Van der Voorn, B.; de Waard, M.; van Goudoever, J.B.; Rotteveel, J.; Heijboer, A.C.; Finken, M.J. Breast-Milk Cortisol and Cortisone Concentrations Follow the Diurnal Rhythm of Maternal Hypothalamus-Pituitary-Adrenal Axis Activity. *J. Nutr.* **2016**, *146*, 2174–2179. [CrossRef]

10. Adam, E.K.; Quinn, M.E.; Tavernier, R.; McQuillan, M.T.; Dahlke, K.A.; Gilbert, K.E. Diurnal cortisol slopes and mental and physical health outcomes: A systematic review and meta-analysis. *Psychoneuroendocrinology* **2017**, *83*, 25–41. [CrossRef]

11. Ruttle, P.L.; Javaras, K.N.; Klein, M.H.; Armstrong, J.M.; Burk, L.R.; Essex, M.J. Concurrent and longitudinal associations between diurnal cortisol and body mass index across adolescence. *J. Adolesc. Health* **2013**, *52*, 731–737. [CrossRef] [PubMed]

12. Wirix, A.J.; Finken, M.J.; von Rosenstiel-Jadoul, I.A.; Heijboer, A.C.; Nauta, J.; Groothoff, J.W.; Chinapaw, M.J.; Kist-van Holthe, J.E. Is There an Association Between Cortisol and Hypertension in Overweight or Obese Children? *J. Clin. Res. Pediatr. Endocrinol.* **2017**, *9*, 344–349. [CrossRef] [PubMed]

13. Patacchioli, F.; Cigliana, G. Maternal plasma and milk free cortisol during the first 3 days of breast-feeding following spontaneous delivery or elective cesarean section. *Gynecolog. Obstet. Invest.* **1992**, *34*, 159–163. [CrossRef] [PubMed]

14. Zigmond, A.S.; Snaith, R.P. The hospital anxiety and depression scale. *Acta Psychiatr. Scand.* **1983**, *67*, 361–370. [CrossRef] [PubMed]

15. Ma, G.; Yao, M.; Liu, Y.; Lin, A.; Zou, H.; Urlando, A.; Wong, W.W.; Nommsen-Rivers, L.; Dewey, K.G. Validation of a new pediatric air-displacement plethysmograph for assessing body composition in infants. *Am. J. Clin. Nutr.* **2004**, *79*, 653–660. [CrossRef] [PubMed]

16. Van der Voorn, B.; Martens, F.; Peppelman, N.S.; Rotteveel, J.; Blankenstein, M.A.; Finken, M.J.; Heijboer, A.C. Determination of cortisol and cortisone in human mother's milk. *Clin. Chim. Acta* **2015**, *444*, 154–155. [CrossRef] [PubMed]

17. Pruessner, J.C.; Kirschbaum, C.; Meinlschmid, G.; Hellhammer, D.H. Two formulas for computation of the area under the curve represent measures of total hormone concentration versus time-dependent change. *Psychoneuroendocrinology* **2003**, *28*, 916–931. [CrossRef]

18. Kyle, U.G.; Schutz, Y.; Dupertuis, Y.M.; Pichard, C. Body composition interpretation. Contributions of the fat-free mass index and the body fat mass index. *Nutrition* **2003**, *19*, 597–604. [CrossRef]

19. TNO. *De Vijfde Landelijke Groeistudie*; TNO: Hague, The Netherlands, 2010.

20. Schonbeck, Y.; Talma, H.; van Dommelen, P.; Bakker, B.; Buitendijk, S.E.; HiraSing, R.A.; van Buuren, S. The world's tallest nation has stopped growing taller: The height of Dutch children from 1955 to 2009. *Pediatr. Res.* **2013**, *73*, 371–377. [CrossRef]

21. Figueiras, A.; Domenech-Massons, J.M.; Cadarso, C. Regression models: Calculating the confidence interval of effects in the presence of interactions. *Stat. Med.* **1998**, *17*, 2099–2105. [CrossRef]

22. De Weerth, C.; Zijl, R.H.; Buitelaar, J.K. Development of cortisol circadian rhythm in infancy. *Early Hum. Dev.* **2003**, *73*, 39–52. [CrossRef]

23. Ackermans, M.T.; Endert, E. LC-MS/MS in endocrinology: What is the profit of the last 5 years? *Bioanalysis* **2014**, *6*, 43–57. [CrossRef] [PubMed]

24. Koontz, M.B.; Gunzler, D.D.; Presley, L.; Catalano, P.M. Longitudinal changes in infant body composition: Association with childhood obesity. *Pediatr. Obes.* **2014**, *9*, e141–e144. [CrossRef]

25. Blair, J.; Adaway, J.; Keevil, B.; Ross, R. Salivary cortisol and cortisone in the clinical setting. *Curr. Opin. Endocrinol. Diabetes Obes.* **2017**, *24*, 161–168. [CrossRef] [PubMed]

26. Savas, M.; Wester, V.L.; de Rijke, Y.B.; Rubinstein, G.; Zopp, S.; Dorst, K.; van den Berg, S.A.A.; Beuschlein, F.; Feelders, R.A.; Reincke, M.; et al. Hair glucocorticoids as biomarker for endogenous Cushing's syndrome: Validation in two independent cohorts. *Neuroendocrinology* **2019**, *109*, 171–178. [CrossRef]

27. Smith, R.E.; Maguire, J.A.; Stein-Oakley, A.N.; Sasano, H.; Takahashi, K.; Fukushima, K.; Krozowski, Z.S. Localization of 11 beta-hydroxysteroid dehydrogenase type II in human epithelial tissues. *J. Clin. Endocrinol. Metab.* **1996**, *81*, 3244–3248. [CrossRef]

28. Macfarlane, A.J.; Blondel, B.; Mohangoo, A.D.; Cuttini, M.; Nijhuis, J.; Novak, Z.; Olafsdottir, H.S.; Zeitlin, J.; Euro-Peristat Scientific Committee. Wide differences in mode of delivery within Europe: Risk-stratified analyses of aggregated routine data from the Euro-Peristat study. *BJOG* **2016**, *123*, 559–568. [CrossRef]

 *nutrients*

Article

Lactoferrin in Human Milk of Prolonged Lactation

Matylda Czosnykowska-Łukacka [1,*,†], **Magdalena Orczyk-Pawiłowicz** [2,*,†], **Barbara Broers** [1] and **Barbara Królak-Olejnik** [1]

[1] Neonatology Department, Wroclaw Medical University, Borowska 213, 50-556 Wroclaw, Poland; barsamut@gmail.com (B.B.); barbara.krolak-olejnik@umed.wroc.pl (B.K.-O.)

[2] Department of Chemistry and Immunochemistry, Wroclaw Medical University, M. Skłodowskiej-Curie 48/50, 50-369 Wrocław, Poland

[*] Correspondence: matylda.czosnykowska-lukacka@umed.wroc.pl (M.C.-Ł.); magdalena.orczyk-pawilowicz@umed.wroc.pl (M.O.-P.)

[†] These authors contributed equally to this work.

Received: 19 July 2019; Accepted: 23 September 2019; Published: 2 October 2019

Abstract: Among the immunologically important bioactive factors present in human milk, lactoferrin (Lf) has emerged as a key player with wide-ranging features that directly and indirectly protect the neonate against infection caused by a variety of pathogens. The concentration of Lf in human milk is lactation-stage related; colostrum contains more than 5 g/L, which then significantly decreases to 2–3 g/L in mature milk. The milk of mothers who are breastfeeding for more than one year is of a standard value, containing macronutrients in a composition similar to that of human milk at later stages. The aim of this study was to evaluate lactoferrin concentration in prolonged lactation from the first to the 48th month postpartum. Lactating women ($n = 120$) up to 48 months postpartum were recruited to the study. The mean value of lactoferrin concentration was the lowest in the group of 1–12 months of lactation (3.39 ± 1.43 g/L), significantly increasing in the 13–18 months group (5.55 ± 4.00 g/L; $p < 0.006$), and remaining at a comparable level in the groups of 19–24 month and over 24 months (5.02 ± 2.97 and 4.90 ± 3.18 g/L, respectively). The concentration of lactoferrin in mother's milk also showed a positive correlation with protein concentration over lactation from the first to the 48th month ($r = 0.3374$; $p = 0.0002$). Our results demonstrate the high immunology potential of human milk during prolonged lactation and that Lf concentration is close to the Lf concentration in colostrum. Evidence of stable or rising immunoprotein levels during prolonged lactation provides an argument for foregoing weaning; however, breastfeeding must be combined with solid foods meet the new requirements of a rapidly growing six-month or older baby.

Keywords: breastfeeding; lactoferrin; prolonged lactation; child nutrition

1. Introduction

Human milk (HM) serves as the gold standard in newborn and infant nutrition. Apart from the main nutrients, HM contains many bioactive proteins, growth factors, cells, and other constituents, which are crucial in the modulation of the development of a competent immune system to protect the term and preterm newborn against pathogens [1]. Among the immunologically important bioactive factors present in human milk, lactoferrin has emerged as a key player with wide-ranging features that directly and indirectly protect the neonate against infection caused by bacteria and other microorganisms.

Lactoferrin (or lactotransferrin, Lf) is an approximately 78 kDa glycoprotein from the transferrin family of proteins. Lf was first identified by Sørensen and Sørensen in 1939 in bovine milk as a red protein in whey [2], and later isolated in 1960 from both human and bovine milk [3–5]. Lf is expressed and secreted by epithelial cells in the main physiological secretions [6–8]. Lf plays a key role in specialized systems to bind iron, affecting cellular proliferation and differentiation, dependent on the degree of iron saturation [9,10]. Clinical trials have demonstrated roles for Lf in the prevention of diarrhea [11,12], neonatal sepsis, and necrotizing enterocolitis in preterm infants [13–15]. Lf may significantly impact development as well as the well-being of infants and later outcomes, which are directly related to breastfeeding. Maternal factors may affect the concentration of Lf in human milk, but affect infant factors as well. Lf concentrations are consistently highest in colostrum, then decrease gradually. However, the specific factors that are related to the Lf concentration in breast milk are unknown. Factors such as geographical location, ethnic origin, genetic polymorphisms, socio-economic status, nutritional status, as well as infants infections, type of delivery, and prematurity have been considered [16–18].

In human milk, Lf is the most abundant protein in the whey fraction [7]. The concentration of Lf in human milk is lactation-stage related: colostrum contains more than 5 g/L, which then significantly decreases to 2–3 g/L in mature milk. Multiple studies have evaluated Lf concentrations in human colostrum and mature milk and in term and preterm milk [7,19,20].

Lf elicits antibacterial effects against bacteria; however, Lf has protective activity due to the high affinity to iron. The iron-free form of Lf (apo-Lf) is a cause of iron deficiency in microorganisms as iron is required for life and growth rate [21]. Research has shown that the direct interaction between Lf and bacteria results in a bactericidal effect [22]. Lf associates with the lipoprotein of bacterial cells and forms receptor complexes. The binding of Lf inhibits the iron intake of bacteria or removes the protective effect of the membrane. Lf causes the release of lipopolysaccharides from the cell's wall, increases permeability of the membrane, and finally kills the Gram-negative bacteria. Lf binds the anionic molecules (e.g., lipoteichoic acid) on the cell surface of Gram-positive bacteria with greater effectiveness than lysozymes and antibiotics [23]. The antiviral effect of Lf is observed in the early phase of infection by inhibition of the growth of viruses. Lf and Lf-derived peptides can effectively act on a broad spectrum of fungal species, such as *Candida albicans, Candida krusei,* and *Aspergillus fumigatus* [24] by increasing membrane permeability, which leads to their death and additionally produces iron deprivation effects. Lf has a positively charged surface, which produces an anti-inflammatory effect. Lf interacts with proteoglycans on the surface of immune cells. This association can trigger signaling pathways that lead to physiological anti-inflammatory responses [25,26].

With reference to the research on the evolution of mammals, the protective function was the first in the ancestral mammary gland, which subsequently evolved to nourish the offspring. The protective components of milk remain highly conserved [27,28].

Breast milk after one year of lactation has nutritional value for children [29]. According to our previous results, the milk of mothers who breastfeed their children over one year or even over two years is standard value milk, and its macronutrient quantity composition is comparable mother's milk up to the first year of lactation [29]. However, further detailed quantitative data on the concentration and variability of the most important bioactive and immunomodulatory human milk proteins are necessary. The aim of this study was to evaluate lactoferrin concentration, the main glycoprotein with anti-pathogen activity, in breast milk during prolonged lactation from 1–48 months of lactation and to identify if any correlation exists between lactoferrin and protein concentrations during a healthy mother's lactation period. This is particularly important because Lf is an element providing innate immunity that is transferred to the breastfed infants and is crucial in the shaping and development of its immature immunological system.

2. Material and Methods

Mothers during lactation were recruited to the study from February 2017 to June 2018 using local groups for breastfeeding women on Facebook. We enrolled 120 participants in the study. The mother's age, socioeconomic status, race, health status, concomitant medications, parity, the mode of delivery, and frequency of breastfeeding were recorded. Milk samples were collected at the Regional Human Milk Bank and University Hospital (Wroclaw, Poland) between 08:00 and 14:00. Providing an interval of a few hours provides greater uniformity of samples. Considering efficiency of milk expression, an electric breast pump was used (Medela Symphony, Baar, Switzerland). Each mother received a sterile set and cup for milk collection. Samples for analyses ware taken immediately after complete emptying of the breast (two aliquots of 2–3 mL each). One session of milk expression should not deprive the infant of their nutritionally required volume. Milk samples were divided into four groups according to months postpartum: the first group was up to 12 months ($n = 24$), second group from 13 to18 months ($n = 33$), third group from 19 to 24 months ($n = 37$) and last group beyond 24 months ($n = 26$). The milk samples after collection were immediately cooled and frozen in −20 °C. This study received ethical approval Nr KB–65/2018 from the University Ethics Committee. Informed and written consent was provided by all participants before sample collection.

2.1. Analysis of the Samples

Each sample was initially heated at 40 °C as recommended by the human milk analyzer producer and homogenized using a sonicator (Sonicator®, MIRIS, Uppsala, Sweden) at 1.5 s/mL, to separate the lipid phase and avoid protein aggregation. Each aliquot was homogenized differently immediately prior to measurement. Analyses were performed in triplicate. Breast milk macronutrient concentration was measured using a human milk analyzer (HMA) (MIRIS, Uppsala, Sweden) calibrated previously with human milk standards. The HMA is based on mid-infrared spectroscopy, enabling the assessment of fat, protein, lactose, energy, and total solids content. Protein is the protein content based on the total amount of nitrogen (N) in a sample. This means that non-protein nitrogen (NPN) compounds are also included in this value. "True protein" is corrected for this and represents only the content of actual protein, hence the denotation "true". The Miris HMA (Uppsala, Sweden) measures both crude protein and true protein to avoid misunderstandings. The obtained results are expressed in g/L. A detailed description of the measurement method was described in previous study [29].

2.2. Skim Milk Sample Preparation

All collected samples of mother's milk were immediately transferred to the dedicated plastic tubes WHAT for freezing and stored at −20 °C until analysis. To obtain the aqueous phase of milk (skim milk), all milk samples were centrifuged at 3500× *g* at 4 °C for 35 min. Next, the milk fat layer and cells were removed [30]. The skim milk samples were aliquoted and stored at −20 °C. All milk samples were handled by the same person. Before the determinations of lactoferrin concentration, the skim milk samples were thawed at room temperature for one hour.

2.3. Analysis of Lactoferrin in Skim Milk

The concentration of lactoferrin in skim milk samples was determined according to a modified procedure reported previously [31]. For testing, 100 µL of 5×10^3, 10×10^3, 25×10^3, and 50×10^3 fold diluted skim milk samples and parallel standard of lactoferrin derived from human milk in concentrations ranging from 0.8 to 50 ng per 100 µL (Sigma Aldrich, St. Louis, MO, USA) in Tris-HCl buffer with sodium chloride (pH = 7.5) were added to the individual microtiter plate wells (Nunc International, Naperville, IL, USA) and incubated for 2 h at 37 °C. The concentration of lactoferrin was analyzed using rabbit anti-human lactoferrin antibodies phosphatase-labeled (Jackson ImmunoResearch Europe Ltd., Ely, UK) in TRIS-HCl buffer (pH = 7.5) with 0.2% Tween 20 for 1 h at 37 °C and then detected by the reaction with phosphatase's substrate, 4-nitrophenyl phosphate, disodium salt (SERVA, Heidelberg, Germany). The absorbance was measured after stopping the enzymatic reaction in a Stat Fax 2100 for Microplate Reader (Awareness Technology Inc., Palm City, FL, USA) using a 405 nm filter and 630 nm as the reference filter. The background absorbance for the ELISA test was low (below 0.04 AU) when Tris-buffered saline was added to the microtiter plate well instead of milk or lactoferrin standard samples. As the negative control, a human albumin (Sigma, St. Louis, MO, USA) was used. All skim milk samples were assayed at four different skim milk sample dilutions in duplicate. The intra-assay and inter-assay coefficients of variation were 4.8% and 5.6%, respectively.

2.4. Statistical Analysis

The statistical analysis was performed using the TIBCO STATISTICA 13.3 software package (StatSoft, Inc., Tulsa, OK, USA). The chi-square test was used for comparison the study population data. Due to the higher interindividual differences reported for many factors of human milk as well as the unequal sample size in the analyzed groups, nonparametric tests were used for analysis. Comparisons between analyzed groups were performed using the Kruskal–Wallis test. The obtained data are presented as the mean ± SD and the median with 25th–75th percentiles. To determine whether any relationships existed among the concentration of lactoferrin and analyzed factors, the correlations were evaluated according to Spearman's coefficient. A p-value lower than 0.05 was considered significant. The charts were prepared using linear regression with a 95% confidence level.

3. Results

The key characteristics of the cohort are detailed in Table 1. All infants were exclusively breastfed up to six months.

Table 1. Characteristics of the study population.

Outcome and Exposure Variables Breastfeeding	Breastfeeding ≤12 Months n = 24% (n/N)	Breastfeeding >12 Months n = 96% (n/N)	χ^2	p-Value	
Maternal age % (n/N)					
25–29	50 (12/24)	33 (31/94)			
30–34	41.7 (10/24)	45 (41/94)	2.672	0.139	NS
35+	8.3 (2/24)	22 (22/94)			
Race/ethnicity					
Caucasian	100 (24/24)	100 (96/96)	-	-	-
Socioeconomic status and education					
Secondary education	12.5 (3/24)	20 (16/96)	0.25	0.617	NS
High education	87.5 (21/24)	83.3 (80/96)			
Parity					
1	79 (19/24)	67.7 (65/96)			
2	21 (5/24)	33.3 (32/96)	1.628	0.22	NS
3	0 (0/24)	2 (2/96)			
Birth weight					
Appropriate for Gestational Age (AGA)	100 (24/24)	100 (96/96)	-	-	-
gestational age					
34–37	4.2(1/24)	1. (1/96)	1.144	0.285	NS
>37	95.8(23/24)	99 (95/96)			
Medicines during lactation					
No medications	71(17/24)	66 (63/96)			
Thyroxine	21(5/24)	29 (28/96)	0.884	0.643	NS
Others	8(2/24)	5 (5/96)			
Maternal diet during lactation					
Vegan/vegetarian	0 (0/24)	8 (8/96)			
Dairy-free diet	21 (5/24)	7 (7/96)	-	-	-
Gluten-free diet	0 (0/24)	3 (3/96)			
Complementary foods introduction					
above 6 months of life	NA	94 (90/96)	-	-	-

χ^2 test at $p < 0.05$; NA—Not assessed, NS—Not significant. -: not calculated

3.1. Lactoferrin

To determine if the concentration of lactoferrin in mother's milk is connected with the month of prolonged lactation, the correlation was calculated; however, the obtained result showed no correlation ($r = 0.1675$; $p = 0.0674$) with prolonged lactation from the first to the 48th month, as shown in Figure 1A.

The mean value of lactoferrin concentration was the lowest value in the group of 12 months of lactation (3.39 ± 1.43 g/L), significantly increasing in the 13–18 months group (5.55 ± 4.00 g/L; $p < 0.006$)

and remaining stable thereafter in the groups of 19–24 months and over 24 months (5.02 ± 2.97 and 4.90 ± 3.18 g/L, respectively) (Table 2).

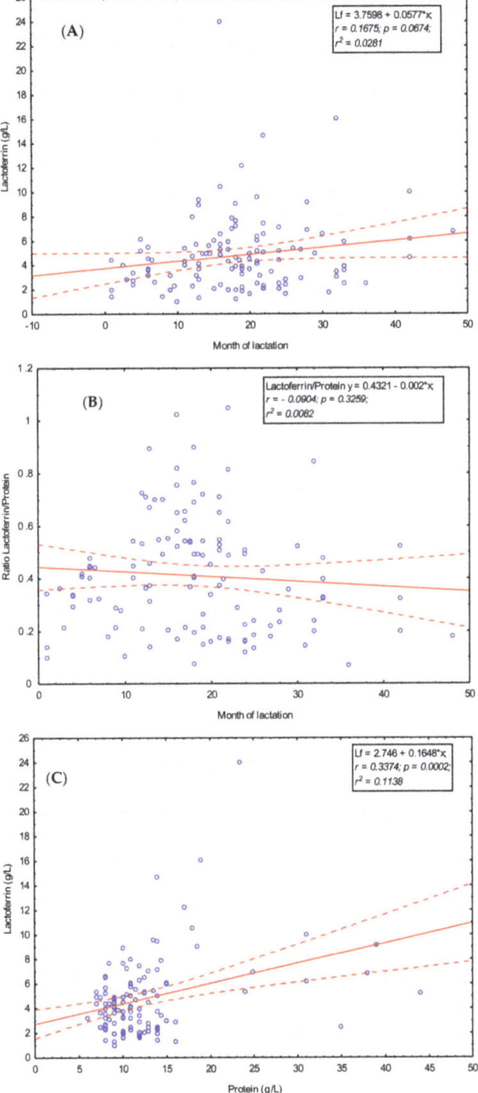

Figure 1. The correlation of (**A**) lactoferrin and (**B**) the ratio of lactoferrin to protein concentration with lactation progression from the 1st to the 48th month and (**C**) the correlation of lactoferrin and protein concentrations for all analyzed samples. A solid line indicates linear regression, and 95% confidence intervals are shown by dotted lines; blue hollow—individual samples; solid circle—2 or more individual samples with the same or very close values. *: multiplied

Table 2. Lactoferrin and protein concentrations in breast milk during prolonged lactation.

Breast Milk, Lactoferrin/Protein Content	Lactation			
	1–12 Months $n = 24$	13–18 Months $n = 33$	19–24 Months $n = 37$	>24 Months $n = 26$
Lactoferrin (g/L)	3.39 ± 1.43 3.24 2.30–4.47	5.55 ± 4.00 * 4.92 3.69–6.08 $p < 0.006$	5.02 ± 2.97 4.42 3.16–6.53	4.90 ± 3.18 4.29 2.49–6.18
Protein (g/L)	10.5 ± 2.3 10.0 8.5–12.8	10.4 ± 3.4 9.0 8.0–11.0	11.2 ± 2.7 ** 11.0 9.5–12.0 $p < 0.05$	19.1 ± 10.7 *** 14.3 12.0–25.0 $p < 0.0001$
Lactoferrin/Protein ratio	0.32 ± 0.12 0.33 0.25–0.41	0.49 ± 0.21 * 0.54 0.36–0.65 $p < 0.002$	0.43 ± 0.21 0.45 0.22–0.54	0.29 ± 0.17 *** 0.24 0.18–0.36 $p < 0.009$

Values are given as the mean ± SD, median and 25th–75th percentiles. The Mann–Whitney U-test was used for statistical calculations, and a p-value lower than 0.05 was considered significant. Significantly different from the milk group of: * 1–12 months of lactation, ** 13–18 months of lactation, *** 19–24 months of lactation.

3.2. Total Protein

The concentration of protein in mother's milk was positively correlated with prolonged lactation during the analyzed period up to 48 months; however, the correlation coefficient value was weak ($r = 0.25$; $p < 0.05$).

The mean values of protein concentrations were comparable in the groups of 1–12 and 13–18 months of lactation (10.5 ± 2.3 and 10.4 ± 3.4 g/L, respectively). In subsequent analyzed lactation periods, i.e., 19–24 and over 24 months of lactation, protein concentrations significantly increased and reached 11.2 ± 2.7 g/L ($p < 0.05$) and 19.1 ± 10.7 g/L ($p < 0.0001$), respectively (Table 2).

3.3. Lactoferrin/Protein Ratio

To identify the existence of any relationship between the concentration of lactoferrin and the concentration of total protein, the correlation coefficient was calculated. Mother's milk lactoferrin was positively correlated with protein concentration ($r = 0.3374$; $p = 0.0002$), as shown in Figure 1B, for the analyzed set of samples derived for 1–48 months of lactation. However, for both lactoferrin and protein concentrations, high interindividual differences among mothers were observed and this result may be characteristic for our particular data set only.

In contrast, the calculated coefficient of lactoferrin to the protein concentration (Lf/Protein) showed no correlation with lactation progression for 1–48 months ($r = -0.0904$; $p = 0.3259$) as shown in Figure 1C. The calculated coefficient of Lf/Protein in the group of 1-12 months of lactation reached 0.32 ± 0.12, significantly increasing in the 13–18-month group (0.49 ± 0.21, $p < 0.002$) and remaining at a comparable level in the 19–24-month group (0.43 ± 0.21), decreasing in the >24 months group to 0.29 ± 0.17 ($p < 0.009$; Table 2, Figure 2).

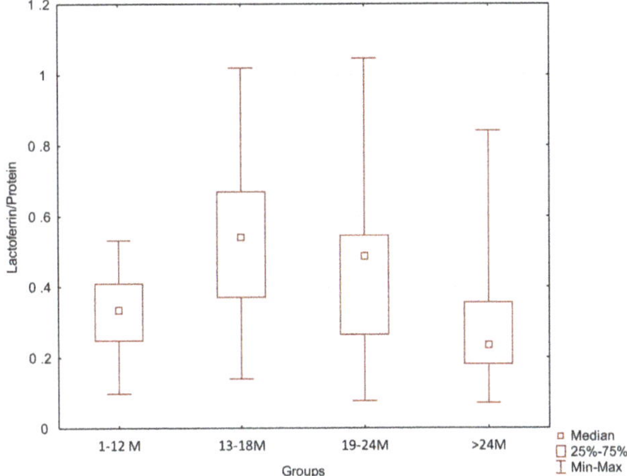

Figure 2. The coefficient of lactoferrin to the protein concentration in the four groups analyzed: Group1 up to 12 months ($n = 24$), group 2 from 13 to 18 months ($n = 33$), group 3 from 19 to 24 months ($n = 37$), and group 2 above 24 months ($n = 26$).

3.4. Lactoferrin Concentration in Relation to Macronutrients

The dependencies between lactoferrin concentration and macronutrients in mother's milk were analyzed throughout the whole lactation period, 1–12, 13–18, 19–24, and >24 months of lactation. The values of the calculated coefficient, which are statistically significant ($p < 0.05$), are given in Table 3. In contrast with the lack of correlation between lactoferrin concentration and month of lactation for 1–48 months of lactation, a positive correlation was found for prolonged lactation, specifically above 24 months. The concentration of lactoferrin over 24 months is negatively correlated negatively with the concentration of carbohydrates ($r = -0.50$) and positively with the concentration of fat ($r = 0.58$), protein ($r = 0.56$), true protein ($r = 0.58$), dry mass ($r = 0.65$), and energy ($r = 0.65$). A statistically significant negative correlation was found between lactoferrin concentration and carbohydrates ($r = -0.32$) and positive correlations between lactoferrin and fat ($r = 0.19$), protein ($r = 0.25$), true protein ($r = 0.24$), dry mass ($r = 0.18$), and energy ($r = 0.19$) in the lactation from month 1 to 48. However, the correlations observed were much stronger in prolonged lactation above the month 24. In contrast, no correlations were found in the other analyzed periods of lactation, namely up to month 12, months 13–18, and 19–24.

Table 3. Correlations between lactoferrin and macronutrients over prolonged lactation from month 1 to 48 and for particular periods.

	Correlation Coefficient (r Value *)				
	Lactoferrin				
Duration of lactation	1–48 months	1–12 months	13–18 months	19–24 months	Over 24 months
	NS	NS	NS	NS	0.39
Carbohydrate	−0.32	NS	NS	NS	−0.50
Fat	0.19	NS	NS	NS	0.58
Protein	0.25	NS	NS	NS	0.56
True protein	0.24	NS	NS	NS	0.58
Dry mass	0.18	NS	NS	NS	0.65
Energy	0.19	NS	NS	NS	0.65

The values of r were calculated according to Spearman's method corresponding to the correlation between the concentrations of lactoferrin and macronutrients over prolonged lactation from month 1 to 48. * all r values are statistically significant with $p < 0.05$. NS—Not significant.

3.5. Lactoferrin Concentration in Breast Milk in Relation to the Number of Feedings

The relationship between lactoferrin concentration of mother's milk and the number of feedings during the analyzed period from month 13 to 48 and the three stages of prolonged lactation, namely months 13–18, 19–24, and >24 of lactation were analyzed. We found no significant correlations for months 13–48 (Figure 3A) as well as for months 13–18 and 19–24; however, for lactation over 24 months, we observed a weak negative tendency between the number of feedings and lactoferrin concentration ($r = -0.26$; $p = 0.2868$; Figure 3B). Additionally, we found no correlation between lactoferrin concentration and maternal age, parity, or time of delivery-gestational age.

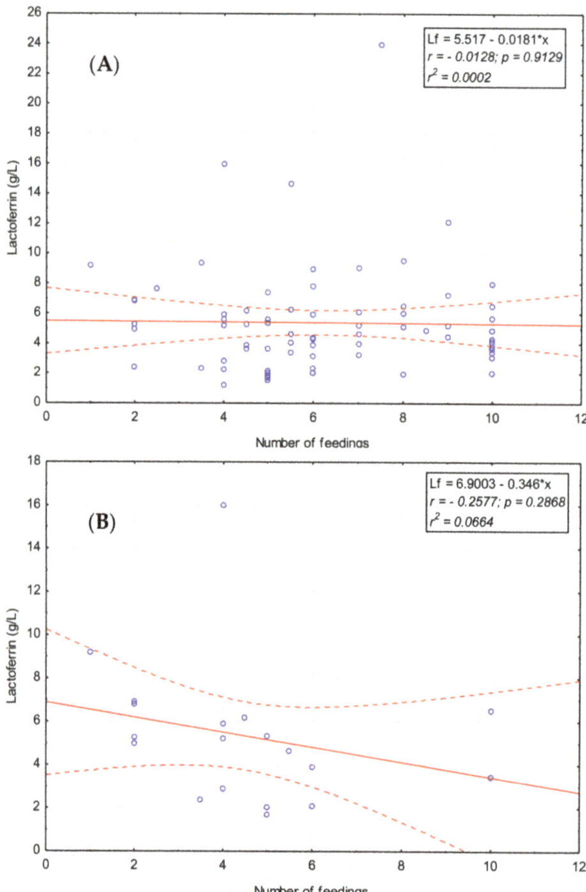

Figure 3. Relationship between the concentration of lactoferrin and the number of feedings mother's milk (**A**) during 13–48 months of lactation, (**B**) over 24 months of lactation. A solid line indicates linear regression, and 95% confidence intervals are shown by dotted lines; blue hollow—individual samples. *: multiplied.

4. Discussion

HM is a dynamic fluid that changes significantly in its composition from the onset of lactation to prolonged lactation above one year postpartum [29,32]. Total milk protein concentration decreases during the lactation period, but this trend is not general for all proteins. In our previous work, we demonstrated that macronutrients concentrations change in prolonged lactation. As a continuation of the previous study, lactoferrin changes in prolonged lactation were assessed in this study. During the first month of lactation, Lf decreases and then remains relatively stable [33]. The highest content of Lf is in colostrum at 5.5 g/L, whereas mature milk contains only 1.5–3.0 g/L, dependent on the stage of lactation in the first year postpartum [32,34,35]. Affolter et al. [36] reported that the concentration of lactoferrin decreases over the lactation period from the fifth day to eight months from 3.30 to 1.17 g/L. However, this decrease was constant during lactation until one to two months (1.24 g/L), and then stabilized until the eighth month. Changes in Lf content may reflect various biological functions of milk during different stages of newborn and infant development [33].

Our results show that the Lf concentration during prolonged lactation ranges from 4.9 to 5.02 g/L. Previous studies have shown that Lf concentration is negatively correlated with the volume of milk expression as well as the stage of lactation and parity [33,37]. The results showed the highest Lf content was recorded between 12 and 24 months of lactation. Above 24 months, concentration decreases, although not significantly, to 4.9 g/L. To the best of our knowledge, this is the first study assessing Lf concentration in human milk during prolonged lactation. These data have shown that Lf content above 12 months of lactation is higher and close to the Lf concentration in colostrum.

The concentration of lactoferrin in mother's milk showed a positive correlation with protein concentration over lactation from the early stage until four years postpartum. The coefficient of Lf concentration to protein did not show a correlation with lactation progression. Lf concentration and milk compositions are influenced by various factors such as health conditions and/or biochemical indicators of blood [33–36].

A notable inter-individual variability of lactoferrin content has also been observed, suggesting a lack of tight regulation in the synthesis and/or blood-to-mammary gland transfer of this protein [34]. Thereby, milk enhances the survival of offspring by promoting immunological competence. The rising concentrations of bioactive proteins, which occur as lactation progresses, are likely a physiological response to the diverse environmental demands of the human baby, which progressively acquires nutritional independence [38].

Raw mother's milk contains a relatively high concentration of Lf, which leads to more effective inhibition of growth of pathogenic bacteria in comparison to pasteurized mother's milk. As shown by Woodman et al. [38], human Lf is much more effective than bovine Lf in the inhibition of bacterial growth, although higher doses of bovine Lf also showed activity against *Bifidobacterium breve* and *Staphylococcus epidermidis*. Given the above, we suggest that human milk from prolonged lactation, which contains a comparable level of Lf, might be considered an alternative source of milk/lactoferrin provided in the feeding of preterm infants, although further detailed analysis is needed. Different methods used to preserve human milk bioactivity significantly impact Lf concentration [31]. Only further studies involving detailed tests of lactoferrin concentration in milk samples from prolonged lactation before and after pasteurization will help answer this question.

The strength of our study is that it is a unique and well-characterized collection of milk samples of mothers with prolonged lactation of up to 48 months, which impacts the credibility of the results obtained. To determine Lf concentration, a specific immunological method, ELISA, was used to prevent obtaining of false positive results. A limitation of our study is the homogeneous test group, which may not enable drawing a causal relationship between the factors that can influence milk Lf concentration, since, as reported by Yang et al. [33], milk Lf concentrations varies amongst different geographical regions in the Chinese population.

Evidence of stable or increasing immunoprotein levels during prolonged lactation provides the additional argument for allowing non-weaning; however, breastfeeding must be combined with solid foods to meet all the new requirements of a rapidly growing six-month or older baby. A number of clinical trials have been conducted to determine the influence of Lf on the protection against neonatal infections [39,40]. Based on published studies of Lf supplementation in preventing infection, determining the Lf concentration during lactation is important not only in the first six months. If the concentration changes during the lactation phase, colostrum has the highest Lf concentration, and decreases significantly during days after delivery. We now know that the milk of long-nursing mothers has a similar concentration to that in colostrum. Physicians should consider this to encourage exclusive breastfeeding, especially in risk groups such as premature babies [41].

To the best of our knowledge, based on the Internet databases such as Web of Science and pubmed, this is the first large-scale study performed on lactoferrin concentration in breast milk during prolonged lactation and the influence of various factors on its concentration. Coincident with the lack of public acceptance of breastfeeding beyond the first or even the second year, these findings have been ignored for decades.

5. Conclusions

Results show that the Lf concentration during prolonged lactation ranges from 4.9 to 5.02 g/L. The highest Lf content was recorded between 12 and 24 months of lactation. Above 24 months, concentration decreases, although not significantly. These data have shown that Lf content above 12 months of lactation is close to the Lf concentration in colostrum.

Author Contributions: M.C.-Ł. and M.O.-P.: These authors contributed equally to this work. Conceptualization, M.C.-Ł., M.O.P. and B.K.-O.; Methodology, M.O.-P. and M.C.-Ł.; Formal Analysis, M.O.-P. and M.C.-Ł.; Writing-Original Draft Preparation, M.C.-Ł. B.K.-O., M.O.-P. and. B.B.; Writing—Review & Editing, B.K.-O., M.O.-P.; Visualization, M.O.-P.; Supervision, B.K.-O.

Funding: This research received no external funding.

Acknowledgments: We would like to thank Zuzanna Sycz and the mothers who came to the Regional Human Milk Bank in University Hospital in Wrocław and expressed breast milk to the study. We would like to also thank Robin and Barbara Royle for their help in the language correction of the text.

Conflicts of Interest: The authors declare no conflict of interest.

References

1. Ballard, O.; Morrow, A.L. Human milk composition: Nutrients and bioactive factors. *Pediatr. Clin. N. Am.* **2013**, *60*, 49–74. [CrossRef] [PubMed]
2. Sorensen, M.; Sorensen, S.P.L. The proteins in whey. *Compte Rendu Trav. Lab. Carlsberg* **1939**, *23*, 55–99.
3. Groves, M.L. The isolation of a red protein from milk. *J. Am. Chem. Soc.* **1960**, *83*, 3345–3350. [CrossRef]
4. Johansson, B. Isolation of an iron-containing red protein from human milk. *Acta Chem. Scand.* **1960**, *14*, 510–512. [CrossRef]
5. Montreuil, J.; Tonnelat, J.; Mullet, S. Preparation and properties of lactosiderophilin (lactotransferrin) of human milk. *Biochim. Biophys. Acta* **1960**, *45*, 413–421. [CrossRef]
6. Rosa, L.; Cutone, A.; Lepanto, M.S.; Paesano, R.; Valenti, P. Lactoferrin: A Natural Glycoprotein Involved in Iron and Inflammatory Homeostasis. *Int. J. Mol. Sci.* **2017**, *18*, 1985. [CrossRef] [PubMed]
7. Liao, Y.; Alvarado, R.; Phinney, B.; Lönnerdal, B. Proteomic characterization of human milk whey proteins during a twelve-month lactation period. *J. Proteome Res.* **2011**, *10*, 1746–1754. [CrossRef]
8. Lönnerdal, B.; Iyer, S. Lactoferrin: Molecular structure and biological function. *Annu. Rev. Nutr.* **1995**, *15*, 93–110. [CrossRef]
9. Vogel, H.J. Lactoferrin, a bird's eye view. *Biochem. Cell Biol.* **2012**, *90*, 233–244. [CrossRef]
10. Lönnerdal, B. Nutritional roles of lactoferrin. *Curr. Opin. Clin. Nutr. Metab. Care* **2009**, *12*, 293–297. [CrossRef]
11. Zavaleta, N.; Figueroa, D.; Rivera, J.; Julia, S.; Segundo, A.; Bo, L. Efficacy of rice-based oral rehydration solution containing recombinant human lactoferrin and lysozyme in Peruvian children with acute diarrhea. *J. Pediatr. Gastroenterol. Nutr.* **2007**, *44*, 258–264. [CrossRef] [PubMed]
12. Ochoa, T.J.; Chea-Woo, E.; Baiocchi, N.; Pecho, I.; Campos, M.; Prada, A.; Valdiviezo, G.; Lluque, A.; Lai, D.; Cleary, T.G. Randomized double-blind controlled trial of bovine lactoferrin for prevention of diarrhea in children. *J. Pediatr.* **2013**, *162*, 349–356. [CrossRef] [PubMed]
13. Manzoni, P.; Rinaldi, M.; Cattani, S.; Pugni, L.; Romeo, M.G.; Messner, H.; Stolfi, I.; Decembrino, L.; Laforgia, N.; Vagnarelli, F.; et al. Bovine lactoferrin supplementation for prevention of late-onset sepsis in very low-birth-weight neonates: A randomized trial. *JAMA* **2009**, *302*, 1421–1428. [CrossRef] [PubMed]

14. Manzoni, P.; Garcia Sanchez, R.; Meyer, M.; Stolfi, I.; Pugni, L.; Messner, H.; Cattani, S.; Betta, P.M.; Memo, L.; Decembrino, L. Exposure to gastric acid inhibitors increases the risk of infection in preterm very low birth weight infants but concomitant administration of lactoferrin counteracts this effect. *J. Pediatr.* **2018**, *193*, 62–67. [CrossRef] [PubMed]

15. Pammi, M.; Suresh, G. Enteral lactoferrin supplementation for prevention of sepsis and necrotizing enterocolitis in preterm infants. *Cochrane Database Syst. Rev.* **2017**, *6*, CD007137. [CrossRef] [PubMed]

16. Mastromarino, P.; Capobianco, D.; Campagna, G.; Laforgia, N.; Drimaco, P.; Dileone, A.; Baldassarre, M.E. Correlation between lactoferrin and beneficial microbiota in breast milk and infant's feces. *Biometals* **2014**, *27*, 1077–1086. [CrossRef] [PubMed]

17. Broadhurst, M.; Beddis, K.; Black, J.; Henderson, H.; Nair, A.; Wheeler, T. Effect of gestation length on the levels of five innate defence proteins in human milk. *Early Hum. Dev.* **2015**, *91*, 7–11. [CrossRef]

18. Villavicencio, A.; Rueda, M.S.; Turin, C.G.; Ochoa, T.J. Factors affecting lactoferrin concentration in human milk: How much do we know? *Biochem. Cell Biol.* **2016**, *95*, 12–21. [CrossRef]

19. Ronayne de Ferrer, P.A.; Baroni, A.; Sambucetti, M.E.; López, N.E.; Cernadas, J.M.C. Lactoferrin levels in term and preterm milk. *J. Am. Coll. Nutr.* **2000**, *19*, 370–373. [CrossRef]

20. Lönnerdal, B. Law infant nutrition: Bioactive proteins of human milk and implications for composition of infant formulas. *Am. J. Clin. Nutr.* **2014**, *99*, 712S–717S. [CrossRef]

21. Law, B.A.; Reiter, B. The isolation and bacteriostatic properties of lactoferrin from bovine milk whey. *J. Dairy Res.* **1977**, *44*, 595–599. [CrossRef] [PubMed]

22. Ostan, N.K.; Yu, R.H.; Ng, D.; Lai, C.C.-L.; Pogoutse, A.K.; Sarpe, V.; Hepburn, M.; Sheff, J.; Raval, S.; Schriemer, D.C.; et al. Lactoferrin binding protein B–a bi-functional bacterial receptor protein. *PLoS Pathog.* **2017**, *13*, e1006244. [CrossRef] [PubMed]

23. Barbiroli, A.; Bonomi, F.; Capretti, G.I.; Iametti, S.; Manzoni, M.; Piergiovanni, L.; Rollini, M. Antimicrobial activity of lysozyme and lactoferrin incorporated in cellulose-based food packaging. *Food Control* **2012**, *26*, 387–392. [CrossRef]

24. Zarember, K.A.; Sugui, J.A.; Chang, Y.C.; Kwon-Chung, K.J.; Gallin, J.I. Human polymorphonuclear leukocytes inhibit Aspergillus fumigatus conidial growth by lactoferrin-mediated iron depletion. *J. Immunol.* **2007**, *178*, 6367–6373. [CrossRef]

25. Legrand, D. Overview of Lactoferrin as a Natural Immune Modulator. *J. Pediatr.* **2016**, *173*, S10–S15. [CrossRef] [PubMed]

26. Wang, B.; Timilsena, Y.P.; Blanch, E.; Adhikari, B. Lactoferrin: Structure, Function, Denaturation and Digestion. *Crit. Rev. Food Sci. Nutr.* **2019**, *59*, 580–596. [CrossRef]

27. Hassiotou, F.; Geddes, D.T. Immune Cell–Mediated Protection of the Mammary Gland and the Infant during Breastfeeding. *Adv Nutr.* **2015**, *6*, 267–275. [CrossRef] [PubMed]

28. Vorbach, C.; Capecchi, M.R.; Penninger, J.M. Evolution of the mammary gland from the innate immune system? *Bioessays* **2006**, *28*, 606–616. [CrossRef]

29. Czosnykowska-Łukacka, M.; Królak-Olejnik, B.; Orczyk-Pawiłowicz, M. Breast Milk Macronutrient Components in Prolonged Lactation. *Nutrients* **2018**, *10*, 1893. [CrossRef]

30. Orczyk-Pawiłowicz, M.; Hirnle, L.; Berghausen-Mazur, M.; Kątnik-Prastowska, I.M. Lactation stage-related expression of sialylated and fucosylated glycotopes of human milk α-1-acid glycoprotein. *Breastfeed Med.* **2014**, *9*, 313–319. [CrossRef]

31. Wesolowska, A.; Sinkiewicz-Darol, E.; Barbarska, O.; Strom, K.; Rutkowska, M.; Karzel, K.; Rosiak, E.; Oledzka, G.; Orczyk-Pawiłowicz, M.; Rzoska, S.; et al. New Achievements in High-Pressure Processing to Preserve Human Milk Bioactivity. *Front. Pediatr.* **2018**, *16*, 323. [CrossRef] [PubMed]

32. Rai, D.; Adelman, A.S.; Zhuang, W.; Rai, G.P.; Boettcher, J.; Lönnerdal, B. Longitudinal changes in lactoferrin concentrations in human milk: A global systematic review. *Crit. Rev. Food Sci. Nutr.* **2014**, *54*, 1539–1547. [CrossRef] [PubMed]

33. Yang, Z.; Jiang, R.; Chen, Q.; Wang, J.; Duan, Y.; Pang, X.; Jiang, S.; Bi, Y.; Zhang, H.; Lönnerdal, B.; et al. Concentration of Lactoferrin in Human Milk and Its Variation during Lactation in Different Chinese Populations. *Nutrients* **2018**, *10*, 1235. [CrossRef] [PubMed]

34. Haschke, F.; Haiden, N.; Thakkar, S.K. Nutritive and bioactive proteins in breastmilk. *Ann. Nutr. Metab.* **2016**, *69*, 16–26. [CrossRef] [PubMed]

35. Garcia-Rodenas, C.L.; De Castro, C.A.; Jenni, R.; Thakkar, S.K.; Beauport, L.; Tolsa, J.-F.; Fischer-Fumeaux, C.J.; Affolter, M. Temporal changes of major protein concentrations in preterm and term human milk. A prospective cohort study. *Clin. Nutr.* **2019**, *38*, 1844–1852. [CrossRef] [PubMed]

36. Affolter, M.; Garcia-Rodenas, C.; Vinyes-Pares, G.; Jenni, R.; Roggero, I.; Avanti-Nigro, O.; de Castro, C.A.; Zhao, A.; Zhang, Y.; Wang, P.; et al. Temporal changes of protein composition in breast milk of Chinese Urban mothers and impact of caesarean section delivery. *Nutrients* **2016**, *8*, 504. [CrossRef] [PubMed]

37. Trend, S.; Strunk, T.; Lloyd, M.; Kok, C.H.; Metcalfe, J.; Geddes, D.T.; Lai, C.T.; Richmond, P.; Doherty, D.A.; Simmer, K.; et al. Levels of innate immune factors in preterm and term mothers' breast milk during the 1st month postpartum. *Br. J. Nutr.* **2016**, *115*, 1178–1193. [CrossRef]

38. Woodman, T.; Strunk, T.; Patole, S.; Hartmann, B.; Simmer, K.; Currie, A. Effects of lactoferrin on neonatal pathogens and Bifidobacterium breve in human breast milk. *PLoS ONE* **2018**, *13*, e0201819. [CrossRef]

39. Ochoa, T.J.; Pezo, A.; Cruz, K.; Chea-Woo, E.; Cleary, T.G. Clinical studies of lactoferrin in children. *Biochem. Cell Biol.* **2012**, *90*, 457–467. [CrossRef] [PubMed]

40. Siqueiros-Cendón, T.; Arévalo-Gallegos, S.; Iglesias-Figueroa, B.F.; García-Montoya, I.A.; Salazar-Martínez, J.; Rascón-Cruz, Q. Immunomodulatory effects of lactoferrin. *Acta Pharmacol. Sin.* **2014**, *35*, 557–566. [CrossRef]

41. Newburg, S.; Walker, W.A. Protection of the neonate by the innate immune system of developing gut and of human milk. *Pediatr. Res.* **2007**, *61*, 2. [CrossRef] [PubMed]

Article

Breastfeeding Disparities between Multiples and Singletons by NICU Discharge

Roser Porta [1], Eva Capdevila [2], Francesc Botet [3], Gemma Ginovart [4], Elisenda Moliner [4], Marta Nicolàs [5], Antonio Gutiérrez [6,7], Jaume Ponce-Taylor [8] and Sergio Verd [9,*]

[1] Neonatal Unit, Dexeus University Hospital, 5 Sabino Arana st, 08028 Barcelona, Spain; roser.porta@quiron.es
[2] Pediatric Unit, Department of Primary Care, Catalonia Health Authority, Balmes st, 08007 Barcelona, Spain; e.capde@gmail.com
[3] Neonatal Unit, University Maternity Hospital, 5 Sabino Arana st. 08028 Barcelona, Spain; fbotet@clinic.cat
[4] Neonatal Unit, Santa Creu i Sant Pau University Hospital, 87 mSant Quinti st. 08041 Barcelona, Spain; gginovart@santpau.cat (G.G.); emoliner@santpau.cat (E.M.)
[5] Neonatal Unit, Germans Trias i Pujol University Hospital, Canyet Road, 08916 Badalona, Spain; mnicolasl.germanstrias@gencat.cat
[6] Department of Hematology, Son Espases University Hospital. IdISBa Balearic Medical Research Council, Valldemossa Road, 79, 07010 Palma de Mallorca, Spain; antoniom.gutierrez@ssib.es
[7] COMIB Advisory, Passeig de Mallorca, 42, 07012 Palma de Mallorca, Spain
[8] Urgent Care Centre, Department of Primary Care, Balearic Health Authority, 1 Illes Balears st. 07014 Palma de Mallorca, Spain; Jaume.ponce@gmail.com
[9] Pediatric Unit, Department of Primary Care, Balearic Health Authority, Matamusinos st. 07013 Palma de Mallorca, Spain
* Correspondence: drsverd@gmail.com; Tel.: +34-600-505-246; Fax: +34-(9)-71-799534

Received: 18 July 2019; Accepted: 6 September 2019; Published: 12 September 2019

Abstract: Multiple pregnancy increases the risk of a range of adverse perinatal outcomes, including breastfeeding failure. However, studies on predictive factors of breastfeeding duration in preterm twin infants have a conflicting result. The purpose of this observational study was to compare feeding practices, at hospital discharge, of twin and singleton very low birth weight infants. The study is part of a prospective survey of a national Spanish cohort of very low birth weight infants (SEN1500) that includes 62 neonatal units. The study population comprised all infants registered in the network from 2002 to 2013. They were grouped into singletons and multiples. The explanatory variables were first analyzed using univariate models; subsequently, significant variables were analyzed simultaneously in a multiple stepwise backward model. During the twelve-year period, 32,770 very low birth weight infants were included in the database, of which 26.957 were discharged alive and included in this analysis. Nine thousand seven hundred and fifty-eight neonates were multiples, and 17,199 were singletons. At discharge, 31% of singleton infants were being exclusively breastfed, 43% were bottle-fed, and 26% were fed a combination of both. In comparison, at discharge, only 24% of multiple infants were exclusively breastfed, 43% were bottle-fed, and 33% were fed a combination of both ($p < 0.001$). On multivariable analysis, twin pregnancy had a statistically significant, but small effect, on cessation of breastfeeding before discharge (OR 1.10; 95% CI: 1.02, 1.19). Risks of early in-hospital breastfeeding cessation were also independently associated with multiple mother-infant stress factors, such as sepsis, intraventricular hemorrhage, retinopathy, necrotizing enterocolitis, intubation, and use of inotropes. Instead, antibiotic treatment at delivery, In vitro fertilization and prenatal steroids were associated with a decreased risk for shorter in-hospital breastfeeding duration. Multiple pregnancy, even in the absence of pathological conditions associated to very low birth weight twin infants, may be an impeding factor for in-hospital breastfeeding.

Keywords: breastfeeding; multiple pregnancy; neonate; premature birth; milk bank; pregnancy outcomes

---

1. Introduction

Historically, formula was provided to very low birth weight (VLBW) infants to ensure rapid growth. However, recent studies show that feeding fortified human milk to preterm infants does not result in lower weight at discharge [1]. Achieving optimal development for preterm infants, whilst minimizing the risk for serious disease in the neonate, remains an important challenge for clinicians. Human milk is recommended for enteral feeding of preterm newborns [2], since it is associated with low rates of necrotizing enterocolitis (NEC) [3], sepsis, retinopathy of prematurity (ROP) [4], and a better neurocognitive development [5]. Despite efforts to offer human milk as the primary source of enteral nutrition during the stay in the neonatal intensive care unit (NICU), infant milk formulas are used in situations in which the mother's milk or donor's milk is not available.

Previous research has identified associations between the provision of breast milk in the NICU and maternal characteristics, infant characteristics, and environmental factors [6–8]. However, studies on predictive factors of breastfeeding duration in premature infants have conflicting results. Some studies have found that higher gestational age is associated with higher rates of breastfeeding [7,9], while others have found that mothers of VLBW infants acknowledge that human milk is a unique factor in improving outcomes for their infant [10] and, consequently, breastfeeding rates are higher among them.

The proportion of infants born preterm is higher for twins, when compared to singletons [11]. Breastfeeding poses unique challenges for the mothers of twins, but little attention has been paid to risk factors for early cessation of breastfeeding for these mothers. Only a few studies have focused on issues related to breastfeeding cessation for full term twins [12,13]. From a physiological standpoint, the World Health Organization (WHO) and other authorities state that mother's milk is sufficient for feeding multiple babies [14]. It provides optimal nutrition for full term or preterm twins. A study comparing singleton and multiple premature, under 30 weeks, gestations, found that the milk volume was higher in mothers of multiples compared with singletons (599 mL/day vs. 430 mL/day) [15]. It has also been reported that the milk yield of individual breasts of mothers of twins is 0.8–2.1 kg per 24 h at six months, whereas the total milk yield of mothers of triplets can reach 3.0 kg per 24 h at 2.5 months [16]. Previous research shows that preterm twin infants may be less likely to be breastfed compared to singleton infants, but the data on this subject remains contradictory [17]. When interpreting the results from these studies, it should be noted that the type of babies included is variable, and that most studies do not provide details on feeding maintenance and complications. One issue that further complicates efforts to address this question is the finding that most preterm complications (NEC, ROP, sepsis, etc.) may be a cause of breastfeeding cessation, and may also be a consequence of breastfeeding cessation. This makes it difficult to explain the relationship between feeding type and severe clinical complications. The aim of this study is to shed light on the influence of major neonatal morbidity, and multiple birth, on in-hospital breastfeeding success. Specifically, we hypothesize that twin birth exerts its own substantial influence on poor breastfeeding outcomes among VLBW infants, irrespective of the clinical situations of these patients.

To test this hypothesis, we have used a multicenter, electronic health record-derived database that represents roughly two-thirds of all Spanish VLBW infants. The objective is to compare breastfeeding rates at hospital discharge of singleton and twin VLBW infants, and to evaluate which neonatal conditions and diseases preclude the provision of breast milk at the time of hospital discharge.

2. Methods

2.1. Design

The study is part of a prospective survey of a national Spanish cohort of VLBW infants (SEN1500 database), based on questionnaires. The SEN1500 network prospectively collects maternal and neonatal data of live-born infants of ≤1500 g who were born in or admitted to the collaborating NICUs within the first 28 days of life. Data were collected using a pre-established form and were submitted electronically using common and specific software. The characteristics, quality control, and data confidentiality systems of this database have been described elsewhere [18]. For the purpose of the present study, data corresponding to all discharged alive newborns within the database over a period of twelve years were retrieved. The SEN1500 database includes 62 neonatal units, which encompasses about two-thirds of all Spanish VLBW infants. The study population comprises all infants registered in the database, during the period 2002–2013. They were grouped into singletons and multiples. The same cohort was used previously to analyze morbidity and mortality associated with multiple birth [19].

2.2. Data Collection

In this study, we assess clinical factors that are known to influence breastfeeding during the hospital stay of singletons and twins. The main variable of interest was any breastfeeding, i.e., not being completely weaned from breastmilk, on the day when hospital discharge was completed. Data on maternal and infant characteristics were obtained from medical records by local health care professionals, using a pretested standardized questionnaire. The study variables were defined according to the variables catalogue of the SEN1500 database. We collected demographic data (birth weight, gender), obstetrical data (fertilization, antenatal care, gestational age, type of delivery, twinning), procedures and treatment (intra partum antibiotics, inotropic therapy, intubation, surfactant administration, ventilation, oxygen therapy, ductus closure, retinopathy surgery), neonatal morbidity (necrotizing enterocolitis, intraventricular haemorrhage, sepsis), length of hospitalization and type of feeding at discharge.

2.3. Outcomes

Exclusive breast milk feeding was defined as the infant receiving no food or drink other than the mother's own milk. Any breast milk feeding was defined as the infant receiving mother's own milk, independent of the addition of formula or other food and/or drink. SEN1500 was set up by the Spanish Association of Neonatologists. It encompasses 62 neonatal units, about two-thirds of all national neonatal units. The operational definitions used to collect feeding data at discharge are: No breastfeeding, any breastfeeding, or exclusive breastfeeding. The database systems are locally controllable. Breastfeeding outcomes are recorded after discharge, in each neonatal unit by the neonatologist responsible for registering obstetric and neonatal information, routinely collected from the Perinatal Health files.

2.4. Statistical Analysis

Data analysis was performed using statistical package SPSS 14.0 for Windows (IBM Company, Chicago, IL, USA). *p*-Values below 0.05 were considered to be of statistical significance. Descriptive statistics were used to outline mother and infant characteristics. Variables following binomial distributions were expressed as frequencies and percentages; and quantitative variables as median and ranges. Comparisons between qualitative variables were made using the Fisher Exact Test or Chi-square. Comparisons between quantitative and qualitative variables were performed through non-parametric tests (U of Mann-Whitney or Kruskal-Wallis).

Logistic regression (both univariate and multivariate) was used to assess the relationship between potential predictor variables of breastfeeding at discharge, a broad range of explanatory variables with a *p*-value under 0.1 was chosen to appear in the final regression model, yet this less restrictive approach is accepted, as well as considering explanatory variables one at a time [20]. Out of the variables with a significant relation to breastfeeding at discharge after the univariate analysis, we excluded those that, from a clinical point of view, are a result rather than a cause of feeding practice (anthropometric data at discharge). Finally, the explanatory variables included in the stepwise backward regression model were birth weight, height at birth, birth head circumference, gender, In vitro fertilization, multiple birth, prenatal care, prenatal administration of corticosteroids or antibiotics, delivery type, NICU use of surfactant or inotropic drugs, intubation, ventilation duration, oxygen therapy duration, CRIB score, duct pharmacological closure, surgical management of ROP, NEC, late onset sepsis (LOS), intraventricular hemorrhage (IVH), and supplementary oxygen at 36 weeks PMA (bronchopulmonary dysplasia). The results are presented as odds ratio (OR) with a 95% confidence interval (CI).

2.5. Statement of Ethics

The study was conducted in accordance with the Declaration of Helsinki. The Local Health District Human Research Ethics Committees approved the collection of the data from the SEN1500 database and subsequent analysis. Data collection in SEN1500 is anonymous and has been previously approved by the Ethics and Investigation Committees of the Collaborative Centers. It is in compliance with the Spanish data protection law. Ethical approval code: IIBSP-LEC-2011-129. Informed consent was obtained a legally authorized person, such as a parent or guardian.

3. Results

3.1. Characteristics of the Study Population

The study included 26,957 VLBW infants, of which 9758 were multiples, and 17,199 were singletons—who were all alive when discharged from the NICU. The characteristics of the study sample are summarized in Table 1. The median gestational age was 29 weeks, with a median birth weight of 1190 g. Female infants made up 50% of the sample. Cesarean delivery was common (72%), and 6% of infants were outborn. Use of antenatal care was observed in 22,830 (88%) patients. Antenatal corticosteroid administration rate was 84%, and median length of hospital stay was 50 days. As we have reported elsewhere [18], most patient characteristics and major morbidities differed between multiples and singletons (see Table 2).

Table 1. Characteristics of 26,957 VLBW infants discharged alive from 62 Spanish neonatal units in 2002–2013.

Characteristic	Number (%)
Maternal Factors	
In vitro fertilization (%)	4747 (19)
Prenatal care (%)	22,830 (88)
Antenatal corticosteroid therapy (%)	22,059 (84)
Intrapartum antibiotic chemoprophylaxis (%)	11,502 (47)
Between hospital transfers	1651 (6)
Caesarean section (%)	19,503 (72)
Multiple birth (%)	9758 (36)
Median gestational age at birth, weeks (range)	29 (20–41)
Infant Factors	
Male sex (%)	13,374 (50)
Apgar score at 1 min (%)	7 (0–10)
Median birth weight, g. (range)	1190 (360–1499)
No breastfeeding at discharge (%)	11,592 (43)
Morbidities	
Early onset sepsis (%)	950 (4)
Late onset sepsis (%)	7672 (29)
Necrotising enterocolitis (any) (%)	1506 (6)
Bronchopulmonary dysplasia (%)	3127 (14)
Surgical management of ROP (%)	951 (4)
Median length of stay, days (range)	50 (1–238)

Data are presented as median (minimum–maximum) or number (percentile), unless specified. Abbreviations: ROP, retinopathy of prematurity; VLBW, very low birth weight; g, gram.

Table 2. Demographic and pathological characteristics of singletons versus multiples.

Variables	Singleton	Multiple	*p*-Value
	n = 17,199	*N* = 9758	
Maternal Factors			
Median gestational age (weeks) (range)	29 (20–41)	30 (20–39)	<0.001
In vitro fertilization	833 (5.2%)	3914 (44%)	<0.001
Outborn	1153 (6.7%)	498 (5.1%)	<0.001
Antenatal care	14,320 (86.3%)	8510 (90.8%)	<0.001
Antenatal steroids	13,741 (81.9%)	8318 (86.9%)	<0.001
Maternal *intra partum* antibiotics	7172 (46.3%)	4330 (49%)	<0.001
Vaginal delivery	5510 (32%)	1944 (19.9%)	<0.001
Infant Factors			
Median birthweight (g) (range)	1160 (360–1499)	1220 (395–1499)	<0.001
Male sex (%)	8728 (50.7)	4646 (47.6)	<0.001
Morbidities			
Days of ventilatory therapy	0 (0–835)	0 (0–672)	<0.001
Days of supplemental oxygen	4 (0–391)	3 (0–286)	<0.001
Surfactant at any time	7689 (45%)	4130 (42.6%)	<0.001
Inotropic therapy	3728 (22.7%)	1959 (21.1%)	0.003
Necrotizing enterocolitis (all grades)	1007 (5.9%)	499 (5.1%)	0.011
Early-onset sepsis	671 (3.9%)	279 (2.9%)	<0.001
Late-onset sepsis	5173 (30.5%)	2499 (26%)	<0.001
IVH (all grades)	3471 (21.7%)	1494 (16.5%)	<0.001
Supplementary Oxygen at discharge	1074 (6.3%)	507 (5.2%)	<0.001
NICU length of stay (days)	19 (0–187)	14 (0–238)	<0.001

Data are presented as mean (SD) or number (percentile), unless specified. Abbreviations: IVH, intraventricular hemorrhage; NICU, neonatal intensive care unit. Comparisons between qualitative variables were made using the Fisher Exact Test or Chi-square. Comparisons between quantitative and qualitative variables were performed through non-parametric tests (U of Mann-Whitney or Kruskal-Wallis).

3.2. Breastfeeding Patterns by Study Factors

At hospital discharge, 15,365 (57%) infants were being breastfed, 28% exclusively, and 29% non-exclusively. Out of the singleton infants, at hospital discharge, 31% were being exclusively breastfed, 43% exclusively bottle-fed, and 26% were being fed a combination of both. In comparison, at hospital discharge, only 24% of multiple infants were being exclusively breastfed, 43% exclusively bottle-fed, and 33% were being fed a combination of both. The distribution of the patients by feeding type at discharge differed in these two groups (*p* < 0.001).

3.3. Determinants of Breastfeeding at Hospital Discharge

The results of the univariate analysis are summarised in Table 3. With regards to mother and delivery characteristics, the unadjusted analysis shows that In vitro fertilization (IVF), prenatal care, administration of antenatal corticosteroids, intra partum antibiotics, and vaginal delivery, were all protective factors for being breastfed at discharge. Infant weight, height and head circumference at birth were also positively associated with breastfeeding rates at discharge. Additionally, infants that were being breastfed at discharge received less surfactant and were less often intubated or put on mechanical ventilation. They spent fewer days using supplemental oxygen, they received fewer inotropes and required less surgical treatment for ROP, and IVH (grades III and IV). NEC and LOS were more prevalent in infants not being breastfed at discharge.

Table 3. Univariate analysis of risk factors for no breastfeeding at NICU discharge.

Characteristics	Twins		Single	
	OR (95% CI)	*p*	OR (95% CI)	*p*
Birth weight	1 (1–1)	0.008	0.99 (0.99–0.99)	<0.001
Height at birth	0.99 (0.99–1)	0.011	0.99 (0.99–0.99)	<0.001
Birth head circumference	0.99 (0.99–1)	0.1	0.99 (0.99–0.99)	<0.001
Male gender	1.08 (0.99–1.17)	0.076	0.99 (0.93–1.05)	0.65
Twin birth	—		—	
In vitro fertilization	0.93 (0.85–1.01)	0.099	0.59 (0.5–0.68)	<0.001
Prenatal care	0.84 (0.72–0.97)	0.022	1.46 (1.33–1.6)	<0.001
Prenatal steroids:				
No		<0.001		<0.001
Partial	0.66 (0.57–0.77)		0.84 (0.76–0.93)	
Complete	0.72 (0.63–0.82)		0.74 (0.68–0.8)	
Antibiotic treatment at delivery	0.72 (0.66–0.79)	<0.001	0.86 (0.81–0.92)	<0.001
Delivery type: C-section	1.12 (1.01–1.24)	0.038	0.87 (0.82–0.93)	<0.001
Intubation:	1.22 (1.1–1.35)	<0.001	1.58 (1.48–1.69)	<0.001
CRIB score	1.06 (1.04–1.08)	<0.001	1.11 (1.09–1.12)	<0.001
Conventional ventilation duration	1.01 (1.01–1.02)	<0.001	0.58 (0.55–0.62)	<0.001
Oxygen therapy duration	1 (1–1.01)	<0.001	0.62 (0.55–0.69)	<0.001
Surfactant administration	1.15 (1.06–1.25)	0.001	1.56 (1.46–1.66)	<0.001
Inotropic therapy	1.53 (1.38–1.7)	<0.001	1.96 (1.82–2.12)	<0.001
Ductus closure:				
None		0.022		<0.001
Indometacin	1.12 (0.98–1.28)		1.51 (1.36–1.67)	
Ibuprofen	0.88 (0.77–1)		1.2 (1.08–1.32)	
ROP surgery	1.42 (1.13–1.78)	0.002	2.43 (2.06–2.88)	<0.001
Necrotizing enterocolitis	1.65 (1.37–1.99)	<0.001	1.88 (1.65–2.15)	<0.001
Late onset sepsis	1.27 (1.15–1.39)	<0.001	1.59 (1.48–1.7)	<0.001
Intraventricular hemorrhage				
No		0.001		<0.001
Grade I–II	1.13 (0.99–1.28)		1.24 (1.14–1.36)	
Grade III–IV	1.48 (1.18–1.85)		1.94 (1.67–2.26)	
Supplemental oxygen *at* 36 weeks *PMA*	1.53 (1.33–1.76)	<0.001	2 (1.82–2.21)	<0.001

Data presented as median (range) or percentile. ABBREVIATIONS: NICU, neonatal intensive care unit; PMA, postmenstrual age; ROP, retinopathy of prematurity.

However, when these factors were put into a multivariable model to determine the effect of a factor, while holding all others constant (Table 4), multiple birth (OR 1.10), surfactant administration (OR 1.10), prenatal care (OR 1.11), intubation (OR 1.15), LOS (OR 1.23), surgical treatment for ROP (OR 1.33), use of inotropes (OR 1.38), IVH grades III or IV (OR 1.42), bronchopulmonary dysplasia (OR 1.43) and NEC (OR 1.47), were significantly associated with a lower chance of any breastfeeding at discharge. On the other hand, the primary findings of this study indicate that the odds of a VLBW infant of being breastfed at the time of hospital discharge correlate to (a) the use of ibuprofen for ductus closure, (b) prenatal steroid treatment, (c) In vitro fertilization, and (d) intrapartum antibiotics.

Table 4. Logistical regression analysis of no breastfeeding at discharge as a function of selected predictor variables.

Variables	Odds Ratio (95% CI)	*p*-Value
	Maternal Factors	
Prenatal care	1.11 (1.01–1.22)	0.036
Partial prenatal steroids	0.79 (0.71–0.88)	<0.001
Complete prenatal steroids	0.82 (0.75–0.89)	<0.001
In vitro fertilization	0.84 (0.77–0.92)	<0.001
Antibiotic treatment at delivery	0.86 (0.81–0.92)	<0.001
Multiple birth	1.10 (1.02–1.19)	0.009
	Infant Factors	
Surfactant administration	1.10 (1.01–1.19)	0.023
Endotracheal intubation	1.15 (1.06–1.25)	0.001
	Morbidities	
Ibuprofen treatment for ductus closure	0.73 (0.66–0.81)	<0.001
Late onset sepsis	1.27 (1.18–1.36)	<0.001
Surgical management of ROP	1.33 (1.13–1.57)	0.001
Inotrope support	1.38 (1.26–1.50)	<0.001
Grade III–IV intraventricular hemorrhage	1.42 (1.21–1.67)	<0.001
Supplemental oxygen at 36 weeks PMA	1.43 (1.3–1.59)	<0.001
Necrotizing enterocolitis	1.47 (1.29–1.69)	<0.001

−2 log likelihood: 23234; DF: 17. Abbreviations: ROP, retinopathy of prematurity.

4. Discussion

The incidence of multiple births has risen since the 1970s [21]. Within this group, the proportion of infants born preterm is higher compared to singletons [22]. There is strong evidence that not being fed breastmilk carries greater risks, with preterm babies having specific additional risks. However, mothers of preterm infants may terminate lactation earlier than mothers of full term infants [23]. Researching the obstacles for breastfeeding in preterm babies is a central issue in perinatology [8]. The low prevalence of breastfeeding among preterm infants may be explained by several factors. First, by difficulties inherent to their higher neonatal morbidity [24,25]. Also, maternal socio-demographic factors, and health-care services related factors may play an important role in the initiation and duration of breastfeeding. Although previous studies have examined a wide range of factors related to breastfeeding during a hospital stay for preterm infants, only a few have focused on the effects of multiple births on breastfeeding rates in this population.

Moreover, the few studies that have taken into account this variable have produced controversial results. The results on this issue are quite diverse depending on the study population and on the time of investigation of the nutritional method. To the best of our knowledge, from 1997 to 2017, four papers have found that the combination of prematurity and twin birth results in lower rates and shorter duration of breastfeeding among preterm multiples. On the other hand, six papers have found no breastfeeding differences between singleton and twin preterm infants. A 2004 study assessed exclusive and non-exclusive breastfeeding at day 1, day 3, week 2, month 1, and month 6 postpartum, among 346 mothers of full term and preterm singleton and multiple babies. They found that at any time point, except week 2 and month 1, the proportion of preterm multiples who fed at least some breast milk was less than in other groups [26]. A 2007 telephone survey on 94 VLBW infants reported that mothers of preterm multiples were 16 times more likely to provide formula before week 12 postpartum than mothers who delivered a singleton [27]. A 2014 Danish cohort of preterm infants (24–36 weeks of gestation) has found that multiple birth is significantly associated with the later establishment of exclusive breastfeeding [28]. A 2016 study on a very large sample of very preterm infants shows that the odds for exclusive breastfeeding at discharge for singletons was significantly higher than for twins [29]. Similarly, we found that singleton preterm infants were more likely to be exclusively breastfed at discharge than multiple preterm infants (31% versus 24%). Our results are in line with the previous research on the prevalence of exclusive breastfeeding for VLBW infants at NICU discharge, ranging from 11.4 to 30.5% in various countries [10,17,30]. All studies on this topic report that rates of breastfeeding (exclusive or non-exclusive) decrease with time. Exclusive breastfeeding at NICU discharge is essential to help women in continuing breastfeeding for durations recommended by the WHO. Accordingly, Kaneko et al. reported that the prevalence of exclusive breastfeeding was 22.6% at NICU discharge, and it had dropped only to 15.7% by the start of complementary feeding. These values are slightly higher or lower than those of recent large Japanese or Italian cohorts, respectively [17,31].

On the other hand, six authors have reported that preterm twins can breastfeed as successfully as preterm singleton infants. A 1997 paper on 15 singleton and 18 twin preterm infants showed no differences in feeding method from birth to discharge [32]. More recently, a Danish paper has reported no differences in exclusive breastfeeding rates at discharge [33], an Israeli paper has reported no differences in breastfeeding beyond six months of age [34], and three other papers from Western countries have shown no differences in any breastfeeding rates at discharge from the NICU [6,35,36].

For the majority of mothers, breastfeeding evolves naturally. However, it is not so when mothers experience the unique stress associated with the NICU environment [37]. Most maternal and NICU characteristics that challenge the success of breastfeeding have been extensively studied. Conversely, scarce attention has been devoted to the influence on breastfeeding of a variety of stress factors that preterm and fragile infants face. These include aggressive procedures, life-threatening and damaging conditions, such as IVH or ROP, without forgetting expected feeding difficulties related to the infant's clinical status at feeding time. Therefore, in this study, we chose to examine the link of common neonatal medical complications with the infant continuing to receive breastmilk upon discharge.

VLBW infants are at risk for a variety of medical complications, any of which further limits the mother-infant dyad progression to breastfeeding competence. However, according to our multivariable analysis, only intubation, NEC, LOS, IVH grades III or IV, surgical ROP, the use of inotropes and twin pregnancy were risk factors for breastfeeding cessation before NICU discharge.

Our findings cannot substantiate previous research in this field, yet we have found only one article that identified absence of history of NEC or LOS as significant predictors of continued provision of mother's milk in VLBW infants through to day of life 30 [38], and we have found no published research that reports that lower rates of breastfeeding among multiple preterm infants may reflect their distinct neonatal morbidity.

Our analysis of a very large sample of babies has reduced the number of variables linked to in-hospital breastfeeding cessation to fifteen independent risk factors. Twin pregnancy is one such risk factor, but its magnitude (OR = 1.10) is small enough to explain contradictory results in previous research on this topic. It is expected that small samples with bigger margins of error cannot detect this effect.

Finally, our research confirms that surgical NEC, ROP, bronchopulmonary dysplasia and LOS are linked to no breastfeeding at discharge. It is likely that this is a matter of reverse causation, since it is well established that lack of breastfeeding is a risk factor for such conditions in preterm infants. Likewise, it is reasonable to speculate that prenatal care, including prenatal steroid treatment, boost successful breastfeeding rates. Conversely, surfactant administration, intubation, inotropes, and IVH have not been previously linked to breastfeeding difficulties, and are, therefore, worth further investigation. On the other hand, the same applies to *ductus* closure, In vitro fertilization or *intra partum* antibiotics.

Limitations

The large sample study, the exclusion of recall bias concerning breastfeeding duration and the diverse patient population, are strengths of our study. We were able to show some variation in maternal breast milk provision between multiple and singleton preterm infants. While these findings are important in broadening our knowledge, they are not without limitations.

Breastfeeding was not the focus of the primary study. Consequently, the questions concerning breastfeeding were less detailed than those that would be used in a prospective study. An additional limitation of this study is the use of maternal breast milk feedings at discharge to represent mothers continued provision of breast milk. We were also unable to quantify the volume of maternal breast milk received.

5. Conclusions

Our study of a very large sample of VLBW infants, shows that singleton pregnancy predicts a discrete increase of in-hospital breastfeeding rates, when compared to twin pregnancy. This study encourages a closer look at breastfeeding in the NICU. We support that breastfeeding among VLBW twin infants at NICU discharge is feasible; however, identification of multiple risk factors for the cessation of breastfeeding reflects the complexity of this process. By understanding predictive maternal or infant factors associated with poor breastfeeding outcomes, mothers at risk can be identified for more successful interventions. In addition, we can work to come up with adequate interventions for twin VLBW infants so that, when compared with singleton VLBW infants, neonatal morbidities will no longer be factors associated with failed breastfeeding. It is possible that early interventions for a few serious complications linked to breastfeeding cessation could have minimized the other problems and made it possible for the women to continue. Interventions in the NICU should enable mothers to transition to breastfeeding as soon as possible to promote success until discharge and beyond.

Going forward, for NICUS to excel at feeding human milk, in-depth work remains to determine barriers to a mother's ability to provide breast milk and to establish evidence-based lactation care.

Author Contributions: All authors contributed extensively to the work presented in this paper. S.V. and R.P. conceived the study; E.C. and F.B. developed the methodology; E.M. and M.N. monitored data collection; A.G. analyzed the data; R.P., G.G., F.B. and J.P.-T. critically reviewed the manuscript for important intellectual content; S.V. and J.P.-T. wrote the paper.

Funding: This research received no external funding.

Acknowledgments: We want to thank all children, mothers and caregivers who participated in this study, as well as the nursing staff participating in the research for their cooperation. This project would not have been possible without the generous support from the Library of the Balearic Health Authority and from SEN1500 network.

Conflicts of Interest: All authors declare: No support from any organisation for the submitted work; no financial relationships with any organisations that might have an interest in the submitted work in the previous three years; no other relationships or activities that could appear to have influenced the submitted work.

Abbreviations

CI, 95% confidence interval; IVH, intraventricular hemorrhage; LOS, late onset sepsis; NEC, necrotizing enterocolitis; NICU, neonatal unit; OR, odds ratio; ROP, retinopathy of prematurity; SEN1500, Spanish cohort of very low birth weight; VLBW, very low birth weight; WHO, World Health Organization.

References

1. Lewis, E.D.; Richard, C.; Larsen, B.M.; Field, C.J. The importance of human milk for immunity in preterm infants. *Clin. Perinatol.* **2017**, *44*, 23–47. [CrossRef] [PubMed]
2. Lechner, B.E.; Vohr, B.R. Neurodevelopmental outcomes of preterm infants fed human milk: A Systematic Review. *Clin. Perinatol.* **2017**, *44*, 69–83. [CrossRef] [PubMed]
3. Lloyd, M.L.; Malacova, E.; Hartmann, B.; Simmer, K. A clinical audit of the growth of preterm infants fed predominantly pasteurised donor human milk v. those fed mother's own milk in the neonatal intensive care unit. *Br. J. Nutr.* **2019**, *121*, 1018–1025. [CrossRef] [PubMed]
4. Kumar, R.K.; Singhal, A.; Vaidya, U.; Banerjee, S.; Anwar, F.; Rao, S. Optimizing nutrition in preterm low birth weight infants-Consensus Summary. *Front. Nutr.* **2017**, *4*, 20. [CrossRef] [PubMed]

5.	Maffei, D.; Schanler, R.J. Human milk is the feeding strategy to prevent necrotizing enterocolitis! *Semin. Perinatol.* **2017**, *41*, 36–40. [CrossRef] [PubMed]

6.	Pineda, R.G. Predictors of breastfeeding and breastmilk feeding among very low birth weight infants. *Breastfeed. Med.* **2011**, *6*, 15–19. [CrossRef] [PubMed]

7.	Powers, N.G.; Bloom, B.; Peabody, J.; Clark, R.; Md, N.G.P.; Md, B.B.; Md, J.P.; Md, R.C. Site of care influences breastmilk feedings at NICU discharge. *J. Perinatol.* **2003**, *23*, 10–13. [CrossRef] [PubMed]

8.	Sisk, P.; Quandt, S.; Parson, N.; Tucker, J. Breast milk expression and maintenance in mothers of very low birth weight infants: Supports and barriers. *J. Hum. Lact.* **2010**, *26*, 368–375. [CrossRef]

9.	Espy, K.A.; Senn, T.E. Incidence and correlates of breast milk feeding in hospitalized preterm infants. *Soc. Sci. Med.* **2003**, *57*, 1421–1428. [CrossRef]

10.	Lee, H.C.; Gould, J.B. Factors influencing breast milk versus formula feeding at discharge for very low birth weight infants in California. *J. Pediatr.* **2009**, *155*, 657–662.e2. [CrossRef]

11.	Branum, A.M.; Schoendorf, K.C. Changing patterns of low birthweight and preterm birth in the United States, 1981–1998. *Paediatr. Perinat. Epidemiol.* **2002**, *16*, 8–15. [CrossRef] [PubMed]

12.	Damato, E.G.; Dowling, D.A.; Standing, T.S.; Schuster, S.D. Explanation for cessation of breastfeeding in mothers of twins. *J. Hum. Lact.* **2005**, *21*, 296–304. [CrossRef]

13.	Östlund, Å.; Nordström, M.; Dykes, F.; Flacking, R. Breastfeeding in preterm and term twins–maternal factors associated with early cessation: A population-based study. *J. Hum. Lact.* **2010**, *26*, 235–241. [CrossRef] [PubMed]

14.	Royal College of Paediatrics and Child Health. *Guidance for Health Professionals on Feeding Twins, Triplets and Higher Order Multiples*; RCPCH: London, UK, 2011.

15.	Lau, C.; Hurst, N.; Burns, P.; Schanler, R.J. Interaction of stress and lactation differs between mothers of premature singletons and multiples. *Adv. Exp. Med. Biol.* **2004**, *554*, 313–316. [PubMed]

16.	Leonard, L.G. Breastfeeding rights of multiple birth families and guidelines for health professionals. *Twin. Res.* **2003**, *6*, 34–45. [CrossRef] [PubMed]

17.	Davanzo, R.; Monasta, L.; Ronfani, L.; Brovedani, P.; Demarini, S.; Breastfeeding in Neonatal Intensive Care Unit Study Group. Breastfeeding at NICU discharge: A multicenter Italian study. *J. Hum. Lact.* **2013**, *29*, 374–380. [CrossRef] [PubMed]

18.	Moro, M.S.; Fernández, C.P.; Figueras, J.A.; Pérez, J.R.; Coll, E.; Doménech, E.M.; Jiménez, R.; Pérez, V.S.; Quero, J.J.; Roques, V.S. Sen1500: Design and implementation of a registry of infants weighing less than 1500 g. at birth in Spain. *An. Pediatr.* **2008**, *68*, 181–188.

19.	Porta, R.; Capdevila, E.; Botet, F.; Verd, S.; Ginovart, G.; Moliner, E.; Nicolàs, M.; Rios, J.; SEN1500 Network. Morbidity and mortality of very low birth weight multiples compared with singletons. *J. Matern. Fetal. Neonatal. Med.* **2019**, *32*, 389–397. [CrossRef] [PubMed]

20.	Lang, T. Documenting research in scientific articles: Guidelines for Authors: 3. Reporting multivariate analysis. *Chest* **2007**, *131*, 628–632. [CrossRef]

21.	Wallis, S.K.; Dowswell, T.; West, H.M.; Renfrew, M.J.; Whitford, H.M. Breastfeeding education and support for women with twins or higher order multiples. *Cochrane Database Syst. Rev.* **2017**, CD012003. [CrossRef]

22.	NICE. Multiple Pregnancy: Twin and Triplet Pregnancies. NICE Quality Standard [QS46]. Available online: http://www.nice.org.uk/guidance/qs46/resources/guidance-multiple-pregnancy-pdf2013 (accessed on 24 January 2019).

23.	Killersreiter, B.; Grimmer, I.; Bührer, C.; Dudenhausen, J.W.; Obladen, M. Early cessation of breastmilk feeding in very low birthweight infants. *Early Hum. Dev.* **2001**, *60*, 193–205. [CrossRef]

24.	Morag, I.; Harel, T.; Leibovitch, L.; Simchen, M.J.; Maayan-Metzger, A.; Strauss, T. Factos associated with breast milk feeding of very preterm infants from birth to 6 months corrected age. *Breastfeed. Med.* **2016**, *11*, 138–143. [CrossRef] [PubMed]

25.	Kirchner, L.; Jeitler, V.; Waldhör, T.; Pollak, A.; Wald, M. Long hospitalization is the most important risk factor for early weaning from breast milk in premature babies. *Acta Paediatr.* **2009**, *98*, 981–984. [CrossRef] [PubMed]

26.	Geraghty, S.R.; Pinney, S.M.; Sethuraman, G.; Roy-Chaudhury, A.; Kalkwarf, H.J. Breast milk feeding rates of mothers of multiples compared to mothers of singletons. *Ambul. Pediatr.* **2004**, *4*, 226–231. [CrossRef] [PubMed]

27. Hill, P.D.; Aldag, J.C.; Zinaman, M.; Chatterton, R.T. Predictors of preterm infant feeding methods and perceived insufficient milk supply at week 12 postpartum. *J. Hum. Lact.* **2007**, *23*, 32–38. [CrossRef] [PubMed]
28. Maastrup, R.; Hansen, B.M.; Kronborg, H.; Bojesen, S.N.; Hallum, K.; Frandsen, A.; Kyhnaeb, A.; Svarer, I.; Hallström, I.K. Breastfeeding progression in preterm infants is influenced by factors in infants, mothers and clinical practice: The results of a national cohort study with high breastfeeding initiation rates. *PLoS ONE* **2014**, *9*, e108208. [CrossRef]
29. Ericson, J.; Flacking, R.; Hellström-Westas, L.; Eriksson, M. Changes in the prevalence of breast feeding in preterm infants discharged from neonatal units: A register study over 10 years. *BMJ Open* **2016**, *6*, e012900. [CrossRef]
30. Maia, C.; Brandão, R.; Roncalli, A.; Maranhão, H. Length of stay in a neonatal intensive care unit and its association with low rates of exclusive breastfeeding in very low birth weight infants. *J. Matern. Fetal. Neonatal Med.* **2011**, *24*, 774–777. [CrossRef]
31. Kaneko, A.; Kaneita, Y.; Yokoyama, E.; Miyake, T.; Harano, S.; Suzuki, K.; Ibuka, E.; Tsutsui, T.; Yamamoto, Y.; Ohida, T. Factors associated with exclusive breast- feeding in Japan: For activities to support child-rearing with breast-feeding. *J. Epidemiol.* **2006**, *16*, 57–63. [CrossRef]
32. Liang, R.; Gunn, A.J.; Gunn, T.R. Can preterm twins breast feed successfully? *N. Z. Med. J.* **1997**, *110*, 209–212.
33. Zachariassen, G.; Faerk, J.; Grytter, C.; Esberg, B.; Juvonen, P.; Halken, S. Factors associated with successful establishment of breastfeeding in very preterm infants. *Acta Paediatr.* **2010**, *99*, 1000–1004. [CrossRef]
34. Pinchevski-Kadir, S.; Shust-Barequet, S.; Zajicek, M.; Leibovich, M.; Strauss, T.; Leibovitch, L.; Morag, I. Direct feeding at the breast Is associated with breast milk feeding duration among preterm infants. *Nutrients* **2017**, *9*, 1202. [CrossRef]
35. Rodrigues, C.; Teixeira, R.; Fonseca, M.J.; Zeitlin, J.; Barros, H.; Couto, A.S.; Lopes, A.; Almeida, A.; Portela, A.; Portuguese EPICE (Effective Perinatal Intensive Care in Europe) Network; et al. Prevalence and duration of breast milk feeding in very preterm infants: A 3-year follow-up study and a systematic literature review. *Paediatr. Perinat. Epidemiol.* **2018**, *32*, 237–246. [CrossRef]
36. Mamemoto, K.; Kubota, M.; Nagai, A.; Takahashi, Y.; Kamamoto, T.; Minowa, H.; Yasuhara, H. Factors associated with exclusive breastfeeding in low birth weight infants at NICU discharge and the start of complementary feeding. *Asia Pac. J. Clin. Nutr.* **2013**, *22*, 270–275. [CrossRef]
37. Lau, C. Breastfeeding challenges and the preterm mother-infant dyad: A conceptual model. *Breastfeed. Med.* **2018**, *13*, 8–17. [CrossRef]
38. Romaine, A.; Clark, R.H.; Davis, B.R.; Hendershot, K.; Kite, V.; Laughon, M.; Updike, I.; Miranda, M.L.; Meier, P.P.; Patel, A.L.; et al. Predictors of prolonged breast milk provision to very low birth weight infants. *J. Pediatr.* **2018**, *202*, 23–30. [CrossRef]

Article

Determinants of Exclusive Breastfeeding Cessation in the Early Postnatal Period among Culturally and Linguistically Diverse (CALD) Australian Mothers

Felix Akpojene Ogbo [1,2,*], Osita Kingsley Ezeh [3], Sarah Khanlari [4,5], Sabrina Naz [1], Praween Senanayake [1], Kedir Y. Ahmed [1], Anne McKenzie [6], Olayide Ogunsiji [7], Kingsley Agho [1,3], Andrew Page [1], Jane Ussher [1], Janette Perz [1], Bryanne Barnett AM [8] and John Eastwood [4,9,10,11,12,13]

[1]   Translational Health Research Institute, School of Medicine, Western Sydney University, Campbelltown Campus, Locked Bag 1797, Penrith, NSW 2571, Australia
[2]   General Practice Unit, Prescot Specialist Medical Centre, Welfare Quarters, Makurdi, Benue State 972261, Nigeria
[3]   School of Science and Health, Western Sydney University, Campbelltown Campus, Locked Bag 1797, Penrith, NSW 2571, Australia
[4]   Department of Community Paediatrics, Sydney Local Health District, Croydon Community Health Centre, 24 Liverpool Street, Croydon, NSW 2132, Australia
[5]   School of Medicine and Public Health, University of Newcastle, University Drive, Callaghan, NSW 2308, Australia
[6]   Child and Family Health Nursing, Primary & Community Health, South Western Sydney Local Health District, Narellan, NSW 2567, Australia
[7]   School of Nursing and Midwifery, Western Sydney University, Liverpool Campus, Locked Bag 1797, Penrith, NSW 2571, Australia
[8]   St John of God Raphael Services, Blacktown, NSW 2148, Australia
[9]   Ingham Institute for Applied Medical Research, 1 Campbell Street, Liverpool, NSW 2170, Australia
[10] School of Women's and Children's Health, The University of New South Wales, Kensington, Sydney, NSW 2052, Australia
[11] Menzies Centre for Health Policy, Charles Perkins Centre, School of Public Health, Sydney University, Sydney, NSW 2006, Australia
[12] School of Public Health, Griffith University, Gold Coast, QLD 4222, Australia
[13] Sydney Institute for Women, Children and their Families, Sydney Local Health District, Camperdown, NSW 2050, Australia
*   Correspondence: f.ogbo@westernsyndey.edu.au

Received: 23 May 2019; Accepted: 11 July 2019; Published: 16 July 2019

Abstract: There are limited epidemiological data on exclusive breastfeeding (EBF) among culturally and linguistically diverse (CALD) Australian mothers to advocate for targeted and/or culturally-appropriate interventions. This study investigated the determinants of EBF cessation in the early postnatal period among CALD Australian mothers in Sydney, Australia. The study used linked maternal and child health data from two local health districts in Australia ($N = 25{,}407$). Prevalence of maternal breastfeeding intention, skin-to-skin contact, EBF at birth, discharge, and the early postnatal period (1–4 weeks postnatal), were estimated. Multivariate logistic regression models were used to investigate determinants of EBF cessation in the early postnatal period. Most CALD Australian mothers had the intention to breastfeed (94.7%). Skin-to-skin contact (81.0%), EBF at delivery (91.0%), and at discharge (93.0%) were high. EBF remained high in the early postnatal period (91.4%). A lack of prenatal breastfeeding intention was the strongest determinant of EBF cessation (adjusted odds ratio [aOR] = 23.76, 95% CI: 18.63–30.30, for mothers with no prenatal breastfeeding intention and aOR = 6.15, 95% CI: 4.74–7.98, for those undecided). Other significant determinants of EBF cessation included a lack of partner support, antenatal and postnatal depression, intimate partner violence, low socioeconomic status, caesarean birth, and young maternal age (<20 years). Efforts to improve breastfeeding among women of CALD backgrounds in Australia should focus on women with vulnerabilities to maximise the benefits of EBF.

Keywords: exclusive breastfeeding; Australia; skin-to-skin; culturally and linguistically diverse (CALD)

---

1. Introduction

Global health organisations (such as the World Health Organization and the United Nations Children's Fund, WHO/UNICEF) recommend exclusive breastfeeding (EBF) for the first six months of life [1]. EBF is defined as providing the infant human breastmilk only, and when needed, oral rehydration solution, or drops/syrups of vitamins, minerals, or medicines [2]. Breast milk has a high proportion of fat, protein, sugar, and water that is required for infant growth and development compared to formula milk, as well as the immunologic substances to effectively protect against infectious diseases for the infant [3–5]. Appropriate EBF protects against childhood obesity [6] and has the potential to increase infant cognitive functioning [7]. EBF not only benefits the infant but is also associated with improved maternal health (e.g., a reduced risk developing of type 2 diabetes mellitus) [8] and improved household productivity due to no cost of human milk [9].

In Australia, culturally and linguistically diverse (CALD) is a term used for communities with diverse ethnic backgrounds, traditions, food, nationality, language, dress, societal structures, art, and religious characteristics [10]. While it may be debatable that such a 'label' exists to characterize such diverse populations, it is nevertheless one measure that can identify a sub-group in the Australian population who are often socioeconomically vulnerable, and for the purpose of focused research and programmatic interventions [11,12]. Past research suggests that CALD subgroups are more likely to experience adverse health outcomes, including depressive symptoms [13] and poor uptake of cancer screening [14], compared to the non-CALD populations.

Many qualitative studies conducted in Australia have reported that most CALD women value optimal breastfeeding practices, but a lack of access to traditional post-birth practices [15], stigma, and shame around public breastfeeding and ambivalence towards breastfeeding support [16,17], were barriers to appropriate breastfeeding. However, there is limited epidemiological data on the determinants of inappropriate EBF among CALD subgroups to inform targeted interventions. Quantitative research in the general Australian population suggests that intimate partner violence [18], a lack of partner support [18], assisted delivery [18,19], low socioeconomic status [20], lower maternal age (<25 years) [19], a mother not having intention to breastfeeding [21–23], and self-reported depressive symptoms [21] were associated with cessation of EBF. Notably, it is unclear whether these factors are relevant to mothers from CALD backgrounds as culturally-appropriate and focused intervention strategies are potentially more effective and less costly than traditional interventional approaches [12].

Understanding the factors that influence a mother's decision to initiate, cease, or continue EBF in the early postnatal period is essential as it can provide important information and opportunities for specific interventions. In addition to providing relevant data on EBF among CALD Australian mothers, this study seeks to provide an evidence base on EBF among CALD mothers for breastfeeding policy advocacy and/or evaluation of relevant social and health services for mothers of CALD backgrounds in Australia. This study aimed to investigate the determinants of EBF cessation in the early postnatal period among CALD Australian mothers in Sydney, New South Wales.

2. Materials and Methods

2.1. Data Source

The study was conducted based on retrospective maternal and child health data of all live births in public health facilities in Sydney Local Health District (SLHD) and South Western Sydney Local Health District (SWSLHD) between 2014 and 2016 ($N = 25{,}407$). These data were routinely collected as part of standard care provided during pregnancy and the postnatal period. Antenatal information that included socio-demographic characteristics, history of any previous pregnancy, probable depression based on the Edinburgh Postnatal Depression Scale (EPDS), and mothers' breastfeeding intention were collected by qualified midwives at the first prenatal care visit. Birth and postnatal data such as information on skin-to-skin contact, EBF at discharge and postnatally, were also obtained immediately after birth and during postnatal visits by qualified child and family health nurses (CFHN). During the first prenatal visit, women were asked to identify whether they belong to CALD, non-CALD, or Aboriginal or Torres Strait Islander subpopulations, and this information was entered into the database. CALD population was defined based on the Australian Bureau of Statistics description for the subgroup [10]. These maternal child health (MCH) data were stored in the local health district's Information Management & Technology Division (IM&TD) database. We obtained the perinatal data from the IM&TD, which were cleaned and linked using individual identifiers, and coded for analysis.

2.2. Study Setting

In Sydney, the SLHD and SWSLHD cover 52% of the metropolitan area, with an estimated population of 1.6 million people of different cultural backgrounds [24,25]. SLHD is located in the centre and inner west of Sydney, while SWSLHD is located in the south-western region of Sydney. A number of maternal and child health services are provided to all communities across both districts, including the most socioeconomically disadvantaged populations.

2.3. Outcome Variables

The main outcome variable was EBF in the early postnatal period (defined as 1–4 weeks post-birth). EBF was measured using the National Health and Medical Research Council (NHMRC) infant feeding guidelines [26], consistent with the WHO/UNICEF definitions for assessing infant and young child feeding practices [1]. EBF was defined as the proportion of infants who received only breast milk (including expressed milk) but allowed oral rehydration solution, syrups of vitamins/medicines. The prevalence of mothers' breastfeeding intention, skin-to-skin contact, EBF at delivery and discharge by the determining factors were also measured in the study.

EBF at delivery was defined as the proportion of infants who received only breast milk in the first 24 h post-birth, while EBF at discharge was measured as the proportion of infants who received only breast milk 24 h preceding discharge from the maternity unit. EBF in the early postnatal visit (1–4 weeks postnatal) referred to the proportion of infants who received only breast milk 24 h prior to this postnatal health visit by the CFHN. To assess mothers' breastfeeding intention, they were asked the following question: "Do you plan to breastfeed your child?" The study also considered skin-to-skin contact as past studies have indicated that skin-to-skin contact is the most effective strategy to promote, protect, and support EBF in early life, irrespective of the mode of birthing (vaginal or caesarean births) [27–29]. Skin-to-skin contact (SSC) was defined as placing the naked baby on the mother's bare chest or abdomen immediately or less than 10 min after birth or soon afterwards [30].

In each local health district, assessment of both the mother and baby is conducted between the first and fourth week post-birth by a CFHN during a universal health home visit. Relevant health information (including anthropometric measurement of the baby and assessment of infant feeding practices of the mother) are obtained and entered into the IM&TD database.

2.4. Study Factors

The exposure variables were broadly categorised into socio-demographic and health factors, which were selected based on previous studies [18,19,22,23] and data availability. Socio-demographic factors included maternal age (categorised as <20 years, 20–34 years or ≥35 years), socioeconomic status (SES, categorised as high, middle or low), maternal cigarette smoking in pregnancy (categorised as yes or no), partner support (categorised as yes, not sure or no), and major nationality groups (categorised as Oceania, North-West Europe, Southern-Eastern Europe, North Africa and The Middle East, South-East Asia, North-East Asia, Southern and Central Asia, Americas or Sub-Saharan Africa).

Health factors included pre-existing maternal health problems (such as diabetes mellitus and/or hypertension, categorised as yes or no), history of intimate partner violence (IPV, categorised as yes or no), type of delivery (categorised as normal vaginal, assisted vaginal or caesarean), and self-reported antenatal and postnatal depressive symptoms (categorised as score ≥13 or score <13 on the EPDS) [31]. Maternal breastfeeding intention was also considered a potential factor for the cessation of EBF in the early post-birth period based on past studies [21–23].

SES was calculated using the Socio-Economic Index for Areas (SEIFA). SEIFA is an indicator created by the Australian Bureau of Statistics using principal component analysis. It ranks areas (including a mother's address provided) in Australia according to relative socio-economic advantage and disadvantage [32]. In this study, deciles of SES were categorised into high (top 10% of the population), middle (middle 80%), and low (bottom 10%) groups, in line with previously published studies [13,31]. In accordance with NSW Health policy [33], IPV information was collected from mothers based on the following questions: (i) "within the last year have you been hit, slapped or hurt in other ways by your partner or ex-partner?", physical IPV; and (ii) "are you frightened of your partner or ex-partner?", psychological IPV.

2.5. Statistical Analysis

The analytical approach followed previously published studies [18,31]. Briefly, preliminary analyses were conducted to calculate frequencies of the study outcomes (i.e., breastfeeding intention, skin-to-skin contact, and EBF at delivery, discharge and postnatally) and cross-tabulations with study factors. This was followed by univariate regression models to examine the association between each study factor and cessation of EBF in the early postnatal period. Multivariate logistic regression analyses that adjusted for confounders was conducted to investigate the potential study factors that were associated with cessation of EBF in the early postnatal period among CALD mothers. Models adjusted for the potential confounding factors of birthing facility, the gender of the baby, maternal alcohol intake, and maternal body mass index, as well as socio-economic and health factors [21,34]. Odds ratios, with 95% confidence intervals, were calculated as the measure of association between the risk factors and cessation of EBF.

Our study also investigated the potential effect of missing data on the estimated odds ratios in sensitivity analyses that used an imputed dataset, based on the original data, which comprised complete information for EBF in the early postnatal period. Multiple imputations by chained equations were employed, which assume that data were missing at random [35]. This analytical approach also assumes that the known characteristics of study respondents can be used to examine the characteristics of participants with missing data [36]. All study factors and the outcome variable in the main analysis were included in the multiple imputation models. Revised odds ratios from the imputed data were generated using the *mim* command, for comparison with the complete case analyses. Sensitivity analyses were conducted based on 25 multiple imputations [37], and all analyses were conducted in Stata (Stata Corp, version 15.0, College Station, TX, USA).

2.6. Ethics

The Sydney Local Health District and South Western Sydney Local Health District Human Research Ethics Committees approved the collection of the data from the IM&TD database and subsequent analysis. Approval numbers HREC: LNR/11/LPOOL/463; SSA: LNRSSA/11/LPOOL/464 and Project No: 11/276 LNR; Protocol No X12-0164 and LNR/12/RPAH/266.

3. Results

3.1. Characteristics of the Study Population

The majority of mothers were from South-East Asia (24.5%) and North Africa and The Middle East (23.0%), while the lowest proportion of mothers were from North-West Europe (1.5%) (Table 1).

Table 1. Characteristics of the study population (N = 25,407).

Variables	N	(%)
Sociodemographic factors		
Maternal age group	25,407	
20–34 years	23,812	93.7
<20 years	124	0.5
≥35 years	1471	5.8
SES category	23,786	
Low	12,548	52.8
Middle	9377	39.4
High	1861	7.8
Smoking status	23,750	
No	23,170	97.6
Yes	580	2.4
Supportive partner	21,672	
Yes	21,061	97.2
No	611	2.8
Major nationality group	25,407	
Oceania	1481	5.8
North-West Europe	384	1.5
Southern-Eastern Europe	1316	5.2
North Africa and The Middle East	5846	23.0
South-East Asia	6222	24.5
North-East Asia	3512	13.8
Southern and Central Asia	5127	20.2
Americas	634	2.5
Sub-Saharan Africa	885	3.5
Health factors		
Antenatal health problems	24,603	
No	20,517	83.4
Yes	4086	16.6
Psychosocial intimate partner violence	21,583	
No	21,238	98.4
Yes	345	1.6
Physical intimate partner violence	21,523	
No	21,240	98.7
Yes	283	1.3
Type of delivery	25,378	
Normal vaginal	15,004	59.1
Assisted vaginal	2924	11.5
Caesarean section	7450	29.4

Table 1. *Cont.*

Variables	N	(%)
Antenatal depressive symptoms	20,560	
EPDS ≤ 9	16,972	82.6
EPDS 10–12	2078	10.1
EPDS ≥ 13	1510	7.3
Postnatal depressive symptoms	19,342	
EPDS ≤ 9	17,425	90.1
EPDS 10–12	1194	6.2
EPDS ≥ 13	723	3.7

N = Sample size; EPDS: Edinburgh Postnatal Depression Scale; SES: Socioeconomic status.

3.2. Breastfeeding Patterns by Study Factors

The proportion of CALD Australian mothers who exclusively breastfed in the early postnatal period varied by global regions, with mothers from South-East Asia (26.9%), North Africa and the Middle East (33.8%), and Oceania (12.2%) representing the highest percentage (Table 2). Mothers from the high SES category had a higher proportion of EBF (63.1%) compared to those from middle and low SES categories (33.4% and 3.5%, respectively).

Table 2. Prevalence of breastfeeding among culturally and linguistically diverse (CALD) mothers from South Western Sydney and Sydney Local Health Districts in Sydney, Australia, 2014–2016 (*N* = 25,407).

Variables	Breastfeeding Intention	Skin-to-Skin Contact	EBF at Delivery	EBF at Discharge	EBF at 1–4 Weeks
	n (%)	n (%)	n (%)	n (%)	n (%)
Study outcomes					
Yes	21,232 (94.7)	16,022 (81.2)	19,355 (90.7)	22,555 (93.3)	19,244 (91.4)
No	629 (2.8)	3697 (18.8)	1973 (9.3)	1629 (6.7)	1809 (8.6)
Undecided	572 (2.5)	-	-	-	-
Socio-demographic factors					
Maternal age group					
20–34 years	21,048 (93.8)	18,651 (94.6)	20,023 (93.9)	22,690 (93.8)	1628 (90.0)
<20 years	103 (0.5)	113 (0.6)	112 (0.5)	1375 (5.7)	14 (9.2)
≥35 years	1282 (5.7)	955 (4.8)	1193 (5.6)	119 (0.5)	167 (0.8)
SES category					
High	10,304 (63.4)	9778 (52.9)	10,411 (52.1)	11,877 (52.5)	1109 (63.1)
Middle	7962 (32.8)	7268 (39.3)	7958 (39.8)	8922 (39.5)	588 (33.4)
Low	1598 (3.8)	1427 (7.7)	1612 (8.1)	1817 (8.0)	62 (3.5)
Smoking status					
No	21,860 (97.6)	17,937 (97.3)	19,520 (97.7)	22,207 (97.6)	1532 (92.9)
Yes	543 (2.4)	437 (2.4)	467 (2.3)	540 (2.4)	117 (7.1)
Supportive partner					
Yes	20,501 (97.2)	16,457 (97.2)	17,773 (97.2)	20,190 (97.3)	1454 (95.7)
No	585 (2.8)	481 (2.8)	508 (2.8)	569 (2.7)	66 (4.3)
Major nationality group					
Oceania	1237 (5.5)	1193 (6.1)	1247 (5.9)	1368 (5.7)	221 (12.2)
North-West Europe	360 (1.6)	326 (1.7)	343 (1.6)	373 (1.5)	18 (1.0)
Southern-Eastern Europe	1177 (5.3)	1028 (5.2)	1146 (5.4)	1262 (5.2)	128 (7.08)
North Africa and The Middle East	5100 (22.7)	4649 (23.6)	4901 (23.0)	5526 (22.8)	612 (33.8)
South-East Asia	5437 (24.2)	4958 (25.1)	5235 (24.6)	5868 (24.3)	487 (26.9)
North-East Asia	3150 (14.0)	2837 (14.4)	3048 (14.3)	3431 (14.2)	150 (8.3)

Table 2. *Cont.*

Variables	Breastfeeding Intention	Skin-to-Skin Contact	EBF at Delivery	EBF at Discharge	EBF at 1–4 Weeks
	n (%)	n (%)	n (%)	n (%)	n (%)
Southern and Central Asia	4636 (20.7)	3638 (18.5)	4158 (19.5)	4914 (20.3)	116 (6.4)
Americas	552 (2.5)	439 (2.2)	526 (2.5)	607 (2.5)	42 (2.3)
Sub-Saharan Africa	784 (3.5)	651 (3.3)	724 (3.4)	835 (3.5)	35 (1.9)
Health factors					
Antenatal health problems					
No	17,847 (82.4)	15,821 (83.0)	17,334 (83.7)	19,514 (83.3)	1496 (85.1)
Yes	3819 (17.6)	3245 (17.0)	3370 (16.3)	3916 (16.7)	263 (14.9)
Psychosocial intimate partner violence					
No	20,657 (98.4)	19,618 (98.4)	17,919 (98.4)	20,360 (98.5)	1517 (97.9)
Yes	335 (1.6)	273 (1.6)	290 (1.6)	316 (1.5)	33 (2.1)
Physical intimate partner violence					
No	20,663 (98.7)	16,616 (98.7)	17,911 (98.7)	20,356 (98.7)	1522 (98.1)
Yes	270 (1.3)	213 (1.3)	243 (1.3)	265 (1.3)	30 (1.9)
Type of delivery					
Normal vaginal	13,356 (59.6)	14,903 (75.7)	14,126 (66.3)	14,432 (59.7)	988 (54.7)
Assisted vaginal	2581 (11.5)	2878 (14.6)	2452 (11.5)	2759 (11.4)	163 (9.0)
Caesarean section	6473 (28.9)	1913 (9.7)	4730 (22.2)	6968 (28.8)	655 (36.3)
Antenatal depressive symptoms					
EPDS ≤ 9	16,543 (72.6)	13,406 (83.3)	14,392 (82.9)	16,259 (82.6)	1164 (79.6)
EPDS 10–12	2013 (10.1)	1577 (9.8)	1751 (10.1)	1992 (10.1)	146 (10.0)
EPDS ≥ 13	1465 (7.3)	1121 (6.9)	1222 (7.0)	1433 (7.3)	152 (10.4)
Postnatal depressive symptoms					
EPDS ≤ 9	15,451 (90.2)	13,538 (90.6)	14,739 (90.5)	16,628 (90.2)	1,489 (88.9)
EPDS 10–12	1055 (6.2)	880 (5.9)	973 (6.0)	1139 (6.2)	94 (5.6)
EPDS ≥ 13	633 (3.7)	523 (3.5)	579 (3.5)	677 (3.6)	92 (5.5)

n: cases; antenatal health problems included diabetes mellitus and/or hypertension; EBF at delivery was defined as infants who received only breast milk within the first 24 h post-delivery; EBF at discharge was measured as infants who received only breast milk in the 24 h preceding discharge from the maternity unit; EPDS: Edinburgh Postnatal Depression Scale; SES: Socioeconomic status.

In the antenatal period, almost all mothers intended to breastfeed their babies (94.7%). Approximately 81% of mothers practised skin-to-skin contact, while 91% and 93% exclusively breastfed at delivery and discharge, respectively. A sub-analysis of the data (i.e., the estimation of the confidence interval and p-value around the estimates, data not shown) suggested that there was no significant difference between EBF prevalence at delivery and discharge. In the early postnatal period, EBF remained high among CALD Australian mothers (91.4%; Table 2).

3.3. Determinants of Exclusive Breastfeeding Cessation in the Early Postnatal Period

The study showed that mothers who indicated no prenatal breastfeeding intention or who were undecided about breastfeeding during pregnancy were more likely to cease EBF in the early postnatal period compared to those who indicated prenatal breastfeeding intention in the complete case analyses (adjusted odds ratio (aOR) = 23.76, 95% CI 18.63–30.30, $p < 0.001$, for those with no prenatal breastfeeding intention and aOR = 6.15, 95% CI 4.74–7.98, $p < 0.001$, for those undecided) (Table 3).

Table 3. Determinants of exclusive breastfeeding cessation in the early postnatal period among CALD mothers in South Western Sydney and Sydney Local Health Districts, 2014–2016 (N = 25,407).

Study Factors	Complete Case Analysis				Multiple Imputation Analysis *			
	Unadjusted OR (95% CI)	p Value	Adjusted OR (95% CI) (a)	p Value	Unadjusted OR (95% CI)	p Value	Adjusted OR (95% CI) (a)	p Value
Antenatal breastfeeding intention								
Yes	1.00		1.00		1.00		1.00	
No	31.11 (25.41–38.10)	<0.001	23.76 (18.63–30.30)	<0.001	31.11 (29.90–32.37)	<0.001	27.13 (26.04–28.27)	<0.001
Undecided	7.01 (5.69–8.64)	<0.001	6.15 (4.74–7.98)	<0.001	7.00 (6.72–7.30)	<0.001	6.38 (6.11–6.66)	<0.001
Sociodemographic factors								
Supportive partner								
Yes	1.00		1.00		1.00		1.00	
No	1.81 (1.38–2.36)	<0.001	1.69 (1.20–2.38)	0.003	1.66 (1.58–1.74)	<0.001	1.60 (1.52–1.69)	<0.00
Socio-economic status								
Low	1.00		1.00					
Middle	0.67 (0.61–0.75)	<0.001	0.85 (0.72–0.99)	0.044	0.67 (0.65–0.68)	<0.001	0.84 (0.82–0.86)	<0.00
High	0.36 (0.28–0.47)	<0.001	0.48 (0.32–0.71)	<0.001	0.36 (0.34–0.38)	<0.001	0.59 (0.56–0.63)	<0.00
Maternal age group								
20–34 years	1.00		1.00		1.00		1.00	
<20 years	1.76 (1.48–2.08)	<0.001	1.72 (1.37–2.15)	<0.001	1.75 (1.69–1.81)	<0.001	1.75 (1.69–1.81)	<0.00
≥35 years	1.79 (1.02–3.15)	0.044	1.91 (0.93–3.93)	0.078	1.78 (1.60–1.99)	<0.001	1.61 (1.44–1.80)	<0.00
Smoking status								
No	1.00		1.00		1.00		1.00	
Yes	4.29 (3.45–5.33)	<0.001	3.39 (2.56–4.49)	<0.001	4.28 (4.10–4.47)	<0.001	3.81 (3.65–3.99)	<0.001
Major nationality group								
Oceania	1.00				1.00			
North-West Europe	0.25 (0.15–0.41)	<0.001	0.49 (0.25–0.97)	0.043	0.25 (0.23–0.28)	<0.001	0.47 (0.43–0.52)	<0.001
Southern-Eastern Europe	0.57 (0.45–0.73)	<0.001	0.73 (0.53–1.00)	0.051	0.57 (0.55–0.60)	<0.001	0.71 (0.68–0.74)	<0.001
North Africa and The Middle East	0.67 (0.56–0.79)	<0.001	0.65 (0.51–0.82)	<0.001	0.67 (0.65–0.69)	<0.001	0.64 (0.62–0.67)	<0.001
South-East Asia	0.45 (0.38–0.53)	<0.001	0.46 (0.36–0.58)	<0.001	0.45 (0.43–0.46)	<0.001	0.47 (0.45–0.49)	<0.001
North-East Asia	0.24 (0.19–0.31)	<0.001	0.41 (0.29–0.56)	<0.001	0.24 (0.23–0.26)	<0.001	0.43 (0.41–0.45)	<0.001
Southern and Central Asia	0.11 (0.09–0.14)	<0.001	0.18 (0.13–0.24)	<0.001	0.11 (0.11–0.12)	<0.001	0.14 (0.14–0.15)	<0.001
Americas	0.37 (0.26–0.53)	<0.001	0.40 (0.24–0.66)	<0.001	0.37 (0.35–0.40)	<0.001	0.41 (0.38–0.44)	<0.001
Sub-Saharan Africa	0.21 (0.14–0.31)	<0.001	0.25 (0.15–0.41)	<0.001	0.21 (0.20–0.23)	<0.001	0.24 (0.23–0.26)	<0.001
Health factors								
Antenatal depressive symptoms								
EPDS ≤ 9	1.00		1.00		1.00		1.00	
EPDS 10–12	1.03 (0.86–1.23)	0.783	1.10 (0.88–1.36)	0.398	1.02 (0.99–1.06)	0.161	1.00 (0.97–1.04)	<0.001
EPDS ≥ 13	1.59 (1.33–1.91)	<0.001	1.50 (1.20–1.89)	<0.001	1.59 (1.53–1.64)	<0.001	1.57 (1.51–1.62)	<0.001

Table 3. *Cont.*

Study Factors	Complete Case Analysis				Multiple Imputation Analysis *			
	Unadjusted OR (95% CI)	*p* Value	Adjusted OR (95% CI) (a)	*p* Value	Unadjusted OR (95% CI)	*p* Value	Adjusted OR (95% CI) (a)	*p* Value
Postnatal depressive symptoms								
EPDS ≤ 9	1.00		1.00		1.00		1.00	
EPDS 10–12	0.911 (0.73–1.13)	0.402	0.97 (0.73–1.30)	0.860	0.92 (0.88–0.96)	<0.001	1.00 (0.96–1.04)	0.885
EPDS ≥ 13	1.61 (1.28–2.01)	<0.001	2.07 (1.55–2.77)	<0.001	1.57 (1.50–1.63)	<0.001	1.66 (1.59–1.73)	<0.001
Psychosocial intimate partner violence								
No	1.00		1.00		1.00		1.00	
Yes	1.47 (1.02–2.12)	0.041	1.66 (1.10–2.53)	0.017	1.46 (1.36–1.57)	<0.001	1.50 (1.39–1.61)	<0.001
Physical intimate partner violence								
No	1.00		1.00		1.00		1.00	
Yes	1.57 (1.06–2.31)	0.023	1.49 (0.94–2.38)	0.093	1.56 (1.45–1.69)	<0.001	1.62 (1.50–1.75)	<0.001
Type of delivery								
Normal vaginal	1.00		1.00		1.00		1.00	
Assisted vaginal	0.79 (0.67–0.94)	0.008	0.90 (0.72–1.14)	0.386	0.79 (0.76–0.82)	<0.001	0.96 (0.88–0.92)	0.078
Caesarean section	1.36 (1.23–1.51)	<0.001	1.35 (1.18–1.55)	<0.001	1.36 (1.33–1.39)	<0.001	1.48 (1.45–1.52)	<0.001
Pre-existing maternal health problems								
No	1.00		1.00		1.00		1.00	
Yes	0.87 (0.76–1.00)	0.054	1.02 (0.83–1.07)	0.790	0.87 (0.85–0.89)	<0.001	0.95 (0.93–0.98)	0.004

(a): adjusted for maternal body mass index, gender of the baby, maternal alcohol intake, Aboriginality and birthing facility location, as well as socioeconomic and health factors.
* Sensitivity analyses following multiple imputations for missing values; pre-existing maternal health problems included diabetes mellitus and/or hypertension.

Mothers who reported not having a supportive partner during pregnancy were more likely to stop EBF in the early post-birth period compared to those who reported receiving support from their partner during pregnancy (aOR = 1.69, 95% CI: 1.20–2.38, p = 0.003). Mothers from higher SES groups were less likely to discontinue EBF in the early postnatal period compared to those from lower SES groups (aOR = 0.48, 95% CI: 0.32–0.71, p < 0.001 for high SES and aOR = 0.85, 95% CI: 0.85–0.99, p = 0.044 for middle). Low maternal age (<20 years) was associated with cessation of EBF in the early postnatal period compared to middle reproductive-aged mothers (20–34 years; aOR = 1.72, 95% CI: 1.37–2.15, p < 0.001). Mothers who reported tobacco smoking during pregnancy were more likely to stop EBF in the early postnatal period compared to their counterparts who reported not smoking (aOR = 3.39, 95% CI: 2.56–4.49, p < 0.001) (Table 3).

The odds of stopping EBF in the immediate postpartum period were higher among mothers who reported antenatal depressive symptoms (EPDS ≥ 13; aOR = 1.50, 95% CI: 1.20–1.89, p < 0.001), and those who reported postnatal depressive symptoms (EPDS ≥ 13; aOR = 2.07, 95% CI: 1.55–2.77, p < 0.001). The likelihood of discontinuing EBF in the immediate postnatal period was higher among mothers who reported a history of psychological intimate partner violence (aOR = 1.66, 95% CI: 1.10–2.53, p = 0.017), and those who had caesarean births (aOR = 1.35, 95% CI: 1.18–1.55, p < 0.001). The odds of ceasing EBF in the immediate postpartum period were lower among mothers from all major nationality groups compared to their counterparts in the reference group (Oceania) (Table 3). The effects of maternal age > 35 years, physical intimate partner violence, and having pre-existing maternal health problems on ceasing EBF in the immediate postnatal period were statistically significant in the imputation data compared to those in the complete case analysis. Furthermore, the 95% CI of the aOR for no maternal breastfeeding intention and major nationality group based on the imputation data is much narrower than the one obtained based on the original data set, possibly reflecting the effect of missing data or small sample size (Table 3).

4. Discussion

Our study indicates that most CALD Australian mothers had the intention to breastfeed (94.7%). Eighty-one percent of mothers practised skin-to-skin contact, while 91% and 93% exclusively breastfed at delivery and discharge, respectively. Notably, EBF remained high in the early postnatal period among CALD Australian mothers (91.4%). A lack of maternal prenatal breastfeeding intention (no prenatal breastfeeding intention or undecided) was the strongest risk factor for the cessation of EBF in the early postpartum period. Other significant factors associated with the cessation of EBF in the early postnatal period included a lack of partner support, antenatal and postnatal depressive symptoms, psychosocial IPV, caesarean birthing, low socioeconomic status, and young maternal age (<20 years).

Consistent with past reports from Australia [21] and internationally [38–40], the present study indicates that a mother's prenatal intention not to breastfeed or undecided were the strongest risk factors for the cessation of EBF in the early postnatal period among CALD Australian women. Studies suggest that a good personal attitude towards breastfeeding, encouraging social norms for breastfeeding at home and work and a personal dislike for formula feeding, were essential to a mother's decision to have a definite intention to breastfeed [38,41]. In addition, the level of social support for the mother at the time of breastfeeding from the partner, grandmothers/in-laws [42], or peers [43,44], and the mother's own attitude to breastfeeding were influential to a mother's breastfeeding behaviour [38]. Our study also suggests that the lack of a supportive partner was associated with cessation of EBF in the early postnatal period. While studies have indicated that partners are willing to provide the required support to improve breastfeeding outcomes for both the mother and baby, a lack of appropriate and/or conflicting information from health practitioners to fathers have been flagged as constraints to fathers' full participation in breastfeeding support [45,46]. The involvement of fathers (in addition to grandmothers/in-laws if present) in prenatal breastfeeding education sessions and postnatal support are key priority areas for improving breastfeeding among mothers [47].

In Australia, many epidemiological studies have indicated that higher maternal SES was associated with EBF [19,20]. Our study shows that CALD mothers from higher SES groups were less likely to cease EBF in the early postnatal period compared to those from lower SES groups. Similarly, evidence has shown that infants whose fathers were from high SES groups were more likely to be breastfed up to 6-weeks post-birth compared to infants with fathers of low SES [47]. Possible reasons for why higher SES mothers (CALD and non-CALD) engage in optimal breastfeeding may include increased uptake of breastfeeding-related information and better skills in negotiating flexible workplace hours, creating opportunities for breastfeeding [48,49]. Research shows that young maternal age (<20 years) was associated with increased risk of mental health issues, with subsequent impacts on employment and socioeconomic position in later life [50]. The present study shows that young maternal age (<20 years) is associated with cessation of EBF in the early postnatal period, in line with previously published studies [18,19]. In many Australian communities, charitable organisations provide parenting support, life support, and educational opportunities for young mothers and their babies to lead healthy and productive lives in their local communities [51]. Integration of these services with antenatal and postnatal breastfeeding services (where practicable) would be essential to improve breastfeeding outcomes among young mothers.

Our study demonstrates an association between self-reported maternal antenatal and postnatal depressive symptoms (EPDS ≥ 13) and cessation of EBF within 4-weeks postnatally. Past studies have indicated that perinatal depression is associated with suboptimal breastfeeding, including cessation of EBF in the early postnatal period [52–54]. Our finding highlights the importance of routine perinatal depression screening in Australia [55]. Detailed information on policy and research implications of Australian perinatal depression screening has been published elsewhere [13,31,55,56]. Furthermore, the present study shows that self-reported psychological intimate partner violence is associated with cessation of EBF in the early postnatal period, in line with previous studies [33,57,58]. In most Australian healthcare centres, routine screening of women in the antepartum and postpartum periods for depression [13,55] and IPV [33,59] is conducted in an effort to promptly identify at-risk mothers, ensure referral for clinical assessment, follow-up, and support. Our findings suggest that sustained advocacy is required for perinatal depression and IPV screening among Australian CALD women to improve EBF.

Past studies conducted in Australia and internationally, have shown that cigarette smoking is associated with suboptimal breastfeeding outcomes [18,60]. Similarly, the present study shows that cigarette smoking is associated with cessation of EBF in the early postnatal period among CALD mothers. Evidence suggests that while there is an association between cigarette smoking and suboptimal breastfeeding, the evidence for a plausible biological mechanism is weak [61]. Nevertheless, a recent review indicated that smoking reduces the protective effects of breast milk and also negatively modifies the infant's response to breastfeeding and breast milk [62]. A possible explanation for this association among smokers may stem from a lack of motivation to breastfeed (i.e., less likely to have breastfeeding intention or to initiate breastfeeding) and/or less desire to seek help with breastfeeding challenges compared to non-smokers [60,61,63]. Australia has legislated a range of anti-smoking strategies (including restrictions on advertising, incremental taxation and regulation) to reduce smoking in the community. These initiatives have been shown to be effective in reducing experimentation and uptake of smoking among young people and overall smoking rates across all socio-demographics in Australia [63]. Sustained implementation of these initiatives will have positive impacts on breastfeeding outcomes among CALD Australian mothers in the short- and long-term.

There is robust evidence in the literature which indicates that health facility birthing, mode of birthing, and professional assistance received during birthing are critical to a mother's breastfeeding initiation and continuation [18,64]. This possibly reflects the key role the Baby Friendly Hospital Initiative (BFHI) plays in promoting, protecting, and supporting optimal breastfeeding in health facilities. Our study found that caesarean birthing was associated with cessation of EBF in the early postpartum period, which is similar to findings from previous studies [18,19]. Among Australian states and territories, evidence indicates that NSW has the lowest number of BFHI certified maternity centres, with implications for maternal breastfeeding experiences [18]. It is anticipated that the Australian National Breastfeeding Strategy 2019 and beyond will provide a roadmap for improving breastfeeding outcomes among CALD mothers in Australia, including increasing BFHI certified maternity centres in NSW [65]. The study showed that the odds of stopping EBF in the immediate postpartum period were lower among mothers from all major nationality groups compared to their counterparts in the reference group (Oceania). Future studies that focus on breastfeeding outcomes among sub-groups within the CALD populations may be warranted, as they would provide more insights for cultural care.

Limitations of this study are discussed. First, the cessation of breastfeeding in the early postnatal period was based on self-report, potentially leading a recall and/or measurement bias. The implication of this is that it may have resulted in an underestimation or overestimation of the relationship between the risk factors and the outcome. Second, the study was unable to assess or adjust for all potential determining factors (e.g., prematurity, level of support received postnatally, multi-parity, or partner education) as these may also affect the observed results. Third, we were unable to distinguish mothers who ceased EBF in the first, second, third, or fourth week postpartum. This analysis would have provided detailed information on early cessation of EBF postnatally among CALD Australian mothers. Fourth, the small sample size of mothers who indicated no prenatal breastfeeding intention or who were undecided may account for the large effect size. Fifth, the non-use of maternal and child health data from private healthcare and other local health districts in Sydney is a limitation in the present study. Finally, the unavailability of longitudinal data on EBF from 1 to 6 months postnatally and the use of secondary data were additional limitations in this study. Despite these limitations, the study provides breastfeeding data from CALD Australian women to inform population-level interventional strategies.

5. Conclusions

Our study suggests that many CALD mothers have definite intention to breastfeed, and most mothers practise skin-to-skin contact, EBF at delivery and at discharge from hospital post-birth. Notably, EBF remained high in the early postnatal period (within 4-weeks postpartum) among CALD Australian mothers. A lack of maternal prenatal breastfeeding intention and partner support, antenatal and postnatal depressive symptoms, psychosocial intimate partner violence, caesarean birthing, low socioeconomic status, and young maternal age (<20 years) were associated with the cessation of EBF in the early postnatal period among CALD Australian mothers. Our study provides insight into breastfeeding practices among CALD Australian mothers to inform targeted initiatives, especially those identified to be at risk of early cessation of EBF.

Author Contributions: Conceptualization, F.A.O.; data curation, F.A.O. and S.N.; formal analysis, F.A.O.; funding acquisition, F.A.O., J.U., J.P., B.B.A.M., and J.E.; investigation, F.A.O.; methodology, F.A.O.; project administration, F.A.O.; resources, F.A.O., A.M., A.P., and J.E.; software, F.A.O.; supervision, F.A.O.; validation, F.A.O., O.K.E., S.K., S.N., P.S., K.Y.A., A.M., O.O., K.A., A.P., B.B.A.M., and J.E.; visualization, F.A.O., O.K.E., S.N., P.S., K.Y.A., A.M., O.O., K.A., A.P., J.U., J.P., B.B.A.M., and J.E.; writing—original draft, F.A.O. and O.K.E.; writing—review and editing, F.A.O., O.K.E., S.K., S.N., P.S., K.Y.A., A.M., O.O., K.A., A.P., B.B.A.M., and J.E.

Funding: This research was funded by the early career research funding grant from the Office of the Deputy Vice-Chancellor (Research and Innovation), Western Sydney University, obtained by F.A.O. The APC was also funded through this grant.

Acknowledgments: The authors are grateful to all the health professionals in South Western Sydney Local Health Districts and Sydney Local Health District who spent time entering the data, and also to personnel (particularly Leslie Cramer for data download) at the Information Management & Technology Division for the time spent in generating the data for the analysis. The authors are also grateful to the Early Years Research Group at the Ingham Institute, South Western Sydney Local Health District and the Healthy Home And Neighbourhood at the Sydney Local Health District for the initial funding. F.A.O. is also grateful to Melinda Wolfenden at THRI, Western Sydney University, for the time and effort in proofreading the grant application and excellent administrative support. The effort of Victoria Blight in securing the research grant is also acknowledged.

Conflicts of Interest: The authors declare no conflict of interest. K.A. is a Guest Editor on the Special Issue "Breastfeeding: Short and Long-Term Benefits to Baby and Mother" for Nutrients but did not play any role in the peer-review and decision-making process for this manuscript. The funder had no role in the design of the study; in the collection, analyses, or interpretation of data; in the writing of the manuscript, or in the decision to publish the results.

References

1. World Health Organization. Indicators for assessing infant and young child feeding practices: Part 1: Definitions. In Proceedings of the Consensus Meeting, Washington, DC, USA, 6–8 November 2007.

2. Gartner, L.M.; Morton, J.; Lawrence, R.A.; Naylor, A.J.; O'Hare, D.; Schanler, R.J.; Eidelman, A.I. American academy of pediatrics section on breastfeeding: Breastfeeding and the use of human milk. *Pediatrics* **2005**, *115*, 496–506. [PubMed]

3. Karlsson, O.; Rodosthenous, R.S.; Jara, C.; Brennan, K.J.; Wright, R.O.; Baccarelli, A.A.; Wright, R.J. Detection of long non-coding RNAs in human breastmilk extracellular vesicles: Implications for early child development. *Epigenetics* **2016**, *11*, 721–729. [CrossRef] [PubMed]

4. Fujita, M.; Lo, Y.J.; Brindle, E. Nutritional, inflammatory, and ecological correlates of maternal retinol allocation to breast milk in agro-pastoral Ariaal communities of northern Kenya. *Am. J. Hum. Biol.* **2017**, *29*, e22961. [CrossRef] [PubMed]

5. Verduci, E.; Banderali, G.; Barberi, S.; Radaelli, G.; Lops, A.; Betti, F.; Riva, E.; Giovannini, M. Epigenetic effects of human breast milk. *Nutrients* **2014**, *6*, 1711–1724. [CrossRef] [PubMed]

6. Mannan, H. Early Infant Feeding of Formula or Solid Foods and Risk of Childhood Overweight or Obesity in a Socioeconomically Disadvantaged Region of Australia: A Longitudinal Cohort Analysis. *Int. J. Environ. Res. Public Health* **2018**, *15*, 1685. [CrossRef] [PubMed]

7. Victora, C.G.; Horta, B.L.; de Mola, C.L.; Quevedo, L.; Pinheiro, R.T.; Gigante, D.P.; Gonçalves, H.; Barros, F.C. Association between breastfeeding and intelligence, educational attainment, and income at 30 years of age: A prospective birth cohort study from Brazil. *Lancet Glob. Health* **2015**, *3*, e199–e205. [CrossRef]

8. Victora, C.G.; Bahl, R.; Barros, A.J.; França, G.V.; Horton, S.; Krasevec, J.; Murch, S.; Sankar, M.J.; Walker, N.; Rollins, N.C. Breastfeeding in the 21st century: Epidemiology, mechanisms, and lifelong effect. *Lancet* **2016**, *387*, 475–490. [CrossRef]

9. León-Cava, N.; Lutter, C.; Ross, J.; Martin, L. *Quantifying the Benefits of Breastfeeding: A Summary of the Evidence*; Pan American Health Organization: Washington, DC, USA, 2002.

10. Australian Bureau of Statistics. Culturally and Linguistically Diversity (CALD) Characteristics. Available online: https://www.abs.gov.au/ausstats/abs@.nsf/Lookup/by%20Subject/4529.0.00.003~{}2014~{}Main%20Features~{}Cultural%20and%20Linguistic%20Diversity%20(CALD)%20Characteristics~{}13 (accessed on 18 January 2019).

11. Brewer, R. *Culturally and Linguistically Diverse Women in the Australian Capital Territory: Enablers and Barriers to Achieving Social Connectedness*; Women's Centre for Health Matters: Canberra, Australia, 2009.

12. Adily, A.; Ward, J. Improving health among culturally diverse subgroups: An exploration of trade-offs and viewpoints among a regional population health workforce. *Health Promot. J. Aust.* **2005**, *16*, 207–212. [CrossRef]

13. Ogbo, F.A.; Eastwood, J.; Hendry, A.; Jalaludin, B.; Agho, K.E.; Barnett, B.; Page, A. Determinants of antenatal depression and postnatal depression in Australia. *BMC Psychiatry* **2018**, *18*, 49. [CrossRef]

14. Aminisani, N.; Armstrong, B.K.; Canfell, K. Cervical cancer screening in Middle Eastern and Asian migrants to Australia: A record linkage study. *Cancer Epidemiol.* **2012**, *36*, e394–e400. [CrossRef]

15. Schmied, V.; Olley, H.; Burns, E.; Duff, M.; Dennis, C.-L.; Dahlen, H.G. Contradictions and conflict: A meta-ethnographic study of migrant women's experiences of breastfeeding in a new country. *BMC Pregnancy Childbirth* **2012**, *12*, 163. [CrossRef] [PubMed]

16. Gallegos, D.; Vicca, N.; Streiner, S. Breastfeeding beliefs and practices of African women living in Brisbane and Perth, Australia. *Mater. Child Nutr.* **2015**, *11*, 727–736. [CrossRef] [PubMed]

17. Joseph, J.; Brodribb, W.; Liamputtong, P. "Fitting-in Australia" as nurturers: Meta-synthesis on infant feeding experiences among immigrant women. *Women Birth* **2018**. [CrossRef] [PubMed]

18. Ogbo, F.A.; Eastwood, J.; Page, A.; Arora, A.; McKenzie, A.; Jalaludin, B.; Tennant, E.; Miller, E.; Kohlhoff, J.; Noble, J. Prevalence and determinants of cessation of exclusive breastfeeding in the early postnatal period in Sydney, Australia. *Int. Breastfeed. J.* **2017**, *12*, 16. [CrossRef] [PubMed]

19. Hauck, Y.L.; Fenwick, J.; Dhaliwal, S.S.; Butt, J. A Western Australian survey of breastfeeding initiation, prevalence and early cessation patterns. *Mater. Child Health J.* **2011**, *15*, 260–268. [CrossRef] [PubMed]

20. Amir, L.H.; Donath, S.M. Socioeconomic status and rates of breastfeeding in Australia: Evidence from three recent national health surveys. *Med. J. Aust.* **2008**, *189*, 254–256. [PubMed]

21. Forster, D.A.; McLachlan, H.L.; Lumley, J. Factors associated with breastfeeding at six months postpartum in a group of Australian women. *Int. Breastfeed. J.* **2006**, *1*, 18. [CrossRef] [PubMed]

22. Scott, J.; Landers, M.; Hughes, R.M.; Binns, C. Factors associated with breastfeeding at discharge and duration of breastfeeding. *J. Paediatr. Child Health* **2001**, *37*, 254–261. [CrossRef]

23. Craig, P.L.; Knight, J.; Comino, E.; Webster, V.; Pulver, L.J.; Harris, E. Initiation and duration of breastfeeding in an aboriginal community in south western Sydney. *J. Hum. Lact.* **2011**, *27*, 250–261. [CrossRef]

24. South Western Sydney Local Health District. *Research Strategy for South Western Sydney Local Health District 2012–2021*; South Western Sydney Local Health District: Sydney, Australia, 2012.

25. Sydney Local Health District. Planning. Available online: https://www.slhd.nsw.gov.au/planning/profiles.html (accessed on 22 June 2016).

26. National Health and Medical Research Council. *Infant Feeding Guidelines*; National Health and Medical Research Council: Canberra, Australia, 2012.

27. Bramson, L.; Lee, J.W.; Moore, E.; Montgomery, S.; Neish, C.; Bahjri, K.; Melcher, C.L. Effect of early skin-to-skin mother—Infant contact during the first 3 h following birth on exclusive breastfeeding during the maternity hospital stay. *J. Hum. Lact.* **2010**, *26*, 130–137. [CrossRef]

28. Crenshaw, J.T. Healthy birth practice# 6: Keep mother and baby together—It's best for mother, baby, and breastfeeding. *J. Perinat. Educ.* **2014**, *23*, 211–217. [PubMed]

29. Moore, E.R.; Bergman, N.; Anderson, G.C.; Medley, N. Early skin-to-skin contact for mothers and their healthy newborn infants. *Cochrane Database Syst. Rev.* **2016**. [CrossRef] [PubMed]

30. World Health Organization. *Guideline: Protecting, Promoting and Supporting Breastfeeding in Facilities Providing Maternity and Newborn Services*; CC BY-NC-SA 3.0 IGO; World Health Organization: Geneva, Switzerland, 2017.

31. Eastwood, J.; Ogbo, F.A.; Hendry, A.; Noble, J.; Page, A. Early Years Research Group. The impact of antenatal depression on perinatal outcomes in Australian women. *PLoS ONE* **2017**, *12*, e0169907. [CrossRef] [PubMed]

32. Australian Bureau of Statistics. *Technical Paper: Socio-Economic Indexes for Areas (SEIFA) 2011*; Commonwealth of Australia: Canberra, Australia, 2013.

33. New South Wales Health Department. *Unless They're Asked: Routine Screening for Domestic Violence in NSW Health: An Evaluation of the Pilot Project*; New South Wales Health: Sydney, Australia, 2001.

34. Binns, C.; Gilchrist, D.; Gracey, M.; Zhang, M.; Scott, J.; Lee, A. Factors associated with the initiation of breast-feeding by Aboriginal mothers in Perth. *Public Health Nutr.* **2004**, *7*, 857–861. [CrossRef] [PubMed]

35. Van Buuren, S.; Boshuizen, H.C.; Knook, D.L. Multiple imputation of missing blood pressure covariates in survival analysis. *Stat. Med.* **1999**, *18*, 681–694. [CrossRef]

36. Sterne, J.A.; White, I.R.; Carlin, J.B.; Spratt, M.; Royston, P.; Kenward, M.G.; Wood, A.M.; Carpenter, J.R. Multiple imputation for missing data in epidemiological and clinical research: Potential and pitfalls. *BMJ* **2009**, *338*, b2393. [CrossRef] [PubMed]

37. Spratt, M.; Carpenter, J.; Sterne, J.A.; Carlin, J.B.; Heron, J.; Henderson, J.; Tilling, K. Strategies for multiple imputation in longitudinal studies. *Am. J. Epidemiol.* **2010**, *172*, 478–487. [CrossRef]

38. Colaizy, T.T.; Saftlas, A.F.; Morriss, F.H. Maternal intention to breast-feed and breast-feeding outcomes in term and preterm infants: Pregnancy Risk Assessment Monitoring System (PRAMS), 2000–2003. *Public Health Nutr.* **2012**, *15*, 702–710. [CrossRef]

39. Wu, Y.; Ho, Y.; Han, J.; Chen, S. The influence of breastfeeding self-efficacy and breastfeeding intention on breastfeeding behavior in postpartum women. *J. Nurs.* **2018**, *65*, 42–50.

40. Stuebe, A.M.; Bonuck, K. What predicts intent to breastfeed exclusively? Breastfeeding knowledge, attitudes, and beliefs in a diverse urban population. *Breastfeed. Med.* **2011**, *6*, 413–420. [CrossRef]

41. Kools, E.J.; Thijs, C.; Vries, H.d. The behavioral determinants of breast-feeding in The Netherlands: Predictors for the initiation of breast-feeding. *Health Educ. Behav.* **2005**, *32*, 809–824. [CrossRef] [PubMed]

42. Negin, J.; Coffman, J.; Vizintin, P.; Raynes-Greenow, C. The influence of grandmothers on breastfeeding rates: A systematic review. *BMC Pregnancy Childbirth* **2016**, *16*, 91. [CrossRef] [PubMed]

43. Cameron, A.J.; Hesketh, K.; Ball, K.; Crawford, D.; Campbell, K.J. Influence of peers on breastfeeding discontinuation among new parents: The Melbourne InFANT Program. *Pediatrics* **2010**, *126*, e601–e607. [CrossRef] [PubMed]

44. Shakya, P.; Kunieda, M.K.; Koyama, M.; Rai, S.S.; Miyaguchi, M.; Dhakal, S.; Sandy, S.; Sunguya, B.F.; Jimba, M. Effectiveness of community-based peer support for mothers to improve their breastfeeding practices: A systematic review and meta-analysis. *PLoS ONE* **2017**, *12*, e0177434. [CrossRef] [PubMed]

45. Susin, L.R.O.; Giugliani, E.R.J. Inclusion of fathers in an intervention to promote breastfeeding: Impact on breastfeeding rates. *J. Hum. Lact.* **2008**, *24*, 386–392. [CrossRef] [PubMed]

46. Tohotoa, J.; Maycock, B.; Hauck, Y.L.; Howat, P.; Burns, S.; Binns, C.W. Dads make a difference: An exploratory study of paternal support for breastfeeding in Perth, Western Australia. *Int. Breastfeed. J.* **2009**, *4*, 15. [CrossRef]

47. Maycock, B.; Binns, C.W.; Dhaliwal, S.; Tohotoa, J.; Hauck, Y.; Burns, S.; Howat, P. Education and support for fathers improves breastfeeding rates: A randomized controlled trial. *J. Hum. Lact.* **2013**, *29*, 484–490. [CrossRef]

48. Mitra, A.K.; Khoury, A.J.; Hinton, A.W.; Carothers, C. Predictors of breastfeeding intention among low-income women. *Mater. Child Health J.* **2004**, *8*, 65–70. [CrossRef]

49. McIntyre, E.; Hiller, J.E.; Turnbull, D. Determinants of infant feeding practices in a low socio-economic area: Identifying environmental barriers to breastfeeding. *Aust. N. Z. J. Public Health* **1999**, *23*, 207–209. [CrossRef]

50. Cooklin, A.R.; Canterford, L.; Strazdins, L.; Nicholson, J.M. Employment conditions and maternal postpartum mental health: Results from the Longitudinal Study of Australian Children. *Arch. Women's Ment. Health* **2011**, *14*, 217–225. [CrossRef]

51. New South Wales Health. Youth Health Resources and Contacts for Young People. Available online: https://www.health.nsw.gov.au/kidsfamilies/youth/Pages/yh-resources-for-young-people.aspx (accessed on 22 April 2019).

52. Grigoriadis, S.; VonderPorten, E.H.; Mamisashvili, L.; Tomlinson, G.; Dennis, C.-L.; Koren, G.; Steiner, M.; Mousmanis, P.; Cheung, A.; Radford, K. The impact of maternal depression during pregnancy on perinatal outcomes: A systematic review and meta-analysis. *J. Clin. Psychiatry* **2013**, *74*, 321–341. [CrossRef] [PubMed]

53. Dennis, C.-L.; McQueen, K. The relationship between infant-feeding outcomes and postpartum depression: A qualitative systematic review. *Pediatrics* **2009**, *123*, e736–e751. [CrossRef] [PubMed]

54. Rahman, A.; Hafeez, A.; Bilal, R.; Sikander, S.; Malik, A.; Minhas, F.; Tomenson, B.; Creed, F. The impact of perinatal depression on exclusive breastfeeding: A cohort study. *Mater. Child Nutr.* **2016**, *12*, 452–462. [CrossRef] [PubMed]

55. Beyondblue. Clinical practice guidelines for depression and related disorders—Anxiety, bipolar disorder and puerperal psychosis—In the perinatal period. In *A Guideline for Primary Care Health Professionals*; Beyondblue: Melbourne, Australia, 2011.

56. Buist, A.E.; Barnett, B.E.; Milgrom, J.; Pope, S.; Condon, J.T.; Ellwood, D.A.; Boyce, P.M.; Austin, M.-P.V.; Hayes, B.A. To screen or not to screen-that is the question in perinatal depression. *Med. J. Aust.* **2002**, *177*, S101–S105. [PubMed]

57. Kendall-Tackett, K.A. Violence against women and the perinatal period: The impact of lifetime violence and abuse on pregnancy, postpartum, and breastfeeding. *Trauma Violence Abus.* **2007**, *8*, 344–353. [CrossRef] [PubMed]

58. Martin-de-las-Heras, S.; Velasco, C.; Luna-del-Castillo, J.; Khan, K. Breastfeeding avoidance following psychological intimate partner violence during pregnancy: A cohort study and multivariate analysis. *BJOG Int. J. Obstet. Gynaecol.* **2018**, *126*, 778–783. [CrossRef] [PubMed]

59. Australian Institute of Health and Welfare. *Screening for Domestic Violence during Pregnancy: Options for Future Reporting in the National Perinatal Data Collection*; AIHW: Canberra, Australia, 2015.

60. Amir, L.H.; Donath, S.M. Does maternal smoking have a negative physiological effect on breastfeeding? The epidemiological evidence. *Birth* **2002**, *29*, 112–123. [CrossRef]

61. Amir, L.H. Maternal smoking and reduced duration of breastfeeding: A review of possible mechanisms. *Early Hum. Dev.* **2001**, *64*, 45–67. [CrossRef]

62. Napierala, M.; Mazela, J.; Merritt, T.A.; Florek, E. Tobacco smoking and breastfeeding: Effect on the lactation process, breast milk composition and infant development. A critical review. *Environ. Res.* **2016**, *151*, 321–338. [CrossRef]
63. Australian National Preventive Health Agency. *Smoking and Disadvantage*; Australian National Preventive Health Agency: Canberra, Australia, 2013.
64. Smith, J.; Cattaneo, A.; Iellamo, A.; Javanparast, S.; Atchan, M.; Gribble, K.; Hartmann, B.; Salmon, L.; Tawia, S.; Hull, N.; et al. *Review of Effective Strategies to Promote Breastfeeding: An. Evidence Check Rapid Review Brokered by the Sax Institute for the Department of Health*; Sax Institute: Sydney, Australia, 2018.
65. Australian Government Department of Health. Breastfeeding. Available online: http://www.health.gov.au/internet/main/publishing.nsf/Content/health-pubhlth-strateg-brfeed-index.htm (accessed on 22 April 2019).

Article

Breastfeeding Status and Duration and Infections, Hospitalizations for Infections, and Antibiotic Use in the First Two Years of Life in the ELFE Cohort

Camille Davisse-Paturet [1,*], Karine Adel-Patient [2], Amandine Divaret-Chauveau [3,4], Juliette Pierson [1], Sandrine Lioret [1], Marie Cheminat [5], Marie-Noëlle Dufourg [5], Marie-Aline Charles [1,5] and Blandine de Lauzon-Guillain [1,*]

1   Université de Paris, CRESS, INSERM, INRA F-75004 Paris, France
2   UMR Service de Pharmacologie et Immunoanalyse, CEA, INRA, Université Paris-Saclay, 91191 Gif-sur-Yvette, France
3   Unité d'allergologie pédiatrique, Hôpital d'enfants, CHRU de Nancy, 54500 Vandoeuvre-lès-Nancy, France
4   EA3450, DevAH-Department of Physiology, Faculty of Medicine, University of Lorraine, 54500 Vandoeuvre-lès-Nancy, France
5   Ined, Inserm, Joint Unit Elfe F-75020 Paris, France
*   Correspondence: camille.davisse-paturet@inserm.fr (C.D.-P.); blandine.delauzon@inserm.fr (B.d.L.-G.); Tel.: +33-145-595-053 (C.D.-P.); +33-145-595-019 (B.d.L.-G.)

Received: 31 May 2019; Accepted: 11 July 2019; Published: 15 July 2019

Abstract: In low- and middle-income countries, the protective effect of breastfeeding against infections is well established, but in high-income countries, the effect could be weakened by higher hygienic conditions. We aimed to examine the association between breastfeeding and infections in the first 2 years of life, in a high-income country with relatively short breastfeeding duration. Among 10,349 young children from the nationwide Etude Longitudinale Française depuis l'Enfance (ELFE) birth cohort, breastfeeding and parent-reported hospitalizations, bronchiolitis and otitis events, and antibiotic use were prospectively collected up to 2 years. Never-breastfed infants were used as reference group. Any breastfeeding for <3 months was associated with higher risks of hospitalizations from gastrointestinal infections or fever. Predominant breastfeeding for <1 month was associated with higher risk of a single hospital admission while predominant breastfeeding for ≥3 months was associated with a lower risk of long duration (≥4 nights) of hospitalization. Ever breastfeeding was associated with lower risk of antibiotic use. This study confirmed the well-known associations between breastfeeding and hospitalizations but also highlighted a strong inverse association between breastfeeding and antibiotic use. Although we cannot infer causality from this observational study, this finding is worth highlighting in a context of rising concern regarding antibiotic resistance.

Keywords: breastfeeding; infections; birth cohort; hospitalizations; antibiotic use

---

1. Introduction

In 2013, infectious diseases were in the four main categories of leading causes of death among children under 5 years of age worldwide [1]. Nonetheless, disparities exist between countries. Infections are the leading causes of death in sub-Saharan African countries but not in high-income countries. The good hygienic conditions and health care system available in some countries significantly reduce the prevalence and fatal issue of such diseases but do not fully prevent them.

The World Health Organization (WHO) recommends exclusive breastfeeding for 6 months, or at least the first 4 months of life [2]. These recommendations were mainly based on the protective effect of breastfeeding against infectious morbidity and mortality [3]. In fact, breast milk components, such as immunoglobulin A (IgA) or maternal leukocytes, can both supplement and promote the newborn's immature immune system [4] and therefore lead to protective effect against infections.

More precisely, recent literature has shown that breastfeeding is related to a reduced rate of hospital admission for diarrhea and respiratory infections as well as a protective effect on otitis media in children up to 2 years old [3,5]. Of note, otitis media studies were mostly from high-income countries, whereas results on diarrhea and respiratory infection studies were mostly found in settings from low- and middle-income countries [6,7]. In high-income countries, the preventive effect of breastfeeding on respiratory tract infections is less consistent across studies [7]. In the cluster-randomized trial on promotion of breastfeeding (PROBIT), which took place in Belarus in the 1990s, breastfeeding was related to a reduced risk of gastrointestinal infections in the first year of life [8].

The aim of this study was to assess the association between breastfeeding duration and several indicators of infectious morbidity, in France, a high-income country with the specificity of low breastfeeding initiation rate (69.7%) and median duration below the guidelines (17 weeks among breastfeeding mothers) [3,9].

2. Materials and Methods

2.1. Study Population

This analysis was based on data from the ELFE (Etude Longitudinale Française depuis l'Enfance) study, a multidisciplinary nationwide birth cohort including 18,329 children born in 2011 in France [10]. The inclusion criteria were as follows: singleton or twins born after 33 weeks of gestation, to mothers aged 18 years or older. Participating mothers had to provide written consent for their own and their child's participation. Fathers signed the consent form for the child's participation when present at inclusion or were informed about their rights to oppose it. The ELFE study was approved by the Advisory Committee for Treatment of Health Research Information (Comité Consultatif sur le Traitement des Informations pour la Recherche en Santé), the National Data Protection Authority (Commission Nationale Informatique et Libertés), and the National Statistics Council.

2.2. Breastfeeding

The feeding method was prospectively collected up to 2 years and the calculation of breastfeeding duration was detailed in a previous paper [11]. As previously described, two breastfeeding definitions were used in the present study: any breastfeeding and predominant breastfeeding. Any breastfeeding was defined as the infant receiving breastmilk. Predominant breastfeeding was defined as the only milk given to the infant being human milk (no animal milk or infant formula).

For each breastfeeding definition, infants were first categorized as ever or never breastfed and then according to their breastfeeding duration (never, <1 month, 1 month to <3 months, 3 months to <6 months, ≥6 months).

2.3. Parental Report of Infections

In the present study, infections were assessed with several indicators: hospitalizations; frequency of bronchiolitis and otitis events; and antibiotic use.

At the 1- and 2-year phone interviews, parents reported hospitalizations in the previous year (date, duration, and main cause). Infants with at least one hospitalization for infectious disease were identified (mainly fever, gastrointestinal infection, and bronchiolitis). Hospitalizations from infectious diseases between birth and 2 years of age were characterized by the number of events (none, 1, ≥2) and cumulative duration (never, 1–3 nights, ≥4 nights).

At the 1-year phone interview, parents both reported whether the child ever had a bronchiolitis event and if the child had had at least 3 bronchiolitis events since birth. At the 2-year phone interview, parents reported whether the child had had 3 bronchiolitis episodes or more since birth. A mixed variable was computed from both interviews resulting in a 3-category variable assessing bronchiolitis events in the first 2 years of life (never, 1 or 2 events, ≥3 events).

At the 2-year phone interview, parents reported otitis events since birth (<3 events, ≥3 events). Unfortunately, it was not possible to distinguish infant with no otitis event from those having 1 or 2 events.

At the 1- and 2-year phone interviews, parents reported the frequency of antibiotic use for their child in the previous 12 months. These frequencies were then combined into a 4-category variable (never, once, 2 or 3 times, >3 times).

2.4. Other Variables

Maternal and household data were collected using face-to-face interviews at the maternity unit and then by phone interview at the 2-month follow-up. Because these data were more thoroughly assessed during the 2-month interview and only marginally changed during these 2 months, we used the data collected at 2 months in our analyses. Socio-demographic characteristics collected during the maternity stay were used only when the 2-month values were missing.

Maternal socio-demographic characteristics included age at the birth of her first child (<25, 25–29, 30–34, ≥35 years), education level (below secondary school, secondary school, high school, 2-year university degree, 3-year university degree, 5-year university degree or higher), place of birth (France, abroad), and employment status (unemployed, employed). Household characteristics included income per consumption unit (≤€750, €751 to 1111, €1112 to 1500, €1501 to 1944, €1945 to 2500, >€2500 /month) and composition (couple with children, single parenthood, step family).

Maternal health-related characteristics included smoking status during pregnancy (never smoked, smoked only before pregnancy, smoked only in early pregnancy, smoked throughout pregnancy) and pre-pregnancy body mass index (BMI) (<18.5 kg/m^2, 18.5 to 24.9 kg/m^2, 25.0 to 29.9 kg/m^2, ≥ 30.0 kg/m^2).

Infant's birth order (first born, second, third, fourth, or higher), caesarean-section delivery, sex, twin birth, and gestational age were collected at birth from medical records. Infant's age at first attendance at a shared childcare facility was computed from the 1-year phone interview as a 5-category variable (≤2 months, >2 to ≤4 months, >4 to ≤6 months, >6 to ≤12 months, never attended in the first year).

2.5. Sample Selection

Infants whose parents withdrew consent during the first year ($n = 57$) and not meeting eligibility criteria ($n = 1$) were excluded, resulting in 17,984 eligible infants. In twin pregnancies, one twin was randomly selected ($n = 287$ exclusions) to avoid family clusters.

We excluded infants without any follow-up at 2 years ($n = 4705$). We then excluded infants with missing data on breastfeeding ($n = 147$) and those with missing data on infections or antibiotic use ($n = 1894$). We also excluded infants with incomplete information for potential confounding variables ($n = 889$). These exclusions lead to a sample of 10,349 infants for the complete case analysis regarding parental reports of infections and antibiotic use (Figure 1).

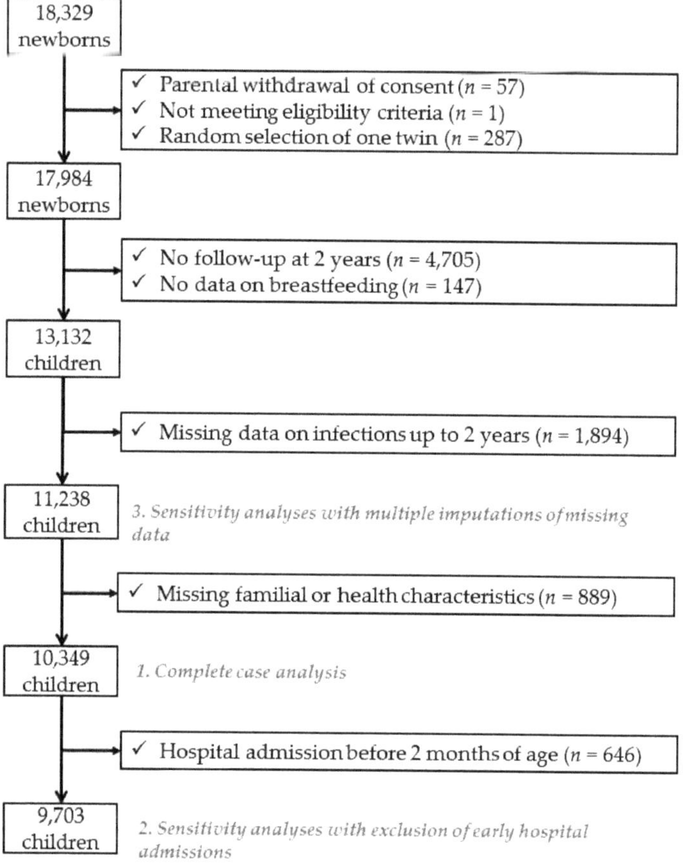

Figure 1. Sample selection.

2.6. Analyses

2.6.1. Main Analyses

To compare selected families to their non-selected ELFE counterparts, we used chi-squared tests for categorical variables and Student *t*-tests for continuous variables.

Associations between breastfeeding and parental reports of infections or antibiotic use were assessed with multinomial logistic regression models. Never-breastfed infants were systematically used as the reference group. In all analyses, we adjusted for potential confounding factors (maternal age at first child, education level, employment status, smoking status during pregnancy and pre-pregnancy BMI, household monthly income per consumption unit, household composition, caesarean section and infant's sex, gestational age, birth order, and age at first attendance at a shared childcare facility), and variables related to study design (recruitment wave, maternity unit size, and mother's region of residence).

2.6.2. Sensitivity Analyses

As early hospitalizations may lead to early breastfeeding cessation, we repeated the main analyses after excluding infants with hospitalizations occurring before the age of 2 months, leading to a sample of 9703 infants.

To deal with missing data for potential confounding factors, we first conducted our analyses on complete cases (Figure 1, n = 10,349). Secondly, we used multiple imputations to deal with these missing data. This method assigned data to missing measurements based on the measurement of infants with similar profiles. We assumed that data were missing at random and generated five independent datasets with the fully conditional specification method (MI procedure, FCS statement, NIMPUTE option), and then calculated pooled effect estimates (SAS MIANALYSE procedure). Further details are available in Table S1, Supplementary Materials. This method allowed us to assess again the association between breastfeeding duration and infections up to 2 years on a sample of 11,238 infants (Figure 1). The model used for these analyses was the same as in our main analysis.

All analyses were carried out with SAS software version 9.4 (SAS Institute, Cary, NC, USA). Statistical significance was defined as $p < 0.05$.

3. Results

The comparison between selected families and their non-selected ELFE counterparts is available in Table S2, Supplementary Materials. Briefly, selected mothers were older, less likely to be single parents, born in France, with a higher education level, and employed and they breastfed longer than non-selected mothers.

The sample's characteristics according to any breastfeeding duration are described in Table 1.

Table 1. Included families' characteristics according to any breastfeeding duration (n = 10,349).

Family Characteristics	Breastfeeding Duration				
	Never (n = 2489)	<1 Month (n = 1704)	1 to <3 Months (n = 1629)	3 to <6 Months (n = 1964)	≥6 Months (n = 2563)
Maternal age at birth (years)	30.4 (4.9)	30.0 (5.0)	30.4 (4.5)	30.9 (4.3)	31.6 (4.6)
Maternal place of birth (France)	96.6% (2405)	94.5% (1611)	92.4% (1505)	90.3% (1774)	82.7% (2120)
Pre-pregnancy body mass index (kg/m^2)	23.9 (5.2)	23.8 (4.9)	23.4 (4.5)	22.8 (4.1)	23.0 (4.3)
Education level					
Below secondary school	7.2% (178)	5.4% (92)	3.8% (62)	3.6% (71)	5.1% (131)
Secondary school	17.2% (428)	15.8% (269)	9.7% (158)	7.3% (143)	7.9% (202)
High school	21.6% (537)	21.4% (365)	18.4% (300)	15.1% (297)	13.7% (352)
2-year university degree	24.5% (611)	25.5% (434)	26.2% (427)	23.7% (465)	21.2% (543)
3-year university degree	15.8% (394)	18.2% (310)	19.6% (320)	22.3% (437)	22.8% (584)
5-year university degree or higher	13.7% (341)	13.7% (234)	22.2% (362)	28.1% (551)	29.3% (751)
Employed before pregnancy	77.9% (1938)	75% (1278)	79.7% (1299)	80.2% (1576)	71.4% (1829)
Traditional household composition	87.9% (2188)	87.9% (1498)	91.3% (1487)	91.6% (1799)	90.2% (2311)
Household monthly income (€)	3379 (3171)	3276 (2750)	3665 (4608)	3738 (3671)	3506 (2663)
Smoking status during pregnancy					
Never smoker	51.9% (1292)	50.1% (853)	57.5% (936)	60.7% (1193)	66.1% (1694)
Only before pregnancy	24% (598)	27.4% (467)	25.9% (422)	25.1% (492)	23.6% (604)
Only in early pregnancy	3.5% (88)	4.7% (80)	3.4% (56)	3.8% (74)	2.9% (75)
Throughout pregnancy	20.5% (511)	17.8% (304)	13.2% (215)	10.4% (205)	7.4% (190)
Caesarean section	19.2% (478)	18% (307)	18% (294)	15.7% (309)	15.2% (389)
Gestational age (weeks)	39.5 (1.5)	39.7 (1.4)	39.7 (1.4)	39.7 (1.4)	39.7 (1.4)
Boys	48.8% (1215)	49.4% (841)	51.6% (841)	50.1% (984)	48.7% (1249)
First born	42.9% (1069)	50.5% (861)	49.7% (809)	45.7% (897)	37.6% (963)
Age at first attendance at a shared childcare facility					
≤2 months	55.7% (1386)	54.2% (924)	48.4% (788)	46.4% (912)	61.6% (1580)
>2 months to 4 months	7.4% (183)	5.8% (99)	8% (130)	3.6% (70)	2.1% (53)
>4 months to 6 months	18.8% (469)	20.4% (347)	26.3% (429)	24.3% (477)	11.2% (286)
>6 months to 12 months	9.1% (227)	9.2% (157)	8.2% (133)	13.7% (269)	10.7% (273)
Never attended in the first year	9% (224)	10.4% (177)	9.1% (149)	12% (236)	14.5% (371)

% (n) or mean (± SD)

3.1. Hospitalizations from Infectious Diseases

Breastfeeding status considered as a binary variable (ever vs. never) was related neither to the number of events (0, 1, ≥2) nor to the cumulative duration of hospitalizations or infectious causes of hospitalizations, whatever the definition of breastfeeding used (any or predominant) (Table 2).

Table 2. Association between breastfeeding and parent reports of hospitalizations from infection: multivariate analyses (*n* = 10,349).

Breastfeeding Status and Duration		Number of Events (Ref = None)			Parental Report of Hospitalizations from Infection								
					Total Duration (Ref = Never)					Causes			
		1	≥2	*p*	1–3 Nights	≥4 Nights	*p*	Fever	*p*	Gastroint. Inf.	*p*	Bronchiolitis	*p*
Number of infants in each group		842	413		470	429		282		397		475	
Any breastfeeding status				0.52			0.33		0.11		0.40		0.98
	Never	1 (Ref)	1 (Ref)		1 (Ref)	1 (Ref)		1 (Ref)		1 (Ref)		1 (Ref)	
	Ever	1.10 (0.93; 1.31)	1.04 (0.82; 1.32)		1.16 (0.92; 1.46)	0.92 (0.73; 1.16)		1.28 (0.95; 1.74)		1.11 (0.87; 1.42)		1.00 (0.80; 1.24)	
Any breastfeeding duration				0.29			0.25		0.17		0.08		0.10
	Never	1 (Ref)	1 (Ref)		1 (Ref)	1 (Ref)		1 (Ref)		1 (Ref)		1 (Ref)	
	<1 month	1.13 (0.90; 1.42)	1.20 (0.89; 1.62)		1.21 (0.90; 1.63)	1.02 (0.76; 1.38)		1.38 (0.94; 2.02)		1.42 (1.05; 1.91)		0.92 (0.68; 1.24)	
	1 to <3 months	1.22 (0.97; 1.53)	1.15 (0.84; 1.58)		1.17 (0.86; 1.60)	1.08 (0.79; 1.46)		1.55 (1.06; 2.28)		0.91 (0.64; 1.29)		1.24 (0.93; 1.65)	
	3 to <6 months	1.13 (0.91; 1.42)	0.94 (0.68; 1.29)		1.28 (0.96; 1.71)	0.85 (0.63; 1.17)		1.13 (0.76; 1.68)		1.03 (0.75; 1.43)		1.07 (0.80; 1.42)	
	≥6 months	0.96 (0.77; 1.20)	0.89 (0.65; 1.21)		1.00 (0.74; 1.34)	0.77 (0.57; 1.04)		1.13 (0.78; 1.65)		1.04 (0.76; 1.42)		0.82 (0.62; 1.10)	
Number of infants in each group		839	413		468	428		281		395		474	
Predominant breastfeeding status				0.42			0.27		0.65		0.73		0.39
	Never	1 (Ref)	1 (Ref)		1 (Ref)	1 (Ref)		1 (Ref)		1 (Ref)		1 (Ref)	
	Ever	1.10 (0.94; 1.29)	0.96 (0.77; 1.19)		1.12 (0.91; 1.39)	0.88 (0.72; 1.09)		1.06 (0.82; 1.39)		1.04 (0.83; 1.30)		1.06 (0.86; 1.30)	
Predominant breastfeeding duration				0.24			0.08		0.25		0.14		0.24
	Never	1 (Ref)	1 (Ref)		1 (Ref)	1 (Ref)		1 (Ref)		1 (Ref)		1 (Ref)	
	<1 month	1.24 (1.01; 1.52)	1.09 (0.82; 1.44)		1.27 (0.97; 1.65)	1.07 (0.81; 1.40)		1.13 (0.80; 1.59)		1.28 (0.97; 1.68)		1.25 (0.96; 1.62)	
	1 to <3 months	1.13 (0.92; 1.39)	0.99 (0.74; 1.32)		1.05 (0.79; 1.38)	1.00 (0.76; 1.31)		1.28 (0.92; 1.79)		1.02 (0.76; 1.37)		1.04 (0.79; 1.37)	
	3 to <6 months	1.02 (0.82; 1.27)	0.79 (0.57; 1.08)		1.08 (0.81; 1.44)	0.67 (0.49; 0.93)		0.85 (0.58; 1.24)		0.84 (0.61; 1.17)		1.00 (0.75; 1.32)	
	≥6 months	0.90 (0.68; 1.19)	0.92 (0.63; 1.34)		1.05 (0.74; 1.50)	0.67 (0.45; 1.00)		0.88 (0.56; 1.40)		0.89 (0.60; 1.34)		0.81 [0.56; 1.18)	

OR (CI 95%) multinomial logistic regressions adjusted for maternal age at first child, education level, employment status, smoking status during pregnancy and pre-pregnancy BMI, household monthly income per consumption unit, household composition, caesarean section and infant's sex, gestational age, birth order, and age at first attendance at a shared childcare facility, recruitment wave, maternity unit size and level, and mother's region of residence. Analyses were performed separately for each breastfeeding definition.

Compared to never-breastfed infants, infants who were predominantly breastfed for <1 month were at higher risk of being hospitalized once, which justifies our sensitivity analysis excluding early infections. The association remained consistent after the exclusion of infants with early hospitalizations (before the age of 2 months) (Table S3, Supplementary Materials).

Compared to never-breastfed infants, infants who were predominantly breastfed for at least 3 months were at lower risk of long duration ($\geq$4 nights) of hospitalizations. These associations remained consistent after the exclusion of infants with early hospitalizations (before the age of 2 months).

Compared to never-breastfed infants, any breastfed for 1 to <3 months infants were at higher risk of hospitalization from fever and any breastfed for <1 month infants were at higher risk of hospitalization from gastrointestinal infections. The first association disappeared after the exclusion of early hospitalization, whereas the second one remained consistent.

3.2. Bronchiolitis Events

Overall, the number of bronchiolitis events was significantly related neither to ever breastfeeding nor to breastfeeding duration. However, ever any breastfeeding tended to be related to a higher risk of 1 or 2 bronchiolitis events but not to frequent bronchiolitis events. In contrast, predominant breastfeeding duration tended to be negatively related to the risk of frequent bronchiolitis events (Table 3). These tendencies remained after the exclusion of early hospitalization events (Table S4, Supplementary Materials).

Table 3. Association between breastfeeding and parent reports of bronchiolitis events, otitis events, and antibiotic use: multivariate analyses ($n = 10,349$).

Breastfeeding Status and Duration		Bronchiolitis Events (Ref = None)			Otitis Events (Ref ≤ 3)		Antibiotic Use (Ref = Never)			
					Parental Report					
		1 or 2	≥3	p	≥3	p	Once	2 or 3 Times	>3 Times	p
Number of infants in each group		6340	1264		2606		1944	1860	4411	
Any breastfeeding status				0.17		0.32				0.02
	Never	1 (Ref)	1 (Ref)		1 (Ref)		1 (Ref)	1 (Ref)	1 (Ref)	
	Ever	1.11 (0.99; 1.24)	1.09 (0.92; 1.28)		1.06 (0.95; 1.18)		0.94 (0.80; 1.09)	0.84 (0.72; 0.98)	0.83 (0.73; 0.94)	
Any breastfeeding duration				0.05		0.37				0.00
	Never	1 (Ref)	1 (Ref)		1 (Ref)		1 (Ref)	1 (Ref)	1 (Ref)	
	<1 month	1.07 (0.93; 1.24)	1.19 (0.96; 1.48)		1.08 (0.94; 1.25)		0.99 (0.80; 1.22)	1.02 (0.83; 1.26)	1.09 (0.92; 1.30)	
	1 to <3 months	1.09 (0.94; 1.26)	1.08 (0.87; 1.35)		1.12 (0.97; 1.29)		1.04 (0.84; 1.28)	0.95 (0.77; 1.18)	0.98 (0.82; 1.17)	
	3 to <6 months	1.15 (1.00; 1.33)	1.20 (0.97; 1.48)		1.06 (0.92; 1.22)		1.00 (0.82; 1.22)	0.78 (0.64; 0.95)	0.79 (0.67; 0.94)	
	≥6 months	1.12 (0.98; 1.29)	0.90 (0.73; 1.12)		0.98 (0.86; 1.13)		0.79 (0.66; 0.95)	0.71 (0.59; 0.85)	0.61 (0.52; 0.71)	
Number of infants in each group		6335	1263		2605		1940	1857	4410	
Predominant breastfeeding status				0.30		0.92				0.00
	Never	1 (Ref)	1 (Ref)		1 (Ref)		1 (Ref)	1 (Ref)	1 (Ref)	
	Ever	1.08 (0.98; 1.19)	1.04 (0.90; 1.21)		0.99 (0.90; 1.10)		0.96 (0.83; 1.10)	0.84 (0.73; 0.96)	0.79 (0.71; 0.89)	
Predominant breastfeeding duration				0.06		0.98				0.00
	Never	1 (Ref)	1 (Ref)		1 (Ref)		1 (Ref)	1 (Ref)	1 (Ref)	
	<1 month	1.09 (0.95; 1.24)	1.19 (0.98; 1.45)		1.02 (0.89; 1.16)		1.11 (0.92; 1.33)	1.05 (0.87; 1.26)	0.97 (0.83; 1.14)	
	1 to <3 months	1.13 (0.99; 1.30)	1.08 (0.89; 1.32)		1.01 (0.88; 1.15)		0.95 (0.79; 1.14)	0.82 (0.68; 0.99)	0.87 (0.74; 1.01)	
	3 to <6 months	1.10 (0.96; 1.270)	0.99 (0.81; 1.22)		0.97 (0.85; 1.11)		0.88 (0.73; 1.06)	0.71 (0.59; 0.85)	0.65 (0.56; 0.76)	
	≥6 months	0.94 (0.80; 1.110)	0.77 (0.59; 1.00)		0.97 (0.82; 1.15)		0.89 (0.71; 1.10)	0.75 (0.60; 0.93)	0.62 (0.51; 0.75)	

OR (CI 95%) multinomial logistic regressions adjusted for maternal age at first child, education level, employment status, smoking status during pregnancy and pre-pregnancy BMI, household monthly income per consumption unit, household composition, caesarean section and infant's sex, gestational age, birth order, and age at first attendance at a shared childcare facility, recruitment wave, maternity unit size and level, and mother's region of residence. Analyses were performed separately for each breastfeeding definition.

3.3. Otitis Events

Both any and predominant breastfeeding were not related to frequent otitis events (at least 3 in the first 2 years) (Table 3).

3.4. Antibiotic Use

Ever breastfeeding was related to a lower risk of frequent antibiotic use (at least 2 times in the first 2 years of life), whatever the definition of breastfeeding used (Table 3).

Compared to no breastfeeding, breastfeeding durations (any and predominant) of at least 3 months were associated with a lower risk of frequent antibiotic use.

All these associations remained after the exclusion of infants with early hospitalizations (Table S4, Supplementary Materials).

3.5. Analyses after Multiple Imputations

Compared to never-breastfed infants, predominantly breastfed for <1 month infants were no longer at higher risk of a single hospitalization event (OR (95% CI) = 1.17 (0.96; 1.42)) (Table S5, Supplementary Materials). Any breastfed for <1 month infants were also no longer at higher risk of hospitalization from gastrointestinal infections (1.32 (0.99; 1.75)).

Compared to never-breastfed infants, any breastfed for ≥6 months infants were at lower risk of longer hospitalizations (≥4 nights) (0.71 (0.53; 0.95)).

Other previously highlighted results remained consistent (Tables S5 and S6, Supplementary Materials).

4. Discussion

In the ELFE study, predominant breastfeeding for over 3 months was related to lower risk of at least 4 nights of hospitalization up to 2 years, while any breastfeeding for over 3 months was related to higher risk of 1 or 2 bronchiolitis events in the first 2 years of age. Finally, both any and predominant breastfeeding durations were negatively associated with frequency of antibiotic use.

We first examined infectious morbidity through common infectious diseases such as bronchiolitis and otitis. Unfortunately, data on diarrhea occurrence were not collected. Contrary to a previous meta-analysis conducted in industrialized countries [12], a lower risk of otitis has not been related to breastfeeding duration. A possible explanation may be the restrictive classification (<3 or ≥3 events) applied in the ELFE questionnaires, not allowing the distinguishing of absence of event from low frequency of events (1 or 2). Consistent with an Italian cohort specifically designed to study respiratory infections [13], we found that, in the main analyses, long duration of predominant breastfeeding (at least 6 months) was associated with a lower risk of frequent bronchiolitis events. Similarly, in the Generation R population-based study, exclusive breastfeeding was related to a lower risk of low respiratory tract infections and, to a lesser extent, upper respiratory tract infections up to 4 years of age [14].

Infections with higher concern can be approximated with hospitalizations. A recent meta-analysis highlighted a protective effect of breastfeeding against hospitalizations from diarrhea and respiratory infections, including lower respiratory tract infection and pneumonia [7]. More recently, the Norwegian MoBa study highlighted a higher risk of hospitalization up to 18 months among infants breastfed for ≤6 months than among those breastfed for at least 12 months [15], but matched sibling analyses, enabled to account for shared maternal characteristics, showed weaker and non-significant associations. We have not observed such associations within the ELFE cohort. Regarding hospitalizations from diarrhea, the meta-analysis included two studies from Europe, highlighting protective associations between breastfeeding and diarrhea [16,17]. Both studies provided data on infants born in the 1990s. Regarding hospitalizations from respiratory infection, the meta-analysis included two studies from Europe, highlighting protective associations between breastfeeding and respiratory infections [18,19], both published before 1995. As infants from the ELFE were born in 2011, they might differ from infants included in these studies. Moreover, in the present analyses only hospitalizations for bronchiolitis could be considered and not all respiratory infections. The low number of hospitalizations for bronchiolitis in our results might prevent any potential association with breastfeeding to arise.

We are unable to provide a biological explanation for the higher risk of 1 or 2 bronchiolitis events in the first 2 years of life related to any breastfeeding but not to predominant breastfeeding. An additional sensitivity analysis adjusting for family history of allergy (parental and/or sibling history of asthma, eczema, and hay fever) did not modify the results (data not shown). A similar unexpected association was found in an Italian case–control study, with a higher breastfeeding rate among infants hospitalized for bronchiolitis than among their control counterparts [20].

Reverse causation bias is a probable hypothesis for the highlighted higher risk of parental reported hospital admission from infection or gastrointestinal infection related to short breastfeeding duration (<1 month) compared to never breastfeeding. As early adverse health events, including hospital admission, can lead to early breastfeeding cessation, the exclusion of early cases of hospitalization allowed us to control this reverse causation bias but not fully as not all adverse health events lead to hospitalization.

Finally, in the present study, antibiotic use was considered as a proxy for bacterial infections. This indicator was strongly related to breastfeeding, a longer duration being related to a lower use of antibiotics up to 2 years of age. Similar results were found in a Czech cross-sectional survey, with lower risk of early exposure to antibiotics among breastfed infants [21]. Likewise, in a Finnish cohort, 1-year breastfed infants were less likely to have been provided with antibiotics during the first year of life than their non-breastfed counterparts [22], and a negative association between breastfeeding duration and antibiotic use was found in a cross-sectional anthropometric and questionnaire study [23]. However, we cannot exclude that health-seeking behaviors could be different among breastfeeding parents and non-breastfeeding parents, leading to the differential use of antibiotics. Moreover, as both breastfeeding and antibiotic use could influence the infant's microbiome [22], microbiome was suggested as a potential mechanism in the association between breastfeeding and lower rates of infections from hospitalization. It would be of great interest to examine these potential mechanisms from stool samples collected in the first months of life.

The ELFE study is a recent nationwide birth cohort aimed at assessing the development of healthy-born children from birth to adulthood from a broad and interdisciplinary point of view. The prospective design limits recall bias for both exposure and outcomes assessment. However, we have to acknowledge the inability to consider exclusive breastfeeding in the present study according to the WHO definition, because the use of water, water-based drinks, and fruit juice in the 0–2-month period were not collected in the ELFE study. While the study may lack specificity when assessing particular outcomes (e.g., antibiotic types), the strength of our approach is the use of complementary indicators of infectious morbidity, with the occurrence of common infectious diseases, hospitalizations, and antibiotic use. Hospitalizations could reflect the most severe cases and allowed for controlling bias due to parental reporting. It is interesting to note that breastfeeding was more related to the duration of hospitalization than to the number of events. When matching with the national health system database will be available, it would be of great interest to conduct similar analyses based on medical care use rather than parental report. The use of antibiotics would be more specific for bacterial infections, but it remains difficult to distinguish a lower need for antibiotics from the reluctance to such use. The very large sample and the collection of detailed socio-demographic or economic data ensure good statistical power and favor control for potential confounders. Exclusion rates due to missing data for these analyses were high, but multiple imputations did not change the results.

5. Conclusions

Even in the context of a high-income country with short breastfeeding duration, we highlighted a lower risk of infectious morbidity related to breastfeeding duration, especially for duration of hospitalization and antibiotic use. The strong association highlighted for antibiotic use would be of great interest in the context of rising concerns regarding antibiotic resistance.

Supplementary Materials: The following are available online at http://www.mdpi.com/2072-6643/11/7/1607/s1. Table S1: Details regarding variables used for the multiple imputations, Table S2: Comparison of included and excluded families (Chi-squared and Student *t*-tests, Table S3: Multivariate adjusted analyses assessing breastfeeding and parent reports of hospitalizations from infection among infants without hospitalization events before 2 months of age (n = 9,703), Table S4: Multivariate adjusted analyses assessing breastfeeding and parent reports of bronchiolitis events, otitis events, and antibiotic use among infants without hospitalization events before 2 months of age (n = 9,703), Table S5: Multivariate adjusted analyses assessing breastfeeding and parent reports of hospitalizations from infection using multiple imputations to deal with missing measurements on familial or health characteristics (n = 11,238), Table S6: Multivariate adjusted analyses assessing breastfeeding and parent reports of bronchiolitis events, otitis events, and antibiotic use using multiple imputations to deal with missing measurements on familial or health characteristics (n = 11,238).

Author Contributions: The corresponding author affirms that all listed authors meet authorship criteria and that no others meeting these criteria have been omitted. C.D.-P. conducted the statistical analyses, interpreted the results, drafted the initial manuscript, and approved the final manuscript as submitted. K.A.-P., A.D.-C., J.P. and S.L. contributed to the interpretation of the results, critically reviewed the manuscript, and approved the final version submitted. M.C. and M.-N.D. contributed to the data acquisition, critically reviewed the manuscript, and approved the final version submitted. M.-A.C. and B.d.L.-G. conceptualized and designed the study, contributed to the interpretation of the results, reviewed and revised the manuscript, and approved the final manuscript as submitted.

Funding: The ELFE survey is a joint project between the French Institute for Demographic Studies (INED) and the National Institute of Health and Medical Research (INSERM), in partnership with the French blood transfusion service (Etablissement français du sang, EFS), Santé publique France, the National Institute for Statistics and Economic Studies (INSEE), the Direction générale de la santé (DGS, part of the Ministry of Health and Social Affairs), the Direction générale de la prévention des risques (DGPR, Ministry for the Environment), the Direction de la recherche, des études, de l'évaluation et des statistiques (DREES, Ministry of Health and Social Affairs), the Département des études, de la prospective et des statistiques (DEPS, Ministry of Culture), and the Caisse nationale des allocations familiales (CNAF), with the support of the Ministry of Higher Education and Research and the Institut national de la jeunesse et de l'éducation populaire (INJEP). Via the RECONAI platform, it received a government grant managed by the National Research Agency under the "Investissements d'avenir" program (ANR-11-EQPX-0038). The funders had no role in the study design, data collection and analysis, decision to publish, or preparation of the manuscript. This research received no external funding.

Acknowledgments: We would like to thank the scientific coordinators (Marie-Aline Charles, Bertrand Geay, Henri Léridon, Corinne Bois, Marie-Noëlle Dufourg, Jean-Louis Lanoé, Xavier Thierry, Cécile Zaros), IT and data managers, statisticians (A. Rakotonirina, R. Kugel, R. Borges-Panhino, M. Cheminat, H. Juillard), administrative and family communication staff, and study technicians (C. Guevel, M. Zoubiri, L. Gravier, I. Milan, R. Popa) of the ELFE coordination team as well as the families who gave their time for the study.

Conflicts of Interest: The authors declare no conflict of interest. The funders of the ELFE survey had no role in the design of the present study; in the collection, analyses, or interpretation of data; in the writing of the manuscript; or in the decision to publish the results.

References

1. Kyu, H.H.; Pinho, C.; Wagner, J.A.; Brown, J.C.; Bertozzi-Villa, A.; Charlson, F.J.; Coffeng, L.E.; Dandona, L.; Erskine, H.E.; Ferrari, A.J.; et al. Global and National Burden of Diseases and Injuries Among Children and Adolescents Between 1990 and 2013: Findings From the Global Burden of Disease 2013 Study. *JAMA Pediatr.* **2016**, *170*, 267–287.

2. World Health Organization. *Feeding and Nutrition of Infants and Young Children, Guidelines for the WHO European Region, with Emphasis on the Former Soviet Countries*; WHO: Geneva, Switzerland, 2003.

3. Victora, C.G.; Bahl, R.; Barros, A.J.; Franca, G.V.; Horton, S.; Krasevec, J.; Murch, S.; Sankar, M.J.; Walker, N.; Rollins, N.C.; et al. Breastfeeding in the 21st century: Epidemiology, mechanisms, and lifelong effect. *Lancet* **2016**, *387*, 475–490. [CrossRef]

4. Field, C.J. The immunological components of human milk and their effect on immune development in infants. *J. Nutr.* **2006**, *135*, 1–4. [CrossRef]

5. Korvel-Hanquist, A.; Djurhuus, B.D.; Homoe, P. The Effect of Breastfeeding on Childhood Otitis Media. *Curr. Allergy Asthma Rep.* **2017**, *17*, 45. [CrossRef]

6. Bowatte, G.; Tham, R.; Allen, K.J.; Tan, D.J.; Lau, M.; Dai, X.; Lodge, C.J. Breastfeeding and childhood acute otitis media: A systematic review and meta-analysis. *Acta Paediatr.* **2015**, *104*, 85–95. [CrossRef]

7. Horta, B.L.; Victora, C.G. *Short-Term Effects of Breastfeeding: A Systematic Review of the Benefits of Breastfeeding on Diarrhoea and Pneumonia Mortality*; World Health Organization: Geneva, Switzerland. 2013.
8. Kramer, M.S.; Chalmers, B.; Hodnett, E.D.; Sevkovskaya, Z.; Dzikovich, I.; Shapiro, S.; Collet, J.P.; Vanilovich, I.; Mezen, I.; Ducruet, T.; et al. Promotion of Breastfeeding Intervention Trial (PROBIT). *JAMA* **2001**, *285*, 413–420. [CrossRef]
9. Wagner, S.; Kersuzan, C.; Gojard, S.; Tichit, C.; Nicklaus, S.; Geay, B.; Humeau, P.; Thierry, X.; Charles, M.A.; Lioret, S.; et al. Breastfeeding duration in France according to parents and birth characteristics. Results from the ELFE longitudinal French Study, 2011. *Bull. Epidemiol. Hebd.* **2015**, *29*, 522–532.
10. Vandentorren, S.; Bois, C.; Pirus, C.; Sarter, H.; Salines, G.; Leridon, H.; Elfe, T. Rationales, design and recruitment for the Elfe longitudinal study. *BMC Pediatr.* **2009**, *9*, 58. [CrossRef]
11. Wagner, S.; Kersuzan, C.; Gojard, S.; Tichit, C.; Nicklaus, S.; Thierry, X.; Charles, M.A.; Lioret, S.; De Lauzon-Guillain, B. Breastfeeding initiation and duration in France: The importance of intergenerational and previous maternal breastfeeding experiences—results from the nationwide ELFE study. *Midwifery* **2019**, *69*, 67–75. [CrossRef]
12. Lodge, C.J.; Bowatte, G.; Matheson, M.C.; Dharmage, S.C. The Role of Breastfeeding in Childhood Otitis Media. *Curr. Allergy Asthma Rep.* **2016**, *16*, 68. [CrossRef]
13. Lanari, M.; Prinelli, F.; Adorni, F.; Di Santo, S.; Faldella, G.; Silvestri, M.; Musicco, M. Maternal milk protects infants against bronchiolitis during the first year of life. Results from an Italian cohort of newborns. *Early Hum. Dev.* **2013**, *89*, 51–57. [CrossRef]
14. Tromp, I.; Kiefte-De Jong, J.; Raat, H.; Jaddoe, V.; Franco, O.; Hofman, A.; De Jongste, J.; Moll, H. Breastfeeding and the risk of respiratory tract infections after infancy: The Generation R Study. *PLoS ONE* **2017**, *12*, e0172763. [CrossRef]
15. Stordal, K.; Lundeby, K.M.; Brantsaeter, A.L.; Haugen, M.; Nakstad, B.; Lund-Blix, N.A.; Stene, L.C. Breast-feeding and Infant Hospitalization for Infections: Large Cohort and Sibling Analysis. *J. Pediatr. Gastroenterol. Nutr.* **2017**, *65*, 225–231. [CrossRef]
16. Quigley, M.A.; Kelly, Y.J.; Sacker, A. Breastfeeding and hospitalization for diarrheal and respiratory infection in the United Kingdom Millennium Cohort Study. *Pediatrics* **2007**, *119*, e837–e842. [CrossRef]
17. Kramer, M.S.; Guo, T.; Platt, R.W.; Sevkovskaya, Z.; Dzikovich, I.; Collet, J.P.; Shapiro, S.; Chalmers, B.; Hodnett, E.; Vanilovich, I.; et al. Infant growth and health outcomes associated with 3 compared with 6 mo of exclusive breastfeeding. *Am. J. Clin. Nutr.* **2003**, *78*, 291–295. [CrossRef]
18. Pisacane, A.; Graziano, L.; Zona, G.; Granata, G.; Dolezalova, H.; Cafiero, M.; Coppola, A.; Scarpellino, B.; Ummarino, M.; Mazzarella, G. Breast feeding and acute lower respiratory infection. *Acta Paediatr.* **1994**, *83*, 714–718. [CrossRef]
19. Howie, P.W.; Forsyth, J.S.; Ogston, S.A.; Clark, A.; Florey, C.D. Protective effect of breast feeding against infection. *BMJ* **1990**, *300*, 11–16. [CrossRef]
20. Nenna, R.; Cutrera, R.; Frassanito, A.; Alessandroni, C.; Nicolai, A.; Cangiano, G.; Petrarca, L.; Arima, S.; Caggiano, S.; Ullmann, N.; et al. Modifiable risk factors associated with bronchiolitis. *Ther. Adv. Respir. Dis.* **2017**, *11*, 393–401. [CrossRef]
21. Parizkova, P.; Dankova, N.; Fruhauf, P.; Jireckova, J.; Zeman, J.; Magner, M. Associations between breastfeeding rates and infant disease: A survey of 2338 Czech children. *Nutr. Diet. J. Dietit. Assoc. Aust.* **2019**. [CrossRef]
22. Korpela, K.; Salonen, A.; Virta, L.J.; Kekkonen, R.A.; De Vos, W.M. Association of Early-Life Antibiotic Use and Protective Effects of Breastfeeding: Role of the Intestinal Microbiota. *JAMA Pediatr.* **2016**, *170*, 750–757. [CrossRef]
23. Krenz-Niedbala, M.; Koscinski, K.; Puch, E.A.; Zelent, A.; Breborowicz, A. Is the Relationship Between Breastfeeding and Childhood Risk of Asthma and Obesity Mediated by Infant Antibiotic Treatment? *Breastfeed. Med.* **2015**, *10*, 326–333. [CrossRef]

 *nutrients* 

Article

Carbohydrates in Human Milk and Body Composition of Term Infants during the First 12 Months of Lactation

Zoya Gridneva [1,*], Alethea Rea [2], Wan Jun Tie [1], Ching Tat Lai [1], Sambavi Kugananthan [1,3], Leigh C. Ward [4], Kevin Murray [5], Peter E. Hartmann [1] and Donna T. Geddes [1]

[1] School of Molecular Sciences, M310, The University of Western Australia, Crawley, Perth, WA 6009, Australia
[2] Centre for Applied Statistics, The University of Western Australia, Crawley, Perth, WA 6009, Australia
[3] School of Human Sciences, The University of Western Australia, Crawley, Perth, WA 6009, Australia
[4] School of Chemistry and Molecular Biosciences, The University of Queensland, St. Lucia, Brisbane, QLD 4072, Australia
[5] School of Population and Global Health, The University of Western Australia, Crawley, Perth, WA 6009, Australia
* Correspondence: zoya.gridneva@uwa.edu.au; Tel.: +61-8-6488-4467

Received: 6 June 2019; Accepted: 26 June 2019; Published: 28 June 2019

Abstract: Human milk (HM) carbohydrates may affect infant appetite regulation, breastfeeding patterns, and body composition (BC). We investigated relationships between concentrations/calculated daily intakes (CDI) of HM carbohydrates in first year postpartum and maternal/term infant BC, as well as breastfeeding parameters. BC of dyads ($n = 20$) was determined at 2, 5, 9, and/or 12 months postpartum using ultrasound skinfolds (infants) and bioelectrical impedance spectroscopy (infants/mothers). Breastfeeding frequency, 24-h milk intake and total carbohydrates (TCH) and lactose were measured to calculate HM oligosaccharides (HMO) concentration and CDI of carbohydrates. Statistical analysis used linear regression/mixed effects models; results were adjusted for multiple comparisons. Higher TCH concentrations were associated with greater infant length, weight, fat-free mass (FFM), and FFM index (FFMI), and decreased fat mass (FM), FM index (FMI), %FM and FM/FFM ratio. Higher HMO concentrations were associated with greater infant FFM and FFMI, and decreased FMI, %FM, and FM/FFM ratio. Higher TCH CDI were associated with greater FM, FMI, %FM, and FM/FFM ratio, and decreased infant FFMI. Higher lactose CDI were associated with greater FM, FMI, %FM, and FM/FFM, ratio and decreased FFMI. Concentrations and intakes of HM carbohydrates differentially influence development of infant BC in the first 12 months postpartum, and may potentially influence risk of later obesity via modulation of BC.

Keywords: human milk carbohydrates; lactose; oligosaccharides; infant; body composition; lactation; daily intake; breastfeeding frequency; milk intake

---

1. Introduction

Nutrition [1] and development of infant body composition (BC) in the early months postpartum [2] are known to play a significant role in the programming of obesity. Breastfeeding and its duration are associated with reduced risk of developing obesity later in life [3]. These observations suggest a dose-dependent effect of breastfeeding on the development of infant BC, but the mechanisms of this effect are not fully elucidated. Although it is well established that added dietary sugar is a risk factor for obesity for children and adults [4] as well as infants [5] and contributes to metabolic diseases [6], little is known how human milk (HM) carbohydrates influence the development of BC during infancy and if they confer protection against the obesity.

The macronutrient composition of HM is remarkably conserved across populations appearing to be largely independent of geographic and ethnic factors [7], maternal nutritional status [8], and adiposity [9]. HM provides a constant supply of carbohydrates to the infant during early life, ensuring appropriate nutrition, maturation and development of their comparatively immature physiological systems. Considering that lactose, the principal carbohydrate, is important for maintenance of constant osmotic pressure in HM [10], it is unlikely that maternal adiposity would have a significant impact on lactose concentration, which is estimated to be approximately 60–78 g/L throughout the lactation, although lactose concentration is still variable between women [11]. As lactose synthesis results in water being drawn into the milk, the rate of lactose synthesis is a major controlling factor of milk production [12], and during the established lactation higher lactose concentration is associated with higher 24-h volumes [13] and higher breastfeeding frequency [14], which is also related to higher 24-h volumes [15].

HM carbohydrates also include small amounts of monosaccharides, such as glucose and galactose, as well as HM oligosaccharides (HMO). Nearly 200 unique oligosaccharides structures that vary from 2 to 22 sugars have been identified [16] and HMO composition and proportion varies between mothers [17], over the course of lactation [18] and is affected by parity, ethnicity, and place of residence, as well as breastfeeding exclusivity [19], and may exert different effects in the infant gut [20]. This suggests that HM carbohydrates may have a dose-dependent effect on infant growth and BC development.

Infant growth rate is associated with the volume of milk consumed [11] and lactose/carbohydrate intakes are lower in breastfed infants at 3 and 6 months compared to formula-fed infants and positively relate to weight and fat-free mass (FFM) but not fat mass (FM) gain [21]. Contradictory, studies of lactose concentration have reported either positive association of HM lactose with infant adiposity at 12 months [22] or no association with infant BC at 6 months [23]. Furthermore, there are differing associations for different HMO (16 HMO investigated) with infant BC, although total concentration was not analyzed [20]. In addition, concentrations of glucose are positively associated with relative weight and both fat and lean mass of breastfed infants [24]. The investigations in this area are extremely limited and need to discern the mechanisms by which HM carbohydrates, concentrations and intakes, may impact development of infant BC, allowing for future targeted interventions that may reduce risk of overweight and obesity later in life.

The purpose of this pilot longitudinal study was to assess the associations of concentration and daily intake of HM carbohydrates (total carbohydrates (TCH), lactose, and HMO) with anthropometric measurements and BC of healthy term infants breastfed on demand and their mothers during the first 12 months of lactation. Relationships of HM carbohydrates with infant 24-h milk intake and breastfeeding frequency were also investigated.

2. Materials and Methods

2.1. Study Overview

We curried out a longitudinal cohort study of mothers with milk supply within a normal range [25], and healthy term infants that were exclusively breastfed [26] at 2 and 5 months and continued breastfeeding on demand until 12 months postpartum without supplementation with formula. The study design, cohort characteristics, data collection procedures, and sample preparation have been described in detail previously, together with the methods for determining maternal and infant anthropometrics and BC [15,27,28], as well as infant 24-h milk intake [29] and breastfeeding frequency [25], and the relationships between the measured parameters.

Briefly, milk samples were collected and BC of healthy dyads (no suspected infectious illness, e.g., a cold or flu, at the time of study visit) was determined when the infants were two and/or five, nine, and 12 months-old utilising bioelectrical impedance spectroscopy (BIS; infants and mothers), and ultrasound for measurements of the skinfolds (ultrasound 2-skinfolds: triceps and subscapular; ultrasound 4-skinfolds: biceps, subscapular, suprailiac, and triceps; infants only) for use in the age and sex-specific equations [15,30]. Infant test-weighing before and after every breastfeeding from each breast over 24–26 h was conducted to measure infant 24-h milk intake and breastfeeding frequency between 2 and 5 months, when milk intake has been reported to be stable [27], and within two weeks of 9 and 12 months. Mothers also estimated and self-reported breastfeeding frequency at the study sessions as the current typical time (h) between the meals.

In addition to standard measured BC parameters (fat-free mass (FFM), fat mass (FM), and percentage FM (%FM)) we have calculated the height-normalized BC indices of mothers and infants: FFM index (FFMI) was calculated as $FFM/length^2$; FM index (FMI) was calculated as $FM/length^2$, both expressed as kg/m^2 [31]. The use of these BC indices is considered less statistically flawed than %FM [32], which includes a single variable (FM) in both numerator and denominator. Furthermore, the dyads' FM to FFM ratios (FM/FFM), which are emerging as values indicative of overall effects of BC (metabolic load and metabolic capacity) [33] were also calculated.

All mothers provided written informed consent for participation in the study, which was approved by the Human Research Ethics Committee of The University of Western Australia (RA/1/4253, RA/4/1/2639) and registered with the Australian New Zealand Clinical Trials Registry (ACTRN12616000368437).

2.2. Measurement of Human Milk Carbohydrates

Lactose concentration was measured in duplicate in pre- and postfeed samples using the enzymatic-spectrophotometric method [11] with recovery of 98.2 ± 4.1% ($n = 10$), detection limit of 30 mM and interassay coefficient of variation (CV) of 3.5%, and averaged for analysis.

Pre- and postfeed samples were pooled for measuring TCH. Skim HM was deproteinized with trichloroacetic acid [34] before dehydration by sulfuric acid [35]. TCH were analyzed in triplicate by UV spectrophotometry at 315nm. Recovery of added lactose was 101.4% ± 2.1% ($n = 7$) with a detection limit of 0.007 g/L and an interassay CV of 3.3% ($n = 7$).

The total HMO concentration (g/L) was estimated by deducting lactose concentration from TCH concentration. On two occasions, where the calculated HMO concentration was less than zero, it was taken to be zero. Glucose and galactose were not measured or accounted for in HMO concentration calculations, as their concentrations in HM are small and thus not detectable in these assays [36].

2.3. Calculated Daily Intakes of Carbohydrates

24-h milk intakes values measured during 24-h test-weighing, and HM carbohydrate concentrations measured in samples taken at the study sessions at corresponding time point were used for calculation of infant daily intakes (CDI). As there are no significant circadian [11] and short-term (weekly) [37] variations in HM carbohydrates concentration during the established lactation, these CDI were considered representative of a typical daily intake.

2.4. Statistical Analyses

This longitudinal study, the details of which, along with power calculation have been described previously [15,27,28], was analyzed using linear mixed models. Fitted models included (a) explanatory variable: maternal BC, responses: carbohydrate concentrations and CDI; (b) explanatory variables: carbohydrate concentration and CDI, response: infant BC; (c) explanatory variable: carbohydrate concentration, responses: breastfeeding parameters (24-h milk intake and breastfeeding frequency); and (d) explanatory variables: carbohydrate concentration, breastfeeding frequency, and 24-h milk intake, response: carbohydrate CDI. The fixed effects were infant age (as a categorical variable) included as an interaction with the explanatory variable and a random effect for each participant. The *p*-value associated with the interaction was examined and if it was below 0.05, the results were reported for a model with fixed effects for infant age and the explanatory variable of interest and a random effect for each participant. Where the interaction was not significant, results were reported for the same model fitted without the interaction.

The analyses for systematic differences between measured parameters at different months after birth used linear mixed model with age as effect factor and participant as a random factor. Differences between each month were analyzed using general linear hypothesis tests (Tukey's all pair comparisons).

Relationships between CDI of carbohydrates measured between 2 and 5, and at 9 and 12 months after birth and changes (Δ) in infant BC and anthropometric parameters between the time points were analyzed using linear regression models.

A false discovery rate (FDR) adjustment [38] was applied to the subgroupings of results to the interaction *p*-value if it was less than 0.05 or to the main effect *p*-value. The adjusted significance levels are reported in the tables or set at the 5% level otherwise. Descriptive statistics are reported as mean ± standard deviation (SD) and range; modeling results as parameters estimates ± standard error (SE). Missing data were dealt with using available case analysis. Statistical analysis and graphics were performed in R 3.1.2.

3. Results

3.1. Subjects

Participant demographic characteristics; determinants of maternal and infant BC, anthropometrics, and breastfeeding parameters (24-h milk intake, 24-h breastfeeding frequency, self-reported breastfeeding frequency); as well as participant attrition and missing data have been reported in detail previously [15,27,28]. The missing data in this analysis also included carbohydrates concentrations (from 80 expected measurements: TCH ($n = 12$); lactose ($n = 14$); HMO ($n = 15$)) and CDI (from 60 expected measurements: lactose ($n = 27$); TCH and HMO ($n = 28$)); missing data were distributed across the time points (Table 1). In addition to previously reported maternal and infant BC, the current analysis also included maternal and infant FM/FFM ratios (Table 1).

3.2. Changes in Body Composition and Human Milk Carbohydrates during First Year of Lactation

Description of the changes in infant and maternal BC and breastfeeding parameters (24-h milk intake, 24-h breastfeeding frequency, self-reported breastfeeding frequency) between 2 and 12 months of lactation have been reported previously [15]. Maternal and infant FM/FFM ratios as well as HM carbohydrates concentrations and CDI over the first year of lactation are detailed in Table 2.

Maternal FM/FFM decreased across the lactation. Infant FM/FFM measured with BIS initially increased and then decreased at the later months of lactation, while infant FM/FFM measured with either ultrasound 2-skinfolds or ultrasound 4-skinfolds did not differ (Table 2).

Concentrations of all, TCH, lactose and HMO did not differ between 2 and 12 months. CDI of TCH and lactose decreased across the lactation, while HMO CDI did not differ (Table 2).

3.3. Maternal Body Composition and Carbohydrates

No associations were seen between maternal BC and concentrations of carbohydrates (data not shown). No associations were seen between maternal BC and CDI of carbohydrates after adjusting for the FDR however, prior to the adjustment higher HMO CDI was associated with higher maternal %FM and FM/FFM between 2 and 5 months, and with lower maternal %FM and FM/FFM at 9 and 12 months (%FM: $p = 0.014$; FM/FFM: $p = 0.013$) (Table A1).

Table 1. Fat mass to fat-free mass ratios and concentrations and 24-h intakes of human milk carbohydrates during first 12 months of lactation.

Characteristics	2 Months	5 Months	9 Months	12 Months
	Mean ± SD (Min–Max)	Mean ± SD (Min–Max)	Mean ± SD (Min–Max)	Mean ± SD (Min–Max)
FM/FFM ratios	(*n* = 14)	(*n* = 20)	(*M* = 18/*I* = 17)	(*M* = 17/*I* = 15)
Maternal FM/FFM	0.56 ± 0.14 [a] (0.35–0.81)	0.51 ± 0.16 (0.30–0.89)	0.48 ± 0.16 (0.25–0.80)	0.43 ± 0.16 (0.24–0.80)
Infant FM/FFM with US 2SF	0.32 ± 0.04 (0.26–0.61)	0.37 ± 0.09 (0.24–0.64)	0.35 ± 0.08 (0.18–0.53)	0.34 ± 0.06 (0.22–0.45)
Infant FM/FFM with US 4SF	0.33 ± 0.05 (0.24–0.44)	0.36 ± 0.07 (0.26–0.56)	0.34 ± 0.07 (0.21–0.45)	0.31 ± 0.06 (0.21–0.41)
Infant FM/FFM with BIS	0.28 ± 0.04 (0.22–0.34)	0.41 ± 0.07 (0.28–0.56)	0.34 ± 0.08 (0.19–0.46)	0.33 ± 0.06 (0.24–0.44)
Concentrations	(*n* = 15)	(*n* = 20)	(*n* = 19)	(*n* = 14)
Total carbohydrates (g/L)	86.7 ± 9.2 (67.1–97.5)	80.7 ± 7.9 (69.3–94.1)	87.8 ± 11.1 (60.9–105.6)	88.4 ± 21.2 (56.9–126.9)
Lactose (g/L)	64.5 ± 4.1 (59.1–77.9)	64.3 ± 5.9 (53.5–70.6)	65.3 ± 5.3 (57.6–79.0)	66.9 ± 4.0 (60.1–79.3)
HMO (g/L)	22.3 ± 10.7 (0–35.8)	16.4 ± 9.9 (2.3–29.9)	22.5 ± 9.2 (0–36.9)	21.4 ± 22.3 (3.0–62.2)
CDI [b]	n/a [b]	(*n* = 17)	(*n* = 8)	(*n* = 8)
Total carbohydrates (g)	n/a	63.2 ± 15.0 (42.9–97.2)	44.8 ± 15.2 (21.2–69.6)	40.7 ± 29.8 (22.2–100.9)
Lactose (g)	n/a	51.2 ± 14.5 (32.6–83.6)	34.0 ± 11.0 (19.6–51.3)	28.7 ± 12.1 (18.0–51.4)
HMO (g)	n/a	12.0 ± 6.0 (2.0–21.6)	10.8 ± 5.4 (0–15.7)	12.0 ± 18.5 (1.5–49.5)

[a] Data are mean ± SD and ranges. [b] Daily intakes (CDI) of carbohydrates were calculated between 2 and 5 months (presented at 5 months here) and within 2 weeks of 9 and 12 months. BIS: bioelectrical impedance spectroscopy; CDI: calculated daily intakes; FFM: fat-free mass; FM: fat mass; FM/FFM: fat mass to fat-free mass ratio; HMO: human milk oligosaccharides; I: infants; M: mothers; n/a: not applicable; US 2SF: ultrasound 2-skinfolds; US 4SF: ultrasound 4-skinfolds.

3.4. Infant Body Composition and Concentrations of Carbohydrates

Higher TCH concentration was associated with increased infant length ($p < 0.001$), weight ($p = 0.003$), FFM (measured with BIS: $p < 0.001$; ultrasound 4-skinfolds: $p = 0.020$), and FFMI (BIS: $p = 0.002$). Higher TCH concentration was associated with an increase in adiposity measured with BIS (FM: $p = 0.016$; FMI: $p = 0.001$; %FM: $p = 0.001$; FM/FFM: $p = 0.003$) at 2 months and decrease at 5, 9, and 12 months (Table 3, Figure 1).

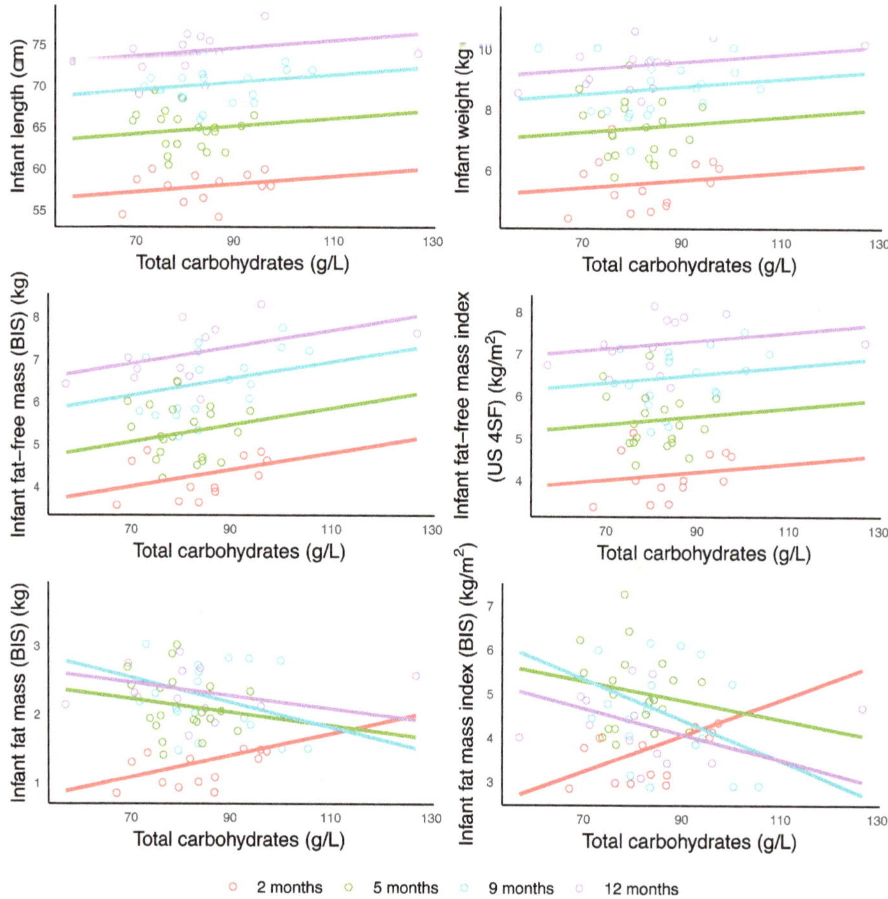

Figure 1. Significant associations between concentration of total carbohydrates and infant anthropometric and body composition measurements. Lines represent linear regression and grouped by the month of lactation. BIS: bioelectrical impedance spectroscopy; US 4SF: ultrasound 4-skinfolds.

Higher HMO concentration was associated with increased FFM (BIS: $p < 0.001$) and FFMI (BIS: $p = 0.008$). Higher HMO concentration was associated with an increase in adiposity measured with BIS (FMI: $p = 0.003$; %FM: $p = 0.005$; FM/FFM: $p = 0.006$) at 2 months and decrease at 5, 9 and 12 months (Table 3, Figure 2).

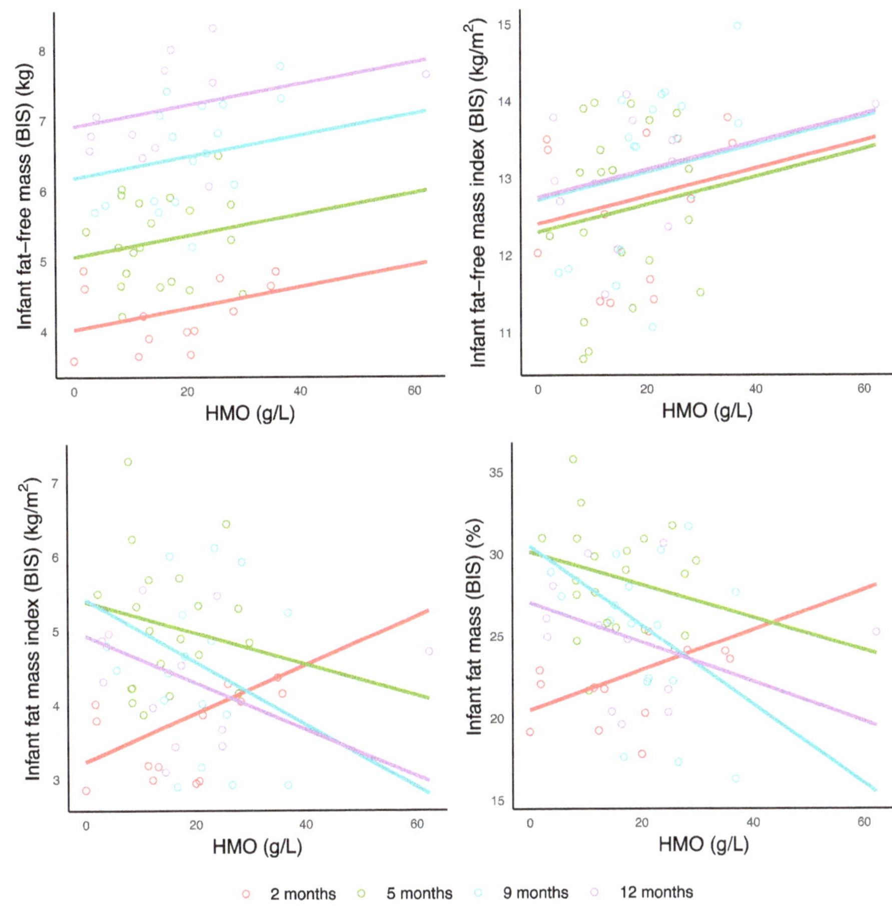

Figure 2. Significant associations between concentration of human milk oligosaccharides (HMO) and infant body composition parameters measured with bioelectrical impedance spectroscopy (BIS). Lines represent linear regression and grouped by the month of lactation.

No further associations were seen after adjusting for the FDR (Table 3).

3.5. Infant Body Composition and Daily Intakes of Carbohydrates

Higher TCH CDI was associated with increased infant FM (ultrasound 2-skinfolds: $p = 0.006$; ultrasound 4-skinfolds: $p < 0.001$), FMI (ultrasound 2-skinfolds: $p = 0.003$), %FM (ultrasound 2-skinfolds: $p = 0.005$), FM/FFM (ultrasound 2-skinfolds: $p = 0.004$), and decreased FFMI (ultrasound 4-skinfolds: $p < 0.001$); and also with BMI (an increase at 5 months and a decrease at 9 and 12 months, $p = 0.019$), and %FM ($p = 0.016$) and FM/FFM ($p = 0.007$) measured with ultrasound 4-skinfolds (an increase at 5 months and 12 months and a decrease at 9 months for both) (Table 4, Figure 3).

Table 2. Differences by infant age/lactation duration within fat mass to fat-free mass ratios and measured human milk carbohydrates [a].

Changes in Characteristics between Time Points	Months after Birth						
	5 and 2 Months	9 and 2 Months	12 and 2 Months	9 and 5 Months	12 and 5 Months	12 and 9 Months	Overall p Value
FM/FFM ratios	**(n = 14)**	**(n = 12)**	**(M = 11/I = 10)**	**(M = 18/I = 17)**	**(M = 18/I = 17)**	**(M = 16/I = 13)**	**(n = 20)**
Maternal FM/FFM	0.01 ± 0.01 [b] (0.98)	−0.02 ± 0.01 (0.63)	−0.04 ± 0.01 [d] (0.002)	−0.02 ± 0.01 (0.27)	−0.05 ± 0.01 [d] (<0.001)	−0.03 ± 0.01 [d] (0.042)	<0.001 [c,d]
Infant FM/FFM with ultrasound 2-skinfolds	0.03 ± 0.03 (0.69)	0.01 ± 0.03 (0.96)	−0.003 ± 0.03 (1.00)	−0.01 ± 0.02 (0.93)	−0.03 ± 0.02 (0.61)	−0.02 ± 0.03 (0.93)	0.59
Infant FM/FFM with ultrasound 4-skinfolds	0.02 ± 0.02 (0.86)	0.002 ± 0.02 (1.00)	−0.03 ± 0.02 (0.54)	−0.01 ± 0.02 (0.87)	−0.05 ± 0.02 (0.095)	−0.03 ± 0.02 (0.41)	0.16
Infant FM/FFM with BIS	**0.12 ± 0.02 (<0.001)**	**0.06 ± 0.02 (0.028)**	0.05 ± 0.02 (0.17)	**−0.06 ± 0.02 (0.006)**	**−0.07 ± 0.02 (0.002)**	−0.01 ± 0.02 (0.95)	**<0.001**
Concentrations	**(n = 15)**	**(n = 14)**	**(n = 8)**	**(n = 19)**	**(n = 14)**	**(n = 14)**	**(n = 20)**
Total carbohydrates (g/L)	−3.0 ± 3.7 (0.86)	1.4 ± 3.8 (0.98)	−1.2 ± 4.1 (0.99)	4.4 ± 3.5 (0.60)	1.8 ± 3.8 (0.97)	−2.6 ± 3.9 (0.91)	0.55
Lactose (g/L)	−1.3 ± 1.8 (0.89)	−1.1 ± 1.8 (0.93)	0.5 ± 1.9 (0.98)	0.2 ± 1.6 (1.00)	1.8 ± 1.8 (0.73)	1.6 ± 1.8 (0.80)	0.70
HMO (g/L)	−2.2 ± 3.9 (0.94)	2.2 ± 3.9 (0.95)	0.7 ± 4.3 (1.00)	4.3 ± 3.6 (0.63)	2.9 ± 3.9 (0.89)	−1.4 ± 4.0 (0.98)	0.69
CDI				**(n = 7)**	**(n = 6)**	**(n = 6)**	**(n = 9)**
Total carbohydrates (g) [e]	n/a [f]	n/a [f]	n/a [f]	**−21.0 ± 5.9 (0.001)**	**−24.7 ± 6.2 (<0.001)**	−3.7 ± 6.6 (0.84)	**0.003**
Lactose (g) [e]	n/a	n/a	n/a	**−19.4 ± 4.4 (<0.001)**	**−23.1 ± 4.4 (<0.001)**	−3.8 ± 4.8 (0.71)	**<0.001**
HMO (g) [e]	n/a	n/a	n/a	−1.4 ± 4.0 (0.94)	−1.5 ± 4.2 (0.94)	−0.05 ± 4.9 (1.00)	0.51

[a] Systematic differences in the measured variables between different months after birth were calculated using general linear hypothesis test (Tukey's all pair comparisons). [b] Data are parameter estimate ± SE of estimate and p-value (in parenthesis). [c] Overall p-value is associated with age as reported in linear mixed model. [d] Bold text indicates significant difference (p < 0.05) between two time points or overall. [e] 24-h milk intake and feeding frequency as meals per 24-h was measured at 24-h milk production and daily intakes (CDI) calculated between 2 and 5 months (n = 17) and within 2 weeks of 9 (n = 8) and 12 months (n = 8). [f] Results are not presented for impractical combinations. BIS: bioelectrical impedance spectroscopy; FFM: fat-free mass; FM: fat mass; FM/FFM: fat mass to fat-free mass ratio; HMO: human milk oligosaccharides; I: infants; M: mothers; n/a: not applicable.

Table 3. Associations between concentrations of human milk carbohydrates and infant characteristics.

Predictor (Concentration, g/L)	2 Months (n = 15)		5 Months (n = 20)		9 Months (n = 19)		12 Months (n = 14)		P-Value (n = 20)		
	Intercept (SE)	Slope (SE)	Intercept (SE)	Slope (SE)	Intercept (SE)	Slope (SE)	Intercept (SE)	Slope (SE)	Predictor	Infant Age (Months)	Interaction
Infant Length (cm)											
TCH	53.90 (1.22) [a]	0.047 (0.013)	60.90 (1.15)	0.047 (0.013)	66.20 (1.21)	0.047 (0.013)	70.40 (1.17)	0.047 (0.013)	**<0.001** [b]	<0.001	0.62 [c]
Lactose	53.60 (2.31)	0.065 (0.034)	60.50 (2.27)	0.065 (0.034)	66.10 (2.26)	0.065 (0.034)	70.0 (2.27)	0.065 (0.034)	0.047	<0.001	0.093
HMO	57.40 (0.62)	0.031 (0.014)	64.30 (0.55)	0.031 (0.014)	69.70 (0.60)	0.031 (0.014)	73.70 (0.60)	0.031 (0.014)	0.036	<0.001	0.67
Infant Weight (kg)											
TCH	4.55 (0.42)	0.013 (0.004)	6.39 (0.40)	0.013 (0.004)	7.65 (0.41)	0.013 (0.004)	8.47 (0.40)	0.013 (0.004)	**0.003**	<0.001	0.54
HMO	5.46 (0.22)	0.009 (0.004)	7.28 (0.21)	0.009 (0.004)	8.54 (0.21)	0.009 (0.004)	9.42 (0.22)	0.009 (0.004)	0.038	<0.001	0.17
Infant Body Mass Index (kg/m^2)											
TCH	11.30 (2.04)	0.063 (0.024)	18.20 (2.38)	−0.007 (0.029)	19.10 (1.83)	−0.016 (0.021)	18.30 (1.29)	−0.013 (0.015)	0.91	<0.001	0.044
HMO	15.70 (0.51)	0.049 (0.020)	17.80 (0.52)	−0.008 (0.026)	17.70 (0.59)	−0.001 (0.023)	17.90 (0.47)	−0.026 (0.016)	0.99	<0.001	0.027
Infant Fat-free Mass with Bioelectrical Impedance Spectroscopy (kg)											
TCH	2.63 (0.35)	0.020 (0.004)	3.69 (0.33)	0.020 (0.004)	4.78 (0.35)	0.020 (0.004)	5.53 (0.33)	0.020 (0.004)	**<0.001**	<0.001	0.28
HMO	4.02 (0.17)	0.015 (0.004)	5.05 (0.15)	0.015 (0.004)	6.17 (0.17)	0.015 (0.004)	6.90 (0.17)	0.015 (0.004)	**<0.001**	<0.001	0.069
Infant Fat-free Mass with Ultrasound 2-skinfolds (kg)											
TCH	3.41 (0.38)	0.009 (0.004)	4.66 (0.36)	0.009 (0.004)	5.63 (0.38)	0.009 (0.004)	6.32 (0.37)	0.009 (0.004)	0.032	<0.001	0.82
Infant Fat-free Mass with Ultrasound 4-skinfolds (kg)											
TCH	3.35 (0.38)	0.010 (0.004)	4.66 (0.36)	0.010 (0.004)	5.64 (0.38)	0.010 (0.004)	6.46 (0.36)	0.010 (0.004)	**0.020**	<0.001	0.86
Infant Fat-free Mass Index with Bioelectrical Impedance Spectroscopy (kg/m^2)											
TCH	10.90 (0.62)	0.022 (0.007)	10.80 (0.58)	0.022 (0.007)	11.20 (0.62)	0.022 (0.007)	11.20 (0.59)	0.022 (0.007)	**0.002**	0.054	0.83
HMO	12.40 (0.28)	0.018 (0.007)	12.30 (0.24)	0.018 (0.007)	12.70 (0.27)	0.018 (0.007)	12.80 (0.27)	0.018 (0.007)	**0.008**	0.030	0.30
Infant Fat Mass with Bioelectrical Impedance Spectroscopy (kg)											
TCH	−0.03 (0.71)	0.016 (0.008)	2.91 (0.83)	−0.010 (0.010)	3.81 (0.63)	−0.018 (0.007)	3.14 (0.44)	−0.010 (0.005)	0.051	<0.001	**0.016**
HMO	1.09 (0.17)	0.012 (0.007)	2.26 (0.17)	−0.009 (0.009)	2.56 (0.20)	−0.016 (0.008)	2.57 (0.15)	−0.011 (0.006)	0.11	<0.001	0.039
Infant Fat Mass with Bioelectrical Impedance Spectroscopy (%)											
TCH	9.84 (6.56)	0.154 (0.078)	37.20 (7.66)	−0.106 (0.010)	46.80 (5.87)	−0.244 (0.067)	35.60 (4.09)	−0.131 (0.049)	0.002	<0.001	**0.001**
HMO	20.50 (1.54)	0.122 (0.069)	30.10 (1.56)	−0.100 (0.087)	30.50 (1.87)	−0.241 (0.080)	27.00 (1.40)	−0.121 (0.055)	0.020	<0.001	**0.005**

Table 3. *Cont.*

Predictor (Concentration, g/L)	2 Months (n = 15)		5 Months (n = 20)		9 Months (n = 19)		12 Months (n = 14)		P-Value (n = 20)		
	Intercept (SE)	Slope (SE)	Intercept (SE)	Slope (SE)	Intercept (SE)	Slope (SE)	Intercept (SE)	Slope (SE)	Predictor	Infant Age (Months)	Interaction
Infant Fat Mass Index with Bioelectrical Impedance Spectroscopy (kg/m²)											
TCH	0.40 (1.47)	0.041 (0.017)	6.81 (1.72)	−0.022 (0.021)	8.59 (1.32)	−0.046 (0.015)	6.72 (0.92)	−0.029 (0.011)	0.008	<0.001	**0.001**
HMO	3.23 (0.36)	0.033 (0.015)	5.39 (0.36)	−0.021 (0.019)	5.41 (0.43)	−0.042 (0.018)	4.93 (0.32)	−0.032 (0.012)	0.033	<0.001	**0.003**
Infant Fat Mass to Fat-free Mass Ratio with Bioelectrical Impedance Spectroscopy											
TCH	0.09 (0.12)	0.003 (0.001)	0.56 (0.14)	−0.002 (0.002)	0.70 (0.10)	−0.004 (0.001)	0.54 (0.07)	−0.003 (0.001)	0.001	<0.001	**0.003**
HMO	0.26 (0.03)	0.002 (0.001)	0.44 (0.03)	−0.002 (0.002)	0.44 (0.03)	−0.004 (0.001)	0.38 (0.03)	−0.002 (0.001)	0.027	<0.001	**0.005**

[a] Parameter estimate ± SE; effects of predictors taken from linear mixed effects models that accounted for month after birth and an interaction between month after birth and predictor with a random effect per participant; if the interaction is not significant parameter estimates are taken from a model with no interaction. [b,c] Results are presented only for interactions or predictors with raw p-values < 0.05; after the false discovery rate adjustment the interaction/predictor p-values were considered to be significant at <0.032 for total carbohydrates (TCH) concentration (indicated by the bold text), at <0.027 for human milk oligosaccharides (HMO) concentration (indicated by the bold text), and at <0.047 for lactose concentration (none are significant).

Table 4. Associations between calculated daily intakes of human milk carbohydrates and infant characteristics.

Predictor (CDI [d], g)	Between 2 and 5 Months (n = 17)		9 Months (n = 8)		12 Months (n = 8)		P-Value (n = 18)		
	Intercept (SE)	Slope (SE)	Intercept (SE)	Slope (SE)	Intercept (SE)	Slope (SE)	Predictor	Infant Age (Months)	Interaction
Infant Body Mass Index (kg/m²)									
TCH	15.60 (1.30) [a]	0.033 (0.019)	18.60 (1.47)	−0.013 (0.029)	18.50 (0.72)	−0.018 (0.013)	0.65 [b]	0.77	**0.019** [c]
Lactose	15.70 (1.00)	0.038 (0.018)	17.90 (1.21)	0.0003 (0.032)	19.00 (0.88)	−0.040 (0.025)	0.38	0.59	**0.011**
Infant Fat-free Mass Index with Ultrasound 4-skinfolds (kg/m²)									
TCH	15.20 (0.56)	−0.034 (0.007)	14.70 (0.47)	−0.034 (0.007)	15.10 (0.45)	−0.034 (0.007)	**<0.001**	0.035	0.37
Lactose	13.30 (0.73)	−0.006 (0.013)	12.90 (0.87)	0.003 (0.023)	15.40 (0.65)	−0.060 (0.018)	0.076	0.13	**0.015**
Infant Fat Mass with Ultrasound 2-skinfolds (kg)									
TCH	1.24 (0.29)	0.011 (0.004)	1.82 (0.25)	0.011 (0.004)	1.94 (0.23)	0.011 (0.004)	**0.006**	<0.001	0.29
Lactose	1.23 (0.31)	0.014 (0.006)	1.93 (0.24)	0.014 (0.006)	2.06 (0.21)	0.014 (0.006)	**0.008**	<0.001	0.19
Infant Fat Mass with Ultrasound 4-skinfolds (kg)									
TCH	1.09 (0.26)	0.014 (0.004)	1.73 (0.22)	0.014 (0.004)	1.64 (0.20)	0.014 (0.004)	**<0.001**	<0.001	0.16
Lactose	1.16 (0.30)	0.015 (0.005)	1.87 (0.23)	0.015 (0.005)	1.80 (0.21)	0.015 (0.005)	**0.004**	<0.001	0.21
HMO	1.73 (0.14)	0.020 (0.008)	2.13 (0.18)	0.020 (0.008)	1.99 (0.17)	0.020 (0.008)	0.010	0.061	0.77

Table 4. *Cont.*

Predictor (CDI [d], g)	Between 2 and 5 Months ($n = 17$)		9 Months ($n = 8$)		12 Months ($n = 8$)		P-Value ($n = 18$)		
	Intercept (SE)	Slope (SE)	Intercept (SE)	Slope (SE)	Intercept (SE)	Slope (SE)	Predictor	Infant Age (Months)	Interaction
				Infant Fat Mass with Ultrasound 2-skinfolds (%)					
TCH	20.10 (2.51)	0.010 (0.036)	21.80 (2.15)	0.010 (0.036)	20.70 (1.98)	0.010 (0.036)	**0.005**	0.64	0.064
Lactose	19.80 (2.80)	0.128 (0.050)	22.30 (2.18)	0.128 (0.050)	21.60 (1.94)	0.128 (0.050)	**0.019**	0.59	0.11
				Infant Fat Mass with Ultrasound 4-skinfolds (%)					
TCH	14.00 (2.91)	0.193 (0.043)	31.10 (4.47)	−0.083 (0.090)	18.80 (2.05)	0.103 (0.042)	<0.001	0.051	**0.016**
Lactose	18.30 (2.66)	0.156 (0.048)	21.60 (2.10)	0.156 (0.048)	18.70 (1.85)	0.156 (0.048)	**0.001**	0.079	0.069
HMO	24.60 (1.29)	0.168 (0.074)	25.10 (1.64)	0.168 (0.074)	21.30 (1.58)	0.168 (0.074)	0.025	0.095	0.51
				Infant Fat Mass Index with Bioelectrical Impedance Spectroscopy (kg/m^2)					
Lactose	3.66 (0.74)	0.027 (0.013)	3.27 (0.58)	0.027 (0.013)	3.55 (0.53)	0.027 (0.013)	0.045	0.62	0.41
HMO	5.37 (0.32)	−0.024 (0.017)	4.51 (0.41)	−0.024 (0.017)	4.64 (0.40)	−0.024 (0.017)	0.049	0.013	0.18
				Infant Fat Mass Index with Ultrasound 2-skinfolds (kg/m^2)					
TCH	3.10 (0.61)	0.025 (0.009)	3.64 (0.52)	0.025 (0.009)	3.36 (0.47)	0.025 (0.009)	**0.003**	0.32	0.10
Lactose	3.01 (0.67)	0.032 (0.012)	3.74 (0.51)	0.032 (0.012)	3.47 (0.47)	0.032 (0.012)	**0.005**	0.18	0.27
				Infant Fat Mass Index with Ultrasound 4-skinfolds (kg/m^2)					
TCH	1.64 (0.74)	0.048 (0.011)	5.22 (1.12)	−0.007 (0.023)	3.21 (0.52)	0.020 (0.011)	<0.001	0.078	0.038
Lactose	2.72 (0.64)	0.038 (0.012)	3.52 (0.50)	0.038 (0.012)	2.84 (0.46)	0.038 (0.012)	**<0.001**	0.065	0.16
HMO	4.28 (0.32)	0.040 (0.018)	4.43 (0.41)	0.040 (0.018)	3.62 (0.39)	0.040 (0.018)	0.034	0.19	0.68
				Infant Fat Mass to Fat-free Mass Ratio with Bioelectrical Impedance Spectroscopy					
HMO	0.43 (0.02)	−0.002 (0.001)	0.34 (0.03)	−0.002 (0.001)	0.36 (0.03)	−0.002 (0.001)	0.024	<0.001	0.095
				Infant Fat Mass to Fat-free Mass Ratio with Ultrasound 2-skinfolds					
TCH	0.23 (0.05)	0.002 (0.001)	0.27 (0.04)	0.002 (0.001)	0.25 (0.04)	0.002 (0.001)	**0.004**	0.68	0.052
Lactose	0.22 (0.06)	0.003 (0.001)	0.28 (0.04)	0.003 (0.001)	0.27 (0.04)	0.003 (0.001)	**0.012**	0.50	0.14
				Infant Fat Mass to Fat-free Mass Ratio with Ultrasound 4-skinfolds					
TCH	0.12 (0.05)	0.004 (0.001)	0.46 (0.08)	−0.002 (0.002)	0.23 (0.04)	0.002 (0.001)	<0.001	0.058	**0.007**
Lactose	0.21 (0.05)	0.003 (0.001)	0.27 (0.04)	0.003 (0.001)	0.22 (0.04)	0.003 (0.001)	**<0.001**	0.084	0.053
HMO	0.33 (0.02)	0.003 (0.001)	0.34 (0.03)	0.003 (0.001)	0.27 (0.03)	0.003 (0.001)	0.027	0.12	0.45

[a] Parameter estimate ± SE; effects of predictors taken from linear mixed effects models that accounted for month after birth and an interaction between month after birth and predictor with a random effect for participant; if the interaction is not significant parameter estimates are taken from a model with no interaction. [b,c] Results are presented only for interactions or predictors with raw *p*-values < 0.05; after the false discovery rate adjustment the interaction/predictor *p*-values were considered to be significant at <0.038 for calculated daily intake (CDI) of total carbohydrates (TCH) (indicated by the bold text), at <0.045 for lactose CDI (indicated by the bold text), and at <0.010 for human milk oligosaccharides (HMO) CDI (none are significant). [d] CDI were measured between 2 and 5 months and within 2 weeks of 9 and 12 months.

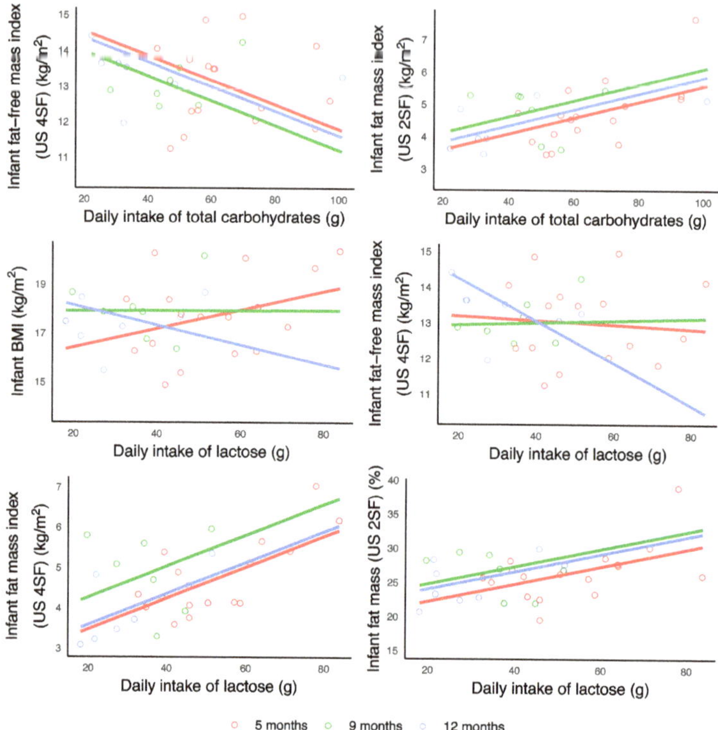

Figure 3. Significant associations between calculated daily intakes of carbohydrates and infant body composition measurements. Lines represent linear regression and grouped by the month of lactation. BMI: body mass index; BIS: US 2SF: ultrasound 2-skinfolds; US 4SF: ultrasound 4-skinfolds.

Higher lactose CDI was associated with increased FM (ultrasound 2-skinfolds: $p = 0.008$; ultrasound 4-skinfolds: $p = 0.004$), FMI (ultrasound 2-skinfolds: $p = 0.005$; ultrasound 4-skinfolds: $p < 0.001$), %FM (ultrasound 2-skinfolds: $p = 0.019$; ultrasound 4-skinfolds: $p = 0.001$), FM/FFM (ultrasound 2-skinfolds: $p = 0.012$; ultrasound 4-skinfolds: $p < 0.001$), and also with BMI (an increase at 5 months, no change at 9 months and a decrease at 12 months, $p = 0.011$) and FFMI (ultrasound 4-skinfolds: no change at 5 and 9 months and a decrease at 12 months, $p = 0.015$) (Table 4, Figure 3).

No further associations were found after adjusting for the FDR (Table 4).

3.6. Associations between Breastfeeding Parameters and Carbohydrates

No significant associations were found between carbohydrates concentrations and 24-h milk intake, 24-h breastfeeding frequency (meals per 24-h), and self-reported breastfeeding frequency (hours between feeds) (data not shown). Higher concentrations of TCH and HMO were associated with higher CDI of these carbohydrates, respectively (TCH: $p = 0.009$; HMO: $p = 0.014$), while no association was seen between lactose concentration and CDI (Table A2).

A higher 24-h milk intake was associated with higher CDI of all measured carbohydrates (TCH: $p < 0.001$; lactose: $p < 0.001$; HMO: $p = 0.016$).

Higher 24-h breastfeeding frequency and lower self-reported breastfeeding frequency were associated with higher CDI of TCH and lactose (24-h breastfeeding frequency: TCH, $p = 0.009$; lactose, $p < 0.001$; self-reported breastfeeding frequency: TCH, $p < 0.001$; lactose, $p = 0.006$) while no associations with HMO CDI were seen (Table A2, Figure 4).

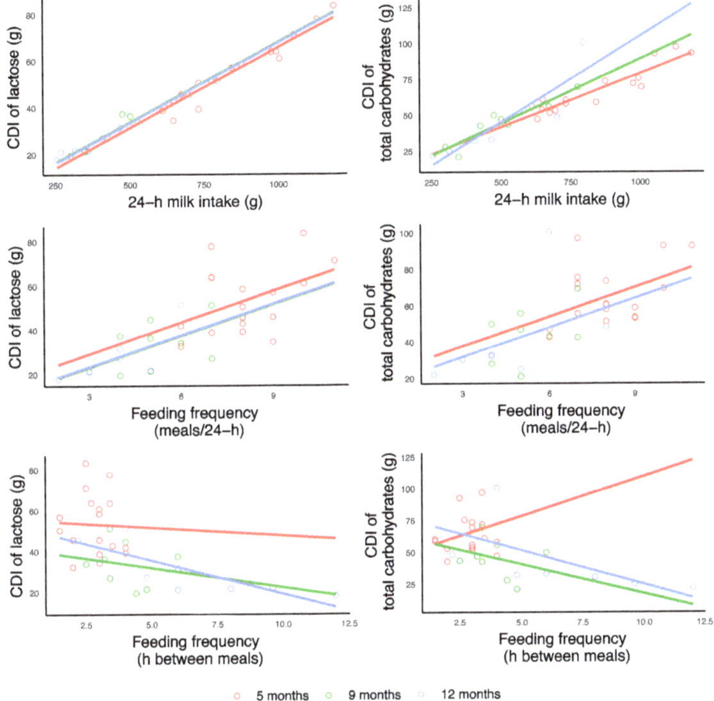

Figure 4. Significant associations between breastfeeding parameters and calculated daily intakes (CDI) of lactose and total carbohydrates measured during 24-h milk production. Lines represent linear regression and grouped by the month of lactation. Where is not seen, the linear regression line for 9 months (green) is under the 12 months line (blue) due to the similar intercept values.

3.7. Changes in Infant Body Composition and Calculated Daily Intakes of Carbohydrates

A significant association was found between changes in infant BC (Δ) between the time points and lactose CDI. Higher lactose CDI at 12 months was associated with greater decrease in FFMI (ultrasound 4-skinfolds: $p = 0.0004$) between 5 and 12 months (Table 5). No further significant associations were found at any time points between changes in infant BC (Δ) between the time points and CDI of both TCH and HMO after adjusting for the FDR (Tables 3 and 4).

Table 5. Associations of calculated daily intakes of lactose at the time points with changes in infant body composition between the time points.

Changes in Infant Characteristic (Response)	Months after Birth					
	5 and 2	9 and 2	12 and 2	9 and 5	12 and 5	12 and 9
Lactose CDI (g) between 2 and 5 months ($n = 17$) [c]						
ΔLength (cm)	−0.019 (0.031) [a] 0.56 [b]	0.059 (0.032) 0.11	**0.100 (0.032)** [d] **0.016**	0.004 (0.026) 0.88	0.021 (0.034) 0.54	0.028 (0.033) 0.41
ΔBMI (kg/m^2)	**0.049 (0.019)** **0.035**	0.015 (0.030) 0.63	−0.006 (0.038) 0.88	−0.015 (0.018) 0.44	−0.033 (0.023) 0.17	−0.017 (0.013) 0.23
ΔFFM with US 2SF (kg)	−0.013 (0.011) 0.27	**0.024 (0.009)** **0.029**	0.015 (0.014) 0.32	0.003 (0.008) 0.76	0.004 (0.010) 0.66	0.0002 (0.005) 0.96
ΔFFM with US 4SF (kg)	0.003 (0.013) 0.84	**0.043 (0.012)** **0.009**	0.044 (0.019) 0.052	0.005 (0.008) 0.51	0.008 (0.009) 0.41	0.004 (0.005) 0.44
ΔFFM with BIS (kg)	0.009 (0.005) 0.12	0.023 (0.016) 0.19	**0.035 (0.012)** **0.025**	0.004 (0.009) 0.63	0.011 (0.011) 0.32	0.012 (0.010) 0.24
ΔFFMI with US 4SF (kg/m^2)	0.041 (0.023) 0.12	0.045 (0.026) 0.13	**0.088 (0.035)** **0.044**	0.006 (0.013) 0.64	−0.002 (0.018) 0.91	−0.004 (0.014) 0.77
ΔFM with US 2SF (kg)	**0.029 (0.010)** **0.014**	0.007 (0.010) 0.53	0.014 (0.015) 0.38	−0.004 (0.008) 0.68	−0.005 (0.008) 0.53	0.001 (0.004) 0.87
ΔFMI with US 2SF (kg/m^2)	**0.057 (0.023)** **0.036**	−0.003 (0.022) 0.91	−0.009 (0.035) 0.80	−0.014 (0.018) 0.46	−0.025 (0.020) 0.23	−0.007 (0.008) 0.40
ΔFMI with BIS (kg/m^2)	0.026 (0.015) 0.11	0.001 (0.029) 0.98	−0.020 (0.023) 0.42	−0.017 (0.017) 0.33	**−0.045 (0.017)** **0.023**	−0.026 (0.019) 0.19
ΔFM/FFM with US 2SF	**0.007 (0.003)** **0.036**	−0.001 (0.002) 0.60	−0.001 (0.005) 0.93	−0.002 (0.002) 0.35	−0.001 (0.001) 0.67	−0.0004 (0.001) 0.65
ΔFM/FFM with BIS	0.001 (0.001) 0.40	−0.001 (0.003) 0.80	**−0.006 (0.002)** **0.032**	−0.002 (0.001) 0.12	**−0.004 (0.002)** **0.034**	−0.002 (0.002) 0.32
Lactose CDI (g) at 9 months ($n = 8$) [c]						
ΔFM with BIS (kg)	n/a [e]	0.053 (0.081) 0.58	−0.033 (0.062) 0.65	−0.020 (0.025) 0.46	**−0.037 (0.014)** **0.045**	−0.017 (0.018) 0.39
Lactose CDI (g) at 12 months ($n = 8$) [c]						
ΔFFMI with US 2SF (kg/m^2)	n/a [e]	n/a [e]	−0.082 (0.036) 0.15	n/a [e]	**−0.097 (0.018)** **0.003**	−0.041 (0.018) 0.072
ΔFFMI with US 4SF (kg/m^2)	n/a	n/a	−0.075 (0.058) 0.33	n/a	**−0.080 (0.007)** **0.0004 *****	−0.042 (0.026) 0.17
ΔFM with US 2SF (kg)	n/a	n/a	**0.036 (0.010)** **0.032**	n/a	0.019 (0.011) 0.13	0.005 (0.006) 0.48

[a] Parameter estimate ± SE and [b] *P*-values for associations between [c] calculated daily intakes (CDI) of lactose (predictor) at given time points and the changes (Δ) in measured variables between different months after birth. [d] Results are presented only for variables with at least one significant raw *P*-value ($p < 0.05$, indicated by the bold text); after the false discovery rate adjustment, the predictor *P*-values were considered to be significant at <0.009 for CDI of lactose between 2 and 5 months (none are significant), at <0.045 for CDI at 9 months (none are significant) and at <0.003 for CDI at 12 months (indicated by the bold text and ***). [e] Results are not presented for impractical time combinations. BIS: bioelectrical impedance spectroscopy; BMI: body mass index; FFM: fat-free mass; FFMI: fat-free mass index; FM: fat mass; FM/FFM: fat mass to fat-free mass ratio; FMI: fat mass index; n/a: not applicable; US 2SF: ultrasound 2-skinfolds; US 4SF: ultrasound 4-skinfolds.

4. Discussion

This proof-of-concept longitudinal study elucidates some of the complex mechanisms by which breastfeeding and HM may affect the development of infant BC and possibly contribute to protection from obesity. Our findings are noteworthy, given the scarcity of data on the effect of HM components, specifically carbohydrates, on infant growth and development. In this study both, concentrations and, more importantly, daily intakes of HM carbohydrates have been found to associate with development of infant BC and are differentially related to infant anthropometry and BC (FM and FFM; Figure 5) throughout the first 12 months of lactation. Moreover, infant breastfeeding frequency was associated with CDI of TCH and lactose, highlighting the important role of breastfeeding in programming of infant appetite regulation and BC in the first year of life.

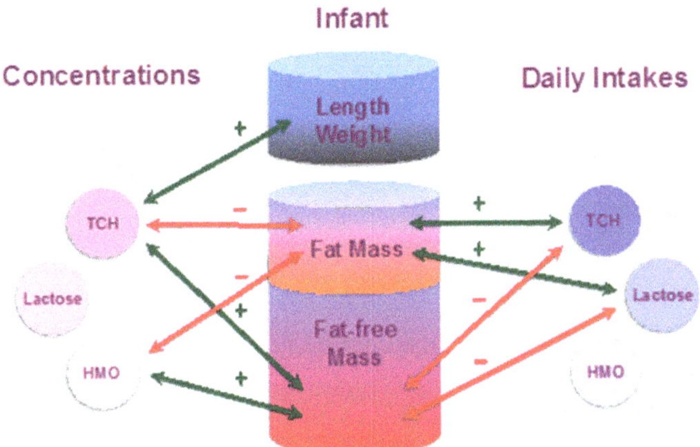

Figure 5. Suggested lactocrine programming of the infant body composition during first 12 months of life. Green arrows indicate positive associations of human milk carbohydrates (concentrations and calculated daily intakes) with infant body composition parameters and red arrows—negative associations. HMO: human milk oligosaccharides; TCH: total carbohydrates.

HM carbohydrates are a major macronutrient which contributes over 40% of energy of the HM [39], predominantly from lactose, and may have greater biological significance compared with other components; thus, it is not surprising that we have established multiple strong associations with infant BC. What is interesting is that the directions of the associations with infant BC were not uniform for the concentration and CDI of the same measured component. This is due to the lack of association between CDI and concentration of lactose, and differences in CDI of TCH and lactose, but not in CDI of HMO, across the time.

Within the normal developmental context of breastfeeding we established that the concentration of lactose, the principal HM carbohydrate, was not related to infant BC or anthropometrics during the first 12 months of lactation (Table 3). Although added dietary sugars have been shown to promote excess adiposity in infants [5], studies of the effect of HM lactose on infant BC are rare. While there has been increased interest in HMO, the importance of lactose for infant development and digestive issues, such as lactose intolerance and malabsorption, is still not well understood. Two recent observational studies found higher HM lactose concentration was related to greater infant weight gain. The first study confirmed a positive association between HM lactose concentration (2–4 weeks lactation) and weight-to-length z-score of 6-month-old infants of mothers with excess weight and obesity [40]. The second study also showed that increased HM lactose concentration (4–8 weeks lactation) was related to increased infant weight gain, BMI, and adiposity (skinfolds) at 12 months [22]. Our contrasting results are likely due to the mixed feeding present in both of the previous studies particularly in the first 5 months, and pooling of HM samples over two weeks in one study [22]. On the other hand, our findings are similar to two studies that reported no relationship of lactose concentration with BC of 6-month-old exclusively breastfed infants [23] and infant growth rate (1–6 months) of exclusively breastfed infants (4 or more months) [11], supporting the notion that exclusivity in the first 4–6 months of life may be important. Alternatively, the lack of associations of lactose concentration could be due to its role in maintaining a constant osmotic pressure [10], and thus being one of the least variable HM components [11,14] during the established lactation.

To account for variation in infant 24-h milk intake, which is known to contribute to infant growth [11], we calculated the CDI of carbohydrates and observed that the absence of associations of lactose concentration with infant BC was mediated by daily intakes. CDI of lactose was positively associated with infant adiposity and negatively with lean body mass (Table 4). Furthermore, lactose CDI at 12 months was associated with a decrease in infant FFMI between 5 and 12 months (Table 5).

Previously daily lactose intake has not been shown to be related to the growth rate of the 4-month-old breastfed infants [11], however the study sample size was very small ($n = 8$). Another study found contradictory results to ours with higher daily intakes of HM lactose being related to increased weight and lean mass gain but not FM in 3–6 months-old breastfed infants [21]. The hypothesis of the growth regulating effect of lactose is supported by porcine studies that show increased weight gain with administered lactose [41], which has been attributed to the prebiotic effect on gut microbiome resulting in growth of lactobacilli and bifidobacteria, improved intestinal health and better minerals absorption. Further, one in vitro HM study reported that in colonic epithelial cells HM lactose induces the gene (*CAMP*) that encodes an antimicrobial peptide, human cathelicidin LL-37, in both a dose- and time-dependent manner [42]. Furthermore, the *CAMP*-inducing capacity of HM lactose increased with time postpartum, while no *CAMP*-inducing effect was found for lactose in commercial formulas. Authors have speculated that small amount of lactose may not be absorbed and, via the *CAMP*-induction, contribute to infant gut homeostasis, protection against pathogens and regulation of gut microbiome [42], which has been implicated in the development of infant weight gain and obesity [43].

The measured HM concentrations of lactose or TCH in our cohort were not related to maternal adiposity, similar to previous findings in a larger cohort [9]. Given that lactose is important for maintaining a constant osmotic pressure in HM [10], maternal adiposity was not expected to have a significant impact on lactose concentration, although negative associations between the lactose concentration and maternal BMI but not %FM have been reported [44]. In contrast, one longitudinal study found positive association between maternal BMI and TCH concentration [45], however this was in a cohort of women with high milk supply which may have influenced the relationship.

Lactose accounts for approximately 85% of the total HM carbohydrates [46] and is often considered akin to TCH despite the significant presence of other carbohydrates. HMO is the third most plentiful component of HM, the amount of which often exceeds total amount of protein [17], while monosaccharides (glucose and galactose) are present in HM in very small quantities [36]. We have used the technique that reliably estimates concentrations and carbon content for monosaccharides, disaccharides as well as polysaccharides with very high molecular weight [34,35] and found that higher concentrations of TCH were associated with greater infant length, weight and lean body mass and decreased adiposity (Table 3). However, TCH CDI was associated with increased infant adiposity and reduced lean mass (Table 4), similar to lactose CDI. Comparable to our findings for TCH, higher concentrations of glucose have been reported to associate with higher relative weight and BMI of 1-month-old breastfed infants [24], while concentration of fructose, thought to be present in HM from maternal diet, was positively associated with weight and lean mass in 6-month-old exclusively breastfed infants [23]. Although present in HM in much lower quantities, together with lactose these small carbohydrates may have a synergistic effect, as previously seen in an in vitro study where stimulation of colonic epithelial cells with combinations of various mono- and disaccharides produced enhanced *CAMP* gene expression compared with the individual effects [42], implying potential roles of multiple HM carbohydrates in infant gut homeostasis in addition to HMO.

Our estimations of HMO also resulted in associations with infant BC. HMO concentration was positively associated with infant lean mass and negatively with adiposity (Table 3), and not HMO CDI (Table 4). Our results are consistent with HMO effects seen in the study of 6-month-old breastfed infants, where concentrations of 16 detected HMO displayed various associations with infant BC, and higher HMO diversity and evenness were associated with lower FM and %FM at 1 month postpartum [20]. Importantly, when examined collectively, the concentration of several HMO in that study explained 33% more of variance in infant adiposity than maternal pre-pregnancy BMI, weight gain during pregnancy or infant sex and age alone. There were discrepancies between HMO concentrations though, with the sum of the 16 individual HMO concentrations (11 g/L at 1 month; 10 g/L at 6 months) [20] being a half of our estimated total HMO concentration at all time points, which could be explained partially by the methodology, with the sulfuric acid–UV method detecting saccharides usually excluded from measurement with HPLC technique, such as acidic oligosaccharides, large oligosaccharides with a degree of polymerization, and small amounts of unidentified saccharides [18].

Other studies of concentrations of individual HMO (2'fucosyllactose) have shown no relationships [47]. Further, a study where 2'fucosyllactose and lacto-N-neotetraose were added to formula showed no relationships with infant anthropometrics over the first year of life [48]. Our results indicate that total HMO as opposed to individual HMO are important, and indeed it has been shown that specific bacteria differentially consume specific HMO [49], with a mixture of HMO feeding the whole microbiome. Thus, HMO likely have an indirect effect on BC by enhancing the growth of beneficial gut bacteria and altering the structure or function of the gut microbiome during the critical periods for the development of obesity [43,50]. Directly measuring the wider variety of HMO would be an advantage in the future research.

The recommendations for breastfeeding are to feed on demand. While we have not seen any associations between concentrations of HM carbohydrates and breastfeeding parameters, we found that infants who fed more frequently and had a higher 24-h milk intake consumed higher daily doses (CDI) of lactose and TCH (Table A2). We have previously reported that frequent feeders consume more milk and have lower FFM and higher FM [15], although the order of the relationship is not clear. Further, lactose is also implicated in gastric emptying of breastfed infants, with higher volumes of higher lactose concentration emptying faster compared to lower volumes of lower concentration [51], supported by previous findings in the group that higher lactose concentration is associated with increased breastfeeding frequency [14]. All of these multiple associations culminate in relationships to infant adiposity and highlight the complexity of pathways by which HM components and the dose that the infant receives impact infant development. In contrast, HMO CDI was not associated with breastfeeding frequency in this study, nor were any strong associations previously seen between total HMO concentration/single dose and gastric emptying of breastfed infants [51], indicating the absence of short-term response to total HMO and their target being the small intestine rather than the stomach.

Before the FDR adjustment, we established that HMO CDI was associated with maternal adiposity (%FM and FM/FFM, Table A1), and associations were contrasting at different stages of lactation; during the exclusive breastfeeding period the relationship was positive, but at 9 and 12 months it was negative. This is in agreement with a study that found negative associations between BMI/adiposity and individual HMO concentrations in maternal serum during pregnancy [52], and with two HM studies, which reported positive and negative correlations between maternal BMI and individual HMO [19,53], indicating maternal factors may contribute to total HMO content. It is plausible that relationships of HMO with maternal adiposity may change when mother returns to pre-pregnancy adiposity, or indicate a possible shift in HMO profile later in lactation when volume of milk produced declines. While the total proportion of HMO has been reported to decline during lactation [18], we did not see the decline in total HMO concentrations over the 12 months of lactation. This emphasizes the importance of HMO for the infant development and accounting for the dose of these carbohydrates.

These findings highlight pitfalls in HM research, with HM components concentrations being the major factor measured with respect to various infant outcomes. Pointedly we see none or opposite relationships for these components when analyzing both, raw concentrations and intakes, rising a question if concentration is a dependable measure. When it comes to HM, concentration of component does not necessarily reflect the dose of the component, considering the variability in volumes of HM consumed, duration of exclusive breastfeeding, as well as in breastfeeding patterns and mixed feeding. When examining the nutritional physiology of the breastfed infant CDI may be a more relevant to understanding the development of BC than concentrations.

Our proof-of-concept study concentrated on women that breastfed on demand for 12 months (exclusively to 5 months and without formula use thereafter), and therefore is more indicative of normative development of BC of the breastfed infant as well as lactation. The strengths of this study also include the longitudinal assessment of HM composition and BC of dyads, the broad variation of maternal adiposity and the analysis of TCH along with lactose. The study limitations are the modest number of dyads related to the multiple measurement time points; the small number of 24-h milk productions at the later months of lactation; an estimation of the total HMO concentration, notwithstanding the technical difficulties with accounting for all HMO. We were unable to collect infant dietary intake data after 5 months, yet it is introduction of and duration of use of formula that alters the BC of breastfed infants, not introduction of solids [21]. Our sample was relatively homogenous (primarily Caucasian term healthy fully-breastfed singletons from mothers of higher social-economic status), thus our results may not be transferable to more diverse populations.

5. Conclusions

This pilot study has provided the methods and design to identify relationships between the maternal BC, HM carbohydrates, 24-h milk intake, breastfeeding frequency, and infant BC. Our results point to a differential effect of concentrations and daily intakes of HM carbohydrates on development of infant lean mass and adiposity during this early time, indicating a potential to improve infant outcome through continuation of breastfeeding during the first 12 months and beyond, which could promote appropriate developmental programming and impact future metabolic health.

Author Contributions: Conceptualization, Z.G., L.C.W., K.M., P.E.H., and D.T.G.; Data curation, Z.G.; Formal analysis, Z.G., A.R., and K.M.; Funding acquisition, D.G.; Investigation, Z.G., W.J.T., C.T.L., S.K., and D.T.G.; Methodology, C.T.L., L.C.W., and D.T.G.; Resources, D.T.G.; Supervision, C.T.L., L.C.W., K.M., P.E.H., and D.T.G.; Visualization, Z.G., A.R., and K.M.; Writing—original draft, Z.G., A.R., and D.T.G.; Writing—review & editing, W.J.T., C.T.L., S.K., L.C.W., K.M., and P.E.H.

Funding: This research was funded by an unrestricted research grant from Medela AG (Switzerland). Z.G. was supported by an Australian Postgraduate Award, The University of Western Australia (Australia). S.K. was supported by a Margaret Lomann-Hall Scholarship from the School of Human Sciences, The University of Western Australia (Australia).

Acknowledgments: The authors are thankful to all the mothers and infants who participated in this study and Anna R. Hepworth for consultation on research design and assistance with data analyses.

Conflicts of Interest: The authors declare that Medela AG provided an unrestricted research grant to D.T.G., from which salaries to D.G., Z.G., W.J.T., and C.T.L. were paid. Medela AG provided a Top-up Scholarship for Z.G. and has provided speaker's fees to D.T.G. for educational lectures. L.C.W. provides consultancy services to ImpediMed Ltd. The funders had no role in the design of the study; in the collection, analyses, or interpretation of data; in the writing of the manuscript, or in the decision to publish the results. The other authors declare no conflicts of interest.

Appendix A

Table A1. Associations between maternal characteristics and calculated daily intakes of human milk oligosaccharides.

Maternal Predictor	Between 2 and 5 Months ($n = 17$)		9 Months ($n = 7$)		12 Months ($n = 7$)		*p*-Value ($n = 18$)		
	Intercept (SE)	Slope (SE)	Intercept (SE)	Slope (SE)	Intercept (SE)	Slope (SE)	Predictor	Infant Age (Months)	Interaction
	Calculated daily intake of human milk oligosaccharides (g)								
Fat mass (%)	1.64 (10.50) [a]	0.322 (0.315)	18.60 (16.00)	−0.209 (0.525)	64.80 (20.10)	−2.01 (0.735)	0.80 [b]	0.90	**0.014** [c]
FM/FFM	6.25 (6.75)	11.80 (12.80)	16.30 (10.80)	−9.10 (23.70)	52.40 (15.40)	−112.0 (40.30)	0.93	0.92	**0.013**

[a] Parameter estimate ± SE; effects of predictors taken from linear mixed effects models that accounted for month after birth and an interaction between month after birth and predictor with a random effect per participant; if the interaction is not significant parameter estimates are taken from a model with no interaction. [b,c] Results are presented only for interactions or predictors with raw *p*-values < 0.05 (indicated by the bold text); after the false discovery rate adjustment, the interaction/predictor *p*-values were considered to be significant at <0.013 for human milk oligosaccharides (none are significant). FF/FFM, fat mass to fat-free mass ratio.

Nutrients **2019**, *11*, 1472

Table A2. Associations of daily intakes of human milk carbohydrates with concentrations and breastfeeding parameters.

Predictor	Between 2 and 5 Months ($n = 17$)		9 Months ($n = 8$)		12 Months ($n = 8$)		P-Value ($n = 18$)		
	Intercept (SE)	Slope (SE)	Intercept (SE)	Slope (SE)	Intercept (SE)	Slope (SE)	Predictor	Infant Age (Months)	Interaction
CDI of TCH (g) [d]									
Concentration of TCH (g)	21.80 (17.10) [a]	0.544 (0.207)	−4.05 (19.30)	0.544 (0.207)	−5.47 (18.60)	0.544 (0.207)	**0.009** [b]	<0.001	0.50 [c]
24-h milk intake (g) [d]	−5.84 (7.05)	0.087 (0.008)	0.92 (5.13)	0.087 (0.008)	0.70 (5.05)	0.087 (0.008)	**<0.001**	0.22	0.059
Feeding frequency (24-h MP) [d]	21.80 (17.30)	5.39 (2.07)	15.80 (12.30)	5.39 (2.07)	15.80 (11.20)	5.39 (2.07)	**0.009**	0.69	0.13
Feeding frequency (SR) [d]	77.30 (5.65)	−4.25 (1.35)	60.50 (7.32)	−4.25 (1.35)	68.40 (10.40)	−4.25 (1.35)	**<0.001**	<0.001	0.052
CDI of lactose (g) [d]									
Concentration of lactose (g)	17.80 (24.00)	0.551 (0.366)	−3.56 (25.40)	0.551 (0.366)	−7.77 (25.70)	0.551 (0.366)	0.14	<0.001	0.31
24-h milk intake (g)	−2.80 (2.54)	0.069 (0.003)	−0.50 (1.86)	0.069 (0.003)	−0.78 (1.72)	0.069 (0.003)	**<0.001**	0.33	0.29
Feeding frequency (24-h MP)	15.10 (11.90)	4.74 (1.42)	8.98 (8.45)	4.74 (1.42)	9.29 (7.38)	4.74 (1.42)	**<0.001**	0.51	0.12
Feeding frequency (SR)	62.00 (4.26)	−3.04 (1.04)	45.70 (5.60)	−3.04 (1.04)	49.80 (7.71)	−3.04 (1.04)	**0.006**	<0.001	0.78
CDI of HMO (g) [d]									
Concentration of HMO (g)	2.62 (1.87)	0.612 (0.105)	2.47 (2.96)	0.388 (0.125)	−4.50 (1.91)	0.799 (0.071)	<0.001	0.001	**0.014**
24-h milk intake (g)	10.00 (8.24)	0.003 (0.010)	−0.04 (10.00)	0.022 (0.019)	−16.10 (8.20)	0.057 (0.016)	0.020	0.50	**0.016**
Feeding frequency (24-h MP)	0.42 (9.60)	1.45 (1.15)	3.00 (6.99)	1.45 (1.15)	4.12 (6.32)	1.45 (1.15)	0.22	0.82	0.76
Feeding frequency (SR)	15.30 (3.62)	−1.10 (1.00)	15.10 (5.16)	−1.10 (1.00)	18.10 (7.54)	−1.10 (1.00)	0.29	0.87	0.36

[a] Parameter estimate ± standard error (SE); effects of predictors taken from linear mixed effects models that accounted for month after birth and an interaction between month after birth and predictor with a random effect per participant; if the interaction is not significant parameter estimates are taken from a model with no interaction. [b,c] After the false discovery rate adjustment, the interaction/predictor *p*-values were considered to be significant at <0.05 for all predictors (indicated by the bold text). [d] 24-h milk intake and breastfeeding frequency as meals per 24-h were measured at 24-h MP; current breastfeeding frequency was also self-reported by mothers at the session as typical hours between meals; CDI were calculated between 2 and 5 months and within 2 weeks of 9 and 12 months.

Table 3. Associations of calculated daily intakes of total carbohydrates at the time points with changes in infant body composition between the time points.

Changes in Infant Characteristic (Response)	Months after Birth					
	5 and 2	9 and 2	12 and 2	9 and 5	12 and 5	12 and 9
Total carbohydrates CDI (g) between 2 and 5 months (*n* = 17) [c]						
ΔLength (cm)	−0.006 (0.026) [a] 0.83 [b]	0.041 (0.028) 0.18	**0.074 (0.026)** [d] **0.023**	0.003 (0.024) 0.91	0.013 (0.030) 0.67	0.016 (0.030) 0.59
ΔWeight (kg)	**0.018 (0.008)** **0.039**	0.025 (0.012) 0.067	0.025 (0.014) 0.11	0.002 (0.007) 0.81	−0.0002 (0.009) 0.98	−0.001 (0.003) 0.81
ΔBMI (kg/m^2)	**0.040 (0.016)** **0.037**	0.013 (0.025) 0.61	−0.008 (0.029) 0.79	−0.009 (0.017) 0.59	−0.022 (0.021) 0.30	−0.013 (0.012) 0.31
ΔFFM with US 4SF (kg)	0.013 (0.012) 0.30	**0.034 (0.013)** **0.038**	**0.045 (0.016)** **0.024**	0.004 (0.007) 0.57	0.007 (0.008) 0.39	0.004 (0.004) 0.33
ΔFFM with BIS (kg)	**0.009 (0.004)** **0.045**	0.021 (0.013) 0.16	0.024 (0.010) 0.051	0.003 (0.008) 0.67	0.006 (0.010) 0.55	0.006 (0.009) 0.54
ΔFM with US 2SF (kg)	**0.024 (0.008)** **0.019**	0.007 (0.008) 0.41	0.011 (0.013) 0.42	−0.003 (0.008) 0.72	−0.003 (0.007) 0.65	0.001 (0.003) 0.85
ΔFMI with US 2SF (kg/m^2)	**0.050 (0.018)** **0.022**	0.001 (0.018) 0.94	−0.004 (0.027) 0.88	−0.013 (0.017) 0.46	−0.020 (0.018) 0.28	−0.006 (0.007) 0.44
ΔFMI with US 4SF (kg/m^2)	0.006 (0.015) 0.73	−0.043 (0.030) 0.19	−0.064 (0.033) 0.10	−0.014 (0.017) 0.41	−0.029 (0.014) 0.071	**−0.015 (0.006)** **0.031**
ΔFM with US 4SF (%)	−0.037 (0.092) 0.70	−0.247 (0.136) 0.11	**−0.371 (0.139)** **0.032**	−0.058 (0.075) 0.45	−0.115 (0.063) 0.089	−0.060 (0.036) 0.12
ΔFM/FFM with US 4SF	−0.0004 (0.002) 0.83	−0.004 (0.002) 0.12	**−0.007 (0.003)** **0.034**	−0.002 (0.002) 0.13	−0.002 (0.001) 0.26	−0.001 (0.001) 0.31
ΔFM/FFM with BIS	0.001 (0.001) 0.37	−0.001 (0.002) 0.71	**−0.005 (0.002)** **0.040**	−0.002 (0.001) 0.14	−0.003 (0.002) 0.15	−0.001 (0.002) 0.58
Total carbohydrates CDI (g) at 9 months (*n* = 8) [c]						
ΔBMI (kg/m^2)	n/a [e]	−0.198 (0.272) 0.54	−0.009 (0.238) 0.97	−0.062 (0.037) 0.16	**−0.081 (0.029)** **0.037**	−0.019 (0.028) 0.52
ΔFFM with US 2SF (kg)	n/a	**0.107 (0.021)** **0.037**	−0.011 (0.032) 0.76	−0.006 (0.017) 0.75	0.002 (0.019) 0.92	−0.008 (0.014) 0.62
ΔFM with BIS (kg)	n/a	−0.029 (0.122) 0.84	−0.095 (0.062) 0.27	−0.022 (0.019) 0.30	**−0.033 (0.010)** **0.018**	−0.011 (0.015) 0.50
ΔFMI with BIS (kg/m^2)	n/a	−0.114 (0.315) 0.75	−0.181 (0.194) 0.45	−0.051 (0.041) 0.27	**−0.077 (0.028)** **0.040**	−0.027 (0.037) 0.51
Total carbohydrates CDI (g) at 12 months (*n* = 8) [c]						
ΔFFMI with US 2SF (kg/m^2)	n/a [e]	n/a [e]	−0.093 (0.030) 0.092	n/a [e]	**−0.038 (0.014)** **0.038**	**−0.021 (0.007)** **0.033**
ΔFFMI with US 4SF (kg/m^2)	n/a	n/a	−0.089 (0.055) 0.25	n/a	**−0.031 (0.009)** **0.029**	−0.022 (0.011) 0.11

[a] Parameter estimate ± SE of estimate and [b] *P*-values for associations between [c] calculated daily intakes (CDI) of total carbohydrates (predictor) at given time points and the changes (Δ) in measured variables between different months after birth. [d] Results are presented only for variables with at least one significant raw *p*-value (*p* < 0.05, indicated by the bold text); after the FDR adjustment, the predictor *p*-values were considered to be significant at <0.019 for CDI of total carbohydrates between 2 and 5 months, at <0.018 for CDI at 9 months and at <0.029 for CDI at 12 months (none are significant). [e] Results are not presented for impractical time combinations. BIS: bioelectrical impedance spectroscopy; BMI: body mass index; FFM: fat-free mass; FFMI: fat-free mass index; FM: fat mass; FM/FFM: fat mass to fat-free mass ratio; FMI: fat mass index; n/a: not applicable; SE: US 2SF: ultrasound 2-skinfolds; US 4SF: ultrasound 4-skinfolds.

Table 4. Associations of calculated daily intakes of total carbohydrates at the time points with changes in infant body composition between the time points.

Changes in Infant Characteristic (Response)	Months after Birth					
	5 and 2	9 and 2	12 and 2	9 and 5	12 and 5	12 and 9
HMO CDI (g) between 2 and 5 months (*n* = 17) [c]						
ΔHead circumference (cm)	0.083 (0.054) [a] 0.16 [b]	−0.007 (0.075) 0.93	0.087 (0.063) 0.21	−0.010 (0.030) 0.74	**0.047 (0.018)** **0.023**	0.047 (0.028) 0.12
HMO CDI (g) at 9 months (*n* = 8) [c]						
ΔBMI (kg/m^2)	n/a [e]	−0.425 (0.231) 0.21	**−0.401 (0.082)** [d] **0.040**	−0.247 (0.129) 0.11	**−0.361 (0.053)** **0.001**	−0.114 (0.096) 0.29
ΔFFMI with US 4SF (kg/m^2)	n/a	−0.130 (0.208) 0.60	−0.211 (0.295) 0.55	−0.024 (0.094) 0.81	−0.185 (0.119) 0.20	**−0.169 (0.058)** **0.032**
ΔFM with BIS [5] (kg)	n/a	−0.200 (0.058) 0.075	−0.048 (0.108) 0.70	−0.115 (0.060) 0.12	**−0.110 (0.043)** **0.049**	0.005 (0.059) 0.94
ΔFMI with BIS (kg/m^2)	n/a	−0.513 (0.176) 0.10	−0.186 (0.257) 0.55	−0.274 (0.119) 0.069	**−0.294 (0.097)** **0.029**	−0.019 (0.143) 0.90

Table 4. *Cont.*

Changes in Infant Characteristic (Response)	Months after Birth					
	5 and 2	9 and 2	12 and 2	9 and 5	12 and 5	12 and 9
	HMO CDI (g) at 12 months (n = 8) [c]					
ΔFMI with BIS (kg/m^2)	n/a [e]	n/a [e]	−0.336 (0.327) 0.41	n/a [e]	**−0.052 (0.017)** 0.032	−0.037 (0.024) 0.17
ΔFM with BIS (%)	n/a	n/a	−1.488 (1.901) 0.52	n/a	**−0.184 (0.070)** 0.047	−0.157 (0.124) 0.26
ΔFM/FFM with BIS	n/a	n/a	−0.024 (0.032) 0.53	n/a	**−0.004 (0.001)** 0.021	−0.003 (0.002) 0.24

[a] Parameter estimate ± standard errors of estimate and [b] P-values for associations between [c] calculated daily intakes (CDI) of human milk oligosaccharides (predictor) at given time points and the changes (Δ) in measured variables between different months after birth. [d] Results are presented only for variables with at least one significant raw p-value ($p < 0.05$, indicated by the bold text); after the false discovery rate adjustment, the predictor p-values were considered to be significant at <0.023 for CDI of human milk oligosaccharides (HMO) between 2 and 5 months, at <0.001 for CDI at 9 months and at <0.021 for CDI at 12 months (none are significant). [e] Results are not presented for impractical time combinations. BIS: bioelectrical impedance spectroscopy; BMI: body mass index; FFM: fat-free mass; FFMI: fat-free mass index; FM: fat mass; FM/FFM: fat mass to fat-free mass ratio; FMI: fat mass index; n/a: not applicable; US 4SF: ultrasound 4-skinfolds.

References

1. Koletzko, B.; Brands, B.; Chourdakis, M.; Cramer, S.; Grote, V.; Hellmuth, C.; Kirchberg, F.; Prell, C.; Rzehak, P.; Uhl, O.; et al. The power of programming and the EarlyNutrition project: Opportunities for health promotion by nutrition during the first thousand days of life and beyond. *Ann. Nutr. Metab.* **2014**, *64*, 187–196. [CrossRef] [PubMed]

2. Wells, J.C.; Chomoto, S.; Fewtrell, M.S. Programming of body composition by early growth and nutrition. *Proc. Nutr. Soc.* **2007**, *66*, 423–434. [CrossRef] [PubMed]

3. Victora, C.G.; Bahl, R.; Barros, A.J.D.; Franca, G.V.A.; Horton, S.; Krasevec, J.; Murch, S.; Sankar, M.J.; Walker, N.; Rollins, N.C.; et al. Breastfeeding in the 21st century: Epidemiology, mechanisms, and lifelong effect. *Lancet* **2016**, *387*, 475–490. [CrossRef]

4. Malik, V.S.; Pan, A.; Willett, W.C.; Hu, F.B. Sugar-sweetened beverages and weight gain in children and adults: A systematic review and meta-analysis. *Am. J. Clin. Nutr.* **2013**, *98*, 1084–1102. [CrossRef] [PubMed]

5. Pan, L.; Li, R.; Park, S.; Galuska, D.A.; Sherry, B.; Freedman, D.S. A longitudinal analysis of sugar-sweetened beverage intake in infancy and obesity at 6 years. *Pediatrics* **2014**, *134*, S29–S35. [CrossRef] [PubMed]

6. Stanhope, K.L. Sugar consumption, metabolic disease and obesity: The state of the controversy. *Crit. Rev. Clin. Lab. Sci.* **2016**, *53*, 52–67. [CrossRef] [PubMed]

7. Butts, C.A.; Hedderley, D.I.; Herath, T.D.; Paturi, G.; Glyn-Jones, S.; Wiens, F.; Stahl, B.; Gopal, P. Human milk composition and dietary intakes of breastfeeding women of different ethnicity from the Manawatu-Wanganui region of New Zealand. *Nutrients* **2018**, *10*, 1231. [CrossRef] [PubMed]

8. Prentice, A. Regional variations in the composition of human milk. In *Handbook of Milk Composition*; Jensen, R.G., Ed.; Academic Press, Inc.: San Diego, CA, USA, 1995; p. 919.

9. Kugananthan, S.; Gridneva, Z.; Lai, C.T.; Hepworth, A.R.; Mark, P.J.; Kakulas, F.; Geddes, D.T. Associations between maternal body composition and appetite hormones and macronutrients in human milk. *Nutrients* **2017**, *9*, 252. [CrossRef] [PubMed]

10. Martin, C.R.; Ling, P.R.; Blackburn, G.L. Review of infant feeding: Key features of breast milk and infant formula. *Nutrients* **2016**, *8*, 279. [CrossRef] [PubMed]

11. Mitoulas, L.R.; Kent, J.C.; Cox, D.B.; Owens, R.A.; Sherriff, J.L.; Hartmann, P.E. Variation in fat, lactose and protein in human milk over 24 h and throughout the first year of lactation. *Br. J. Nutr.* **2002**, *88*, 29–37. [CrossRef] [PubMed]

12. Arthur, P.G.; Smith, M.; Hartmann, P.E. Milk lactose, citrate, and glucose as markers of lactogenesis in normal and diabetic women. *J. Pediatr. Gastroenterol. Nutr.* **1989**, *9*, 488–496. [CrossRef] [PubMed]

13. Nommsen, L.A.; Lovelady, C.A.; Heinig, M.; Lonnerdal, B.; Dewey, K.G. Determinants of energy, protein, lipid, and lactose concentrations in human milk during the first 12 mo of lactation: The Darling study. *Am. J. Clin. Nutr.* **1991**, *53*, 457–465. [CrossRef] [PubMed]

14. Khan, S.; Hepworth, A.R.; Prime, D.K.; Lai, C.T.; Trengove, N.J.; Hartmann, P.E. Variation in fat, lactose, and protein composition in breast milk over 24 hours: Associations with infant feeding patterns. *J. Hum. Lact.* **2013**, *29*, 81–89. [CrossRef] [PubMed]

15. Gridneva, Z.; Rea, A.; Hepworth, A.R.; Ward, L.C.; Lai, C.T.; Hartmann, P.E.; Geddes, D.T. Relationships between breastfeeding patterns and maternal and infant body composition over the first 12 months of lactation. *Nutrients* **2018**, *10*, 45. [CrossRef] [PubMed]

16. Ninonuevo, M.R.; Park, Y.; Yin, H.; Zhang, J.; Ward, R.E.; Clowers, B.H.; German, J.B.; Freeman, S.L.; Killeen, K.; Grimm, R.; et al. A strategy for annotating the human milk glycome. *J. Agric. Food Chem.* **2006**, *54*, 7471–7480. [CrossRef] [PubMed]

17. Bode, L. The functional biology of human milk oligosaccharides. *Early Hum. Dev.* **2015**, *91*, 619–622. [CrossRef] [PubMed]

18. Chaturvedi, P.; Warren, C.D.; Altaye, M.; Morrow, A.L.; Ruiz-Palacios, G.; Pickering, L.K.; Newburg, D.S. Fucosylated human milk oligosaccharides vary between individuals and over the course of lactation. *Glycobiology* **2001**, *11*, 365–372. [CrossRef] [PubMed]

19. Azad, M.B.; Robertson, B.; Atakora, F.; Becker, A.B.; Subbarao, P.; Moraes, T.J.; Mandhane, P.J.; Turvey, S.E.; Lefebvre, D.L.; Sears, M.R.; et al. Human milk oligosaccharide concentrations are associated with multiple fixed and modifiable maternal characteristics, environmental factors, and feeding practices. *J. Nutr.* **2018**, *148*, 1733–1742. [CrossRef] [PubMed]

20. Alderete, T.A.; Autran, C.; Brekke, B.E.; Knight, R.; Bode, L.; Goran, M.; Fields, D. Associations between human milk oligosaccharides and infant body composition in the first 6 mo of life. *Am. J. Clin. Nutr.* **2015**, *102*, 1381–1388. [CrossRef] [PubMed]

21. Butte, N.; Wong, W.; Hopkinson, J.; Smith, E.; Ellis, K. Infant feeding mode affects early growth and body composition. *Pediatrics* **2000**, *16*, 1355–1366. [CrossRef] [PubMed]

22. Prentice, P.; Ong, K.K.; Schoemaker, M.H.; van Tol, E.A.F.; Vervoort, J.; Hughes, I.A.; Acerini, C.L. Breast milk nutrient content and infancy growth. *Acta Paediatr.* **2016**, *105*, 641–647. [CrossRef] [PubMed]

23. Goran, M.I.; Martin, A.A.; Alderete, T.A.; Fujiwara, H.; Fields, D. Fructose in breast milk is positively associated with infant body composition at 6 months of age. *Nutrients* **2017**, *9*, 146. [CrossRef] [PubMed]

24. Fields, D.; Demerath, E. Relationship of insulin, glucose, leptin, IL-6 and TNF-a in human breast milk with infant growth and body composition. *Pediatr. Obes.* **2012**, *7*, 304–312. [CrossRef] [PubMed]

25. Kent, J.C.; Mitoulas, L.R.; Cregan, M.D.; Ramsay, D.T.; Doherty, D.A.; Hartmann, P.E. Volume and frequency of breastfeedings and fat content of breast milk throughout the day. *Pediatrics* **2006**, *117*, e387–e395. [CrossRef] [PubMed]

26. Binns, C.W.; Fraser, M.L.; Lee, A.H.; Scott, J. Defining exclusive breastfeeding in Australia. *J. Paediatr. Child Health* **2009**, *45*, 174–180. [CrossRef] [PubMed]

27. Gridneva, Z.; Kugananthan, S.; Rea, A.; Lai, C.T.; Ward, L.C.; Murray, K.; Hartmann, P.E.; Geddes, D.T. Human milk adiponectin and leptin and infant body composition over the first 12 months of lactation. *Nutrients* **2018**, *10*, 1125. [CrossRef] [PubMed]

28. Gridneva, Z.; Tie, W.J.; Rea, A.; Lai, C.T.; Ward, L.C.; Murray, K.; Hartmann, P.E.; Geddes, D.T. Human milk casein and whey protein and infant body composition over the first 12 months of lactation. *Nutrients* **2018**, *10*, 1332. [CrossRef]

29. Arthur, P.; Hartmann, P.; Smith, M. Measurement of the milk intake of breast-fed infants. *J. Pediatr. Gastroenterol. Nutr.* **1987**, *6*, 758–763. [CrossRef]

30. Gridneva, Z.; Hepworth, A.R.; Ward, L.C.; Lai, C.T.; Hartmann, P.E.; Geddes, D.T. Determinants of body composition in breastfed infants using bioimpedance spectroscopy and ultrasound skinfolds-Methods comparison. *Pediatr. Res.* **2016**, *81*, 423–433. [CrossRef]

31. Van Itallie, T.B.; Yang, M.U.; Heymsfield, S.B.; Funk, R.C.; Boileau, R.A. Height-normalized indices of the body's fat-free mass and fat mass: Potentially useful indicators of nutritional status. *Am. J. Clin. Nutr.* **1990**, *52*, 953–959. [CrossRef]

32. Wells, J.C.K.; Cole, T.J. Adjustment of fat-free mass and fat mass for height in children aged 8 y. *Int. J. Obes.* **2002**, *26*, 947–952. [CrossRef] [PubMed]

33. Xiao, J.; Purcell, S.A.; Prado, C.M.; Gonzalez, M.C. Fat mass to fat-free mass ratio reference values from NHANES III using bioelectrical impedance analysis. *Clin. Nutr.* **2018**, *37*, 2284–2287. [CrossRef] [PubMed]

34. Euber, J.; Brunner, J. Determination of lactose in milk products by high-performance liquid chromatography. *J. Dairy Sci.* **1979**, *62*, 685–690. [CrossRef]

35. Albalasmeh, A.; Berhe, A.; Ghezzehei, T. A new method for rapid determination of carbohydrate and total carbon concentrations using UV spectrophotometry. *Carbohydr. Polym.* **2013**, *97*, 253–261. [CrossRef] [PubMed]

36. Newburg, D.; Neubauer, S. Carbohydrates in milks: Analysis, quantities, and significance. In *Handbook of Milk Composition*; Jensen, R., Ed.; Academic Press, Inc.: San Diego, CA, USA, 1995; pp. 273–349.

37. Khan, S.; Prime, D.K.; Hepworth, A.R.; Lai, C.T.; Trengove, N.J.; Hartmann, P.E. Investigation of short-term variations in term breast milk composition during repeated breast expression sessions. *J. Hum. Lact.* **2013**, *29*, 196–204. [CrossRef] [PubMed]

38. Curran-Everett, D. Multiple comparisons: Philosophies and illustrations. *Am. J. Physiol. Regul. Integr. Comp. Physiol.* **2000**, *279*, R1–R8. [CrossRef] [PubMed]

39. Hambraeus, L. Human milk composition. *Nutr. Abstr. Rev. Clin. Nutr.* **1984**, *54*, 219–236.

40. Young, B.E.; Patinkin, Z.W.; Pyle, L.; de la Houssaye, B.; Davidson, B.S.; Geraghty, S.; Morrow, A.L.; Krebs, N. Markers of oxidative stress in human milk do not differ by maternal BMI but are related to infant growth trajectories. *Matern. Child Health J.* **2017**, *21*, 1367–1376. [CrossRef] [PubMed]

41. Pierce, K.; Callan, J.; McCarthy, P.; O'Doherty, J. Effects of high dietary concentration of lactose and increased soyabean meal inclusion in starter diets for piglets. *Anim. Sci.* **2004**, *79*, 445–452. [CrossRef]

42. Cederlund, A.; Kai-Larsen, Y.; Printz, G.; Yoshio, H.; Alvelius, G.; Lagercrantz, H.; Strömberg, R.; Jörnvall, H.; Gudmundsson, G.H.; Agerberth, B. Lactose in human breast milk, an inducer of innate immunity with implications for a role in intestinal homeostasis. *PLoS ONE* **2013**, *8*, e53876:1–e53876:12. [CrossRef]

43. Koleva, P.T.; Bridgman, S.L.; Kozyrskyj, A.L. The infant gut microbiome: Evidence for obesity risk and dietary intervention. *Nutrients* **2015**, *7*, 2237–2260. [CrossRef] [PubMed]

44. Quinn, E.A.; Largado, F.; Power, M.; Kuzawa, C.W. Predictors of breast milk macronutrient composition in Filipino mothers. *Am. J. Hum. Biol.* **2012**, *24*, 533–540. [CrossRef] [PubMed]

45. Michaelsen, K.F.; Skafte, L.; Badsberg, J.H.; Jorgensen, M. Variation in macronutrients in human bank milk: Influencing factors and implications for human milk banking. *J. Pediatr. Gastroenterol. Nutr.* **1990**, *11*, 229–239. [CrossRef] [PubMed]

46. Coppa, G.V.; Gabrielli, O.; Pierani, P.; Catassi, C.; Carlucci, A.; Giorgi, P.L. Changes in carbohydrate composition in human milk over 4 months of lactation. *Pediatrics* **1993**, *91*, 637–641. [PubMed]

47. Sprenger, N.; Lee, L.Y.; De Castro, C.A.; Steenhout, P.; Thakkar, S.K. Longitudinal change of selected human milk oligosaccharides and association to infants' growth, an observatory, single center, longitudinal cohort study. *PLoS ONE* **2017**, *12*, e0171814:1–e0171814:15. [CrossRef] [PubMed]

48. Puccio, G.; Alliet, P.; Cajozzo, C.; Janssens, E.; Corsello, G.; Sprenger, N.; Wernimont, S.; Egli, D.; Gosoniu, L.; Steenhout, P. Effects of infant formula with human milk oligosaccharides on growth and morbidity: A randomized multicenter trial. *J. Pediatr. Gastroenterol. Nutr.* **2017**, *64*, 624–631. [CrossRef] [PubMed]

49. Yu, Z.-T.; Chen, C.; Newburg, D.S. Utilization of major fucosylated and sialylated human milk oligosaccharides by isolated human gut microbes. *Glycobiology* **2013**, *23*, 1281–1292. [CrossRef] [PubMed]

50. Bode, L. Human milk oligosaccharides: Every baby needs a sugar mama. *Glycobiology* **2012**, *22*, 1147–1162. [CrossRef]

51. Gridneva, Z.; Kugananthan, S.; Hepworth, A.R.; Tie, W.J.; Lai, C.T.; Ward, L.C.; Hartmann, P.E.; Geddes, D.T. Effect of human milk appetite hormones, macronutrients, and infant characteristics on gastric emptying and breastfeeding patterns of term fully breastfed infants. *Nutrients* **2017**, *9*, 15. [CrossRef]

52. Jantscher-Krenn, E.; Aigner, J.; Reiter, B.; Köfeler, H.; Csapo, B.; Desoye, G.; Bode, L.; van Poppel, M.N.M. Evidence of human milk oligosaccharides in maternal circulation already during pregnancy—A pilot study. *Am. J. Physiol. Endocrinol. Metab.* **2019**, *316*, E347–E357. [CrossRef]

53. McGuire, M.K.; Meehan, C.L.; McGuire, M.A.; Williams, J.E.; Foster, J.; Sellen, D.W.; Kamau-Mbuthia, E.W.; Kamundia, E.W.; Mbugua, S.; Moore, S.E.; et al. What's normal? Oligosaccharide concentrations and profiles in milk produced by healthy women vary geographically. *Am. J. Clin. Nutr.* **2017**, *105*, 1086–1100. [CrossRef] [PubMed]

Article

Comprehensive Preterm Breast Milk Metabotype Associated with Optimal Infant Early Growth Pattern

Marie-Cécile Alexandre-Gouabau [1,*], Thomas Moyon [1], Agnès David-Sochard [1], François Fenaille [2], Sophie Cholet [2], Anne-Lise Royer [3], Yann Guitton [3], Hélène Billard [1], Dominique Darmaun [1,4], Jean-Christophe Rozé [1,4] and Clair-Yves Boquien [1,5]

[1] INRA, UMR1280, Physiopathologie des Adaptations Nutritionnelles, Institut des maladies de l'appareil digestif (IMAD), Centre de Recherche en Nutrition Humaine Ouest (CRNH), Nantes F-44093, France; Thomas.Moyon@univ-nantes.fr (T.M.); agnes.david@univ-nantes.fr (A.D.-S.); helene.billard@univ-nantes.fr (H.B.); dominique.darmaun@chu-nantes.fr (D.D.); jeanchristophe.roze@chu-nantes.fr (J.-C.R.); clair-yves.boquien@univ-nantes.fr (C.-Y.B.)

[2] Service de Pharmacologie et d'Immunoanalyse, Laboratoire d'Etude du Métabolisme des Médicaments, CEA, INRA, Université Paris Saclay, MetaboHUB, F-91191 Gif-sur-Yvette, France; francois.fenaille@cea.fr (F.F.); sophie.cholet@cea.fr (S.C.)

[3] LUNAM Université, ON;IRIS, Laboratoire d'Etude des Résidus et Contaminants dans les Aliments (LABERCA), USC INRA 1329, Nantes F-44307, France; anne-lise.royer@oniris-nantes.fr (A.-L.R.); yann.guitton@oniris-nantes.fr (Y.G.)

[4] CHU, Centre Hospitalo-Universitaire Hôtel-Dieu, Nantes F-44093, France

[5] EMBA, European Milk Bank Association, Milano I-20126, Italy

[*] Correspondence: Marie-Cecile.Alexandre-Gouabau@univ-nantes.fr; Tel.: +33-(0)2-53-48-20-12; Fax: +33-(0)2-53-48-20-03

Received: 28 January 2019; Accepted: 25 February 2019; Published: 28 February 2019

Abstract: Early nutrition impacts preterm infant early growth rate and brain development but can have long lasting effects as well. Although human milk is the gold standard for feeding new born full-term and preterm infants, little is known about the effects of its bioactive compounds on breastfed preterm infants' growth outcomes. This study aims to determine whether breast milk metabolome, glycome, lipidome, and free-amino acids profiles analyzed by liquid chromatography-mass spectrometry had any impact on the early growth pattern of preterm infants. The study population consisted of the top tercile-Z score change in their weight between birth and hospital discharge ("faster grow", $n = 11$) and lowest tercile ("slower grow", $n = 15$) from a cohort of 138 premature infants (27–34 weeks gestation). This holistic approach combined with stringent clustering or classification statistical methods aims to discriminate groups of milks phenotype and identify specific metabolites associated with early growth of preterm infants. Their predictive reliability as biomarkers of infant growth was assessed using multiple linear regression and taking into account confounding clinical factors. Breast-milk associated with fast growth contained more branched-chain and insulino-trophic amino acid, lacto-N-fucopentaose, choline, and hydroxybutyrate, pointing to the critical role of energy utilization, protein synthesis, oxidative status, and gut epithelial cell maturity in prematurity.

Keywords: breast milk metabolome; glycome; lipidome; free amino acid; preterm infant; growth trajectory

---

1. Introduction

A growing body of evidence supports the impacts on lifelong health of exposure to multiple factors in early life [1]. Therefore, studying the influence of intrauterine environments and perinatal exposure are keys to understanding early growth and development and health throughout life. Indeed, putative benefits of breastfeeding in new born full-term infants are, at least in part, due to its complex composition in various macronutrients, micronutrients, and other bioactive compounds [2–4]. Maternal breast milk is the recommended nutrition for feeding pre-mature infants [5], due to its reported health benefits such as (i) a significant decrease in the risk of developing prematurity-related morbidities [6,7], including necrotizing enterocolitis [8] and infection [8,9]; (ii) a significant decrease in the feeding intolerance [8,10]; and (iii) an improvement in neurodevelopmental outcomes [8,11]. However, feeding unfortified human milk may lead to insufficient or inadequate postnatal nutritional intake for many preterm infants in the first few weeks of extra-uterine life, particularly the very preterm infants born with a low birth weight and before 28 weeks gestation. Additionally, it is often associated with extra-uterine growth restriction [11,12], which could have severe adverse consequences in term of developmental delay [13–15]. Fortification of human milk is therefore recommended by the European Society for Paediatric Gastroenterology Hepatology and Nutrition (EPSGHAN) [16]. Yet, even among preterm infants receiving protein-fortified human milk, a large range of variation is observed in the early postnatal growth patterns [17].

The host of low-molecular-weight metabolites present in breast milk fully justifies the application of metabolomics/lipidomics, a promising holistic approach in neonatology used, by our [18–20] and other laboratories [21–23]. Metabolomics have been shown to generate new insights when investigating human milk [20,24–26] during the first month of lactation [27] or pre-term and full-term human milk metabolomes over a full lactation period [28]. We also reported, for the first time, the association of early growth trajectory with a specific lipidomic signature in the human milk of mothers delivering preterm infants over the first month of lactation [20]. Human milk oligosaccharides (HMO) are other unique components known to affect the gut microbiota and may contribute to the reduced incidence of necrotizing enterocolitis [28–30], improved brain development [31], and growth patterns observed in breastfed infants [32,33]. Additionally, amino acid [34] and fatty acid [35] metabolism by mammary gland were suggested to affect milk production and infant growth, leading to a metabolic imprinting, which may persist into adulthood. To the best of our knowledge, this is the first study to explore in depth the relationships between the metabolome, lipidome, and glycome of human milk, and the early preterm infant growth during hospital stays in neonatal intensive care units. To fill this gap, we tested the potential of the liquid chromatography-mass spectrometry-based phenotypic approach to investigate the composition of human breast milk from mothers delivering a preterm newborn during the early course of lactation. More in detail, the current study aims at shedding light on the relationships between breast milk composition, characterized using targeted free amino acid pattern and non-targeted metabolomic, lipidomics and glycomic signatures, and the early growth of preterm infants nourished by their own mother's milk. As in our earlier reported pilot study [20], the present work was conducted within a larger prospective-monocentric-observational early birth LACTACOL cohort in which we selected two groups of preterm infants presenting very different growth trajectories during hospital stays. We previously reported in details [20] the breast milk lipidome in link with both infant growth groups, during the first month of lactation. The aims of the present work therefore are three-fold: (i) to assess metabolome, glycome and free amino acids pattern in the breast milk provided to preterms infants from week two to week four of lactation; (ii) to evaluate, initially and in week three-expressed breast milk samples, the interactions between human breast milk metabolome, lipidome and glycome and their association with the weight gain of infants between birth and time of discharge; and (iii) to identify a set of breast milk biomarkers with predictive ability on the postnatal weight growth trajectory of the preterm infants, taking into account confounding clinical factors. We hypothesized that our holistic approach, incorporating data from multiple breast milk compartments (i.e., metabolome, glycome, lipidome, and free amino acids), would considerably enhance our understanding of the molecular mechanisms linking breast milk composition to optimal early-growth of preterm infants.

2. Materials and Methods

2.1. Study Design and Population

The present pilot study was conducted within a larger prospective study of the previously published LACTACOL birth cohort of preterm infant mother dyads [20], whose primary objective was to explore the impact of breast milk protein content received by preterm infants during hospital stays, on neurodevelopmental outcomes at 2 years of age. A total of 118 mothers and 138 infants born between 27–34 weeks of gestational age with no severe congenital pathology and no major diseases, except prematurity and who received, for a minimum of 28 days, their own mother's breast milk only, were finally enrolled in the LACTACOL cohort (Figure 1). The current data were obtained on both sub-groups of infants selected among infants enrolled in the LACTACOL cohort and in the ancillary study, whose aim was to assess the relationship between breast milk composition (metabolome, lipidome, glycome, and amino acids) and preterm infant's growth pattern during the first month of life. These 26 selected infants presented no severe neonatal morbidity or necrotizing enterocolitis or retinopathy of prematurity.

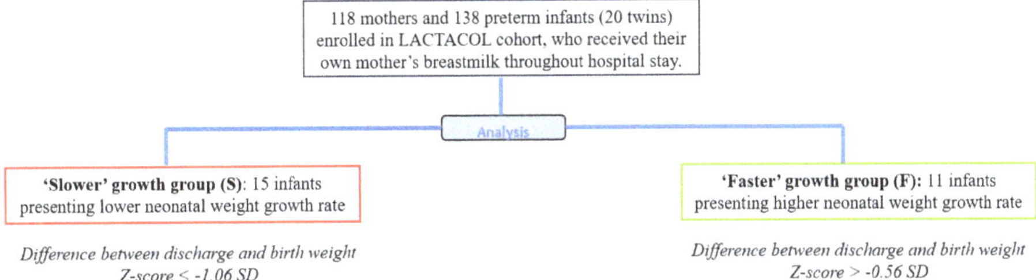

Figure 1. Study flowchart of infants enrolled in the ancillary study of the mono-centric prospective population-based LACTACOL (for global study flowchart of LACTACOL, see [20]). Among the 138 infants included in the LACTACOL cohort, no infant presented necrotizing enterocolitis (NEC), 4 infants had retinopathy of prematurity (ROP) of light severity, 3 presented intraventricular hemorrhage (IVH) of grade 2, 8 displayed bronchopulmonary dysplasia (BPD) at 28 days and 6 at 36 weeks' postmenstrual age. The 26 selected infants did not have NEC, ROP, or BPD.

Clinical characteristics were collected both on mothers and infants, including: maternal age, educational level, pre-gravid body mass index (BMI), adverse events during pregnancy and delivery, infants' gestational ages, birth weight, and head circumference, growth trajectory through hospital discharge, and events during hospital stays in neonatology. According to the EPSGHAN recommendations [16], preterm infants received parenteral nutrition and minimal enteral feeding with expressed breast milk predominantly provided by their own mother and fortified using Eoprotine® (Milupa, 1564 Domdidier, Suisse) and FortiPré® (Guigoz, 77186 Noisiel, France) for protein and carbohydrate intakes, and Liquigen® (Nutricia, 93406 Saint-Ouen, France) for lipid intakes, as previously detailed [20].

2.2. Ranking Infants according to Early Growth Trajectory

Infants enrolled in the LACTACOL cohort were ranked according to their change in weight Z-score (expressed in units of Standard Deviation (SD) and calculated as previously described [20] between birth and hospital discharge). For the first time, we chose to limit our longitudinal analysis of human breast milk composition to a small number of preterm mother infant dyads with no formal sample size calculation due to the exploratory nature of this pilot study. Then, the present study population consisted of the top tercile-Z score change in their weight between birth and hospital discharge ("faster grow", *n* = 11 infants) and lowest tercile ("slower grow", *n* = 15 infants), from our population of 138 enrolled preterm babies (born before 32 weeks gestation) (Figure 1).

2.3. Ethics

This research study was approved by the National Data Protection Authority (Commission Nationale de l'Informatique et des Libertés, N° 8911009) and the appropriate ethics Committee for the Protection of People Participating in Biomedical Research (CPP-Ouest I, reference CPP RCB-2011-AOO292-39). The current data were obtained in the ancillary study number three of the LACTACOL cohort registered at www.clinicaltrials.gov under #NCT01493063. The milk biobank was approved by the Committee for the Protection of Persons in medical research (CPP CB-2010-03). Parents received oral and written information in the maternity ward or neonatal unit and lactation support and training on proper sample collection from the study lactation consultant. A written consent was obtained from all parents at enrolment.

2.4. Human Milk Collection and Targeted Free Amino Acid (FAA) Analysis

Weekly representative 24-h breast milk expression was performed manually by mothers at home during all the lactation periods corresponding to their infant hospital stay, then processed, aliquoted, and frozen at -80 °C until analysis, as previously described [20]. FAA concentrations were determined in expressed breast milk samples collected from week two to week four of lactation using Ultra Performance Liquid Chromatography-High-Resolution-Mass Spectrometry (UPLC-HR-MS), as previously described [36]. Briefly, following a delipidation step by centrifugation and a deproteinization step by addition of sulfosalicylic acid and centrifugation, free amino acids (FAA) from supernatant were derivatized using AccQ®TagTM Ultra reagent (Waters Corporation, Milford, MA, USA)), separated on an Acquity H-Class® UPLC system (Waters Corporation, Milford, MA, USA), combined with a Xevo TQD® mass spectrometer (Waters Corporation, Milford, MA, USA), then identified and quantified, using the Waters TargetLinks™ software (Waters Corporation, Milford, MA, USA).

2.5. Breast Milk Liquid Chromatography-High-Resolution-Mass Spectrometry (LC-HRMS)–Based Glycomic Profiling

The extraction and reduction of oligosaccharides in human milk collected from week two to week four of lactation were performed as previously described [37]. Briefly, 10 μL of human milk were diluted by adding 450 μL of water and then delipidated by centrifugation. The lower phase was reduced with an NaBH4 solution and loaded onto a porous graphitized PGC cartridge (Hyperseb Hypercarb®, Thermo Scientific, San Jose, CA, USA). Milk reduced oligosaccharides were separated on a Hypercarb® column (2.1 mm i.d. $\times$ 100 mm, 3 μm particle size, Thermo Scientific, San Jose, CA, USA) on an Ultimate 3000 HPLC system (Thermo Scientific, San Jose, CA, USA). HMO chromatographic separation was performed at 30 °C with a flow rate of 300 μL/min using the gradient conditions with mobile phases A (water containing 0.1% formic acid) and B (acetonitrile containing 0.1% formic acid), as described by Oursel et al. [37]. Column effluent was directly introduced into the electrospray source of a hybrid quadruple time-of-flight (Q-TOF Impact HD) instrument (Bruker Daltonics, Bremen, Germany) operating in the positive ion mode. The source parameters were the following: 3700 V for the capillary voltage, 8.0 L/min for the dry gas, and 200 °C for the dry heater.

2.6. Breast Milk Liquid Chromatography-High-Resolution-Mass Spectrometry (LC-HRMS)–Based Lipidomic and Metabolomic Profiling

The organic and aqueous layers, following Bligh-Dyer extraction [38] of the same milk samples, were collected from week two to week four of lactation, dried separately, and subsequently reconstituted in acetonitrile-isopropanol-water (ACN: IPA: H_2O 65:30:5, $v/v/v$) and in water-acetonitrile (H_2O: ACN 95:5, v/v) for lipid and polar species, respectively. Then, lipidomic and metabolomic profilings were performed using separation on a 1200 infinity series® HPLC-system (Agilent Technologies, Santa Clara, CA, USA) coupled to an Exactive Orbitrap® MS (Thermo Fisher Scientific, Bremen, Germany) equipped with a heated electrospray (H-ESI II) source (operating in polarity switch mode), as previously described [20]. Concerning lipidomic profiling, a reverse phase

CSH® C_{18} (100 × 2.1 mm² i.d., 1.7 μm particle size) column (Waters Corporation, Milford, MA, USA) was used for lipid species separation using ACN:H$_2$O (60:40) and IPA:ACN:H$_2$O (88:10:2) as solvent A and B, respectively, with both containing 10 mM ammonium acetate and 0.1% acetic acid [39]. Concerning metabolomics fingerprinting, polar species separation was performed on the same LC-HRMS system on a reverse phase with a Hypersil GOLD C18 column (1.9 μm particle size, 100 × 2.1 mm) using a mobile phase of water (95%) and acetonitrile (5%), each containing 0.1% acetic acid according to Courant et al. [40]. The precision associated with sample preparation and LC-HRMS measurement was determined on the basis of a quality control (QC) consisting of a pool of 10 mothers' milk provided by the milk bank of Nantes Hospital Center.

2.7. Lipidomic, Metabolomic and Glycomic Data Treatment and Metabolites Annotation

Lipidomics and metabolomics raw data files were preprocessed and converted to the *.mzXML open file format using Xcalibur 2.2® (Thermo Fisher Scientific, San Jose, CA, USA) and MSConvert® (http://proteowizard.sourceforge.net/), respectively [41]. Then, lipidomics and metabolomics data were extracted using (i) pre-processing with the open-source XCMS® [42] within Workflow4Metabolomics® (W4M) (http://workflow4metabolomics.org) [43] for nonlinear retention time alignment and automatic integration for each detected features combined with CAMERA® [44] for annotation of isotopes and adducts, and (ii) normalization of intra- and inter-batch effects using Quality Control (QC) samples [45]. A manual curation, for the quality of integration and a filtration of the resulting XCMS (m/z; Retention Time (RT)) features by a 30% relative SD cutoff within the repeated pooled QC injections [46] were performed. Thereafter, accurate mass measurement of each putative metabolite was submitted to LIPID Metabolites and Pathways Strategy (LipidMaps®, www.lipidmaps.org), Human Metabolite Data base (HMDB®, www.hmdb.ca), Biofluid Metabolites Database (MetLin®, metlin.scripps.edu), and Milk Metabolome Database (MCDB®, www.mcdb.ca) annotation. Moreover, the lipids and metabolites of interest were identified with the use of the (pseudo) tandem mass spectrometry spectrum generated by all ion fragmentation [39] combined with the use of in-house reference databanks [47]. Metabolite's identification level was level one, for metabolites definitively annotated with our home data base (i.e., based upon characteristic physicochemical properties of a chemical reference standard (m/z, RT) and their M/MS spectra compared to those of breastmilk QC) or level two, for metabolites putatively annotated (i.e., without chemical reference standards, based upon physicochemical properties and MS/MS spectral similarity with public/commercial spectral libraries, e.g., LipidMaps®, MetLin®, and MCDB®). Monosaccharide compositions of HMOs were deduced from accurately measured masses (<5 ppm on average) and previously determined retention times were obtained through the use of some commercial HMO molecules [37]. Complementary MS/MS experiments were then performed to confirm putative structures. When it was not possible to clearly determine HMO structures, HMOs were named according to their monosaccharide compositions and denoted as hexose (Hex), N-acetylhexosamine (HexNAc), fucose (Fuc), and N-acetylneuraminic acid (NeuAc) numbers. In addition, isomeric forms were distinguished by a lower-case letter added after the monosaccharide composition (e.g., 4230a and 4230b). Overall, 89 (45 monosaccharide compositions) distinct HMOs were detected. Relative HMO abundances were calculated by dividing absolute HMO peak area by each sample's total HMO peak areas.

2.8. Statistical Analyses

In Tables 1–4, values were reported as medians and 25% and 75% percentiles. Statistical analyses were carried out using GraphPad Prism® software version 6.00 (La Joya, CA, USA), SIMCA P® version 14 (Umetrics AB, Sweden) and R version 3.4. (R Development Core Team, 2013; http://www.R-project.org). For all data analyses, the significance level (α) was set to 5%. Multivariate statistical models were applied separately on each glycomic, lipidomic, metabolomics, fatty acids and free amino acids data matrix considering the *a priori* structure into "faster" vs. "slower" infants' growth groups. We chose to take into account the higher (compared to glycomic data) variability in magnitude for lipidomic and metabolomic features; this is the reason why a Log Pareto scaling [48] was performed. Lipidomic or metabolomic data were submitted to the statistical workflow previously used with success on lipidomic profiling [20] in order to: (i) select the lipid/metabolic species providing a clear separation between the two infant postnatal growth subgroups from week two to week four of lactation, using the Analysis of Variance-PLS (AoV-PLS) combined with a Fisher's Linear Discriminant Analysis (LDA) procedure [49]; (ii) check the selected biomarkers predictive ability for infant weight growth, using Mann-Whitney U-test combined with multiple testing filtering (FDR); and (iii) confront them to the various confound clinical variables (mother's body mass index, birth weight, gestational age, complementary parenteral and enteral nutrition with the protein, lipid and energy intakes, duration of parenteral feeding and ventilation, and length of hospital-stay) and, in turn, test their reliability as biomarkers of infant's growth, by using multiple linear regression (MLR) combined to FDR on the remaining variables candidates as biomarkers, i.e., 80, 60, or 35 models for metabolomic, lipidomic or glycomic data, respectively. Moreover, we hypothesized that high-level data fusion, resulting in a meaningful synthesis, was expected to provide a holistic picture of the preterm breast milk composition. In order to integrate multiple–omics analytical sources and chemometrics for a comprehensive metabolic profiling of human preterm milk associated with an optimal infant weight growth, we used clustering or classification methods aiming at discriminating groups of milks using "omics" data. In order to simplify the model, we discarded the time lactation point factor of the present study and focused on "omics" data provided at week three of lactation, which had previously been shown to display the higher discriminating effect on preterm breast milk lipidome [20]. Additionally, as including an excessive amount of irrelevant variables would deteriorate the models, and in order to ovoid overfitting, all variables provided by AoV-PLS-DA scores with variables of importance in the projection (VIP)-index below 1.5 were removed in both metabolomic and lipidomic data and only the annotated representative metabolites and lipid species were kept. The input resulting metabolomic and lipidomic Log Pareto-scaled blocks were concatenated with mean and deviation standard-scaled blocks (i.e., glycomic profiling, fatty acid, and free amino acid patterns). Then, we tested on the super-matrix thus obtained an unsupervised unfold principal components analysis (UPCA-clustering method) [50] and supervised multi-block partial least squares analysis (MB-PLS- classification method) [51] strategies that searched for directions of similar sample distributions in the multidimensional spaces defined by each block of "omics" data, i.e., common components. Variables of interest for the discrimination of milk metabotype were selected according to their coordinates on the common components axes in the MB-PLS model.

3. Results

3.1. Subject Characteristics

The median difference between discharge and birth weight Z-score was −0.479 SD and −1.538 SD, for infants with "faster growth" and "slower growth", respectively. Two sets of twins belonged to the "slower" growth group, and two others sets of twins followed opposite trajectories regarding their weight Z-score difference between birth and hospital discharge, i.e., one twin belonged to the "faster" growth group, whereas the other twin belonged to the "slower" growth group. Table 1 displays the median maternal and infants' characteristics. Despite similar gestational age and hospital stay-lengths, the group of infants with 'faster' growth presented a 25% lower birth weight and a 69% greater gain in weight Z-score compared to the group of infants with "slower" growth. This negative correlation between birth weight and weight Z-score at time of discharge was previously reported in the large LIFT cohort of 2277 preterm infants by our team [52] and in another cohort [53].

Table 1. Maternal and preterm infants' characteristics.

Characteristics	"Faster" Growth Rate	"Slower" Growth Rate	*p*-Value
Maternal characteristics	11	11	
Age (years)	29.00 ± 4.52 (25.00; 35.00)	30.00 ± 4.12 (26.00; 33.00)	0.908
BMI before gestation (kg/m^2)	22.32 ± 5.26 (19.14; 28.91)	24.00 ± 5.11 (20.83; 30.80)	0.789
Infants characteristics at birth	11 (7 males and 4 females)	15 (10 males and 5 females)	
Neonatal Morbidity (number of events) *	0	0	
Gestational age (weeks)	31.00 ± 1.37 (30.0; 32.00)	30.00 ± 1.68 (29.00; 32.00)	0.288
Length of hospital stay (days)	51.50 ± 3.16 (37.25; 56.25)	49.50 ± 4.21 (36.75; 54.75)	0.849
Birth weight (kg)	1.200 ± 0.293 (1.020; 1.445)	1.605 ± 0.211 (1.465; 1.705)	**0.005**
Birth weight Z-score (SD)	-1.592 ± 0.958 (-2.079; -0.571)	0.564 ± 0.718 (-0.290; 0.842)	**0.000**
BMI at birth (kg/m^2)	7.694 ± 1.573 (7.139; 9.884)	9.455 ± 0.857 (8.843; 9.900)	0.161
Discharge weight (kg)	2.340 ± 0.320 (2.029; 2.520)	2.565 ± 0.270 (2.355; 2.720)	**0.041**
Discharge weight Z-score (SD)	-1.878 ± 0.857 (-2.264; -1.127)	-1.142 ± 0.682 (-1.552; -0.953)	0.146
BMI at Discharge (kg/m^2)	11.98 ± 0.485 (11.66; 12.28)	12.67 ± 0.955 (11.78; 13.36)	**0.047**
Difference between discharge and birth weight Z-score (SD)	-0.479 ± 0.189 (-0.668; -0.294)	-1.538 ± 0.417 (-1.953; -1.230)	**<0.001**

*: The development of comorbidities was clearly described in the same pilot study [20]. Values are medians and 25% and 75% percentiles. *p* values for comparison between "faster" and "slower" growth groups were derived using Mann-Whitney *U* test. Parameters in bold presented a significant *p*-value < 0.05.

Initially, time course breast milk compositional changes were detected, during the first month of lactation in our two sub-groups of 11 mothers delivering preterm newborns, who presented very different growth trajectories during their hospital stays using (i) targeted free amino-acids quantification combined with (ii) metabolomic (and lipidomic) and (iii) glycomic signatures. Then, multi- and univariate statistical models were applied to identify significant changes in metabolites that are associated with early postnatal infant growth. For second time, we focused (iv) our "omics" data fusion models on one representative time of lactation (week 3) to identify similar expression changes in various molecules and, in turn, highlight a few biological pathways of interest associated with optimal preterm infant growth.

3.2. Targeted Free Amino Acid Quantification

In the present pilot study, breast milk provided to the "faster" growth group presented a slightly higher essential amino acid content combined with a significantly higher content of branched-chain, insulinotrophic and gluconeogenic amino acids, as well as a decrease in sulfur amino acid content (with only taurine and methionine quantified). More specifically, breast milk arginine and tyrosine concentrations were significantly higher in the "faster" growth group than that in the "slower" growth group, whereas glycine and taurine levels were lower in the "faster" growth group with a trend toward lower glutamate and glutamine concentrations. Considering the predictive ability of free amino acid for infant weight growth during hospital stay, branched-chain and insulinotrophic amino acids were significant using multiple linear regression combined with multiple correction and taking into account maternal and infant clinical variables, whereas sulfur amino acids (taurine and methionine) presented only a trend (Table 2).

Table 2. Concentration levels of free amino acids in breast milk provided to preterm infants with "faster" or "slower" growth during the W2 to W4 lactation period.

Free Amino Acids (μM)	W2 to W4 Median (25% and 75% Percentile)		Mann-Whitney p-Value from W2 to W4	FDR Corrected MW q-Value from W2 to W4	MLR p-Value From W2 to W4	FDR Corrected MLR q-Value from W2 to W4
	"Slower" Growth ($n = 38$)	"Faster" Growth ($n = 29$)				
EAA	234.4 (216.9–278.5)	277.1 (221.4–343.5)	0.0675 [t]	0.18	0.06 #	0.08 #
Arginine	11.02 (7.56–21.84)	18.34 (11.45–28.13)	0.0079 **	0.05 *	0.10 #	0.08 #
Isoleucine	10.12 (7.56–14.51)	10.70 (8.12–18.01)	0.2224	0.29	0.08 #	0.44
Leucine	30.23 (19.93–35.46)	32.25 (24.00–39.02)	0.2407	0.29	0.05 #	0.40
Proline	30.21 (26.56–37.13)	30.32 (26.16–38.80)	0.9425	0.49	0.85	0.83
Methionine	5.29 (3.11–8.35)	6.27 (3.88–8.90)	0.6038	0.43	0.44	0.74
Phenylalanine	12.43 (8.81–15.47)	12.42 (7.25–16.43)	0.9277	0.48	0.48	0.77
Threonine	86.71 (74.34–115.7)	78.62 (66.76–107.5)	0.3197	0.31	0.39	0.71
Tryptophan	2.48 (1.89–4.23)	2.73 (1.97–4.22)	0.7676	0.45	0.59	0.21
Valine	51.13 (38.90–55.06)	52.09 (37.75–56.30)	0.8034	0.45	0.41	0.71
NEAA	2610 (2146–3280)	2512 (1753–3182)	0.2592	0.32	0.78	0.25
Alanine	206.7 (186.9–332.4)	201.0 (166.5–254.3)	0.5466	0.43	0.78	0.83
Aspartic acid & asparagine	66.93 (39.53–90.83)	57.39 (29.92–80.49)	0.3382	0.31	0.86	0.83
Glutamine	455.9 (211.4–902.3)	375.0 (134.4–572.0)	0.0838 [t]	0.17	0.14	0.52
Glutamic acid	1319 (898.5–1449)	1220 (906.3–1480)	0.6567	0.43	0.27	0.52
Glx	1838 (1381–2259)	1754 (1177–2098)	0.0613 [t]	0.14	0.52	0.21
Glycine	89.25 (68.99–105.2)	69.60 (54.38–103.09)	0.0126 *	0.05 *	0.18	0.56
Serine	86.01 (68.78–116.2)	86.01 (63.81–112.3)	0.8034	0.46	0.23	0.59
Tyrosine	11.64 (6.70–15.56)	14.49 (11.47–21.00)	0.0349 *	0.10	0.11	0.49
Taurine	313.9 (275.5–428.2)	270.0 (174.0–313.2)	0.0031 **	0.03 *	0.14	0.51
BCAA	85.15 (71.48–93.3)	101.1 (84.99–121.5)	0.0075 **	0.04 *	0.06 #	0.08 #
Insulino-trophic amino acid	182.0 (166.6–219.2)	224.9 (175.7–275.0)	0.0427 *	0.10 [t]	0.07 #	0.08 #
SAA	327.9 (297.0–430.7)	227.9 (168.6–355.1)	0.0019 **	0.03 *	0.13	0.09 #

Values are medians (25% and 75% percentiles) from amino acid concentrations from week 2 to week 4 of lactation period. EAA: essential amino acids; NEAA: non-essential amino acids; FAA: free amino acid; Glx: glutamine + glutamic acid; BCAA: branched chain amino-acids (valine + leucine + isoleucine); Insulinotrophic and glycemic amino acids = valine + leucine + isoleucine + threonine + arginine. Sulfur amino acids (SAA): taurine and methionine. Variables were considered as significantly modified between the two groups of infants' growth (Mann-Whitney U test) when their multiple comparisons adjusted P-values (i.e., False Discovery Rate (FDR) corrected-MW q-value) was < 0.05. *: MW p-value or FDR-corrected MW q-value < 0.05; **: MW p-value or FDR-corrected MW q-value < 0.01; t: MW p-value or FDR-corrected MW q-value < 0.1. Multiple linear regression (MLR) for infant weight Z-score (p-value) was also combined to FDR and predictive ability for infant weight growth was considered reliable when MLR q-value was < 0.1 (#).

3.3. Lipidomics and Metabolomcs Profiling

Lipidomic analysis of the human breast milk provided (from week two to week four of lactation) to the 26 infants selected in the present pilot study was previously reported [20]. The most discriminant features associated with infant growth during hospital stays corresponded to a cluster of 1256 VIP-based lipid species. Among the 50 AoV-PLS/LDA- and FDR-selected annotated lipid biomarkers, nine lipid species appeared of paramount interest due to their significant (10% threshold) MLR q-value for delta weight Z-score (data from Table 4, [20]). Similarly, metabolomic LC-HRMS (ESI^+/ESI^-) data obtained on preterm breast milk from week two to week four of lactation, were processed using AoV-PLS procedure [49] to assess the association between the metabolites and the *a priori* grouping structure ("faster" vs. "slower" infant growth). The score plots clearly highlighted the separation between breast milk metabotypes associated with 'faster' or 'slower' infant growth in both positive (Figure 2a) and negative (supplementary Figure S1) ionization modes with the breast milk metabolomic profiles, corresponding to the four sets of twins plotted between both clusters (depicted with blue symbols in Figure 2a and Figure S1a). Then, the selected appropriate components of AoV-PLS (for both ionization modes) were subjected to a Fisher's linear discriminant analysis (LDA) to test the significance of growth factor (Figure 2b and Figure S1b). Their cross-validation error rates of the LDA canonical variables for positive and negative mode were both equal to 7.14%. The most discriminant features associated with infant growth during hospital stay corresponded to a cluster of 125 (resp. 119) VIP-based metabolites species (VIP-index above 1.5) in the positive (resp. negative) ionization mode leading to 68 features that could be annotated (Table 3).

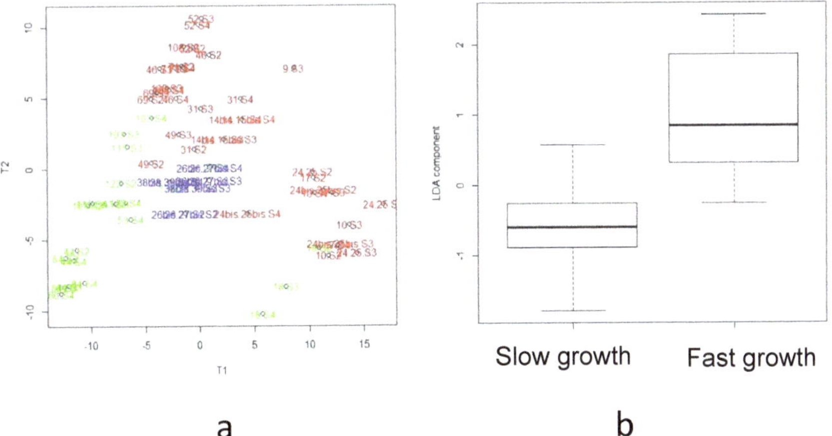

Figure 2. Analysis of Variance (AoV)-PLS and linear discriminant analysis (LDA) models based on the LC-ESI^+-HRMS metabolomics profiles of human preterm milk on the factor weight Z-score (discharge-birth): AoV-PLS score plot with 56% of variance (R2Y = 34%) on components 1–2 (**a**) and LDA (built on components of AoV-PLS) with a *p*-value = 0) (**b**). Breast milk provided to preterm infants who experienced "faster" (green) or "slower" (red) growth and to twin infants with discordant growth rate, one twin with high growth rate and the other one with low growth rate, (blue).

Table 3. Abundance (10^6) of annotated metabolites that discriminated metabotypes of breast milk provided to preterm infants with "faster" or "slower" growth during the W2 to W4 lactation period.

Metabolites (Annotation Level)	a, b, c	mz	Abundance of Metabolites (10^6)			
			Median (25% and 75% Percentile), W2 to W4			
			"Slower" Growth (n = 38)	"Faster" Growth (n = 29)	Mann-Whitney p-Value (FDR-Corrected MW q-Value in Exposant)	MLR p-Value (FDR-Corrected MLR q-Value in Exposant)
Amino acid						
Hippuric acid [1]	a	180.0654 (M + H)$^+$	1.36 (0.79–1.81)	1.36 (0.88–2.20)	0.86	0.33
2-hydroxyhippuric acid [2]	a, b, c	194.0459 (M – H)$^-$	0.07 (0.05–0.126)	0.08 (0.05–0.15)	0.25	0.04
Valine [1]	a	118.0865 (M + H)$^+$	2.81 (1.72–3.23)	2.67 (0.91–3.17)	0.79	0.17
Leucine [1]	a, c	130.0872 (M – H)$^-$	1.92 (1.45–2.87)	2.99 (1.61–4.66)	0.02 **	0.92
N-Carbamoylsarcosine [2]	a, c	133.0609 (M + H)$^+$	0.96 (0.66–1.45)	1.69 (1.18–2.38)	0.0003 **	0.96
Tryptophan metabolism						
Tryptophan [1]	a, c	205.0970 (M + H)$^+$	4.20 (3.59–4.77)	4.76 (3.51–7.95)	0.18	0.79
Kynurenine [1]	a, c	192.0653 (M–NH3 + H)$^+$	0.97 (0.64–1.55)	0.72 (0.59–0.93)	0.06 ^{t}	0.80
1H-Indole-3-carboxaldehyde [2]	a, b	146.0599 (M + H)$^+$	2.18 (1.55–2.81)	2.26 (1.66–3.63)	0.27	0.98
Indole-3-ethanol [2]	a	184.0732 (M + Na)$^+$	9.66 (6.11–13.01)	9.59 (6.62–12.06)	0.82	0.14
3-Methylindole [2]	a	132.0806 (M + H)$^+$	0.43 (0.35–0.55)	0.47 (0.39–0.63)	0.10	0.70
Tyrosine metabolism						
hydroxyphenylacetic acid [1]	a, c	151.0399 (M – H)$^-$	0.32 (0.25–0.37)	0.30 (0.21–0.37)	0.61	0.08
p-Cresol (4-methylphenol) [2]	a, b, c	107.0501 (M – H)$^-$	1.84 (1.16–2.85)	1.37 (0.09–2.21)	0.04 *	0.20
p-Cresol sulfate [2]	a	187.0070 (M – H)$^-$	5.28 (4.14–9.27)	4.89 (2.67–6.79)	0.13 ^{t}	0.17
Sulphur metabolism						
Cystathionine [2]	a	240.1015 (M + NH$_4$)$^+$	0.67 (0.51–0.80)	0.87 (0.66–1.35)	0.02 *	0.90
Methionin [1]	a, c	150.0580 (M + H)$^+$	1.82 (1.49–2.19)	1.81 (1.63–2.69)	0.36	0.89
Se-Adenosylselenohomocysteine [2]	a	228.0314 (M + H + Na)$^+$	0.22 (0.17–0.25)	0.19 (0.13–0.24)	0.15 ^{t}	0.09
S-Adenosylhomocysteine [2]	a	365.1048 (M – H$_2$O– H)$^+$	6.53 (5.02–7.63)	4.48 (3.92–6.84)	0.007 *	0.20
Hydrogen sulfite [2]	a, b, c	79.9573 (M – H)$^-$	0.52 (0.38–0.85)	0.44 (0.31–0.67)	0.07 ^{t}	0.23
Thiocyanic acid [2]	a, c	150.0018 (M – H)$^-$	0.75 (0.64–0.83)	0.57 (0.51–0.76)	0.02 *	0.69
Aromatic compound						
Benzoic acid [1]	a	121.0294 (M – H)$^-$	3.14 (2.27–4.18)	2.49 (2.09–2.95)	0.01 *	0.16
Hydroxyphenyllactic acid [2]	a	241.0730 (M + Hac-H)$^-$	0.14 (0.11–0.16)	0.11 (0.08–0.18)	0.17	0.96
Pyridines and Derivatives/Nucleosides						
Niacinamide [1]	a	123.0554 (M + H)$^+$	6.45 (5.27–10.82)	11.00 (6.17–13.84)	0.029 *	0.16

Table 3. *Cont.*

Metabolites (Annotation Level)	a, b, c	mz	Abundance Des Ions (10^6)			
			Median (25% and 75% Percentile), W2 to W4			
			"Slower" Growth ($n = 38$)	"Faster" Growth ($n = 29$)	Mann-Whitney p-Value (FDR-Corrected MW q-Value in Exposant)	MLR p-Value (FDR-Corrected MLR q-Value in Exposant)
Energy metabolism						
Hydroxyhexanoylcarnitine [2]	a	276.1803 $(M + H)^+$	0.11 (0.04–0.22)	0.08 (0.05–0.21)	0.84	0.52
Oxoicosanoyl-CoA [2]	a, b	547.2129 $(M + H+ NH4)^+$	0.66 (0.53–0.89)	0.81 (0.63–1.20)	0.04 *	0.87
3-Hydroxypimelyl-CoA [2]	a	943.2103 $(M + NH4)^+$	1.77 (1.66–2.02)	1.79 (1.61–2.04)	0.98	0.91
Hexanoylglycine [2]	a, b	174.1123 $(M + H)^+$	0.20 (0.12–0.26)	0.13 (0.08–0.19)	0.02 *	0.15
Heptanoylglycine [2]	a	229.1544 $(M + H)^+$	0.64 (0.50–0.97)	0.65 (0.49–0.81)	0.58	0.69
Gamma-Butyrolactone/	a	85.0293 $(M – H)^-$	42.9 (25.08–58.13)	48.37 (34.67–65.61)	0.32	0.68
But-2-enoic/Isocrotonic acid [2]	a, b	631.3089 $(M – H)^-$	0.01 (0.00–0.03)	0.02 (0.01–0.06)	0.01 *	0.63
butyl 2-dodecanoic acid/ 5-Tetra dodecanoic acid [2]	a	225.1859 $(M – H)^-$	0.09 (0.04–0.19)	0.06 (0.01–0.10)	0.05 [t]	0.81
caproic acid [1]	a	115.0763 $(M – H)^-$	0.13 (0.11–0.15)	0.14 (0.11–0.18)	0.46	0.26
3-hydroxycapric acid [2]	a, b, c	187.1339 $(M – H)^-$	0.43 (0.26–0.81)	0.62 (0.50–1.11)	0.01 *	0.97
Geranic acid [2]		167.1077 $(M – H)^-$	0.09 (0.07–0.22)	0.08 (0.06–0.16)	0.13	0.03
Sebacic acid [1]	a, c	261.1345 (M-CH3COO)$^-$	0.10 (0.07–0.18)	0.13 (0.08–0.19)	0.18 [t]	0.72
3-Hydroxysebacic acid [2]	a, c	217.1081 $(M – H)^-$	0.04 (0.03–0.06)	0.06 (0.03–0.07)	0.07 [t]	0.68
3,4-Methylenesebacic acid [2]	a	225.1132 $(M – H)^-$	0.04 (0.03–0.06)	0.03 (0.02–0.04)	0.12 [t]	0.82
2-Hydroxybutyric acid [1]	a, b, c	103.0399 $(M – H)^-$	2.93 (2.35–3.72)	3.85 (3.05–4.81)	0.005 *	0.77
2-hydroxy-3-methylbutyric acid [1]	a, b, c	117.0555 $(M – H)^-$	0.99 (0.81–1.82)	1.52 (1.07–2.07)	0.01 *	0.06
pyridosine [2]	a, c	253.1195 $(M – H)^-$	0.20 (0.09–0.28)	0.11 (0.07–0.22)	0.06 [t]	0.17
Glycerophosphorylcholine [2]	a, c	292.0724 $(M – H)^-$	0.68 (0.49–0.96)	0.58 (0.41–0.71)	0.07 [t]	0.40
N-Heptanoylglycine [2]	a, b	186.1135 $(M – H)^-$	0.93 (0.47–1.83)	0.61 (0.40–1.16)	0.07 [t]	0.14
Butyryl glycine/Saccharopine [2]	a	335.1455 * $(M + Fa – H)^-$	0.22 (0.11–0.53)	0.28 (0.08–0.51)	0.58	0.62
2-Phenylglycine [2]	a, c	150.0559 $(M – H)^-$	0.14 (0.09–0.22)	0.17 (0.10–0.29)	0.11 [t]	0.08
Cis-aconitic acid [1]	a	154.9983 (M-H_2O – H) $^-$	1.33 (0.58–2.27)	1.73 (1.28–2.49)	0.09 [t]	0.85
Pyruvic acid [1]	a, b, c	147.0297 (M-CH3COO)$^-$	2.15 (1.27–3.49)	3.55 (1.94–6.10)	0.03 *	0.95
Citraconic [1]	a, b, c	129.0192 $(M – H)^-$	13.82 (9.49–26.19)	21.43 (13.82–29.91)	0.02 *	0.95
2-Keto-glutaramic acid [2]	a, c	144.0302 $(M – H)^-$	0.38 (0.30–0.44)	0.37 (0.31–0.58)	0.21	0.44
Panthothenic acid [1]	a, c	200.0929 (M-H_2O – H)$^-$	0.09 (0.06–0.11)	0.10 (0.08–0.11)	0.51	0.007
4-Heptenal [2]	a, b	111.0814 $(M – H)^-$	0.43 (0.33–0.50)	0.35 (0.26–0.45)	0.04 *	0.55
2-Methylpentanal [2]	a	99.0814 $(M – H)^-$	0.12 (0.09–0.13)	0.10 (0.09–0.12)	0.06 [t]	0.98
Undecenal [2]	a, b, c	167.1440 $(M – H)^-$	0.08 (0.06–0.12)	0.06 (0.04–0.13)	0.006 *	0.02
Methyl 2-octynoate [2]	a	153.0919 $(M – H)^-$	0.10 (0.07–0.13)	0.07 (0.06–0.11)	0.09 [t]	0.06

Table 3. *Cont.*

Metabolites (Annotation Level)	a, b, c	mz	Abundance Des Ions (10^6)			
			Median (25% and 75% Percentile), W2 to W4			
			"Slower" Growth ($n = 38$)	"Faster" Growth ($n = 29$)	Mann-Whitney p-Value (FDR-Corrected MW q-Value in Exposant)	MLR p-Value (FDR-Corrected MLR q-Value in Exposant)
4-Methylphenyl-acetaldehyde [2]	a	133.0658 (M − H)$^-$	0.07 (0.05–0.11)	0.06 (0.05–0.09)	0.04 *	0.14
4-Hydroxynonenal [2]	a, c	155.1077 (M − H)$^-$	0.06 (0.04–0.09)	0.07 (0.05–0.10)	0.13 [t]	0.79
cis-4-Decenedioic acid [2]	a, b, c	199.0973 (M − H)$^-$	0.07 (0.05–0.11)	0.10 (0.07–0.16)	0.01 *	0.94
Tetradecanedioic acid [1]	a, c	257.1761 (M − H)$^-$	0.08 (0.05–0.12)	0.12 (0.08–0.19)	*0.07* [t]	0.62
Dodecanedioic acid [2]	a, c	229.1445 (M − H)$^-$	0.41 (0.30–0.56)	0.35 (0.30–0.54)	0.42	0.04
Heptanoic acid [2]	a, c	129.0920 (M − H)$^-$	0.17 (0.14–0.21)	0.14 (0.10–0.19)	*0.06* [t]	0.74
2-benzyloctanoic acid [2]	a, b, c	233.1544 (M − H)$^-$	1.45 (0.78–2.31)	0.87 (0.61–1.45)	0.007 *	0.81
N-methylethanolaminium phosphate [2]	a, b, c	136.0165 (M–H$_2$O − H)$^-$	0.36 (0.27–0.51)	0.26 (0.19–0.30)	<0.0001 **	0.8
Phosphorylcholine [1]	a,	206.0551 (M + Na)$^+$	5.78 (4.26–6.63)	4.03 (0.68–5.67)	0.001 *	0.2
Glycerophosphocholin [2]	a	280.0917 (M + Na)$^+$	18.88 (9.66–30.18)	16.98 (3.54–24.57)	0.12 [t]	0.57
Choline [1]	a,	105.11080 (M + H)$^+$	2.67 (2.25–3.66)	3.30 (2.74–4.65)	0.008 *	0.02
Glucuronide/oligosides						
Dihydrocaffeic acid 3-O-glucuronide [2]	a, b, c	383.0763 (M + Na)$^+$	1.15 (0.87–1.29)	1.05 (0.86–1.23)	0.49	0.5
2-Fucosyllactose [2]	a, c	511.1629 (M + H)$^+$	106.6 (82.1–152.5)	122.6 (98.5–148.7)	0.34	0.79
N-acetyl-D-glucosamine [2]	a	244.0788 (M + Na)$^+$	2.94 (2.71–3.51)	3.20 (2.55–3.81)	0.74	0.68
Lacto-N-fucopentaose-2 [2]	a, b	876.2936 (M + Na)$^+$	9.29 (7.62–12.68)	10.56 (8.07–15.04)	0.15 [t]	0.72
Saccharopine [2]	a, c	335.1455 (M-CH3COO)$^-$	0.22 (0.11–0.53)	0.28 (0.08–0.51)	0.58	0.62

Values are medians (25% and 75% percentiles) from metabolites abundances from week 2 to week 4 of lactation period. Metabolites' annotation level in brackets: 1: identification level, definitively annotated with our home data base (i.e., based upon characteristic physicochemical properties of chemical reference standards (m/z, RT) and their MS/MS spectra compared to those of breastmilk QC); 2: putatively annotated compounds (i.e., without chemical reference standards, based upon physicochemical properties and MS/MS spectral similarity with public/commercial spectral libraries). (**a**) VIP in AoV-PLS/LDA ESI$^+$ or ESI$^-$ model, (**b**) loadings in MB-PLS model, (**c**) loadings in ACC model; when the letters (a, b or c) are in italic, that means that the significance of metabolites, as VIP or loadings in statistical models, is just a trend. Variables were considered as significantly modified between the two groups of infants' growth (Mann-Whitney U test) when their multiple comparisons adjusted *P*-values (i.e., FDR corrected-MW *q*-value) was < 0.05. FDR-corrected MW *q*-value was labelled in exposant: *: FDR-corrected *q*-value < 0.05; **: FDR-corrected MW *q*-value < 0.01; t: FDR-corrected MW *q*-value < 0.1. Predictive ability of metabolites for infant weight growth was considered reliable when FDR-corrected MLR *q*-value was < 0.1 (and labelled in exposant as #).

We identified (Table 3) the association between our two groups of preterm infant's growth and breast milk metabolites (VIP in AoV-PLS models) that manually map to pathways such as the arginine-creatinine pathway, aromatic amino acid metabolism (including intermediate products of tryptophan, tyrosine, and/or phenylalanine catabolism), nicotinamide adenine dinucleotide precursors (such as nicotinamide and tryptophan), sulfur metabolism, oligosaccharides (e.g., 2'-Fucosyllactose and Lacto-N-FucoPentaose), mitochondrial fatty acid beta-oxidation, (including metabolites such as acylglycine and several analytes of the tricarboxylic acid cycle), pyruvic, citraconic, and aconitic acids, and choline metabolism. Many metabolites were significantly different between both groups of infants' growth when their multiple comparisons adjusted *P*-values (i.e., *q*-value after false discovery rate (FDR)) was <0.1, such as higher levels in orotic acid, nicotinamide, hydroxybutyric acid, pyruvic and citraconic acids, and choline in the "faster" group besides lower abundance in cresol and benzoic acid, for example. Among these metabolites, only a few metabolites predicted early infant growth (significant MLR *p*-value but unsuccessful for multiple correction testing, FDR-corrected MLR *q*-value), including: hydroxy-3-methylbutyric acid, undecenal, dodecanedioic acid and choline.

3.4. Glycomics Profiling

Milk samples were analyzed for changes in their HMO profiles from W2 to W4 of lactation in association with postnatal weight growth trajectory during hospital stay. On the PLS-DA score plot of PLS components (PCs 1 and 2) (Figure 3a), as expected, there was a clear difference in milk HMO profiles depending on secretor status in preterm milk samples, as the milk samples were separated into secretor (21 mothers) or non-secretor (five mothers) groups. The overal percentage of non-secretor mothers in our sub-cohort of LACTACOL was consistent with the proportions in most human populations, i.e., approximately 20% [28], but was significantly higher for mothers of infants ranked in the "slower" growth group (36%, i.e., 4 non-secretor mothers) versus those ranked in the "faster" growth group (10%, i.e., only one non-secretor mother). In this study, secretor or non-secretor status (i.e., mothers expressing or not the 1,2-fucosyl-transferase 2) was essentially defined by either high or low 2'Fucosyllactose (2'FL) levels, respectively, as measured by LC-MS [28,54]. Milk from mothers that was classified as non-secretors showed very low or even no detectable levels of 2'-FL, whereas milk from mothers classified as secretors contained high amounts of 2'-FL. Interestingly, exclusion of the five HMO profiles from non-secretor mothers in the statistical PLS-DA model improved the separation between the milks of the "faster" and "slower" infant growth groups (Figure 3b). As shown in Table 4, relative abundances of several measured HMOs differed significantly between breast milks provided to preterm infants with "faster" or "slower" growth during the W2 to W4 lactation period with higher overall levels of HMOs in milk given to fast-growing infants. In detail, and regardless of the maternal secretor status, breast milk provided to infants with optimal growth contained more total fucosylated HMOs (essentially due to the mono-fucosylayed, Lacto-N-FucoPentaose I (LNFPI percentage), both di-fucosylated HMOs, an isomer of Lacto-N-difucosyl-hexaose (LNDFH with the following monosaccharides structure: 3210, i.e., 3 Hex//1 HexNac/2 Fuc/0 NeuAc) 4210d, and neutral HMOs such as pLNH (lacto-N-hexaose), but contained less neutral HMO 3000 and di-fucosylated HMO 4210b. Many of these HMOs were found to be variables of interest for breast milk glycome discrimination (VIP-PLS-DA index above 1.0, as reported in Table 4). Of note, breast milk presented no between-group differences in any sialylated HMOs. As reported in Table 4, many HMOs were significantly predictive of infant weight Z-score, as they successfully passed the multiple linear regression test (MLR *p*-value), following adjustment for maternal and infant confounding factors and after multiple correction testing (FDR-corrected MLR *q*-value significant, i.e., <0.1). Indeed, the fucosylated LNFP I, an isomer of LNDFH 3'FL (3'-fucosyllactose) and the neutral pLNH were predictive of infant weight Z-score for secretor mothers only, whereas, the four minor di-fucosylated HMOs (4230c, 4230b, 4240b, and 4210d) remained predictive of infant weight Z-score regardless of the secretor or non-secretor status of mothers.

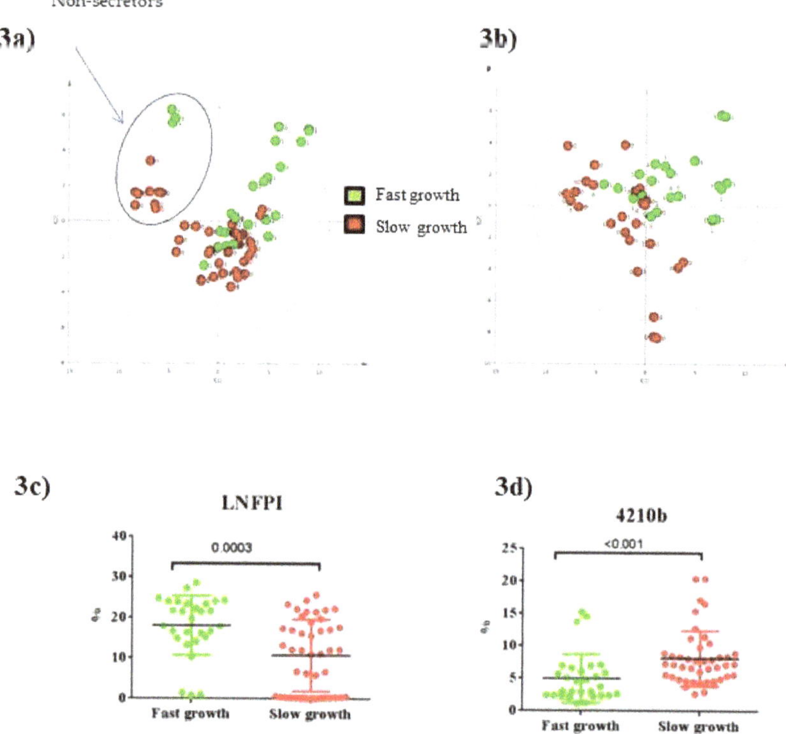

Figure 3. PLS-DA score plot based on the LC- HRMS glycomics profiles of human preterm milk of all mothers (**a**) (with non-secretor mothers circled on the basis of on the concentration of 2'-FL in their milks) or only of secretor mothers (**b**), on the factor weight *Z*-score (discharge-birth) with 45–39% of variance (R2Y = 37–45%), respectively, on components 1–2. Scatter plot (median) from W2 to W4 of lactation period (with secretor and non-secretor mothers) for two representative HMOs selected among VIP of interest: LNFPI (**c**) and 4210b (**d**), respectively. *P* values for comparison between "faster" and "slower" growth groups were derived using Mann-Whitney *U* test.

Table 4. Major HMOs detected in breast milk glycome and provided to preterm infants with "faster" or "slower" growth during the W2 to W4 lactation period.

HMOs	Composition						Median (25% and 75% Percentile) from W2 to W4 (Secretors and Non Secretors Mothers)		Mann-Whitney p-Value (FDR-Corrected MW q-Value in Exposant)		VIP–PLS-DA (C1–C2)		MLR-p-Value (FDR-Corrected MLR q-Value in Exposant)	
	mz	RT	Hex	HexNac	Fuc	NeuAc	"Slower" Growth (n = 38)	"Faster" Growth (n = 29)	Secretors and Non Secretors	Secretors Only	Secretors and Non Secretors	Secretors Only	Secretors and Non Secretors	Secretors Only
Fucosylated							61.46 (50.28–65.13)	62.82 (60.20–65.19)	0.1847 [t]	0.1370				0.73
Sialylated							8.47 (6.97–9.40)	7.45 (6.70–9.11)	0.2545	0.2773				0.37
Fucosylated./Sialylated							1.95 (1.50–2.32)	1.63 (1.33–2.30)	0.0973 [t]	0.1422				0.35
Neutral							28.30 (26.78–38.23)	26.87 (25.92–29.57)	0.0352 *	0.6966				0.95
Mono Fucosylated							28.49 (19.06–34.74)	36.95 (31.83-39.21)	0.0020	0.0669				0.19
Di Fucosylated							26.64 (23.83–28.53)	23.12 (19.14–26.77)	0.0026	0.0054 *				0.10
Tri Fucosylated							3.52 (2.96–4.00)	2.85 (2.63–3.26)	0.0061	0.0883				0.43
Tetra Fucosylated							3.20 (2.54–4.07)	2.94 (1.77–3.83)	0.1520	0.0243 *				0.004 *
LNFPI≠	856.3280	10.2	3	1	1	0	11.99 (0.36-18.65)	19.86 (15.00–23.69)	0.0003 **	0.0045 *	1.22	1.26	0.63	0.03 *
pLNH≠	1075.4023	18	4	2	0	0	0.26 (0.14–0.35)	0.38 (0.26–0.54)	0.0014 **	0.0013 *	1.44	1.75	0.50	0.10 *
2'-FL≠	491.1958	8.5	2	0	1	0	11.22 (0.10–13.87)	12.58 (10.29–15.09)	0.0882 [t]	0.9433	1.10	1.08	0.63	0.22 *
6'-SL	636.2333	9.6	2	0	0	1	2.50 (2.30–3.06)	2.47 (1.81–3.05)	0.3219	0.9433	0.65	0.29	0.26	0.31 *
LNnH≠	1075.4023	15.1	4	2	0	0	0.82 (0.32–1.44)	0.95 (0.58–2.31)	0.0689 [t]	0.8116	1.25	1.22	0.67	0.52 [t]
LNT/LNnT≠	710.2701	10.9	3	1	0	0	22.52 (20.29–35.65)	22.46 (20.04–25.21)	0.1991 [t]	0.2591	1.00	0.81	0.48	0.62 [t]
LSTc/b	1001.3655	18.4	3	1	0	1	2.59 (2.00–3.57)	2.63 (2.25–3.41)	0.7012	0.1422	0.89	0.87	0.78	0.65 [t]
LNDFH I	1002.3859	5.7	3	1	2	0	4.26 (0.44–6.25)	5.10 (0.51–6.96)	0.3986	0.8903	0.55	0.74	0.22	0.50 [t]
3' FL	491.1958	1.8	2	0	1	0	0.03 (0.00–0.08)	0.00 (0.00–0.03)	0.0133 *	0.1697	0.94	0.70	0.12	0.09 *
3'SL	636.2333	18.9	2	0	0	1	0.15 (0.13–0.21)	0.14 (0.11–0.18)	0.1555 [t]	0.9433	0.46	0.48	0.26	0.16 *
LNDFH ≠	1002.3859	9.6	3	1	2	0	0.26 (0.14-0.36)	0.36 (0.27-0.45)	0.0020 **	0.1050	1.12	1.05	0.43	0.04 *
LNDFHx	1002.3859	6.8	3	1	2	0	0.26 (0.11–1.11)	0.14 (0.07–0.24)	0.0293 *	0.0726 [t]	0.97	1.04	0.15	0.06 *
4230c≠	1513.5760	13.5	4	2	3	0	0.11 (0.00–0.18)	0.07 (0.04–0.14)	0.5405	0.0002 **	1.90	1.55	0.0062 *	0.07 *
4210d≠	1221.4602	17.4	4	2	1	0	0.18 (0.00–0.39)	0.47 (0.27–1.31)	0.0005 **	0.0142 *	1.69	1.66	0.10	0.04 *
4220e≠	1367.5181	13	4	2	2	0	1.59 (0.37–2.43)	1.75 (1.35–2.77)	0.1799 [t]	0.6128	1.28	1.76	0.86	0.78 [t]
4230b≠	1513.576	8.2	4	2	3	0	0.65 (0.00–1.49)	0.61 (0.06–1.20)	0.7967	0.0893 [t]	1.38	1.33	0.0095 *	0.09 *
3000≠	507.1907	6.4	3	0	0	0	0.26 (0.19–0.31)	0.16 (0.12–0.23)	<0.0001 ***	0.0022 *	2.06	0.90	0.53	0.22 *
5300 (2+)	720.7709	18.1	5	3	0	0	0.15 (0.10–0.23)	0.16 (0.14–0.32)	0.0771 [t]	0.6305	1.41	1.39	0.72	0.37 *
6420c (2+)≠	1049.3949	18.1	6	4	2	0	0.44 (0.09–0.64)	0.50 (0.32–0.72)	0.1893 [t]	0.5220	1.11	0.67	0.30	0.76 [t]
6430d (2+)≠	1122.4238	16.8	6	4	3	0	0.14 (0.00–0.26)	0.15 (0.10–0.24)	0.2102 [t]	0.2663	1.18	1.00	0.34	0.87 [t]
5310c	1586.5924	18	5	3	1	0	0.29 (0.22–0.38)	0.38 (0.25–0.46)	0.0428 *	0.4094	0.90	1.02	0.95	0.16 *
2110a	694.2752	3.6	2	1	1	0	0.01 (0.00–0.27)	0.24 (0.00–0.51)	0.0811 [t]	0.6772	0.90	1.49	0.48	0.60 [t]
4240b≠	1659.6339	13.9	4	2	4	0	0.04 (0.00–0.11)	0.04 (0.00–0.08)	0.9654	0.0457 *	1.29	0.68	0.0007 **	0.05 *
2020a≠	637.2537	11.1	2	0	2	0	0.49 (0.04–0.61)	0.55 (0.45–0.66)	0.0840 [t]	0.9700	1.10	1.23	0.72	0.20 *
5310b	1586.5924	17	5	3	1	0	0.24 (0.13–0.31)	0.16 (0.09–0.27)	0.1091 [t]	0.0420 *	0.93	0.97	0.84	0.40 *

Table 4. *Cont.*

HMOs	Composition						Median (25% and 75% Percentile) from W2 to W4 (Secretors and Non Secretors Mothers)		Mann-Whitney *p*-Value (FDR-Corrected MW *q*-Value in Exposant)		VIP –PLS-DA (C1–C2)		MLR-*p*-Value (FDR-Corrected MLR *q*-Value in Exposant)	
	mz	RT	Hex	HexNAc	Fuc	NeuAc	"Slower" Growth ($n = 38$)	"Faster" Growth ($n = 29$)	Secretors and Non Secretors	Secretors Only	Secretors and Non Secretors	Secretors Only	Secretors and Non Secretors	Secretors Only
4220a	684.2627	7.7	4	2	2	0	0.20 (0.14–0.29)	0.11 (0.07–0.25)	0.0216 *	0.0911 *t*	0.49	0.52	0.10	0.82 *t*
4220b≠	1367.5181	8.8	4	2	2	0	0.31 (0.09–1.78)	0.23 (0.07–0.33)	0.0895 *t*	0.3352	1.02	0.99	0.15	0.14 *
4210b≠	1221.4602	12.6	4	2	1	0	7.19 (5.03–8.90)	3.75 (2.37–6.33)	<0.0001 ***	0.0003 **	1.05	1.84	0.30	0.07 *
5330b	1878.7082	13.9	5	3	3	0	0.23 (0.00–0.39)	0.29 (0.00–0.38)	0.3749	0.4106	0.96	1.10	0.05	0.26 *
5320a	1732.6503	12.9	5	3	2	0	0.27 (0.13–0.70)	0.13 (0.05–0.22)	0.0004 **	0.0045 *	1.06	1.37	0.47	0.16 *

Relative HMO abundances (%) were calculated by dividing absolute HMO peak area by each sample's total HMO peak areas. Values are medians (25% and 75% percentiles) from relative HMOs abundances from week 2 to week 4 of lactation period. P values for comparison between "faster" and "slower" growth groups were derived using Mann-Whitney *U* test. Variables were considered as significantly modified between the two groups of infants' growth when their multiple comparisons adjusted *P*-values (e.g., q-value after false discovery rate) was < 0.05. *: FDR-corrected MW *q*-value < 0.05; **: FDR-corrected MW *q*-value < 0.01; ***: FDR-corrected MW *q*-value < 0.001; *t*: FDR-corrected MW *q*-value < 0.1. Multiple Linear Regression (MLR) for infant weight Z-score (p-value) was also combined with FDR, and predictive ability for infant weight growth was considered reliable when MLR *q*-value was < 0.05. *: FDR-corrected MLR *q*-value < 0.05; *t*: FDR-corrected MLR *q*-value < 0.1. ≠: variables of importance for PLS-DA model (VIP > 1.0). LNDFH, Lacto-N-difucosyl-hexaose; LNT, lacto-N-tetraose; pLNH, p-Lacto-N-Hexaose; LNFP, lacto-N-fucopentaose; LNnT, lacto-N-neotetraose; LNT, lactoN-tetraose; LST, sialyl-lacto-N-tetraose; 2′-FL, 2′-fucosyllactose; 3-FL, 3-fucosyllactose; 3′-SL, 3′-sialyllactose; Hex, hexose; HexNac, N-acetylhexosamine; Fuc, fucose and NeuAc, N-acetylneuraminic acid. Fucosylation was further investigated to determine the differences in the abundance of mono, di, tri, and tetrafucosylation (based on the number of fucose residues).

3.5. Integration of Multi-Omics Data sets

The aim of the current pilot study was to determine whether multi-block modeling could be applied for relating MS-based metabolomic, lipidomic, glycomic, fatty acids, and free amino acids data with regard to the predictive component that is the infant growth trajectory. Due to the complexity of longitudinal study by extracting the relevant information from multiple "omics" data sources, we chose to focus on "omics" data obtained on one representative time of lactation (week three of lactation). More specifically, the horizontal concatenation of annotated VIP provided by the AoV-PLS-DA model applied on MS-based metabolomic (i.e., 68 metabolites), lipidomic (i.e., 143 lipid species previously selected in [20]), and glycomic (79 HMOs) data with all metabolites–species issues from free amino acid and total fatty acid quantification was a straightforward solution to providing an extended analytical coverage of the biochemical diversity characterizing the breast milk samples. Concerning the glycomics data set, we had to overcome the problem due to the breast milk clustering based on maternal secretor or non-secretor status. As the relatively substantial variation in HMOs between the high and low 2'FL levels clusters were recently reported not to impact term infant growth of either sex up to four months [54], we chose to perform, on glycomic data, a mean-centered scaling of HMOs abundances on maternal secretor or non-secretor status before the "omics" data fusion. Then, on the super-matrix thus obtained, we tested UPCA or MBPLS multi-block strategies. Following the fusion of 340 selected annotated variables resulting from the combination of five data sources (blocks), the unsupervised UPCA score plot (Figure 4a) showed—on the principal components PC 3–4 that reported 21.4% variance—a clustering of breast milk samples and indicated a specific metabotype in the milk provided to preterm infants with "faster" growth versus that fed to infants with "slower" growth. The supervised MB-PLS score plot (Figure 4b) clearly highlighted on components 1–2 (18.5% of variance for the five blocks data), two breast milk metabotypes associated with 'faster' or 'slower' infant growth. The breast milk profiles, corresponding to the four sets of twins and depicted with blue symbols in Figure 4 were plotted between both "faster" and "slower" clusters. The most discriminant features associated with infant growth during hospital stay corresponded to a cluster of 87 metabolites species selected according to their loadings on the first MB-PLS components PC1 and PC2, and including: (i) many HMOs, such as LNFPI and 4210d, positively correlated with PC1 (i.e., associated with "faster" growth) or 4230b and 4230c, negatively correlated with PC1 (i.e., associated with "slower" growth), (ii) a few free amino acids (such as valine and glycine associated with "slower" growth), (iii) several lipidomic-species that were associated with "faster" growth, such as medium-chain saturated fatty acids (MCSAT, e.g., pentadecanoic and myristic acid), triglycerides (TG(46:0) and TG(50:2)), phospholipids (PS(38:4) and PE(38:3)), or that were associated with "slower" growth, such as oleic acid, plasmalogen-derivatives (PC(P-34:2) and PE(P-36:0)), lyso-phosphatidylethanolamine-containing arachidonic acid (LysoPE(20:4)), ceramide (Cer(18:1/22:0)), and very long-chain TG (TG(54:4) and TG(58:7)) and finally, (iv) few metabolomic-species positively (3-hydroxycapric acid, dihydrocaffeic acid 3-O-glucuronide, LNFPII) or negatively (9-undecenal, heptanoyl- and hexanoyl-glycine, 3-hydroxy-adipic acid, valerenic acid) correlated to PC1. Among these metabolites, MCSAT and oleic acid were previously shown to be predictive of optimal early growth [20].

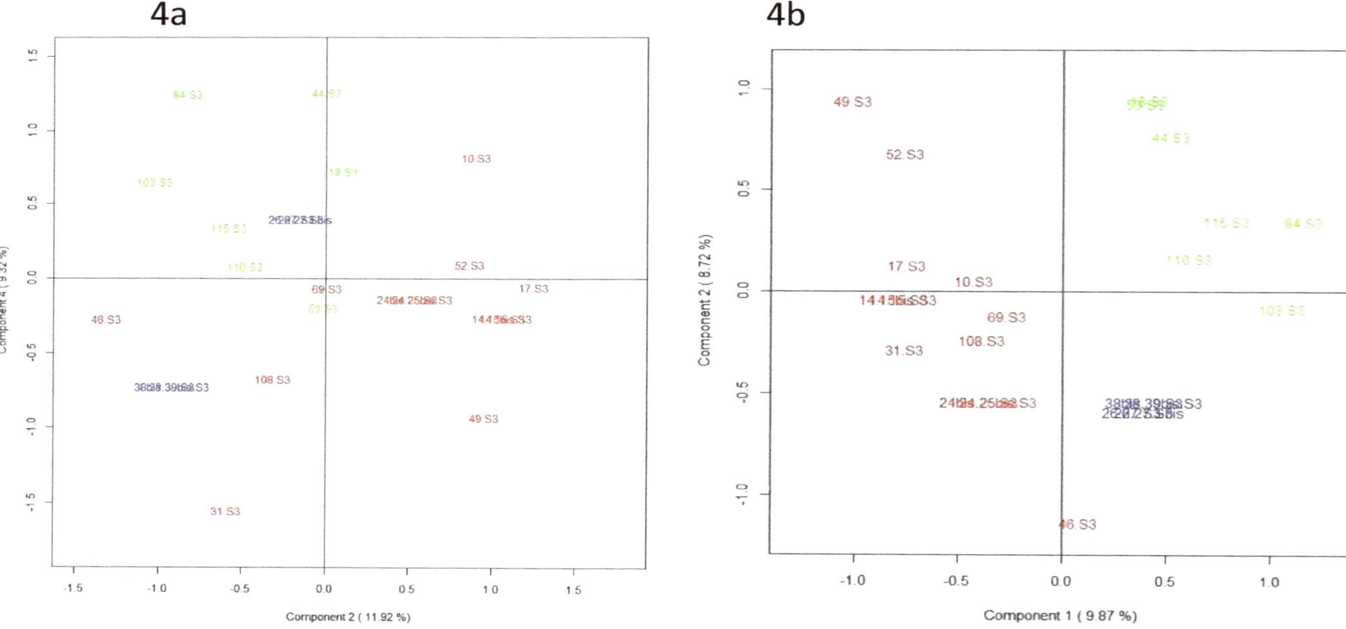

Figure 4. Score plot generated from UPCA (**a**) (with 21.4% of variance on components 3–4) and MB-PLS-DA (**b**) (with 18.5% of variance on components 1–2) based on the lipidomics/metabolomics/glycomics/FA/FAA profiles of human preterm milk at week 3 of lactation on the factor weight Z-score (discharge minus birth). Breast milk provided to preterm infants who experienced "faster" (green) or "slower" (red) growth and to twin infants with discordant growth rate, one twin with high growth rate and the other one with low growth rate (blue).

4. Discussion

To date, only a few metabolomics studies have been reported on preterm human milk in the first few weeks of lactation [27,28]. To the best of our knowledge, the current pilot study is the first to comprehensively characterize and compare the preterm human milk lipidome (previously reported in [20]), metabolome, glycome, total fatty acid, and free amino acid profiles in relation to the growth velocity of preterm infant early in life in a group of 26 breastfeeding mother infant dyads. Our findings strongly suggest that molecular species other than "classic" macronutrients in human breast milk might affect infant growth early in life.

4.1. Higher Breast Milk Content in Branched-Chain and Insulino-Trophic Amino Acid and in Tyrosine Associated to Optimal Infant Growth

The higher levels of total branched-chain amino acids (BCAA) in breast milk provided to infants who experienced a "faster" early growth was consistent with the muscle protein anabolic effects of BCAA reported in adult humans [55]. In a previous study conducted on obese mothers of full-term infants, we found a 20% higher BCAA content in the breast milk obtained from obese mothers compared with control, lean mothers [56]. Several authors argued that the higher BCAA content of most formulas compared to human milk may contribute to the higher early weight gain observed in bottle-fed (compared to breastfed) full-term infants [57,58]. However, whether a high milk BCAA content directly impacts the growth of the breast-fed child remains to be explored. Among the insulino-trophic amino acid, arginine, the sole endogenous source of nitric oxide (NO), and a precursor of polyamines and creatine may regulate angiogenesis, mammary gland development, enhance protein synthesis, and decrease protein degradation in mammary epithelial cells [59], thereby improving lactation performance [60,61]. In our study, the optimal early infant growth associated with higher breast milk contents of arginine and carbamoylsarcosine, an intermediate in the creatine-arginine pathway, might reflect the positive effects of arginine and polyamine on muscle protein synthesis and on immune response [61], its beneficial effects on the intestinal mucosa itself, the prevention of necrotizing enterocolitis [62–64] and its protective effect for the nervous system [61]. Higher availability of creatine may serve as a source for phosphorylation in muscle tissue and fat metabolism. The higher amounts of tyrosine combined with lower abundances of tyrosine catabolism products (hydroxyphenylacetic acid and cresol) [23] observed in the "faster" breast milk are consistent with the key role of tyrosine in infant growth as a precursor of thyroxine, a hormone involved in energy metabolism, and of dopamine, a neurotransmitter.

4.2. Enhanced Milk Fat Availability by Infants with an Early "Faster" Growth Velocity

Our data highlighted lower concentrations of two free amino acids involved in bile acid conjugation, glycine and taurine, in the "faster" growth group. These data were observed along with an enhanced total lipid content previously reported in "faster" milk [20]. We speculate that the lower milk content in glycin and taurine, under their free form, along with the higher fat content could be explained by enhanced bile acid conjugation in "faster" breast milk, which could, consequently, facilitate the solubilization of lipids, sterols, and fat-soluble vitamins by forming mixed micelles, and, in turn, the uptake of these nutrients into enterocytes despite gastrointestinal immaturity. This also is in agreement with higher coefficients of fat absorption reported in breast-fed vs. formula-fed new borns, and related to the bile salt-dependent lipase (BSL) present in human milk [65]. The "faster" breast milk also presented decreased glycine derivatives, including hexanoyl- and heptanoylglycine, which are products of the mitochondrial fatty acid beta-oxidation [66]. These findings suggest low rates of fatty acids utilization in mammary gland and/or breast milk, which could enhance fat availability for breast-fed preterm infants. Moreover, the "faster" breast milk contained higher amounts in beta-hydroxybutyrate, a ketone body involved in the mammary gland synthesis of triglycerides, leading to the formation of milk fat globules [34]. Apart from its pivotal role as an energy fuel for extrahepatic tissues like brain, heart, or skeletal muscle in newborns, hydroxybutyrate is reported to play a role as a signaling mediator, a driver of protein post-translational modification, and a modulator of inflammation and oxidative stress [67]; such mechanisms could explain the link between beta-hydroxybutyrate availability and optimal early growth in a context of prematurity. Moreover, preterm breast milk metabolome showed decreased levels of few odd medium unbranched (heptanoic and undecenal) or branched-chain (BCFAs, as 2-benzyloctanoic acid and methyl-2-octynoic acids) fatty acids but enhanced levels in one specific BCFA, the citraconic acid (known as methylmaleate, an odd and methyl-branched short chain unsaturated dicarboxylic fatty acid) in the "faster" group. Interestingly, mother's milk BCFAs [68] have been reported to reduce the occurrence of necrotizing enterocolitis in a mouse model due to their active roles in reducing inflammation and altering microbiota in the gut [69]. However, the potential effects of BCFAs on the infant health and development remain to be examined [68], as Wongtangtintharn et al. [70] suggested that BCFA lowers fatty acid synthesis [71]. Indeed, the higher lipid content we previously reported in the "faster" breast milk lipidome [20] seems to be consitent with the lower levels of many BCFA observed in the present study. Additionally, the fact that odd chain saturated fatty acid are reported to pass into the milk of lactating cows [72] suggests that these fatty acids may also cross the blood-brain barrier and act on early postnatal brain development. Finally, the higher availability of niacinamide in "faster" milk suggests niacinamide could act, through its role as a precursor of co-enzymes NAD and NADP, and impact fatty acid utilization since NAD^+ is used as a coenzyme in energy production (glycolysis, mitochondrial respiration) and improves gastrointestinal tract repair after damage as well as immune response [73].

Taken together, our findings suggest that the "faster" milk supplied (i) a higher content in energy, with calories supplied under a more digestible form of the preterm newborns, to overcome the immaturity of their digestive tractus, (ii) larger amounts or amino acids that promoted protein anabolism, and (iii) bioactive molecule, that played critical roles in energy homeostasis and gastrointestinal function, contributing to an optimal early-growth in infants born preterm. These findings also suggest that the higher lipid content in breast milk metabotype associated with preterm infants with optimal weight growth during hospitalization is used for enhanced fat oxidation rather than fat deposition, promoting tissue growth and likely a preferential fat-free mass deposition, which might in turn contribute to the recovery of the body composition and optimization of neurodevelopmental outcomes [74]. Moreover, the high BCAA and arginine intakes in breast milk provided to infants who belong to the "faster" group are also in agreement with this putative fat-free mass accretion in preterm infants, which could explain an optimal weight gain.

Additionnally, the higher content in phosphatidylcholine and sphingomyelin previously reported in the "faster" milk of the same pilot study [20] is consistent with the higher amounts of choline observed in breast milk metabolome provided to infants with optimal early-growth trajectory. Choline plays a key role as a precursor for acetylcholine (a neuromediator) and betaine (a source of labile methyl groups), which could increase the availability of methionine and choline and also enhance liver glycogenesis [75]. Indeed, this higher choline content may have a profound benefic impact upon the homestatic mechanisms and upon the physiological function in preterm infants, leading to an optimal growth during their hospitalization.

4.3. Di-Fucosylated HMOs Associated to Early Preterm Infant Growth

We found similar proportions of fucosylated and sialylated HMOs in preterm milk in good agreement with an earlier study from another team [31] but with slight differences in HMO-fucosylation level (according to the number of fucose residues carried) in both growth groups, such as a higher abundance of mono-(LNFPI) and a lower abundance of di- and tri-fucosylated oligosaccharides in breast milk provided to infants with "faster" growth. Given the small number of non-secretor mothers in this pilot study (five among 26 mothers) and because only one preterm milk sample was collected from a non-secretor mother who delivered a preterm infant with a "faster" growth, meaningful conclusions regarding differences in "faster" and "slower" growth milk on the basis of maternal secretor status cannot be drawn. There are scarce data to clarify the putative relationship between the maternal milk fucosyl-transferase 2 (FUT2)-dependent oligosaccharide status and infant anthropometry [32,54]. Recently, a specific FUT2-dependent HMO, LNFP I, but not 2′FL, was reported to be associated with the weight of full term infants at six months of age [32], but that association was not observed on growth curves over the first four months of life in groups of healthy, full term infants [54]. In our cohort, we observed consistent effects of the minor di-fucosylated HMOs (4230c, 4230b, 4240b, and 4210d) but not 2′FL and of the mono-fucosylated LNFP I (considering only secretors mother infant dyads) concentrations in breast milk on preterm infant growth up to 4 weeks of lactation. After multiple linear regression combined with multiple corrections and adjustment for confounding maternal and infant factors, the ability of the latter HMOs to predict early weight gain persisted. These di-fucosylated HMOs were also selected as discriminant metabolites in our MB-PLS model built on the five blocks of milk components. However, given the small sample size included in this exploratory study, future work would be needed in larger samples with longer follow-up to identify the exact contribution of specific HMOs to preterm infant growth. The impact of HMOS may involve effects on intestinal epithelium and gut microbiome, as fucosylated HMOs (and particularly LNFPI) support increased *bifidobacteria*, which dominate the microbiota of breastfed infants [76] but also on immune development and/or protection from infection through systemic effects [32], thereby likely affecting infant growth and body composition.

5. Conclusions

Previous studies have shown breastfeeeding is associated with many health benefits in pre-term infants. However, whether such benefits are due to mother's milk constituents, *per se*, remains unclear. Breastmilk may not always be adequate for pre-term infants with high nutrition density requirements, which may lead to insufficient weight gain. We believe this to be the first study to document the changes in the metabolomics/lipidomics/glycomics/amino acids profiles of pre-term human milk during the first month of lactation, related to preterm infant growth during hospital stays. We showed that specific differences in milk metabolites exist between breast milk provided to preterm infants with optimal or non-optimal early growth, and a set of few milk metabolites were identified as predictive of infant growth parameters in the present pilot study, pointing to the critical role of energy utilization, protein synthesis, oxidative status, gut epithelial cell maturity, or more indirectly, gut microbiome in the context of prematurity. In particular, our findings highlighted robust biomarkers, i.e., arginine, tyrosine, hydroxybutyrate, niacinamide, choline, and lacto-N-fucopentaose I, that displayed a good ability to predict weight gain during hospital stays. Moreover, this preterm milk metabolomic signature suggests that the optimal early growth trajectory during hospital stays of preterm infants could be combined with preferential fat utilization and fat-free mass accretion. A clear limitation of our pilot study is its small population sample size and slight differences in birth weight between the groups. Although it has long been known that infants born with a lower birth weight grow faster [36,37], matching groups for birth weight would have been nearly impossible. We therefore have to admit that it cannot be ascertained whether the difference in breastmilk composition is programmed by infant antenatal growth or, alternatively, is one among the many determinant factors of postnatal growth trajectory. This study is also limited by the statistical method we applied because the Mann-Whitney *U* test has a lower power than the parametric method. Before the metabolites identified here can be considered valid biomarkers, they need to be quantified in the entire preterm newborn LACTACOL cohort and, ideally, validated in other longitudinal birth cohorts. Nevertheless, this pioneer integrative analysis for human breast milk might open the way to novel strategies using human milk as a tool to improve the outcome of a frail population of newborn infants. Indeed, providing optimal individual nutrition remains a daunting challenge and the subject of heated debate between neonatalogists on the issue of "individualized" fortification of human milk to improve growth and long term outcomes for all preterm infants.

Supplementary Materials: The following are available online at http://www.mdpi.com/2072-6643/11/3/528/s1. Figure S1: AoV-PLS and LDA models, based on the LC-ESI–HRMS metabolomics profiles of human preterm milk, on the factor weight Z-score (discharge-birth): AoV-PLS score plot with 63% of variance (R2Y = 218%) on components 1–2 (Figure S1a) and LDA (built on components of AoV-PLS) with a *p*-value = 0) (Figure S1b). Breast milk provided to preterm infants who experienced 'faster' (green) or 'slower' (red) growth and to twin infants with discordant growth rate, one twin with high growth rate and the other one with low growth rate, (blue).

Author Contributions: M.-C.A.-G., C.-Y.B., J.-C.R.: had full access to all of the data in the study and took responsibility for the integrity of the data and the accuracy of the data analysis designed research; C.-Y.B., J.-C.R.: conceived the LACTACOL study and design; H.B. and C.-Y.B.: managed the LACTACOL cohort; M.-C.A.-G.: conceived and supervised the present metabolomics study conducted on a subset of infants of LACTACOL cohort; T.M.: conceived the appropriate statistical tools and/or performed the statistical analysis; A.D.-S., A.-L.R., Y.G., F.F., S.C.: conducted metabolomics analysis and/or metabolites identification using MS; M.-C.A.-G.: drafted the manuscript; T.M., F.F., C.-Y.B. and D.D.: critically revised the manuscript for important intellectual content; and all authors: acquired, analyzed, or interpreted the data, and read and approved the final manuscript.

Acknowledgments: This work was part of the LACTACOL project. The LACTACOL projet was funded by Région Pays-de-la-Loire (2011–2012) and by the 'Fonds Européen de Développement Economique et Régional » (FEDER, Grant 38395, Project 6226, 2013–2015). We thank lactating mothers for having accepted donation of breast milk for the LACTACOL protocol, andCécile Boscher, Laure Simon, and Evelyne Gauvard for caring mothers and their infants as well as guarantying the collection of maternal milk. We also thank the Biological Resource Centre for biobanking (CHU Nantes, Hôtel Dieu, Centre de Ressources Biologiques (CRB), Nantes, F-44093, France (BRIF: BB-0033-00040)) for storing human milk samples in the milk biobank (DC-2009-982). Metabolomics LC-HRMS experiments, bioinformatics tool development and data treatment were performed on the LABERCA, a technical plateau from CORSAIRE Metabolomics platform (Biogenouest network) with the collaborative portal dedicated to metabolomic data processing, analysis and annotation 'Workflow4Metabolomics'. Glycomics LC-HRMS experiments were performed on the CEA-Saclay plateau (MetaboHUB, France). The authors would like to acknowledge Lauriane Rambaud (LABERCA, Nantes, France) for her technical assistance in breast milk metabobolome annotation.

Conflicts of Interest: No funder/sponsor had any role in the design and conduct of the study; collection, management, analysis, and interpretation of the data; the preparation, review, or approval of the manuscript; or the decision to submit the manuscript for publication.

Abbreviations

AoV-PLS	analysis of variance combined to partial least squares regression
BCFA	branched-chain fatty acids
ESI	electrospray ionization
FDR	false discovery rate
GA	gestational age
HMO	human milk oligosaccharides
LC-HR-MS	Liquid-Chromatography-High-Resolution-Mass-Spectrometry
MG PLS-DA	multi-group partial least squares discriminant analysis
MLR	multiple linear regression
MCSAT	medium-chain saturated fatty acid
U-PCA	unfold principal component analysis
SD	standard deviation
W4M	Workflow4Metabolomics®

References

1. Victora, C.G.; Bahl, R.; Barros, A.J.; França, G.V.; Horton, S.; Krasevec, J.; Murch, S.; Sankar, M.J.; Walker, N.; Rollins, N.C.; et al. Breastfeeding in the 21st century: Epidemiology, mechanisms, and lifelong effect. *Lancet* **2016**, *387*, 475–490. [CrossRef]

2. Mills, S.; Ross, R.P.; Hill, C.; Fitzgerald, G.F.; Stanton, C. Milk intelligence: Mining milk for bioactive substances associated with human health. *Int. Dairy J.* **2011**, *21*, 377–401. [CrossRef]

3. Andreas, N.J.; Kampmann, B.; Le-Doare, K.M. Human breast milk: A review on its composition and bioactivity. *Early Hum. Dev.* **2015**, *91*, 629–635. [CrossRef] [PubMed]

4. Mosca, F.; Giannì, M.L. Human milk: Composition and health benefits. *La Pediatria Medica e Chirurgica* **2017**, *39*. [CrossRef] [PubMed]

5. Boquien, C.Y. Human milk: An ideal food for nutrition of preterm newborn. *Front. Pediatr.* **2018**, *6*. [CrossRef] [PubMed]

6.   Johnson, T.J.; Patel, A.L.; Bigger, H.R.; Engstrom, J.L.; Meier, P.P. Economic benefits and costs of human milk feedings: A strategy to reduce the risk of prematurity-related morbidities in very-low-birth-weight infants. *Adv. Nutr. Int. Rev. J.* **2014**, *5*, 207–212. [CrossRef] [PubMed]

7.   Sankar, M.J.; Sinha, B.; Chowdhury, R.; Bhandari, N.; Taneja, S.; Martines, J.; Bahl, R. Optimal breastfeeding practices and infant and child mortality: A systematic review and meta-analysis. *Acta Paediatr.* **2015**, *104*, 3–13. [CrossRef] [PubMed]

8.   Quigley, M.; McGuire, W. Formula versus donor breast milk for feeding preterm or low birth weight infants. *Cochrane Database Syst. Rev.* **2014**. [CrossRef] [PubMed]

9.   Corpeleijn, W.E.; de Waard, M.; Christmann, V.; van Goudoever, J.B.; Jansen-van der Weide, M.C.; Kooi, E.M.; Koper, J.F.; Kouwenhoven, S.M.; Lafeber, H.N.; Mank, E.; et al. Effect of donor milk on severe infections and mortality in very low-birth-weight infants: The early nutrition study randomized clinical trial. *JAMA Pediatr.* **2016**, *170*, 654–661. [CrossRef] [PubMed]

10. Boyd, C.A.Q.; Quigley, M.A.; Brocklehurst, P. Donor breast milk versus infant formula for preterm infants: Systematic review and meta-analysis. *Arch. Dis. Child. Fetal Neonatal Ed.* **2007**, *92*, F169–F175. [CrossRef] [PubMed]

11. Rozé, J.C.; Darmaun, D.; Boquien, C.Y.; Flamant, C.; Picaud, J.C.; Savagner, C.; Claris, O.; Lapillonne, A.; Mitanchez, D.; Branger, B.; et al. The apparent breastfeeding paradox in very preterm infants: Relationship between breast feeding; early weight gain and neurodevelopment based on results from two cohorts; EPIPAGE and LIFT. *BMJ Open* **2012**, *2*, e000834. [CrossRef] [PubMed]

12. Curtis, M.; Rigo, J. Extrauterine growth restriction in very-low-birthweight infants. *Acta Paediatr.* **2004**, *93*, 1563–1568. [CrossRef] [PubMed]

13. Ehrenkranz, R.A.; Dusick, A.M.; Vohr, B.R.; Wright, L.L.; Wrage, L.A.; Poole, W.K. Growth in the neonatal intensive care unit influences neurodevelopmental and growth outcomes of extremely low birth weight infants. *Pediatrics* **2006**, *117*, 1253–1261. [CrossRef] [PubMed]

14. Larroque, B.; Ancel, P.Y.; Marret, S.; Marchand, L.; André, M.; Arnaud, C.; Pierrat, V.; Rozé, J.C.; Messer, J.; Thiriez, G.; et al. Neurodevelopmental disabilities and special care of 5-year-old children born before 33 weeks of gestation (the EPIPAGE study): A longitudinal cohort study. *Lancet* **2008**, *371*, 813–820. [CrossRef]

15. Frondas-Chauty, A.; Simon, L.; Branger, B.; Gascoin, G.; Flamant, C.; Ancel, P.Y.; Darmaun, D.; Rozé, J.C. Early growth and neurodevelopmental outcome in very preterm infants: Impact of gender. *Arch. Dis. Child. Fetal Neonatal Ed.* **2014**, *99*, F366–F372. [CrossRef] [PubMed]

16. Agostoni, C.; Buonocore, G.; Carnielli, V.P.; De Curtis, M.; Darmaun, D.; Decsi, T.; Domellöf, M.; Embleton, N.D.; Fusch, C.; Genzel-Boroviczeny, O.; et al. Enteral nutrient supply for preterm infants: Commentary from the European Society of Paediatric Gastroenterology; Hepatology and Nutrition Committee on Nutrition. *J. Pediatr. Gastr. Nutr.* **2010**, *50*, 85–91. [CrossRef] [PubMed]

17. Henriksen, C.; Westerberg, A.C.; Rønnestad, A.; Nakstad, B.; Veierød, M.B.; Drevon, C.A.; Iversen, P.O. Growth and nutrient intake among very-low-birth-weight infants fed fortified human milk during hospitalisation. *Brit. J. Nutr.* **2009**, *102*, 1179–1186. [CrossRef] [PubMed]

18. Alexandre-Gouabau, M.C.; Courant, F.; Le Gall, G.; Moyon, T.; Darmaun, D.; Parnet, P.; Coupé, B.; Antignac, J.P. Offspring metabolomic response to maternal protein restriction in a rat model of intrauterine growth restriction (IUGR). *J. Proteome Res.* **2011**, *10*, 3292–3302. [CrossRef] [PubMed]

19. Alexandre-Gouabau, M.C.; Courant, F.; Moyon, T.; Küster, A.; Le Gall, G.; Tea, I.; Antignac, J.P.; Darmaun, D. Maternal and cord blood LC-HRMS metabolomics reveal alterations in energy and polyamine metabolism, and oxidative stress in very-low birth weight infants. *J. Proteome Res.* **2013**, *12*, 2764–2778. [CrossRef] [PubMed]

20. Alexandre-Gouabau, M.-C.; Moyon, T.; Cariou, V.; Antignac, J.-P.; Qannari, E.M.; Croyal, M.; Soumah, M.; Guitton, Y.; Billard, H.; Legrand, A.; et al. Breast milk lipidome is associated with early growth trajectory in preterm infants. *Nutrients* **2018**, *10*, 164. [CrossRef] [PubMed]

21. Fanos, V.; Atzori, L.; Makarenko, K.; Melis, G.B.; Ferrazzi, E. Metabolomics application in maternal-fetal medicine. *BioMed Res. Int.* **2013**, *9*. [CrossRef] [PubMed]

22. Demmelmair, H.; Koletzko, B. Variation of metabolite and hormone contents in human milk. *Clin. Perinatol.* **2017**, *44*, 151–164. [CrossRef] [PubMed]

23. Carraro, S.; Baraldi, E.; Giordano, G.; Pirillo, P.; Stocchero, M.; Houben, M.; Bont, L. Metabolomic Profile of Amniotic Fluid and Wheezing in the First Year of Life—A Healthy Birth Cohort Study. *J. Pediatr.* **2018**, *196*, 264–269. [CrossRef] [PubMed]

24. Marincola, F.C.; Noto, A.; Caboni, P.; Reali, A.; Barberini, L.; Lussu, M.; Murgia, F.; Santoru, M.L.; Atzori, L.; Fanos, V.J. A metabolomic study of preterm human and formula milk by high resolution NMR and GC/MS analysis: Preliminary results. *Matern. Fetal Neonatal Med.* **2012**, *25*, 62–67. [CrossRef] [PubMed]

25. Wu, J.; Domellöf, M.; Zivkovic, A.M.; Larsson, G.; Öhman, A.; Nording, M.L. NMR-based metabolite profiling of human milk: A pilot study of methods for investigating compositional changes during lactation. *Biochem. Biophys. Res. Comm.* **2016**, *469*, 626–632. [CrossRef] [PubMed]

26. George, A.; Gay, M.; Trengove, R.; Geddes, D. Human Milk Lipidomics: Current Techniques and Methodologies. *Nutrients* **2018**, *10*, 1169. [CrossRef] [PubMed]

27. Spevacek, A.R.; Smilowitz, J.T.; Chin, E.L.; Underwood, M.A.; German, J.B.; Slupsky, C.M. Infant Maturity at Birth Reveals Minor Differences in the Maternal Milk Metabolome in the First Month of Lactation. *J. Nutr.* **2015**, *45*, 1698–1708. [CrossRef] [PubMed]

28. Sundekilde, U.K.; Downey, E.; O'Mahony, J.A.; O'Shea, C.A.; Ryan, C.A.; Kelly, A.L.; Bertram, H.C. The effect of gestational and lactational age on the human milk metabolome. *Nutrients* **2016**, *8*, 304. [CrossRef] [PubMed]

29. Maffei, D.; Schanler, R.J. Human milk is the feeding strategy to prevent necrotizing enterocolitis! *Semin. Perinatol.* **2017**, *41*, 36–40. [CrossRef] [PubMed]

30. Bode, L. Human Milk Oligosaccharides in the Prevention of Necrotizing Enterocolitis: A journey from in vitro and in vivo models to mother-infant cohort studies. *Front Pediatr.* **2018**, *6*, 385. [CrossRef] [PubMed]

31. De Leoz, M.L.A.; Gaerlan, S.C.; Strum, J.S.; Dimapasoc, L.M.; Mirmiran, M.; Tancredi, D.J.; Smilowitz, J.T.; Kalanetra, K.M.; Mills, D.A.; German, J.B.; et al. Lacto-N-tetraose; fucosylation; and secretor status are highly variable in human milk oligosaccharides from women delivering preterm. *J. Proteome Res.* **2012**, *11*, 4662–4672. [CrossRef] [PubMed]

32. Alderete, T.L.; Autran, C.; Brekke, B.E.; Knight, R.; Bode, L.; Goran, M.I.; Fields, D.A. Associations between human milk oligosaccharides and infant body composition in the first 6 mo of life. *Am. J. Clin. Nutr.* **2015**, *102*, 1381–1388. [CrossRef] [PubMed]

33. Charbonneau, M.R.; O'Donnell, D.; Blanton, L.V.; Totten, S.M.; Davis, J.C.; Barratt, M.J.; Cheng, J.; Guruge, J.; Talcott, M.; Bain, J.R.; et al. Sialylated milk oligosaccharides promote microbiota-dependent growth in models of infant undernutrition. *Cell* **2016**, *164*, 859–871. [CrossRef] [PubMed]

34. Rezaei, R.; Wu, Z.; Hou, Y.; Bazer, F.W.; Wu, G. Amino acids and mammary gland development: Nutritional implications for milk production and neonatal growth. *J. Anim. Sci. Biotechnol.* **2016**, *7*, 20. [CrossRef] [PubMed]

35. Metrustry, S.J.; Karhunen, V.; Edwards, M.H.; Menni, C.; Geisendorfer, T.; Huber, A.; Reichel, C.; Dennison, E.M.; Cooper, C.; Spector, T.; et al. Metabolomic signatures of low birthweight: Pathways to insulin resistance and oxidative stress. *PLoS ONE* **2018**, *13*, e0194316. [CrossRef] [PubMed]

36. Ferchaud-Roucher, V.; Desnots, E.; Naël, C.; Martin Agnoux, A.; Alexandre-Gouabau, M.C.; Darmaun, D.; Boquien, C.Y. Use of UPLC-ESI-MS/MS to quantitate free amino acid concentrations in micro-samples of mammalian milk. *SpringerPlus* **2013**, *2*, 622. [CrossRef] [PubMed]

37. Oursel, S.; Junot, C.; Fenaille, F. Comparative analysis of native and permethylated human milk oligosaccharides by liquid chromatography coupled to high resolution mass spectrometry. *J. Chromatogr. B* **2017**, *1071*, 49–57. [CrossRef] [PubMed]

38. Bligh, E.G.; Dyer, W.J. A rapid method of total lipid extraction and purification. *Can. J. Biochem. Physiol.* **1959**, *37*, 911–917. [CrossRef] [PubMed]

39. Gallart-Ayala, H.; Courant, F.; Severe, S.; Antignac, J.P.; Morio, F.; Abadie, J.; Le Bizec, B. Versatile lipid profiling by liquid chromatography-high resolution mass spectrometry using all ion fragmentation and polarity switching. Preliminary application for serum samples phenotyping related to canine mammary cancer. *Anal. Chim. Acta* **2013**, *796*, 75–83. [CrossRef] [PubMed]

40. Courant, F.; Royer, A.L.; Chéreau, S.; Morvan, M.L.; Monteau, F.; Antignac, J.P.; Le Bizec, B. Implementation of a semi-automated strategy for the annotation of metabolomic fingerprints generated by liquid chromatography-high resolution mass spectrometry from biological samples. *Analyst* **2012**, *137*, 4958–4967. [CrossRef] [PubMed]

41. Kessner, D.; Chambers, M.; Burke, R.; Agus, D.; Mallick, P. ProteoWizard: Open source software for rapid proteomics tools development. *Bioinformatics* **2008**, *24*, 2534–2536. [CrossRef] [PubMed]

42. Smith, C.A.; Want, E.J.; O'Maille, G.; Abagyan, R.; Siuzdak, G. XCMS: Processing mass spectrometry data for metabolite profiling using nonlinear peak alignment, matching, and identification. *Anal. Chem.* **2006**, *78*, 779–787. [CrossRef] [PubMed]

43. Giacomoni, F.; Le Corguillé, G.; Monsoor, M.; Landi, M.; Pericard, P.; Pétéra, M.; Duperier, C.; Tremblay-Franco, M.; Martin, J.F.; Jacob, D.; et al. Workflow4Metabolomics: A collaborative research infrastructure for computational metabolomics. *Bioinformatics* **2014**, *31*, 1493–1495. [CrossRef] [PubMed]

44. Kuhl, C.; Tautenhahn, R.; Bottcher, C.; Larson, T.R.; Neumann, S. CAMERA: An integrated strategy for compound spectra extraction and annotation of liquid chromatography/mass spectrometry data sets. *Anal. Chem.* **2011**, *84*, 283–289. [CrossRef] [PubMed]

45. Van Der Kloet, F.M.; Bobeldijk, I.; Verheij, E.R.; Jellema, R.H. Analytical error reduction using single point calibration for accurate and precise metabolomic phenotyping. *J. Prot. Res.* **2009**, *8*, 5132–5141. [CrossRef] [PubMed]

46. Dunn, W.B.; Broadhurst, D.; Begley, P.; Zelena, E.; Francis-McIntyre, S.; Anderson, N.; Brown, M.; Knowles, J.D.; Halsall, A.; Haselden, J.N.; et al. Procedures for large-scale metabolic profiling of serum and plasma using gas chromatography and liquid chromatography coupled to mass spectrometry. *Nat. Protoc.* **2011**, *6*, 1060. [CrossRef] [PubMed]

47. Ferchaud-Roucher, V.; Croyal, M.; Krempf, M.; Ouguerram, K. Plasma lipidome characterization using UHPLC-HRMS and ion mobility of hypertriglyceridemic patients on nicotinic acid. *Atherosclerosis* **2015**, *241*, e123–e124. [CrossRef]

48. Van den Berg, R.A.; Hoefsloot, H.C.; Westerhuis, J.A.; Smilde, A.K.; van der Werf, M.J. Centering, scaling, and transformations: Improving the biological information content of metabolomics data. *BMC Genom.* **2006**, *7*, 142. [CrossRef] [PubMed]

49. El Ghaziri, A.; Qannari El, M.; Moyon, T.; Alexandre-Gouabau, M.-C. ANOVA-PLS: A new method for the analysis of multivariate data depending on several factors. *Electron. J. Appl. Stat. Anal.* **2015**, *8*, 214–235. [CrossRef]

50. Henrion, R. N-way principal component analysis theory; algorithms and applications. *Chemometr. Intell. Lab.* **1994**, *25*, 1–23. [CrossRef]

51. Westerhuis, J.A.; Kourti, T.; MacGregor, J.F. Analysis of multiblock and hierarchical PCA and PLS models. *J. Chemometr.* **1998**, *12*, 301–321. [CrossRef]

52. Simon, L.; Frondas-Chauty, A.; Senterre, T.; Flamant, C.; Darmaun, D.; Rozé, J.C. Determinants of body composition in preterm infants at the time of hospital discharge. *Am. J. Clin. Nutr.* **2014**, *100*, 98–104. [CrossRef] [PubMed]

53. Steward, D.K.; Pridham, K.F. Growth patterns of extremely low-birth-weight hospitalized preterm infants. *J. Obstet. Gynaecol. Neonat. Nurs.* **2002**, *31*, 57–65. [CrossRef]

54. Sprenger, N.; De Castro, C.A.; Steenhout, P.; Thakkar, S.K. Longitudinal change of selected human milk oligosaccharides and association to infants' growth, an observatory, single center, longitudinal cohort study. *PLoS ONE* **2017**, *12*, e0171814. [CrossRef] [PubMed]

55. Lynch, C.J.; Adams, S.H. Branched-chain amino acids in metabolic signalling and insulin resistance. *Nat. Rev. Endocrinol.* **2014**, *10*, 723. [CrossRef] [PubMed]

56. De Luca, A.; Hankard, R.; Alexandre-Gouabau, M.C.; Ferchaud-Roucher, V.; Darmaun, D.; Boquien, C.Y. Higher concentrations of branched-chain amino acids in breast milk of obese mothers. *Nutr. J.* **2016**, *32*, 1295–1298. [CrossRef] [PubMed]

57. McCormack, S.E.; Shaham, O.; McCarthy, M.A.; Deik, A.A.; Wang, T.J.; Gerszten, R.E.; Clish, C.B.; Mootha, V.K.; Grinspoon, S.K.; Fleischman, A. Circulating branched-chain amino acid concentrations are associated with obesity and future insulin resistance in children and adolescents. *Pediatr. Obes.* **2013**, *8*, 52–61. [CrossRef] [PubMed]

58. Kirchberg, F.F.; Harder, U.; Weber, M.; Grote, V.; Demmelmair, H.; Peissner, W.; Rzehak, P.; Xhonneux, A.; Carlier, C.; Ferre, N.; et al. Dietary protein intake affects amino acid and acylcarnitine metabolism in infants aged 6 months. *J. Clin. Endocr. Metab.* **2015**, *100*, 149–158. [CrossRef] [PubMed]

59. Ma, Q.; Hu, S.; Bannai, M.; Wu, G. L-Arginine regulates protein turnover in porcine mammary epithelial cells to enhance milk protein synthesis. *Amino Acids* **2018**, *50*, 621–628. [CrossRef] [PubMed]

60. Kim, S.W.; Wu, G. Regulatory role for amino acids in mammary gland growth and milk synthesis. *Amino Acids* **2009**, *37*, 89–95. [CrossRef] [PubMed]

61. Tan, C.; Zhai, Z.; Ni, X.; Wang, H.; Ji, Y.; Tang, T.; Ren, W.; Long, H.; Deng, B.; Deng, J.; et al. Metabolomic profiles reveal potential factors that correlate with lactation performance in sow milk. *Sci. Rep. UK* **2018**, *8*. [CrossRef] [PubMed]

62. Bode, L. Recent advances on structure; metabolism; and function of human milk oligosaccharides. *J. Nutr.* **2006**, *136*, 2127–2130. [CrossRef] [PubMed]

63. Plaza-Zamora, J.; Sabater-Molina, M.M.; Rodríguez-Palmero, M.; Rivero, M.; Bosch, V.; Nadal, J.M.; Zamora, S.; Larqué, E. Polyamines in human breast milk for preterm and term infants. *Br. J. Nutr.* **2013**, *3*, 1–5. [CrossRef] [PubMed]

64. Shah, P.S.; Shah, V.S.; Kelly, L.E. Arginine supplementation for prevention of necrotising enterocolitis in preterm infants. *Cochrane Database Syst. Rev.* **2017**. [CrossRef] [PubMed]

65. Blazquez, A.M.; Cives-Losada, C.; Iglesia, A.; Marin, J.J.; Monte, M.J. Lactation during cholestasis: Role of ABC proteins in bile acid traffic across the mammary gland. *Sci. Rep. UK* **2017**, *7*, 7475. [CrossRef] [PubMed]

66. Luan, H.; Liu, L.F.; Tang, Z.; Zhang, M.; Chua, K.K.; Song, J.X.; Mok, V.C.; Li, M.; Cai, Z. Comprehensive urinary metabolomic profiling and identification of potential noninvasive marker for idiopathic Parkinson's disease. *Sci. Rep. UK* **2015**, *5*. [CrossRef] [PubMed]

67. Puchalska, P.; Crawford, P.A. Multi-dimensional roles of ketone bodies in fuel metabolism; signaling; and therapeutics. *Cell Metab.* **2017**, *25*, 262–284. [CrossRef] [PubMed]

68. Dingess, K.A.; Valentine, C.J.; Ollberding, N.J.; Davidson, B.S.; Woo, J.G.; Summer, S.; Peng, Y.M.; Guerrero, M.L.; Ruiz-Palacios, G.M.; Ran-Ressler, R.R.; et al. Branched-chain fatty acid composition of human milk and the impact of maternal diet: The Global Exploration of Human Milk (GEHM) Study. *Am. J. Clin. Nutr.* **2016**, *105*, 177–184. [CrossRef] [PubMed]

69. Ran-Ressler, R.R.; Khailova, L.; Arganbright, K.M.; Adkins-Rieck, C.K.; Jouni, Z.E.; Koren, O.; Ley, R.E.; Brenna, J.T.; Dvorak, B. Branched chain fatty acids reduce the incidence of necrotizing enterocolitis and alter gastrointestinal microbial ecology in a neonatal rat model. *PLoS ONE* **2011**, *6*. [CrossRef] [PubMed]

70. Wongtangtintharn, S.; Oku, H.; Iwasaki, H.; TODA, T. Effect of branched-chain fatty acids on fatty acid biosynthesis of human breast cancer cells. *J. Nutr. Sci. Vitaminol.* **2004**, *50*, 137–143. [CrossRef]

71. Vlaeminck, B.; Fievez, V.; Cabrita, A.R.J.; Fonseca, A.J.M.; Dewhurst, R.J. Factors affecting odd-and branched-chain fatty acids in milk: A review. *Anim. Feed Sci. Tech.* **2006**, *131*, 389–417. [CrossRef]

72. Jenkins, B.; West, J.; Koulman, A. A review of odd-chain fatty acid metabolism and the role of pentadecanoic acid (C15: 0) and heptadecanoic acid (C17: 0) in health and disease. *Molecules* **2015**, *20*, 2425–2444. [CrossRef] [PubMed]

73. Gill, B.D.; Indyk, H.E. Development and application of a liquid chromatographic method for analysis of nucleotides and nucleosides in milk and infant formulas. *Int. Dairy J.* **2007**, *17*, 596–605. [CrossRef]

74. Gianni, M.L.; Roggero, P.; Mosca, F. Human milk protein vs. formula protein and their use in preterm infants. *Cur. Opin. Clin. Nutr.* **2019**, *22*. [CrossRef] [PubMed]

75. Zeisel, S.H. Choline: Critical role during fetal development and dietary requirements in adults. *Annu. Rev. Nutr.* **2006**, *26*, 229–250. [CrossRef] [PubMed]

76. Newburg, D.S.; Morelli, L. Human milk and infant intestinal mucosal glycans guide succession of the neonatal intestinal microbiota. *Pediatr. Res.* **2015**, *77*, 115–120. [CrossRef] [PubMed]

Review

Role of Vitamin A in Mammary Gland Development and Lactation

M. Teresa Cabezuelo [1,2], **Rosa Zaragozá** [3,*], **Teresa Barber** [4] and **Juan R. Viña** [4]

1   Department of Physiology, Universitat de València, Avda. Blasco Ibañez, 15, 46010 Valencia, Spain; tecabar@alumni.uv.es
2   University Hospital Doctor Peset, Gaspar Aguilar, 90, 46017 Valencia, Spain
3   Department of Human Anatomy and Embryology-INCLIVA Biomedical Research Institute, Universitat de València, 46010 Valencia, Spain
4   Department of Biochemistry and Molecular Biology-INCLIVA Biomedical Research Institute, Universitat de València, 46010 Valencia, Spain; teresa.barber@uv.es (T.B.); juan.r.vina@uv.es (J.R.V.)
*   Correspondence: rosa.zaragoza@uv.es; Tel.: +34-96386-4187

Received: 25 October 2019; Accepted: 19 December 2019; Published: 27 December 2019

Abstract: Vitamin A (all-*trans*-retinol), its active derivatives retinal and retinoic acid, and their synthetic analogues constitute the group of retinoids. It is obtained from diet either as preformed vitamin A or as carotenoids. Retinal plays a biological role in vision, but most of the effects of vitamin A are exerted by retinoic acid, which binds to nuclear receptors and regulates gene transcription. Vitamin A deficiency is an important nutritional problem, particularly in the developing world. Retinol and carotenoids from diet during pregnancy and lactation influence their concentration in breast milk, which is important in the long term, not only for the offspring, but also for maternal health. In this study, we review the role of vitamin A in mammary gland metabolism, where retinoid signaling is required not only for morphogenesis and development of the gland and for adequate milk production, but also during the weaning process, when epithelial cell death is coupled with tissue remodeling.

Keywords: vitamin A; retinoic acid; vitamin A deficiency; lactating mammary gland; weaning; involution

1. Introduction

Vitamin A, an isoprenic-derived micronutrient, strictly refers to the alcoholic form all-*trans*-retinol, although in a broad sense includes both its active derivatives as well as other synthetic analogues that exhibit its biological actions. All of these molecules constitute the group of retinoids [1]. Their important role has been demonstrated through many scientific approaches and clinical observations. Retinoids are required for embryonic development and exert important effects on postnatal physiological events, including cell differentiation and proliferation, immunity, ocular function, and reproduction, and are also important antioxidants [2–9]; however, it must also be considered that treatment with retinoic acid (RA) can increase oxidative stress [10].

Retinol is a liposoluble micronutrient essential in our diet. It is available in foods either as preformed vitamin A; as retinyl ester (RE), abundant in some animal-derived sources such as liver, eggs, dairy products, and fatty fish; or as provitamin A carotenoids, mainly β-carotene, abundant in dark colored fruits and vegetables such as green leaves, carrots, ripe mangos, and other orange–yellow vegetables. This heterogeneous group of carotenoids presents other healthy effects as well, including antioxidant, anti-inflammatory, and immunomodulatory properties [11]. Dietary vitamin A is absorbed in the small intestine in the form of retinol and transported in blood attached to retinol-binding protein (RBP). Inside the cells, retinol is oxidized into its main biologically active derivatives, first retinaldehyde (retinal), which plays a role in vision, and then retinoic acid (RA), which regulates the expression of multiple target genes. Most functions of vitamin A are mediated through the RA activation of two types of transcription factors, retinoic acid receptors (RARs) and retinoid X receptors (RXRs), each with three subtypes (α, β, and γ) and each of those with different isoforms [12]. All-*trans*-RA (atRA) is a high-affinity ligand for RARs, while 9-*cis*-RA binds to RARs and RXRs. Upon binding to RA, RAR–RXR heterodimers bind to the RA response element (RARE) in the promoter regions of target genes and stimulate their transcription. Moreover, RA also has extranuclear, nontranscriptional effects, which directly influence the expression of several genes through phosphorylation processes. Different transcriptional effects of retinol and retinal have also been extensively described [12–15].

According to the World Health Organization (WHO), vitamin A deficiency (VAD) is a public health problem in 50% of countries; it is the most common nutritional disorder in the world, along with protein malnutrition. The leading cause of VAD in humans is deficient dietary intake, which occurs particularly in developing countries and may be exacerbated by high rates of infection, especially diarrhea and measles. In low- and middle-income countries, where intake of animal-based foods is low, dietary provitamin A carotenoids are the main source of vitamin A, and higher amounts are needed to meet the requirements of this vitamin. VAD might be aggravated during the stages of life in which nutritional demands are greater, such as pregnancy, lactation, and early childhood [16]. Dietary deficiency can start early in life, with the colostrum being discarded or with inadequate breastfeeding, which means that babies are deprived of their first important source of vitamin A. Mothers from poorer populations, where intake of vitamin A and carotenoids is marginal, have a much lower content of these micronutrients in their milk compared to mothers from developed countries. It is important to consider that nutrients provided through maternal milk derive not only from the diet during pregnancy and lactation, but also from maternal reserves. In this sense, the potential long-term effects of nutrient depletion during any of these periods are important not only to fetal and child development, but also to maternal health [11,16–19].

This review focuses on the role of vitamin A in the mammary gland along each developmental stage. First, it covers the importance of retinoids in the morphogenesis and development of the gland. Next, we note that during lactation, it is essential to meet the requirements to fulfill an adequate milk production. Finally, we also consider the role of this micronutrient in the weaning process, when extensive tissue remodeling occurs and epithelial tissue regresses to return the gland to a structure similar to the pre-pregnant state, thus the tissue is prepared for the next pregnancy and lactation. We also cover the impact of vitamin A deficiency on both the mother and her offspring's health during the period of pregnancy and lactation, when the most important stages of development in human life occur.

2. Vitamin A Dietary Sources, Availability, and Requirements

Dietary vitamin A is obtained in mainly two forms, as preformed vitamin A or as provitamin A carotenoids. Preformed vitamin A is mainly found as long-chain fatty acid esters of retinol, particularly retinyl esters (REs), which are easily hydrolyzed endogenously to form retinol. Preformed vitamin A is found in animal sources such as eggs, fish oils, organ meats, and dairy products and their derivatives, although liver is considered the source with the greatest quantitative importance. On the other hand, the provitamin A carotenoids, polyisoprenoid pigments of vegetable origin, are partially converted to vitamin A in the intestinal mucosa. To date, more than 750 carotenoids have been identified, although only about 40 of them are consumed in substantial amounts in the human diet, the most abundant being β-carotene, lycopene, lutein, β-cryptoxanthin, α-carotene, and zeaxanthin [20]. Not all carotenoids exhibit provitamin A properties; only 10% of natural carotenoids can be metabolically cleaved to produce at least one molecule of retinol. The main structural requirement of a carotenoid for its provitamin A activity is the presence of at least one ring of unsubstituted β-ionone with a polyene chain of 11 or more carbon atoms [21,22]. As previously mentioned, only a few of these precursors with provitamin A activity are found in significant amounts in the human diet, mainly β-carotene, with minor contributions from α-carotene, β-cryptoxanthin, β-canthaxanthin, and β-echinenone.

The dietary sources of these carotenoids are colored fruits (orange, apricot, mango, etc.) and some vegetables and their formulations (carrot, tomato, sweet potato, pumpkin, broccoli, cabbage, spinach, red palm oil, etc.) [23–25]. Other carotenoids commonly found in foods, such as lycopene, lutein, and zeaxanthin, do not meet these structural requirements and are not precursors of vitamin A.

Dietary sources of vitamin A differ among countries, due to not only economic but also cultural factors. In Western countries, approximately 75% of vitamin A is ingested as preformed REs and comes mainly from dairy products, or fortified products such as breakfast cereals, butter or margarine, and infant formula. On the other hand, in developing countries, 70–90% of vitamin A is consumed as provitamin A carotenoids. Therefore, in populations in low-income countries where vitamin A is mainly consumed in the form of carotenoids, the risk of VAD increases [16,26]. Moreover, in low-income countries with a rice-based diet, VAD is aggravated, since rice lacks this vitamin. Consequently, in some developing countries, vitamin A is added to certain products, such as sugar, thus improving the vitamin A status in the global population. Several successful efforts have been made to improve the provitamin A content in major crops such as wheat, rice, and potato, which are poor in β-carotene (e.g., "golden" rice or "golden" potato tubers that overexpress three bacterial genes for β-carotene synthesis) [27–30]. In spite of these modifications, the prevalence of subclinical deficiency is increasing [25]. However, it also has to be taken into account that VAD is not exclusive to developing countries, as it can also occur in Western societies depending on different factors such as a highly restrictive diet [16,31–36].

The bioavailability of vitamin A in food is defined as the ingested fraction available for use and storage, and depends on the capacity of the digestive process to release it from its original matrix in the food and on the fraction, which, once absorbed, is converted into retinol. Vitamin A bioavailability is higher in foods of animal origin (retinyl esters, preformed vitamin A) compared to vegetable origin (provitamin A carotenoids), although some of the latter group are considered good sources of vitamin A. Based on the efficiency of absorption and conversion into vitamin A, 1 μg of all-*trans*-retinol is equivalent to 12 μg of β-carotene and 24 μg of α-carotene or β-cryptoxanthin. However, the efficiency of β-carotene to be converted into retinoids is probably worse than previously thought, particularly in developing countries where vitamin A status and several other factors can reduce its conversion efficiency [31,36]. Accordingly, the use of a 21:1 ratio for mixed diets (12:1 for fruits and 26:1 for vegetables) has been suggested [25,36]. Thus, compared with preformed vitamin A, provitamin A carotenoids are relatively poor but important sources of vitamin A for the global population, as a higher amount is needed to meet vitamin A needs. On the other hand, it is important to consider for all groups of populations that several factors can influence the bioavailability of vitamin A from the diet, including the presence and severity of infections and parasites, intestinal or liver disease, iron and zinc status, xenobiotics, levels of dietary fat, protein malnutrition, alcohol intake, dietary source (preformed or provitamin A carotenoids), and food processing [11,16,31,35,37–40].

The requirements for vitamin A are expressed in retinol activity equivalent (RAE): 1 μg of RAE is equal to 1 μg of all-*trans*-retinol. In terms of international units (IUs), 1 IU is equal to 0.3 μg of all-*trans*-retinol or 0.3 μg of RAE. The recommended dietary allowances (RDAs) for children, men, and women are 300–600, 900, and 700 mg of RAE/day, respectively. However, the demand for micronutrients increases to 750 mg of RAE/day during pregnancy and 1300 mg of RAE/day during lactation. The gestational period is nutritionally relevant for maternal, fetal, and infant health. During the breastfeeding period, lactating mothers are susceptible to vitamin deficiency as the neonate feeds on her stores through the milk. Vitamin A is one of the most critical micronutrients in this period, affecting lung function and maturation, and thus susceptibility to infection. Inadequate maternal intake of vitamin A translates to an inadequate supply to the fetus during pregnancy and to the neonate during lactation through breast milk [37,41–44]. It is noteworthy that this deficiency during embryogenesis and the early months of life cannot be compensated in the postnatal period. Furthermore, in humans, an insufficient supply of vitamin A to the fetus is associated with organ malformations, preterm birth, and low neonatal stores, which increase the risk of infectious diseases and could exert a negative effect on health later in life [36,45–47]. Pregnant and breastfeeding women are considered to be at risk due to their higher demand for micronutrients, and should be advised to consume enriched products with β-carotene or, even better, vitamin A plus β-carotene, in order to avoid nutritional deficits [16,19,25,31–34,48].

In contrast to what was established in the case of vitamin A, there is no dietary intake recommendation for carotenoids for any population group. The average intake of β-carotene in Western societies is 3.9 mg/day and the total intake of provitamin A carotenoids is approximately 5.2 mg/day, higher in woman than men. The maximum daily amount of β-carotene that the adult intestine can cleave is approximately 2.5 mg, and the minimum amount of fat required for optimal absorption of carotenoids is about 3–5 g per meal [31,34,41,45].

3. Intertissue Flux and Metabolic Transformations of Vitamin A

Vitamin A, mainly found in the diet as REs, is hydrolyzed in the lumen of the small intestine into retinol molecules. This free retinol enters the enterocytes by a carrier-mediated process and by passive diffusion. On the other hand, vitamin A precursors, carotenoids, cross the epithelial cells by membrane-bound transporters and are metabolized within the cells to retinal, which can then be converted to retinol. Absorption of vitamin A appears to be fairly high and more significant than that of carotenoids. However, due to their hydrophobic nature, the absorption of both preformed and vitamin A precursors depends on micellar solubilization and thus on the fat content of the diet [25].

After absorption, retinol is re-esterified to REs, incorporated into chylomicrons together with a small fraction of carotenoids, and transported for a short time in plasma. Finally, the action of lipoprotein lipase (LPL) in different tissues affects chylomicron remnants that contain REs. Those chylomicron remnants are directed mainly to the liver in a receptor-mediated process (low-density lipoprotein (LDL) receptor, or via LDL-related protein, LRP). REs are then hydrolyzed back to retinol in the liver, which is secreted into plasma associated with retinol binding protein (RBP4), oxidized to RA, or transferred to stellate cells for storage mainly as esters of palmitic acid. Thus, the liver serves as the main pool for vitamin A (Figure 1) [23,40,49–52]. When needed, the vitamin A reserve is mobilized and secreted as the all-*trans*-retinol–RBP (holo-RBP4) complex to circulating blood. This complex associates with transthyretin (1:1) in plasma, reducing the glomerular filtration of retinol in the kidneys.

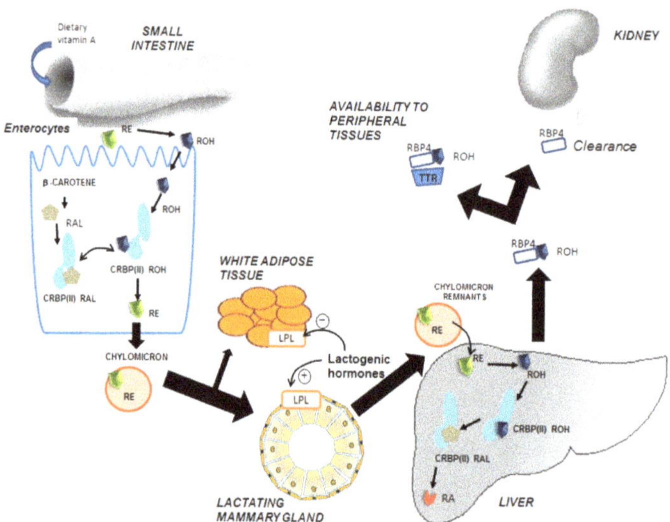

Figure 1. Intertissue flux of vitamin A during lactation. Dietary intake of vitamin A occurs mainly in

the form of precursor β-carotene from vegetables and fruits, or retinyl esters (REs) from sources of animal origin. Both molecules are absorbed by enterocytes in the small intestine, although REs must be hydrolyzed into retinol (ROH) within the intestinal lumen by means of hydrolases. Within enterocytes, ROH and β-carotene are metabolized into REs, which in turn are secreted into the lymph in chylomicrons. Circulating chylomicrons, before reaching the liver, are hydrolyzed in tissues with lipoprotein lipase (LPL) activity and REs released into these extrahepatic tissues. During lactation, increased levels of lactogenic hormones regulate LPL activity, inducing it in the mammary tissue, whereas it is reduced in white adipose tissue. The effect of these hormones redirects REs into the mammary epithelial cells during lactation. Finally, chylomicron remnants are taken up by hepatocytes, where REs are hydrolyzed into ROH, stored, or oxidized to retinoic acid (RA). When needed, ROH is released from hepatic cells into plasma bound to retinol-binding protein (RBP4). Once in plasma, it associates with the stabilizing protein transthyretin (TTR), which decreases the renal clearance of vitamin A [49–55].

Retinol is delivered from plasma to extrahepatic cells by free diffusion or a receptor-mediated process (stimulated by RA 6, STRA6) [56]. Once inside the cells, retinol is reversibly oxidized to retinal by enzymes of the retinol dehydrogenase complex and short-chain dehydrogenase/reductase (SDR/RDH) families. Retinal is irreversibly oxidized to RA by cytosolic aldehyde dehydrogenase 1 (ALDH1) isoenzymes. RA regulates gene transcription by binding to nuclear retinoid receptors (RARs and RXRs), which act as transcription factors and can be further oxidized for excretion, catalyzed by mono-oxygenases of the CYP family. There are cellular binding proteins for retinol and retinal (CRBP) and RA (CRABP), which enable their metabolism and action [25,40,49,56].

4. Role of Vitamin A during Development and Regression of the Mammary Gland

Vitamin A has several pleiotropic effects in cell differentiation, embryogenesis, apoptosis, etc. Retinoic acid and its precursor retinol are known to be involved in the maintenance, differentiation, and function of many epithelial tissues, among which the mammary gland is included.

4.1. An Overview of Mammary Gland Development

The mammary gland structure consists of a compound tubulo-alveolar gland embedded within irregular connective tissue. This complex organ undergoes a series of changes from conception to senescence, which are regulated by hormonal activity, mainly prolactin, estrogen, and progesterone. The peculiarity of this highly dynamic glandular tissue lies in the fact that it reaches full development only after birth. During the lifetime of the female, the mammary gland will suffer profound changes in structure and function to adapt to the physiological processes of pregnancy, lactation, and involution. Moreover, with each menstrual cycle, the mammary gland undergoes a round of expansion, proliferation, and regression under the influence of sexual hormones. Along these different stages, the subpopulations of cells that form mammary tissue will proliferate, differentiate, or even die, giving rise to significant remodeling of the glandular architecture [57].

At birth, a rudimentary embryonic ductal tree is present and will remain quiescent until puberty. At this point, under the control of sexual hormones, ductal epithelium develops and grows, invading the mammary fat pad in a process known as branching morphogenesis. Finally, the gland reaches a bilayered ductal tree structure, where luminal epithelial cells form the alveoli, with their apical side toward the lumen and basal side surrounded by myoepithelial contractile cells. Throughout adulthood, the epithelium proliferates and regresses during each estrous/menstrual cycle [58]. However, it is during pregnancy and lactation when the most dramatic changes occur: Proliferation of the alveolar epithelium concomitant with differentiation into secretory alveoli capable of producing milk. During lactation, oxytocin released by suckling stimuli will trigger the contraction of myoepithelial basal cells, and thus milk ejection toward the nipple. Then, after weaning, secretory epithelial cells are no longer needed and there is regression of the epithelial compartment, together with re-differentiation of adipocytes. This process, termed involution, includes programmed cell death and remodeling of the mammary gland and extracellular matrix (ECM), mainly driven by metalloproteases, returning the gland to the pre-pregnant state. Finally, during the postmenopausal period, the breast undergoes a second process of involution, characterized by atrophy of the parenchymal structures [59].

The mammary gland is a unique experimental model to study all of the physiological changes that occur along the different developmental stages: Proliferation, branching morphogenesis, differentiation, cell death, regression, and adipocyte differentiation. In this sense, several murine models have been extensively used to decipher the underlying mechanisms controlling all of these processes [60–70].

4.2. Retinoids and Mammary Gland Development

RA is an important regulator of cell differentiation and plays a major role in embryonic development and tissue remodeling [71]. The biological action of RA is mediated by three nuclear receptors, RARα, β, and γ, which are ubiquitously expressed in the breast. Moreover, several in vivo and in vitro studies have shown that RA is required for mammary gland morphogenesis. RA signaling is involved in initiating mammary gland formation, and precise levels of RA are crucial for normal mammary epithelium development during embryogenesis. Indeed, the expression of several genes involved in RA signaling, such as retinaldehyde dehydrogenase 2 (*Raldh2*) or RARβ, has been detected in mouse embryos at day 10.5. In cultured mouse flanks, the expression of genes involved in mammary gland development could be modulated by retinoic acid signaling. It is interesting to point out that during embryonic development, the mammary epithelium was inhibited by both increased and decreased levels of RA, reinforcing the importance of a narrow range for this molecule to exert its physiological function [60].

Nevertheless, the role of retinoids within the mammary gland is not restricted to the early embryonic stages when rudimentary mammary gland is formed. On the contrary, retinoid signaling regulates almost all developmental stages of the mammary tissue. Several studies have shown that retinoic acid signaling is required for development of the mammary gland after birth. The effects of retinoids are mediated through RARs and RXRs, which are differently expressed in distinct phases of mammary gland development [61]. Indeed, RARα controls mammary development before and during puberty, and loss of this nuclear receptor has a major effect on expansion of the neonatal ductal tree and branching morphogenesis in pubertal mice [61,62]. Similarly, in adult human breast tissue, RARα is also ubiquitously expressed in the nuclei of mammary epithelial cells, and the effects of RA seem to be controlled by the expression of *Raldh1*, which converts retinol to RA. In fact, in primary cultures from human mammary epithelial cells, when this gene was knocked down, mammosphere formation was significantly reduced. This effect could be rescued by treating the cells with RA, suggesting that the retinoic acid pathway is essential for mammosphere formation and thus for branching morphogenesis [63].

Analogous results were obtained in murine models of pregnancy and lactation. Abrogation of the RA signaling pathway in transgenic mouse models led to mammary gland hyperplasia [61,64] and excessive branching morphogenesis, and these mice displayed lactational deficiency. These effects of VAD deficiency could be reverted by RA supplementation, confirming that vitamin A is essential for mammary gland development at both nulliparous and pregnancy/lactation stages [61]. Further reinforcing these results, it was shown that RA induced lumen formation in vitro in a two-dimensional (2D) model where mammary epithelial cells were grown in collagen gels [65], and treatment of MCF-10A cells with RA induced re-differentiation of these cells [66]. The role of vitamin A in lumen morphogenesis is by mediation through the RARα signaling pathway, extending beyond the intracellular effects, since extracellular matrix remodeling is required for lumen formation. Moreover, RA deficiency in mice shows stromal hypercellularity and increased collagen, altering homeostasis in the ECM [64]. All of these findings therefore indicate that retinoids play an important role in the development and maintenance of glandular epithelial structures.

4.3. Retinoid Signaling During Mammary Gland Involution

At the end of lactation, following weaning of the offspring, the mammary gland undergoes massive cell death and tissue remodeling with regression of the epithelial structures in order to return to a virgin-like state. Involution is characterized by three major events: Programmed cell death of epithelial cells, degradation of the basement membrane and ECM remodeling, and, finally, adipocyte repopulation with a concomitant loss of the lobular–alveolar structure (Figure 2) [68–70]. Several signaling pathways are known to modulate mammary gland involution. Milk stasis within the gland activates programmed cell death of epithelial secretory cells that are no longer needed; this process is ruled by STAT3 and NF-κB transcription factors, which are rapidly activated upon cessation of suckling stimuli [68,72,73].

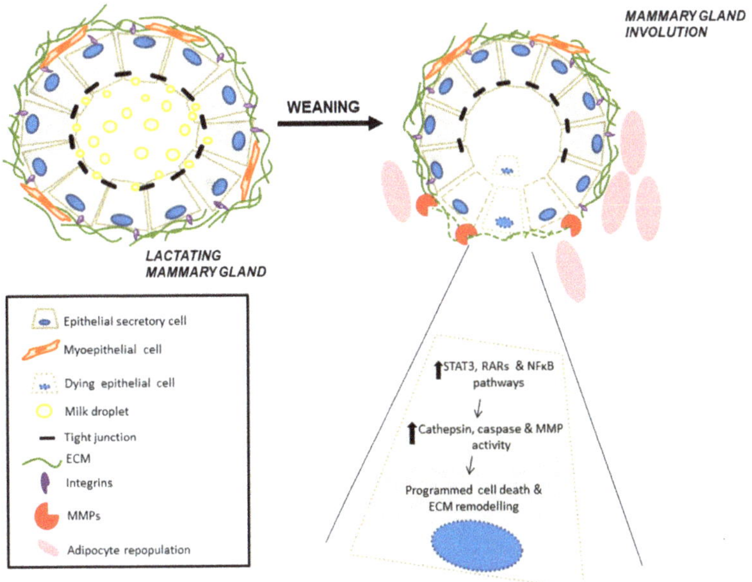

Figure 2. Schematic model of mammary gland involution. Lactation is characterized by an alveolar architecture lined by secretory mammary epithelial cells and contractile myoepithelial cells underneath. After weaning, milk stasis within the lumen and other local factors initiate involution of the mammary gland. At this point, cell shedding and cell death (cells with dotted lines) begins by 12 h after milk stasis, but lactation is still reversible until 48 h of weaning. Beyond this time point, epithelial tight junctions are disrupted, followed by extracellular matrix (ECM) remodeling and adipocyte re-differentiation [68,70,72,73]. MMP, metalloprotease; RAR, retinoic acid receptor; STAT3, signal transducer and activator of transcription 3; NF-kB, nuclear factor kappa B.

One could hypothesize that since milk production has ceased during this stage and blood flow to the gland also decreases, vitamin A recruitment to the gland would also be reduced, thus inhibiting the whole signaling pathway. Surprisingly, vitamin A plays an important role during weaning, not as a micronutrient to be secreted within milk, but as a signaling molecule instead. The RA signaling pathway is fully activated during mammary gland involution. Cytoplasmic transporter CRABP II and CRBP-I carrier proteins, together with several isoforms of RAR and RXRα, are significantly increased during involution. Besides, as previously described, RARα exerts its effect by binding to gene promoters such as the metalloprotease 9 (MMP-9) promoter, which induces MMP-9 mRNA and protein levels and, finally, proteinase activity [67]. The role of retinoids in matrix remodeling is reinforced by the fact that administration of an acute dose of retinol palmitate to control lactating rats also induced MMP-9 expression. This emphasizes the importance of retinoids in vivo to regulate mammary gland involution [67,68].

5. Vitamin A and Carotenoid Uptake by the Mammary Gland during Lactation

Vitamin A is transferred to the offspring by limited placental transfer during gestation, and mostly through maternal milk during the neonatal period. Because of the limited transplacental transfer, mammal newborns have low stores of vitamin A, depending significantly on the lactation period to acquire enough reserves and maintain adequate growth and development [37,44,45,74–79].

Lactation is characterized by widespread changes in the metabolism of different tissues to ensure a sufficient supply of nutrients to the mammary gland for milk production [80]. Among these changes, LPL in adipose tissue is inhibited; meanwhile, this enzyme is activated in the mammary tissue, therefore triglycerides are redirected to the latter [53–55]. Regarding amino acid and glutathione metabolism, there is also redistribution from the liver to the mammary gland, together with a nitrogen-sparing mechanism [81–85]. In addition, several physiological changes are required to ensure proper milk production, such as hyperphagia, liver and mammary gland hypertrophy, increased cardiac output, and increased blood flow to the gland. In this sense, substrate uptake by the lactating mammary gland is determined by several factors: (1) Increased availability of the substrates, which is met by hyperphagia and increased blood flow and substantial redistribution from different tissues to the mammary gland [81,82]; (2) transport mechanisms with increased expression of several carriers [85]; and (3) increased metabolism of nutrients within the gland [83,84]. All of these factors are under tight regulation by local and general effectors and contribute to milk production (Figure 3).

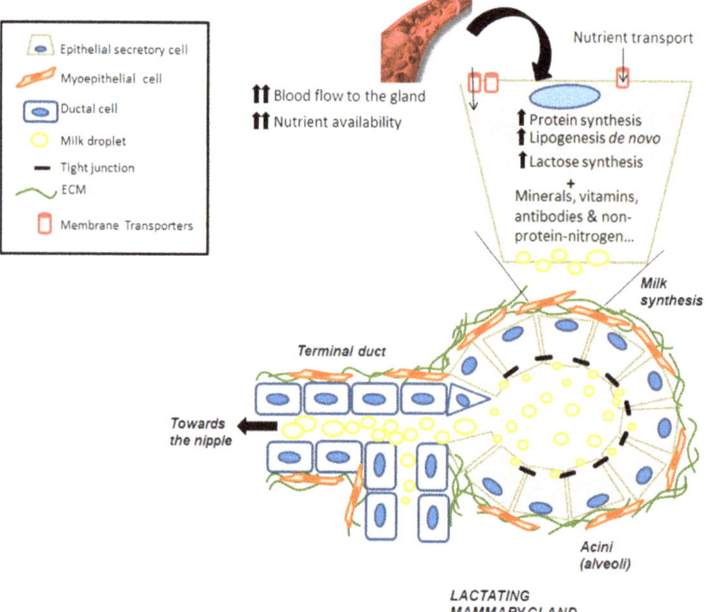

Figure 3. Metabolic events that lead to milk synthesis in alveolar cells from lactating mammary gland. During lactation, increased blood flow and intertissue redistribution of nutrients toward the mammary gland, together with increased membrane transporters, ensure bioavailability of precursors for milk synthesis within the epithelial cells. Both lipid and carbohydrate anabolism are increased, as is protein synthesis. Within the apex of the alveoli, milk-lipid droplets are formed and released toward the lumen with other milk components such as vitamins, minerals, and salts [81–85].

Vitamin A sources for the mammary gland comprise REs and carotenoids packed in the chylomicrons synthesized from dietary intake, and serum retinol (RBP4–retinol), which does not vary over a wide range of vitamin A ingestion and will be esterified within the mammary gland. During lactation, hyperphagia allows for higher vitamin A and carotenoid intake, which is associated with increased redistribution to the mammary tissue and thus to the secreted milk. REs associated with chylomicrons from the diet travel through the lymph and plasma, reaching extrahepatic tissues before getting to the liver. During breastfeeding, a great proportion of this vitamin A from the diet is redirected, preferably to lactating mammary tissue, as the activity of LPL in the mammary gland is a determinant for this highest uptake by the gland [53–55]. Indeed, the higher activity of LPL in thr mammary gland and the depressed activity in other tissues, such as adipose tissue, triggers the transfer of fatty acids, partial triglycerides, some unesterified cholesterol, and fat-soluble vitamins to the parenchymal cells of the gland during lactation (Figure 1). Once inside the mammary secretory epithelial cells, retinol is found to be bound to CRBPs, which will facilitate metabolism or excretion of this compound. In fact, it has been reported that CRBP-I and CRBP-III are both expressed in the mammary gland at different developmental stages. CRBP-III-null mice showed reduced RE levels in milk, suggesting that this isoform plays a role in retinol metabolism within lactating mammary tissue [86]. Human-milk vitamin A is present almost exclusively as REs, mainly retinyl palmitate, in the lipid fraction of the milk.

Hormonal changes are responsible for this nutrient redistribution and higher uptake within the mammary tissue. Preceding parturition, a number of hormones (insulin, steroids, and prolactin) play a role in preparing the gland for lactation [76]. Several studies using animal models have elucidated the role of these hormones in the intertissue flux of nutrients, together with the metabolic adaptations that occur during lactation [53–55,80,83–85]. During breastfeeding, the regulation of metabolism in lactating rats, and in particular milk production, appears to depend mainly on two hormones, prolactin and insulin. The insulin/glucagon ratio in lactation is decreased, but the number of insulin receptors, especially those with high affinity for the hormone, are increased in lactating mammary gland, contrary to what happens in adipocytes and hepatocytes, where the number of insulin receptors does not change. In this context, it is important to highlight that mammary tissue lacks glucagon receptors, and consequently, glucagon cannot affect mammary gland metabolism. On the other hand, prolactin levels in serum, which are already increased at the end of gestation, enhance LPL activity within the mammary gland and negatively regulate LPL activity within adipocytes, allowing for increased metabolism of RE-chylomicrons in mammary tissue. Overall, these studies highlight that high plasma prolactin levels are mainly responsible for the cooperative changes in the metabolism of different tissues during lactation that preferably redirect the circulating substrates to the mammary gland for milk synthesis; insulin is also likely to play a role in this intertissue flux of nutrients [25,53–55,76,80,83,85].

6. Vitamin A Concentration in Milk

Maternal milk is considered to be the optimal source of nutrition for infants and contains all essential vitamins. Accordingly, international organizations (e.g., WHO, UNICEF, American Academy of Pediatrics (AAP), European Society for Paediatric Gastroenterology Hepatology and Nutrition (ESPGHAN)) recommend exclusive breastfeeding during the first 6 months of life [17,41,44]. It is species-specific and a complete food adapted to the requirements for the survival and healthy development of offspring, by providing macronutrients (fats, carbohydrates, proteins, and free amino acids), micronutrients (vitamins and minerals), protective factors, and other important components for growth, such as cytokines, oligosaccharides, growth factors, and hormones [17,42,75–77]. A recent meta-analysis concluded that breast milk is not only a perfect nutritional supply for the infant, but is also probably the most specific personalized medication, at a time in life when genetic expression is being achieved. Breastfeeding provides protection against childhood infections and malocclusion, and is probably related to reductions in instances of overweight and diabetes, and globally could prevent 823,000 deaths of children younger than 5 years annually [87]. However, even in low- and middle-income countries, where breastfeeding duration is longer than in high-income countries, only 37% of children under 6 months of age are exclusively breastfed [87].

The milk produced in the first days after birth (colostrum) is higher in proteins, vitamins A, B12, and K, and immunoglobulins and lower in fat content than mature milk [42,76,88]. Rapid changes occur in breast milk composition during the first week postpartum, and it can be considered mature approximately 15 days after birth [76]. The volume and composition of breast milk are not homogeneous; the milk differs over time and between species, and it also depends on the demands made by the infant and the nutritional status of the mother in the same species. Maternal intake influences the macro- and micronutrient content of breast milk, as well as its immunological properties [44,47,89]. Indeed, the amount of retinol concentration in milk seems to be independent of serum retinol levels. On the contrary, several reports have revealed a strong correlation between vitamin A and carotenoid content in the diet during pregnancy and lactation and the amount of these micronutrients secreted in breast milk [17,43,90,91]. Due to the positive relationship between dietary intake of carotenoids and breastmilk concentration, breastfeeding mothers should have a diet abundant in this micronutrient. Therefore, the recommended dietary allowance (RDA) is higher for pregnant and lactating women than for non-pregnant women [17,19,31,41–43,48,75,90,92].

The concentration of vitamin A in human milk decreases over the course of lactation; it is maximal in the colostrum and reaches a plateau in mature milk. In healthy mothers, the vitamin A concentration varies from 5 to 7 µM in colostrum, 3 to 5 µM in transitional milk, and 1.4 to 2.6 µM in mature milk. The higher concentrations of retinol in colostrum allow tissue stores in the newborn to be rapidly replenished after limited placental transfer during gestation [25,76,88,90,92]. The newborn has a small amount of retinol pools in the liver (range 0–0.34 µmol/g in normal birth weight babies), achieved through the placental route, with maternal milk as the most important source of vitamin A [37,44,45,74–79]. Preterm birth aggravates the situation, since vitamin A liver stores are even lower, as the placental route has been cut off. These babies show lower plasma levels of retinol and retinol binding protein (RBP) at birth compared with full-term newborns. To compensate this deficit in retinol liver pools, it was observed that transitional milk in prematurity had a significant increase in micronutrient levels [91]. Milk retinol concentration in colostrum is lower in mothers of preterm babies (~3 µM) than in mothers of full-term babies [25,91]; however, it increases in transitional milk to values similar to those observed in milk from mothers of full-term babies (3.5 µM). There are no differences in mature milk between preterm and full-term, with milk retinol concentration of ~2 µM in mothers of preterm babies [91]. In this context, it is noteworthy that retinol level in preterm milk seems to be independent of the degree of prematurity [91].

During the first 6 months of life, infants from well-nourished mothers who are exclusively breastfed will accumulate around 310 µmol of vitamin A in the liver, approximately 60-fold higher than vitamin A accumulated throughout gestation by the transplacental pathway [25,88,90]. In fact, it has been reported that partial breastfeeding continues to provide a significant proportion of the recommended intake in infants after 6 months: 42% from 6 to 12 months and 61% during the second year in cases where vitamin A levels are low in the diet after weaning [78].

7. Deficiency of Vitamin A and Carotenoids during Pregnancy and Lactation

Vitamin A is necessary to maintain epithelial tissues, vision, and immune function. Breastfeeding from vitamin A-depleted mothers predisposes infants to the consequences of deficiency. Furthermore, VAD cannot be compensated by postnatal supplementation. As previously mentioned, pregnant and breastfeeding women are considered to be at risk of VAD in developing countries, where vitamin A in the diet is provided mainly as precursor carotenoids with lower bioavailability and efficiency to be converted into retinol. In this sense, several recent studies have shown that low dietary intake correlates with lower concentrations of vitamin A in milk of nursing mothers in Brazil [93,94], African countries [95–99], and Southeast Asia [26]. Thus, vitamin A supplementation for lactating mothers in these areas is regarded as a practical public health strategy to support maternal breastfeeding of newborns, more so of premature babies [25,40,90,92,100,101].

Serum retinol concentration is used to assess vitamin A status, with values below a cutoff of 0.70 µmol/L representing VAD, and below 0.35 µmol/L representing severe VAD. For pregnancy or lactation, the cutoff value is higher, and serum retinol concentration below 1.05 µmol/L reflects low vitamin A status among pregnant and lactating women [16]. The WHO Global Database estimated that clinical VAD (with night blindness and Bitot's spots) and biochemical VAD (with serum retinol concentration <0.70 µmol/L) affected 7 and 219 million preschool-aged children, respectively [16]. Regarding estimations in pregnant women, 19.8 million had low vitamin A status (serum retinol or breast milk concentration < 1.05 µmol/L), of whom 7.2 million were deficient in vitamin A (<0.70 µmol/L) and 6.2 million experienced gestational night blindness. These estimates found that nearly two-thirds of the women with night blindness lived in South and Southeast Asia [16,77,102].

However, it is important to point out that even in Western countries, where vitamin A- and β-carotene-rich foods are normally available, there are groups at risk for low vitamin A levels [16,45, 77,103,104]. One-fifth of the population in developed countries does not get the full recommended intake, with plasma and liver concentrations of vitamin A lower than those accepted as normal [104]. This situation can be aggravated by the increasingly common tendency to reduce fat intake in the diet and to engage in uncontrolled weight loss diets. Therefore, in these countries, it has been considered as subclinical vitamin A deficiency, which has increased markedly worldwide in the last decades [16,32,33,40]. Although calorie and protein content are largely unaffected by a woman's diet, there are some micronutrients in breast milk that will only be at adequate levels if there is sufficient dietary intake. Nutritional supplements should be given to these mothers to fulfill the requirements of micronutrients such as vitamin A in newborns [19,48,103]. It is well established that maternal VAD in pregnancy leads to placental dysfunction, fetal loss, congenital malformations, and preterm birth. Vitamin A-deficient mothers will produce breast milk that is low in the vitamin, consequently triggering VAD in the offspring. During lactation, if breast milk does not provide the neonate with appropriate vitamin A levels, the immune system might be affected, raising the risk for infectious diseases. Indeed, respiratory tract infections and complications during viral infections (e.g., measles) are increased in children with VAD [45]. Also, lung development could be compromised, increasing the risk for bronchopulmonary disease [45]. Indeed, recommended vitamin A during pregnancy, and especially during lactation, increases considerably, an almost 2-fold increase compared to non-pregnancy in terms of RDA [18,31,37,45,74,77,92,102]. A low neonatal store not only increases the risk of infectious diseases in the perinatal period, but also could have a negative effect on health later on in life [45–47]. The most specific clinical effect and one of the first manifestations of VAD is xerophthalmia with nyctalopia. If untreated, this can progress to night blindness; depressed immunity; squamous metaplasia of mucous epithelium in several organs; hyperkeratosis; disturbances in cell differentiation, organ development, growth, and reproduction; increased risk of infection; and mortality [16,32,33,38,43,46,74,102]. Similarly, VAD during pregnancy in experimental animals led to the formation of hypoplastic organs or alterations in fetal organs, such as eyes, brain, kidneys, and lungs, and even to the death and resorption of fetuses, depending on the severity of the deficiency [105–107].

8. Conclusions

Vitamin A plays a role in mammary gland metabolism during lactation. On the one hand, RA is essential for mammary gland development and, later on, in secretory epithelia to achieve adequate milk production. On the other hand, retinoids, through the RARα-dependent signaling pathway, have also been shown to regulate, at least in part, the weaning process, where epithelial cell death is coupled with tissue remodeling. Nevertheless, RA should not be considered exclusively as a signaling molecule controlling the development of mammary tissue, but also as a micronutrient essential for fetal and neonatal development. In this sense, the fetal liver stores only a small amount of vitamin A during gestation; therefore, the neonate depends on the external supply from the mother's milk.

An appropriate dietary intake of retinol and β-carotene during pregnancy and lactation together with the intertissue flux of different nutrients in the gland during lactation influence vitamin A concentration in breast milk. This is important in the long term, not only for maternal health, but also for the offspring, because vitamin A is involved in postnatal physiological functions.

Author Contributions: J.R.V. and T.B. were responsible for the general structure of the paper. J.R.V., T.B., M.T.C., and R.Z. performed the literature search, compiled and analyzed the data, drafted the manuscript, and designed the drawings. The final version was discussed by all authors, who gave their approval before submission. All authors have read and agreed to the published version of the manuscript.

Funding: This review and the APC were funded by Generalitat Valenciana, grant PROMETEO-2018-167 to J.R.V. and T.B.

Conflicts of Interest: The authors have no relevant interests to disclose. The funders had no role in the design of the study; in the collection, analyses, or interpretation of data; in the writing of the manuscript; or in the decision to publish the results.

References

1. McLaren, D.S.; Kraemer, K. Vitamin A in nature. *World Rev. Nutr. Diet* **2012**, *103*, 7–17. [CrossRef] [PubMed]
2. De Luca, L.M. Retinoids and their receptors in differentiation, embryogenesis and neoplasia. *FASEB J.* **1991**, *5*, 2924–2933. [CrossRef] [PubMed]
3. Livrea, M.A.; Tesoriere, L. Antioxidant activity of vitamin A within lipid environments. In *Fat-Soluble Vitamins*; Subcellular Biochemistry; Springer: Boston, MA, USA, 1998; Volume 30, pp. 113–143. [CrossRef]
4. Barber, T.; Borrás, E.; Torres, L.; García, C.; Cabezuelo, F.; Lloret, A.; Pallardó, F.V.; Viña, J.R. Vitamin A deficiency causes oxidative damage to liver mitochondria in rats. *Free Radic. Biol. Med.* **2000**, *29*, 1–7. [CrossRef]
5. Clagett-Dame, M.; Knutson, D. Vitamin A in reproduction and development. *Nutrients* **2011**, *3*, 385–428. [CrossRef] [PubMed]
6. Gimeno, A.; Zaragozá, R.; Vivó-Sesé, I.; Viña, J.R.; Miralles, V.J. Retinol, at concentration greater than the physiological limit, induces oxidative stress and apoptosis in human dermal fibroblasts. *Exp. Dermatol.* **2004**, *13*, 45–54. [CrossRef]
7. Rhinn, M.; Dollé, P. Retinoic acid signaling during development. *Development* **2012**, *139*, 843–858. [CrossRef]
8. Ross, A.C. Vitamin A and retinoic acid in T cell-related immunity. *Am. J. Clin. Nutr.* **2012**, *96*, 1166S–1172S. [CrossRef]
9. Sommer, A.; Vyas, K.S. A global clinical view on vitamin A and carotenoids. *Am. J. Clin. Nutr.* **2012**, *96*, 1204S–1206S. [CrossRef]
10. Esteban-Pretel, G.; Marín, M.P.; Renau-Piqueras, J.; Barber, T.; Timoneda, J. Vitamin A deficiency alters rat lung alveolar basement membrane: Reversibility by retinoic acid. *J. Nutr. Biochem.* **2010**, *21*, 227–236. [CrossRef]
11. Zielinska, M.A.; Wesołowska, A.; Pawlus, B.; Hamułka, J. Health Effects of Carotenoids during Pregnancy and Lactation. *Nutrients* **2017**, *9*, 838. [CrossRef]
12. Chambon, P. A decade of molecular biology of retinoic acid receptors. *FASEB J.* **1996**, *10*, 940–954. [CrossRef] [PubMed]
13. Ziouzenkova, O.; Orasanu, G.; Sharlach, M.; Akiyama, T.E.; Berger, J.P.; Viereck, J.; Hamilton, J.A.; Tang, G.; Dolnokowski, G.G.; Vogel, S.; et al. Retinaldehyde represses adipogenesis and diet-induced obesity. *Nat. Med.* **2007**, *13*, 695–702. [CrossRef] [PubMed]
14. Berry, D.C.; Noy, N. Signaling by Vitamin A and Retinol-Binding Protein in Regulation of Insulin Responses and Lipid Homeostasis. *Biochim. Biophys. Acta* **2012**, *1821*, 168–176. [CrossRef] [PubMed]
15. Al Tanoury, Z.; Piskunov, A.; Rochette-Egly, C. Vitamin A and Retinoid Signaling: Genomic and Nongenomic Effects. *J. Lipid Res.* **2013**, *54*, 1761–1775. [CrossRef] [PubMed]
16. WHO, World Health Organization. Global prevalence of vitamin A deficiency in populations at risk 1995–2005. In *WHO Global Database on Vitamin A Deficiency*; World Health Organization: Geneva, Switzerland, 2009; ISBN 978-92-4-159801-9.
17. Emmett, P.M.; Rogers, I.S. Properties of human milk and their relationship with maternal nutrition. *Early Hum. Dev.* **1997**, *9*, S7–S28. [CrossRef]

18. Strobel, M.; Tinz, J.; Biesalski, H.K. The importance of β-carotene as a source of vitamin A with special regard to pregnant and breastfeeding women. *Eur. J. Nutr.* **2007**, *46*, 1–20. [CrossRef]

19. Cruz, S.; Pereira da Cruz, S.; Ramalho, A. Impact of Vitamin A Supplementation on Pregnant Women and on Women Who Have Just Given Birth: A Systematic Review. *J. Am. Coll. Nutr.* **2018**, *37*, 243–250. [CrossRef]

20. Borel, P.; Desmarchelier, C. Genetic Variations Associated with Vitamin A Status and Vitamin A Bioavailability. *Nutrients* **2017**, *9*, 246. [CrossRef]

21. Das, B.C.; Thapa, P.; Karki, R.; Das, S.; Mahapatra, S.; Liu, T.C.; Torregroza, I.; Wallace, D.P.; Kambhampati, S.; Van Veldhuizen, P.; et al. Retinoic Acid Signaling Pathways in Development and Diseases. *Bioorg. Med. Chem.* **2014**, *22*, 673–683. [CrossRef]

22. Harrison, E.H.; Curley, R.W., Jr. Carotenoids and Retinoids: Nomenclature, Chemistry, and Analysis. In *The Biochemistry of Retinoid Signaling II*; Subcellular Biochemistry; Springer: Dordrecht, The Netherlands, 2016; Volume 81, pp. 1–19. [CrossRef]

23. Eroglu, A.; Harrison, E.H. Carotenoid metabolism in mammals, including man: Formation, occurrence, and function of apocarotenoids. *J. Lipid Res.* **2013**, *54*, 1719–1730. [CrossRef]

24. Kim, Y.K.; Wassef, L.; Chung, S.; Jiang, H.; Wyss, A.; Blaner, W.S.; Quadro, L. β-Carotene and its Cleavage Enzyme β-Carotene-15,15'-oxygenase (CMOI) Affect Retinoid Metabolism in Developing Tissues. *FASEB J.* **2011**, *25*, 1641–1652. [CrossRef] [PubMed]

25. Combs, G.F., Jr.; McClung, J.P. *The Vitamins: Fundamental Aspects in Nutrition and Health*; Academic Press, Elsevier: New York, NY, USA, 2017. [CrossRef]

26. Wieringa, F.T.; Dijkhuizen, M.A.; Berger, J. Micronutrient deficiencies and their public health implications for South-East Asia. *Curr. Opin. Clin. Nutr. Metab. Care* **2019**, *22*, 479–482. [CrossRef] [PubMed]

27. Tang, G.; Qin, J.; Dolnikowski, G.G.; Russell, R.M.; Grusak, M.A. Golden rice is an effective source of vitamin A. *Am. J. Clin. Nutr.* **2009**, *89*, 1776–1783. [CrossRef] [PubMed]

28. Diretto, G.; Al-Babili, S.; Tavazza, R.; Scossa, F.; Papacchioli, V.; Migliore, M.; Beyer, P.; Giuliano, G. Transcriptional-metabolic networks in beta-carotene-enriched potato tubers: The long and winding road to the golden phenotype. *Plant Physiol.* **2010**, *154*, 899–912. [CrossRef]

29. Moghissi, A.A.; Pei, S.; Liu, Y. Golden Rice: Scientific, Regulatory and Public Information Processes of a Genetically Modified Organism. *Crit. Rev. Biotechnol.* **2016**, *36*, 535–541. [CrossRef]

30. Govender, L.; Pillay, K.; Siwela, M.; Modi, A.T.; Mabhaudhi, T. Improving the Dietary Vitamin A Content of Rural Communities in South Africa by Replacing Non-Biofortified white Maize and Sweet Potato with Biofortified Maize and Sweet Potato in Traditional Dishes. *Nutrients* **2019**, *11*, 1198. [CrossRef]

31. NAS, National Academy of Sciences. Dietary References Intakes: The Essential Guide to Nutrient Requirements. In *Food and Nutrition Board*; National Academic Press: Washington, DC, USA, 2006; Available online: http://www.nap.edu/catalog.php?record_id511537#toc (accessed on 25 June 2019).

32. Underwood, B.A. Vitamin A Deficiency Disorders: International Efforts to Control a Preventable "Pox". *J. Nutr.* **2004**, *134*, 231S–236S. [CrossRef]

33. Sommer, A. Vitamin A deficiency and clinical disease: An historical overview. *J. Nutr.* **2008**, *138*, 1835–1839. [CrossRef]

34. Weber, D.; Grune, T. The contribution of β-carotene to vitamin A supply of humans. *Mol. Nutr. Food Res.* **2012**, *56*, 251–258. [CrossRef]

35. Priyadarshani, A.M. A Review on Factors Influencing Bioaccessibility and Bioefficacy of Carotenoids. *Crit. Rev. Food Sci. Nutr.* **2015**, *13*. [CrossRef]

36. Barber, T.; Esteban-Pretel, G.; Marín, M.P.; Timoneda, J. Vitamin A: An Overview. In *Vitamin A and Carotenoids: Chemistry, Analysis, Function and Effects*; Preedy, V.R., Ed.; The Royal Society of Chemistry: London, UK, 2012; pp. 396–416. ISBN 978-1-84973-368-7. [CrossRef]

37. McCauley, M.E.; van den Broek, N.; Dou, L.; Othman, M. Vitamin A supplementation during pregnancy for maternal and newborn outcomes. *Cochrane Database Syst. Rev.* **2015**, *10*. [CrossRef] [PubMed]

38. McLaren, D.S.; Kraemer, K. *Manual on Vitamin Deficiency Disorders (VADD)*, 3rd ed.; Sight and Life Press: Basel, Switzerland, 2012; ISBN 978-3-906412-58-0.

39. Lieber, C.S. Alcohol: Its metabolism and interaction with nutrients. *Annu. Rev. Nutr.* **2000**, *20*, 395–430. [CrossRef] [PubMed]

40. Timoneda, J.; Rodríguez-Fernández, L.; Zaragozá, R.; Marín, M.P.; Cabezuelo, M.T.; Torres, L.; Viña, J.R.; Barber, T. Vitamin A Deficiency and the Lung. *Nutrients* **2018**, *10*, 1132. [CrossRef] [PubMed]

41. Zielinska, M.A.; Hamulka, J.; Wesolowska, A. Month of Lactation and Its Associations with Maternal Dietary Intake and Anthropometric Characteristics. *Nutrients* **2019**, *11*, 193. [CrossRef]

42. Allen, L.H. Multiple micronutrients in pregnancy and lactation: An overview. *Am. J. Clin. Nutr.* **2005**, *81*, 1206S–12125. [CrossRef]

43. Bravi, F.; Wiens, F.; Decarli, A.; Dal Pont, A.; Agostoni, C.; Ferraroni, M. Impact of maternal nutrition on breast-milk composition: A systematic review. *Am. J. Clin. Nutr.* **2016**, *104*, 646–662. [CrossRef]

44. Erick, M. Breast milk is conditionally perfect. *Med. Hypotheses* **2018**, *111*, 82–89. [CrossRef]

45. Biesalski, H.K.; Nohr, D. The importance of vitamin A during pregnancy and childhood: Impact on lung function. In *Vitamin A and Carotenoids: Chemistry, Analysis, Function and Effects*; Preedy, V.R., Ed.; The Royal Society of Chemistry: London, UK, 2012; pp. 532–554. ISBN 978-1-84973-368-7. [CrossRef]

46. Biesalski, H.K.; Black, R.E. Hidden Hunger. Malnutrition and the First 1000 Days of Life: Causes, Consequences and Solutions. *World Rev. Nutr. Diet* **2016**, *115*, 134–141. [CrossRef]

47. Davisse-Paturet, C.; Adel-Patient, K.; Divaret-Chauveau, A.; Pierson, J.; Lioret, S.; Cheminat, M.; Dufourg, M.-N.; Charles, M.-A.; de Lauzon-Guillain, B. Breastfeeding Status and Duration and Infections, Hospitalizations for Infections, and Antibiotic Use in the First Two Years of Life in the ELFE Cohort. *Nutrients* **2019**, *11*, 1607. [CrossRef]

48. Soares, M.M.; Silva, M.A.; Garcia, P.P.C.; Silva, L.S.D.; Costa, G.D.D.; Araújo, R.M.A.; Cotta, R.M.M. Effect of vitamin A suplementation: A systematic review. *Cien. Saude Colet.* **2019**, *24*, 827–838. [CrossRef]

49. Blomhoff, R.; Blomhoff, H.K. Overview of retinoid metabolism and function. *J. Neurobiol.* **2006**, *66*, 606–630. [CrossRef] [PubMed]

50. D'Ambrosio, D.N.; Clugston, R.D.; Blaner, W.S. Vitamin A Metabolism: An Update. *Nutrients* **2011**, *3*, 63–103. [CrossRef] [PubMed]

51. Blaner, W.S.; Li, Y.; Brun, P.J.; Yueng, J.J.; Lee, S.A.; Clugston, R.D. Vitamin A Absorption, Storage and Mobilization. In *The Biochemistry of Retinoid Signaling II*; Subcellular Biochemistry; Springer: Dordrecht, The Netherlands, 2016; Volume 81, pp. 95–126. [CrossRef]

52. Saeed, A.; Dullaart, R.P.F.; Schreuder, T.C.M.A.; Blokzijl, H.; Faber, K.N. Disturbed Vitamin A Metabolism in Non-Alcoholic Fatty Liver Disease (NAFLD). *Nutrients* **2018**, *10*, 29. [CrossRef] [PubMed]

53. Hang, J.; Rillema, J.A. Prolactin's effects on lipoprotein lipase (LPL) activity and on LPL mRNA levels in cultured mouse mammary gland explants. *Proc. Soc. Exp. Biol. Med.* **1997**, *214*, 161–166. [CrossRef]

54. Zhao, W.S.; Hu, S.L.; Yu, K.; Wang, H.; Wang, W.; Loor, J.; Luo, J. Lipoprotein Lipase, Tissue Expression and Effects on Genes Related to Fatty Acid Synthesis in Goat Mammary Epithelial Cells. *Int. J. Mol. Sci.* **2014**, *15*, 22757–22771. [CrossRef]

55. Jensen, D.R.; Gavigan, S.; Sawicki, V.; Witsell, D.L.; Eckel, R.H.; Neville, M.C. Regulation of Lipoprotein lipase activity and mRNA in the mammary gland of the lactating mouse. *Biochem. J.* **1994**, *298*, 321–327. [CrossRef]

56. Noy, N. Vitamin A transport and cell signaling by the retinol-binding protein receptor STRA6. In *The Biochemistry of Retinoid Signaling II*; Subcellular Biochemistry; Springer: Dordrecht, The Netherlands, 2016; Volume 81, pp. 77–94. [CrossRef]

57. Inman, J.L.; Robertson, C.; Mott, J.D.; Bissell, M.J. Mammary gland development: Cell fate specification, stem cells and the microenvironment. *Development* **2015**, *142*, 1028–1042. [CrossRef]

58. Fata, J.E.; Chaudhary, V.; Khokha, R. Cellular turnover in the mammary gland is correlated with systemic levels of progesterone and not 17beta-estradiol during the estrous cycle. *Biol. Reprod.* **2001**, *65*, 680–688. [CrossRef]

59. Stingl, J.; Raouf, A.; Emerman, J.T.; Eaves, C.J. Epithelial progenitors in the normal human mammary gland. *J. Mammary Gland Biol. Neoplasia* **2005**, *10*, 49–59. [CrossRef]

60. Cho, K.W.; Kwon, H.J.; Shin, J.O.; Lee, J.M.; Cho, S.W.; Tickle, C.; Jung, H.S. Retinoic acid signaling and the initiation of mammary gland development. *Dev. Biol.* **2012**, *365*, 259–266. [CrossRef]

61. Wang, Y.A.; Shen, K.; Wang, Y.; Brooks, S.C. Retinoic acid signaling is required for proper morphogenesis of mammary gland. *Dev. Dyn.* **2005**, *234*, 892–899. [CrossRef] [PubMed]

62. Cohn, E.; Ossowski, L.; Bertran, S.; Marzan, C.; Farias, E.F. RARα1 control of mammary gland ductal morphogenesis and wnt1-tumorigenesis. *Breast Cancer Res.* **2010**, *12*, R79. [CrossRef] [PubMed]

63. Honeth, G.; Lombardi, S.; Ginestier, C.; Hur, M.; Marlow, R.; Buchupalli, B.; Shinomiya, I.; Gazinska, P.; Bombelli, S.; Ramalingam, V.; et al. Aldehyde dehydrogenase and estrogen receptor define a hierarchy of cellular differentiation in the normal human mammary epithelium. *Breast Cancer Res.* **2014**, *16*, R52. [CrossRef] [PubMed]

64. Pierzchalski, K.; Yu, J.; Norman, V.; Kane, M.A. CrbpI regulates mammary retinoic acid homeostasis and the mammary microenvironment. *FASEB J.* **2013**, *27*, 1904–1916. [CrossRef]

65. Montesano, R.; Soulié, P. Retinoids induce lumen morphogenesis in mammary epithelial cells. *J. Cell Sci.* **2002**, *115*, 4419–4431. [CrossRef]

66. Arisi, M.F.; Starker, R.A.; Addya, S.; Huang, Y.; Fernandez, S.V. All trans-retinoic acid (ATRA) induces re-differentiation of early transformed breast epithelial cells. *Int. J. Oncol.* **2014**, *44*, 1831–1842. [CrossRef]

67. Zaragozá, R.; Gimeno, A.; Miralles, V.J.; García-Trevijano, E.R.; Carmena, R.; García, C.; Mata, M.; Puertes, I.R.; Torres, L.; Viña, J.R. Retinoids induce MMP-9 expression through RAR alpha during mammary gland remodeling. *Am. J. Physiol. Endocrinol. Metab.* **2007**, *292*, E1140–E1148. [CrossRef]

68. Zaragozá, R.; García-Trevijano, E.R.; Lluch, A.; Ribas, G.; Viña, J.R. Involvement of Different networks in mammary gland involution after the pregnancy/lactation cycle: Implications in breast cancer. *IUBMB Life* **2015**, *67*, 227–238. [CrossRef]

69. McNally, S.; Stein, T. Overview of Mammary Gland Development: A Comparison of Mouse and Human. *Methods Mol. Biol.* **2017**, *1501*, 1–17. [CrossRef]

70. Jena, M.K.; Jaswal, S.; Kumar, S.; Mohanty, A.K. Molecular mechanism of mammary gland involution: An update. *Dev. Biol.* **2019**, *445*, 145–155. [CrossRef]

71. Zile, M.H. Function of vitamin A in vertebrate embryonic development. *J. Nutr.* **2001**, *131*, 705–708. [CrossRef] [PubMed]

72. Torres, L.; Serna, E.; Bosch, A.; Zaragozá, R.; García, C.; Miralles, V.J.; Sandoval, J.; Viña, J.R.; García-Trevijano, E.R. NF-KB as a node for signal amplification during weaning. *Cell Physiol. Biochem.* **2011**, *28*, 833–846. [CrossRef] [PubMed]

73. Hughes, K.; Watson, C.J. The Multifaceted Role of STAT3 in Mammary Gland Involution and Breast Cancer. *Int. J. Mol. Sci.* **2018**, *19*, 1695. [CrossRef] [PubMed]

74. Underwood, B.A. The Role of Vitamin A in Child Growth, Development and Survival. In *Nutrient Regulation during Pregnancy, Lactation, and Infant Growth*; Advances in Experimental Medicine and Biology; Allen, L., King, J., Lönnerdal, B., Eds.; Springer: Boston, MA, USA, 1994; Volume 352, pp. 201–208. [CrossRef]

75. Report on Health and Social Subjects. 32. Present day practice in infant feeding: Third report. Report of a working party of the Panel on Child Nutrition Committee on Medical Aspects of Food Policy. *Rep. Health Soc. Subj.* **1988**, *32*, 1–66.

76. De Vries, J.Y.; Pundir, S.; Mckenzie, E.; Keijer, J.; Kussmann, M. Maternal Circulating Vitamin Status and Colostrum Vitamin Composition in Healthy Lactating Women—A Systematic Approach. *Nutrients* **2018**, *10*, 687. [CrossRef] [PubMed]

77. WHO, World Health Organization. *Essential Nutrition. Actions: Improving Maternal, Newborn, Infant and Young Child Health and Nutrition*; World Health Organization: Geneva, Switzerland, 2013. Available online: https://www.who.int/nutrition/publications/infantfeeding/essential_nutrition_actions/en/ (accessed on 20 May 2019).

78. Ross, J.S.; Harvey, P.W. Contribution of breastfeeding to vitamin A nutrition of infants: A simulation model. *Bull. World Health Organ.* **2003**, *81*, 80–86.

79. Da Silva, A.G.C.L.; de Sousa Rebouças, A.; Mendonça, B.M.A.; Silva, D.C.N.E.; Dimenstein, R.; Ribeiro, K.D.D.S. Relationship between the dietary intake, serum, and breast milk concentrations of vitamin A and vitamin E in a cohort of women over the course of lactation. *Matern. Child Nutr.* **2019**, *15*, e12772. [CrossRef]

80. Williamson, D.H. Regulation of metabolism during lactation in the rat. *Reprod. Nutr. Dev.* **1986**, *26*, 597–603. [CrossRef]

81. Barber, T.; García de la Asunción, J.; Puertes, I.R.; Viña, J.R. Amino acid metabolism and protein synthesis in lactating rats fed on a liquid diet. *Biochem. J.* **1990**, *270*, 77–82. [CrossRef]

82. Barber, T.; Triguero, A.; Martínez-López, I.; Torres, L.; García, C.; Miralles, V.J.; Viña, J.R. Elevated expression of liver gamma-cystathionase is required for the maintenance of lactation in rats. *J. Nutr.* **1999**, *129*, 928–933. [CrossRef]

83. Canul-Medina, G.; Fernandez-Mejia, C. Morphological, hormonal, and molecular changes in different maternal tissues during lactation and post-lactation. *J. Physiol. Sci.* **2019**. [CrossRef] [PubMed]

84. Zaragozá, R.; Garcia, C.; Rus, A.D.; Pallardo, F.V.; Barber, T.; Torres, L.; Miralles, V.J.; Viña, J.R. Inhibition of liver trans-sulphuration pathway by propargylglycine mimics gene expression changes found in the mammary gland of weaned lactating rats: Role of glutathione. *Biochem. J.* **2003**, *373*, 825–834. [CrossRef] [PubMed]

85. De la Asunción, J.G.; Devesa, A.; Viña, J.R.; Barber, T. Hepatic amino acid uptake is decreased in lactating rats. In vivo and in vitro studies. *J. Nutr.* **1994**, *124*, 2163–2171. [CrossRef] [PubMed]

86. Piantedosi, R.; Ghyselinck, N.; Blaner, W.S.; Vogel, S. Cellular retinol-binding protein type III is needed for retinoid incorporation into milk. *J. Biol. Chem.* **2005**, *280*, 24286–24292. [CrossRef] [PubMed]

87. Victora, C.G.; Bahl, R.; Barros, A.J.; França, G.V.; Horton, S.; Krasevec, J.; Murch, S.; Sankar, M.J.; Walker, N.; Rollins, N.C.; et al. Breastfeeding in the 21st century: Epidemiology, mechanisms, and lifelong effect. *Lancet* **2016**, *387*, 475–490. [CrossRef]

88. Dror, D.K.; Allen, L.H. Retinol-to-Fat Ratio and Retinol Concentration in Human Milk Show Similar Time Trends and Associations with Maternal Factors at the Population Level: A Systematic Review and Meta-Analysis. *Adv. Nutr.* **2018**, *9* (Suppl. 1), 332S–346S. [CrossRef]

89. Van de Pavert, S.A.; Ferreira, M.; Domingues, R.G.; Ribeiro, H.; Molenaar, R.; Moreira-Santos, L.; Almeida, F.F.; Ibiza, S.; Barbosa, I.; Goverse, G.; et al. Maternal Retinoids Control Type 3 Innate Lymphoid Cells and Set the Offspring Immunity. *Nature* **2014**, *508*, 123–127. [CrossRef]

90. Debier, C.; Larondelle, Y. Vitamins A and E: Metabolism, roles and transfer to offspring. *Br. J. Nutr.* **2005**, *93*, 153–174. [CrossRef]

91. Lima, M.S.; da Silva Ribeiro, K.D.; Pires, J.F.; Bezerra, D.F.; Bellot, P.E.; de Oliveira Weigert, L.P.; Dimenstein, R. Breast milk retinol concentration in mothers of preterm newborns. *Early Hum. Dev.* **2017**, *106–107*, 41–45. [CrossRef]

92. Bastos Maia, S.; Rolland Souza, A.S.; Costa Caminha, M.F.; Lins da Silva, S.; Callou Cruz, R.S.B.L.; Carvalho Dos Santos, C.; Batista Filho, M. Vitamin A and Pregnancy: A Narrative Review. *Nutrients* **2019**, *11*, 681. [CrossRef]

93. Machado, M.R.; Kamp, F.; Nunes, J.C.; El-Bacha, T.; Torres, A.G. Breast Milk Content of Vitamin A and E from Early- to Mid-Lactation is Affected by Inadequate Dietary Intake in Brazilian Adult Women. *Nutrients* **2019**, *11*, 2025. [CrossRef] [PubMed]

94. Bezerra, D.S.; Ribeiro, K.D.S.; Lima, M.S.R.; Pires Medeiros, J.F.; da Silva, A.G.C.L.; Dimenstein, R.; Osório, M.M. Retinol status and associated factors in mother-newborn pairs. *J. Hum. Nutr. Diet.* **2019**. [CrossRef] [PubMed]

95. Hanson, C.; Lyden, E.; Anderson-Berry, A.; Kocmich, N.; Rezac, A.; Delair, S.; Furtado, J.; Van Ormer, M.; Izevbigie, N.; Olateju, E.K.; et al. Status of Retinoids and Carotenoids and Associations with Clinical Outcomes in Maternal-Infant Pairs in Nigeria. *Nutrients* **2018**, *10*, 1286. [CrossRef] [PubMed]

96. Abebe, Z.; Haki, G.D.; Schweigert, F.J.; Henkel, I.M.; Baye, K. Low breastmilk vitamin A concentration is prevalent in rural Ethiopia. *Eur. J. Clin. Nutr.* **2019**, *73*, 1110–1116. [CrossRef] [PubMed]

97. Wessells, K.R.; Young, R.R.; Ferguson, E.L.; Ouédraogo, C.T.; Faye, M.T.; Hess, S.Y. Assessment of dietary intake and nutrient gaps, and development of food-based recommendations, among pregnant and lactating women in Zinder, Niger: An optifood linear programming analysis. *Nutrients* **2019**, *11*, 72. [CrossRef] [PubMed]

98. Kaliwile, C.; Michelo, C.; Titcomb, T.J.; Moursi, M.; Donahue Angel, M.; Reinberg, C.; Bwembya, P.; Alders, R.; Tanumihardjo, S.A. Dietary intake patterns among lactating and non-lactating women of reproductive age in rural Zambia. *Nutrients* **2019**, *11*, 288. [CrossRef] [PubMed]

99. Petry, N.; Jallow, B.; Sawo, Y.; Darboe, M.K.; Barrow, S.; Sarr, A.; Ceesay, P.O.; Fofana, M.N.; Prentice, A.M.; Wegmüller, R.; et al. Micronutrient Deficiencies, Nutritional Status and the Determinants of Anemia in Children 0–59 Months of Age and Non-Pregnant Women of Reproductive Age in The Gambia. *Nutrients* **2019**, *11*, 2275. [CrossRef]

100. Meyer, S.; Gortner, L.; NeoVitaA Trial investigators. Up-date on the NeoVitaA Trial: Obstacles, challenges, perspectives, and local experiences. *Wien. Med. Wochenschr.* **2017**, *167*, 264–270. [CrossRef]

101. Atalhi, N.; El Hamdouchi, A.; Barkat, A.; Elkari, K.; Hamrani, A.; El Mzibri, M.; Haskell, M.J.; Mokhtar, N.; Aguenaou, H. Combined consumption of a single high-dose vitamin A supplement with provision of vitamin A fortified oil to households maintains adequate milk retinol concentrations for 6 months in lactating Moroccan women. *Appl. Physiol. Nutr. Metab.* **2019**. [CrossRef]

102. WHO, World Health Organization. *Micronutrients Deficiencies*; World Health Organization: Geneva, Switzerland, 2018; Available online: http://www.who.int/nutrition/topics/vad/en/ (accessed on 5 May 2019).

103. Copp, K.; DeFranco, E.A.; Kleiman, J.; Rogers, L.K.; Morrow, A.L.; Valentine, C.J. Nutrition Support Team Guide to Maternal Diet for the Human-Milk-Fed Infant. *Nutr. Clin. Pract.* **2018**, *33*, 687–693. [CrossRef]

104. Hanson, C.; Schumacher, M.; Lyden, E.; Furtado, J.; Van Ormer, M.; McGinn, E.; Rilett, K.; Cave, C.; Johnson, R.; Weishaar, K.; et al. Status of vitamin A and related compounds and clinical outcomes in maternal-infant pairs in the midwestern United States. *Ann. Nutr. Metab.* **2017**, *71*, 175–182. [CrossRef] [PubMed]

105. Wolbach, S.B.; Howe, P.R. Tissue changes following deprivation of fat-soluble A vitamin. *J. Exp. Med.* **1925**, *42*, 753–777. [CrossRef] [PubMed]

106. Wilson, J.G.; Roth, C.B.; Warkany, J. An analysis of the syndrome of malformations induced by maternal vitamin A deficiency. Effects of restoration of vitamin A at various times during gestation. *Am. J. Anat.* **1953**, *92*, 189–217. [CrossRef] [PubMed]

107. Takahashi, Y.I.; Smith, J.E.; Winick, M.; Goodman, D.S. Vitamin A deficiency and fetal growth and development in the rat. *J. Nutr.* **1975**, *105*, 1299–1305. [CrossRef] [PubMed]

Review

What Evidence Do We Have for Pharmaceutical Galactagogues in the Treatment of Lactation Insufficiency?—A Narrative Review

Luke E. Grzeskowiak [1,2,*], **Mary E. Wlodek** [3] **and Donna T. Geddes** [4]

1   Adelaide Medical School, Robinson Research Institute, The University of Adelaide, Adelaide, SA 5005, Australia
2   SA Pharmacy, Flinders Medical Centre, SA Health, Bedford Park, Adelaide, SA 5042, Australia
3   Department of Physiology, The University of Melbourne, Melbourne, VIC 3010, Australia; m.wlodek@unimelb.edu.au
4   School of Molecular Sciences, The University of Western Australia, Crawley, Perth, WA 6009, Australia; donna.geddes@uwa.edu.au
*   Correspondence: luke.Grzeskowiak@adelaide.edu.au; Tel.: +61-8-8313-1687

Received: 29 March 2019; Accepted: 24 April 2019; Published: 28 April 2019

Abstract: Inadequate breast milk supply is a frequently reported reason for early discontinuation of breastfeeding and represents a critical opportunity for intervening to improve breastfeeding outcomes. For women who continue to experience insufficient milk supply despite the utilisation of non-pharmacological lactation support strategies, pharmacological intervention with medications used to augment lactation, commonly referred to as galactagogues, is common. Galactagogues exert their pharmacological effects through altering the complex hormonal milieu regulating lactation, particularly prolactin and oxytocin. This narrative review provides an appraisal of the existing evidence regarding the efficacy and safety of pharmaceutical treatments for lactation insufficiency to guide their use in clinical practice. The greatest body of evidence surrounds the use of domperidone, with studies demonstrating moderate short-term improvements in breast milk supply. Evidence regarding the efficacy and safety of metoclopramide is less robust, but given that it shares the same mechanism of action as domperidone it may represent a potential treatment alternative where domperidone is unsuitable. Data on remaining interventions such as oxytocin, prolactin and metformin is too limited to support their use in clinical practice. The review provides an overview of key evidence gaps and areas of future research, including the impacts of pharmaceutical galactagogues on breast milk composition and understanding factors contributing to individual treatment response to pharmaceutical galactagogues.

Keywords: galactagogues; low milk supply; lactation; breast milk; breast feeding

---

1. Mothers' Own Breast Milk is Best

Breast milk is considered the optimal source of enteral nutrition to support the growth and development of all infants [1]. Exclusive breastfeeding is advocated for the first six months, with the continuation of breastfeeding supported until the age of two years or longer [2]. Infants who are not breastfed or are breastfed for shorter durations are at an increased risk of early and later-life morbidities and mortality [1–3]. Specifically, the reduced provision of breast milk is associated with increased risks of gastrointestinal, urinary, respiratory and middle-ear infections, together with a range of non-communicable diseases such as asthma, allergies, obesity and diabetes, and some childhood cancers [4]. Breastfeeding also confers a range of maternal benefits, which include reduced risks of breast, endometrial and ovarian cancer, together with faster return to pre-pregnancy weight and improvements in the psychological wellbeing [4,5].

The provision of mothers' own breast milk is particularly important in the feeding of vulnerable preterm infants, where superiority of mothers' own milk to donor human milk and infant formula with respect to the composition and bioactivity are critically important [3]. The use of mothers' own breast milk during hospitalisation reduces the incidence and severity of preventable morbidities, including necrotizing enterocolitis (NEC), late onset sepsis, chronic lung disease, retinopathy of prematurity, rehospitalisation after discharge, and neurodevelopmental problems in infancy and childhood [3]. For example, a recent Cochrane review showed that preterm infants receiving infant formula are three times more likely to develop NEC, which has a mortality rate of 20–40% [6]. The benefits are further exemplified by evidence from a cohort of very low birth weight infants where for every 10 mL/kg/day increase in the ingestion of mothers' own breast milk, the odds of rehospitalisation up to 30 months of age decreased by 5% [7]. These benefits extend long-term, with predominant breast milk feeding in the first 28-days of life is associated with better IQ, academic achievement, working memory, and motor function at seven years of age in very preterm infants [8]. In addition, the ability for a mother to provide her own breast milk provides psychological benefits including greater feelings of attachment, empowerment, and confidence [9].

The reality, however, is that many mothers face challenges related to initiating and sustaining breastfeeding. One of the key challenges related to breast milk production, with perceived inadequate breast milk supply is a frequently reported reason for early discontinuation of breastfeeding or decreased exclusivity in women who have initiated breastfeeding [10]. As an example of the potential magnitude of the problem, a large survey of 1323 mothers in the US highlighted that 50% of women stopped breastfeeding earlier than they had intended due to insufficient breast milk supply [11]. These surveys however, suffer a lack of objective evidence to confirm whether the low milk supply is real or perceived. Further, they often do not include information about early breastfeeding practices during the establishment of lactation such as the frequent emptying of the breast. Nevertheless, even if true lactation insufficiency is a fraction of previously reported figures, a significant opportunity exists for the development and implementation of strategies targeting improvements in breast milk supply to improve breastfeeding outcomes and subsequent maternal and infant health. In light of this, there continues to be widespread interest in and use of medications to augment lactation, commonly referred to as galactagogues. Prior to appraising the evidence regarding the role of pharmaceutical galactagogues in the treatment of lactation insufficiency, it is necessary to consider the physiology of lactation.

2. Physiology of Lactation

The development of functional lactation is a multi-stage event and is regulated by a complex hormonal milieu that includes numerous reproductive hormones (e.g., oestrogen, progesterone, prolactin, oxytocin) and metabolic hormones (e.g., glucocorticoids, insulin, insulin-like growth factor 1 (IGF-I), growth hormone, and thyroid hormone) [12]. Secretory differentiation occurs in mid- to late pregnancy when the differentiation of mammary epithelial cells into lactocytes occurs, conferring the ability to synthesize and secrete key components of human milk. This process is largely driven by circulating levels of oestrogen and progesterone that not only stimulate milk duct development, but also suppress prolactin's action on milk production during pregnancy. Secretory activation is triggered by the sudden drop in progesterone following delivery of the placenta, accompanied by high levels of circulating prolactin. Secretory activation is confirmed by the onset of copious milk secretion and typically occurs between 24–102 h (average 60 h) after birth. This is often described as the milk 'coming in'. Milk ejection is primarily regulated by oxytocin, which stimulates the contraction of myoepithelial cells around the alveoli and facilitates milk release [13]. Once lactation is established, an autocrine control (local feedback) mechanism regulates ongoing milk production, meaning supply is largely based on the effective removal of milk from the breast and is less sensitive to circulating levels of prolactin [13].

Given the important roles of prolactin and oxytocin in regulating milk secretion and milk ejection, these hormones have become common pharmacological targets for influencing breast milk supply. Figure 1 provides an overview of endogenous and exogenous factors influencing prolactin and oxytocin, while also highlighting mechanisms of action of common pharmaceutical galactagogues. A more detailed review of the neuroendocrine regulation of prolactin and oxytocin secretion and function can be found elsewhere and is beyond the scope of this review [14]. In brief, the production of prolactin in the anterior pituitary is regulated by the presence of prolactin inhibiting factors (PIF) and prolactin releasing factors (PRF), which are controlled by the hypothalamus. The concentrations of these inhibitory and stimulatory factors is influenced by a variety of external stimuli such as infant suckling, sounds of the infant crying, and stress. The key PIF is dopamine, which exerts an inhibitory effect on prolactin production. Positive external stimuli lead to an inhibitory effect on dopamine release and a resultant increase in prolactin levels. Similarly, medications that block dopamine, such as the dopamine antagonists domperidone and metoclopramide, block its inhibitory effects and result in an increase in prolactin secretion. In contrast, a range of hormones act to stimulate prolactin production, including thyrotropin releasing hormone, cortisol, and oxytocin. Other hormones that exert direct or indirect effects on the mammary gland to influence milk synthesis include the growth hormone, prolactin and insulin. Milk contains a small whey protein called the feedback inhibitor of lactation (FIL). The accumulation of milk in the breast leads to milk stasis and an increased concentration of FIL, which provides a negative feedback to slow milk production by inhibiting the action of prolactin on lactocytes.

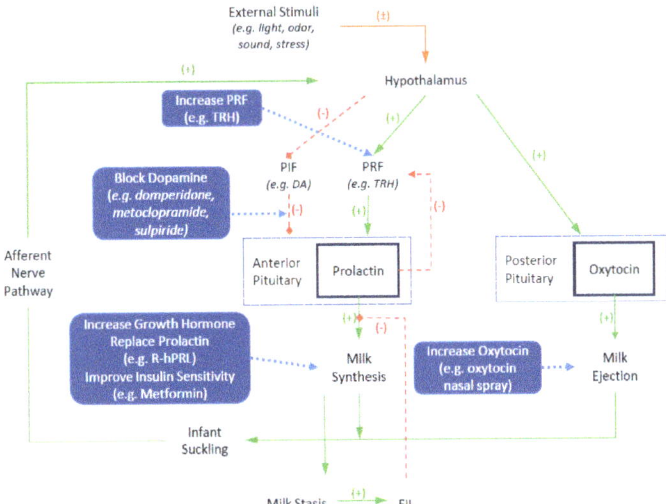

Figure 1. Physiology of lactation and mechanisms of action of various pharmaceutical galactagogues (Adapted from [14–16]). Abbreviations: TRH, thyrotrophin releasing hormone; PIF, prolactin inhibitory factor; PRF, prolactin releasing factor; DA, dopamine; R-hPRL, recombinant human prolactin; FIL, feedback inhibitor of lactation.

3. Risk Factors for Lactation Insufficiency

Three key aspects are required for establishing an adequate breast milk supply including sufficient mammary tissue, normal hormone levels, and regular effective removal of milk [17]. Insufficient mammary tissue can be the result of hypoplastic breasts, breast surgery such as mastectomy or breast reduction, or cyst removal [18]. Numerous maternal factors can influence hormone levels including retained placenta, severe postpartum haemorrhage, hypothyroidism, high levels of stress or anxiety, certain medications, anaemia, diabetes, obesity, polycystic ovary syndrome, cigarette smoking, or alcohol consumption [17,18]. Lastly, the effective and regular removal of milk can be influenced by the method and frequency of expressing if indirectly feeding the infant, or the inability for infants to effectively remove milk from the breast if feeding directly as a result of issues such as poor attachment, low infant intraoral vacuums and delivery complications. Factors such as the use of infant formula, early introduction of solids, infant sedation, or infant medical problems (e.g., low birth weight, congenital abnormalities, illness) may also exacerbate poor feeding and ineffective removal of breast milk [18].

In particular, mothers of preterm infants face many challenges in establishing and maintaining an adequate supply of breast milk during their infant's prolonged hospitalisation. This is driven by multiple factors including; physiological immaturity of the breast associated with preterm birth, maternal morbidities, inability for the preterm infant to breastfeed directly from the breast, and the stress of having an infant admitted to the Neonatal Intensive Care Unit (NICU). Each of these factors has the ability to interfere with the establishment of normal milk supply, with one previous study demonstrating that 82% of women birthing preterm experienced delayed secretory activation [19]. While longer-term breastfeeding outcomes were not collected for this cohort, other studies in mothers of term infants have demonstrated that delayed secretory activation is associated with an increased risk of early cessation of breastfeeding [20].

4. Use of Galactagogues in Clinical Practice

Relatively little evidence exists regarding the prevalence of pharmaceutical galactagogue use. A recent clinical practice survey of Australian Neonatal Units (NNUs) identified that 100% of NNUs reported domperidone as their first line pharmacological agent of choice in the management of lactation insufficiency [21]. While 75% of NNUs reported having a clinical guideline on domperidone, significant variability was evident across these guidelines with respect to the dose and duration of treatment.

The preference towards the use of domperidone for lactation insufficiency is supported by a recent audit of phone calls to an Australian pregnancy and lactation counselling service regarding the use of galactagogues during lactation highlighted that domperidone accounted for >90% of phone calls [22]. In contrast, the discussion of herbal galactagogues increased from 0% in 2001 to 23% in 2014. Overall, these findings highlight the popularity of domperidone use in the Australian setting, together with the increasing popularity of herbal galactagogues.

Three studies have examined patterns of domperidone use over the past 10–15 years. An Australian study undertaken at a single tertiary level maternity hospital identified the increasing prevalence of use from 0.1% of all deliveries in 2000 to 5% in 2010 [23]. A recognised limitation of this study lies in the fact that it likely underestimates total domperidone use, as it could not take into account domperidone dispensed from community pharmacies. In comparison, a Canadian study investigated the number of women that dispensed domperidone in the first six months postpartum among 320,351 live births occurring between 2002 and 2011 [24]. Over this time period the prevalence of domperidone use increased from 7% to 18% in women delivering at term and 17% to 32% in women delivering preterm. The most recent study used data from the Clinical Practice Research Datalink in England and demonstrated a 3.8-fold increase in postpartum domperidone prescribed from 0.56 per 100 person-years in 2002–2004 to 2.1 per 100 person-years in 2011–2013, with overall use much lower than that reported in previous studies [25]. Risk factors commonly associated with domperidone use in these studies included preterm birth, delivery by caesarean section, maternal obesity, diabetes or gestational diabetes, and primiparity [24–26].

5. Efficacy and Safety of Galactagogues

In order to evaluate the efficacy and safety of galactagogues we searched four electronic databases from inception to January 2019: Ovid Medline, EMBASE, Web of Science, and SCOPUS. Medical subject headings (e.g., MeSH headings) and free word combinations using Boolean logic of the following search items were used: 'galactagogues' OR individual medication names (e.g., 'domperidone') AND 'lactation' OR 'low milk supply'. Previous reviews, bibliographies of published trials and cross references were also searched. No language restrictions were applied. For the evaluation of galactagogue efficacy, studies were restricted to those investigating the use of galactagogues in the treatment of established lactation insufficiency and using controlled study designs. Studies investigating herbal galactagogues were excluded from this review.

An overview of controlled intervention trials evaluating the efficacy of various galactagogues is provided in Table 1. Domperidone [27–37] and metoclopramide [28,29,38–40] appear to be the most commonly studied galactagogues in the setting of established lactation insufficiency, with limited studies evaluating the effects of other galactagogues including sulpiride [41,42], growth hormone [43,44], human recombinant prolactin [45], thyrotropin releasing hormone [46,47], and metformin [48]. No studies have evaluated the use of oxytocin in the treatment of lactation insufficiency. The focus of this review is on the use of galactagogues in the treatment of lactation insufficiency, studies that included women without an established diagnosis of lactation insufficiency were not considered [49–59].

5.1. Domperidone

Domperidone is a dopamine antagonist and is thought to act as a galactagogue by increasing serum prolactin (Figure 1). The majority of studies evaluating the effects of domperidone have been undertaken among mothers of preterm infants. Treatment durations (five to 28 days) and treatment initiation (>1 to >3 weeks post-partum) varied widely between studies. The most commonly used interventional dose consisted of 10 mg three times daily, with two studies evaluating a higher dose of 20 mg three times daily [27,34].

All placebo-controlled studies demonstrate treatment effects in favor of domperidone. Five studies involving mothers of preterm infants have recently been combined in a meta-analysis, demonstrating that the administration of 30 mg/day led to a modest increase in maternal breast milk volume of 88 mL daily [60]. Missing from this meta-analysis was the recent study by Fazilla et al., which demonstrated a similar improvement to the previous pooled estimate of 109 mL/day (95%CI 69 to 149 mL/day) [32]. Effects of domperidone on mothers of term infants is limited to two studies [33,35], with both demonstrating treatment effects in favor of domperidone compared with placebo. Inam et al. utilised a treatment target of 50 mL per single expression episode (both breasts) with 36 of 50 (72%) women in the domperidone arm achieving this target compared with 11 of 40 (22%) women in the placebo arm ($p = 0.002$) [33]. Petraglia et al. evaluated changes in breast milk supply by weighing infants before and after breast feeding, demonstrating greater improvement in daily breast milk supply among women receiving domperidone (347 ± 36 to 673 ± 44 mL/day), compared with placebo (335 ± 30 to 398 ± 45 mL/day; $p < 0.01$) [35].

Table 1. Overview of controlled studies evaluating the efficacy of pharmaceutical galactagogues for the treatment of lactation insufficiency.

Study (Country)	Intervention (n/N) and Dose	Duration (Days)	Delivery Gestation (Weeks)	Infant Age (Weeks)	Effect on Milk Production
Domperidone					
Asztalos 2017 (Canada) [37]	*Domperidone* (45): 10 mg TDS *Placebo* (40/45)	14	<30	>1 to ≤3	Change in milk supply: *Domperidone*: 121 (±96) to 267 (±189) mL/day; *Placebo*: 115 (±95) to 217 (±168) mL/day; $p = 0.10$
Campbell-Yeo 2010 (Canada) [30]	*Domperidone* (21/22): 10 mg TDS *Placebo* (24)	14	<31	≥3	Change in milk supply: *Domperidone*: 184 (±167) to 380 (±202) mL/day; *Placebo*: 218 (±155) to 251 (±172) mL/day
da Silva 2001 (Canada) [31]	*Domperidone* (6/11): 10 mg TDS *Placebo* (8/9)	7	<37 †	Any	Difference from baseline: *Domperidone*: 49.5 (±29.4 mL/day; *Placebo*: 8.0 (±39.5) mL/day; $p < 0.05$
Fazilla 2017 (Indonesia) [32]	*Domperidone* (25): 10 mg TDS *Placebo* (25):	7	<37	>1	Difference from baseline: *Domperidone*: 181.6 (±80.2) mL/day; *Placebo*: 72.4 (±57.8) mL/day; $p = 0.0001$
Inam 2013 (Pakistan) [33]	*Domperidone* (50): 10 mg TDS *Placebo* (50)	7	Term	1	Number reaching target of ≥ 50 mL per single expression: *Domperidone*: 36/50 (76%); *Placebo*: 11/50 (22%); $p = 0.002$
Knoppert 2013 (Canada) [34]	*Domperidone* (5/8): 10 mg TDS *Domperidone* (6/7): 20 mg TDS	28	<33	≥2 to ≤3	Change in milk supply: 10 mg TDS: 150 to 420 mL/day; 20 mg TDS: 300 to 720 mL/day
Petraglia 1985 (Italy) [35]	*Domperidone* (9): 10 mg TDS *Placebo* (8)	10	Term	>2	Change in milk supply: *Domperidone*: 347 (±36) to 573 (±44) mL/day; *Placebo*: 335 (±30) to 398 (±45) mL/day; $p < 0.01$
Rai 2016 (India) [36]	*Domperidone* (14/16): 10 mg TDS *Placebo* (16)	7	<37†	≥1 to ≤2	Difference from baseline: Domperidone: Median 136.0 (IQR 126.5 to 240) mL/day; Placebo: Median 70 (IQR 49.5 to 97); $p = 0.004$
Wan 2008 (Australia) [27]	*Domperidone* (4/7): 10 mg TDS & 20 mg TDS	7–14	<37	>2	Change in milk supply: Baseline: 208.8 (±182.3) mL/day; 30 mg: 566.4 (±229.3) mL/day; 60 mg: 705.6 (±388.1) mL/day
Metoclopramide					
Kauppila 1981 (Finland) [39]	*Metoclopramide* (37/45): 5 mg TDS, 10 mg TDS, 15 mg TDS & *placebo*	14	Term	>1	Difference from baseline (single feed): 5 mg TDS: 11.2 ± 28.1 mL; 10 mg TDS: 42.5 ± 34.7 mL; 15 mg TDS: 50.0 ± 35.9 mL; *Placebo*: 4.0 ± 27.5 mL
Kauppila 1985 (Finland) [38]	*Metoclopramide* (8): 10 mg TDS *Placebo* (5)	21	Term	≥4 to ≤20	Change in milk supply: *Metoclopramide*: 285 (±75) to 530 (±162) mL/day ($p < 0.01$); *Placebo*: individual results not stated
Sakha 2008 (Iran) [40]	*Metoclopramide* (10): 10 mg TDS *Placebo* (10)	15	Term	NR	Infant weight gain: *Metoclopramide*: 328.5 g; *Placebo*: 351.5 g ($p = 0.68$)
Domperidone/Metoclopramide					
Blank 2000 (Australia) [28]	*Domperidone* (9): 10 mg TDS *Metoclopramide* (11): 10 mg TDS *Placebo* (9)	5	<34	≥1	Change in milk supply: *Domperidone*: 120 (±81) to 239 (±105) mL/day; *Metoclopramide*: 100 (±53) to 184 (±100) mL/day; *Placebo*: 143 (±57) to 172 (±117) mL/day
Ingram 2012 (Canada) [29]	*Domperidone* (31/38): 10 mg TDS *Metoclopramide* (34/42): 10 mg TDS	10	<37†	NR	Change in milk supply: *Domperidone*: 174 (±126) to 285 ±158) mL/day; *Metoclopramide*: 133 (±115) to 212 (±154) mL/day; *Difference*: 31.0 (−5.67 to 67.6) mL/day
Sulpiride					
Ylikorkala 1982 (Finland) [41]	*Sulpiride* (14): 50 mg TDS *Placebo* (12/14)	28	NS	≤16	Difference from baseline: *Sulpiride*: 265 mL/day; *Placebo*: −50 mL/day
Ylikorkala 1984 (Finland) [42]	*Sulpiride* (13/14): 50 mg TDS *Placebo* (11/14)	14	NS	≤16	Difference from baseline: *Sulpiride*: 646 (±67) mL/day; *Placebo*: 428 (±71) mL/day ($p < 0.05$)

Table 1. *Cont.*

Study (Country)	Intervention (n/N) and Dose	Duration (Days)	Delivery Gestation (Weeks)	Infant Age (Weeks)	Effect on Milk Production
Human Growth Hormone (hGH)					
Gunn 1996 (New Zealand) [43]	*hGH* (9/10): 0.2 IU/kg/day SC (max 16 IU/kg/day) *Placebo* (9/10)	7	26–34	NR	Change from baseline: *hGH*: 139 (±49) to 175 (±46) mL/day; $p < 0.01$ *Placebo*: 93 (±50) to 102 (±69) mL/day; $p = $ NS
Milsom 1998 (New Zealand) [44]	*hGH* (5): 0.05 IU/kg/day SC *hGH* (5): 0.1 IU/kg/day SC *hGH* (6): 0.2 IU/kg/day SC	7	Term	<16	Percentage increase in breast milk volume: High dose group (0.2 IU): 36 ± 12.6 Low/Mid dose group (0.05/0.1 IU): 4.7 ± 9.7; $p < 0.04$
Recombinant Human Prolactin (R-hPRL)					
Powe 2010 (USA) [45]	*R-hPRL* (3): 60 mcg/kg SC BD *R-hPRL* (3): 60 mcg/kg SC daily *Placebo* (4)	7	24–32	≥1 to ≤4	Percentage change in milk supply: *R-hPRL* (Twice Daily): 429 (±338%); *R-hPRL* (Once Daily): 44 (±28%); *Placebo*: −12 (±27%)
Thyrotrophin-Releasing Hormone (TRH)					
Zarate 1975 (Mexico) [46]	*TRH* (5): 20 mg TDS PO *Placebo* (4)	7	NR	NR	No change in milk production (actual volume not reported)
Peters 1991 (Germany) [47]	*TRH* (10): 1 mg QID IN *Placebo* (9)	10	Term	Day 6	Change from baseline: *TRH*: 142.0 (±33.9) to 253.0 (±105.3) g/day ($p = 0.014$); *Placebo*: 150.0 (±46.2) to 140.6 (±57.7) g/day ($p = 0.87$)
Metformin					
Nommsen-Rivers 2019 (USA) [48]	*Metformin* (10): Day 1–7 – 750 mg/day; Day 8–14 – 1500 mg/day; Day 14–28 – 2000 mg/day *Placebo* (5)	28	Term	1–8	Difference from baseline: *Metformin*: Median 8 (IQR −23 to 33) mL/day; *Placebo*: Median −58 (IQR −62 to −1) mL/day

Abbreviations: TDS, three times daily; BD, twice daily; SC, subcutaneous; IN, intranasal; QID, four times daily; NR, not reported. † denotes infants admitted to neonatal unit.

Two underpowered studies have compared the effects of different doses of domperidone [27,34]. One of these was a small case-crossover study where women were randomised to receive an initial dose of 30 mg/day or 60 mg/day, then crossover to the other arm of the study after 7–14 days of treatment. Daily milk volume production was only reported for four of six women who were considered treatment responders, with daily breast milk volume increasing from 208.8 ± 182.3 mL/day at baseline to 566.4 ± 229.3 mL/day at a dose of 30 mg/day and 705.6 ± 388.1 mL/day at a dose of 60 mg/day [27]. The reported increases were statistically significantly higher compared with baseline values, but the small sample size of four women prohibited the identification of statistically significant differences between the two different doses. In contrast, Knoppert et al. randomised 15 women to either receive 30 mg/day or 60 mg/day for four weeks [34]. Daily breast milk volume increased from a median of 150 to 420 mL/day among those receiving the 30 mg/day dose, and 300 to 720 mL/day among those receiving the 60 mg/day dose, but the differences between the groups were not statistically significant [34]. Collectively, while these two studies provide promising evidence of potential further improvements in daily breast milk volume with the use of a higher dose of domperidone (i.e., 60 mg/day compared to 30 mg/day), further research is required to determine if a higher dose can be considered superior.

Maternal adverse events most commonly associated with the use of domperidone in clinical trials include headache, dry mouth, and gastrointestinal disturbances. That said, compared to placebo, no increased risk of maternal adverse events (RR 1.05; 95%CI 0.65–1.71) has been identified in clinical trials. Similarly, no trials have reported an increased risk of adverse neonatal events [58,59]. Indeed the concentrations of domperidone in breast milk are low with the absolute infant dose only 0.04 (95%CI 0.03–0.07) micrograms/kg/day for a maternal dose of 30 mg/day, and 0.07 (95%CI 0.05–0.11) micrograms/kg/day for a maternal dose of 60 mg/day [27]. One small case-crossover study demonstrated a higher prevalence of adverse events with a higher dose of domperidone [27]. Among the seven study participants, there were a total of five adverse events reported at 30 mg/day (abdominal cramping [$n = 1$], dry mouth [$n = 3$], headache [$n = 1$]), and 12 adverse events reported at 60 mg/day (abdominal cramping [$n = 2$], constipation [$n = 1$], dry mouth [$n = 5$], depressed mood [$n = 1$], headache [$n = 3$]) [27]. Also, one woman withdrew from the study as a result of experiencing severe abdominal cramps while receiving 60 mg/day [27]. In contrast, a more recent double blind randomised controlled trial comparing 30 mg/day to 60 mg/day identified no adverse events in either group [34]. Therefore, while it currently remains unclear, it is possible that a higher dose of domperidone may be associated with a greater risk of maternal adverse events.

The use of domperidone in lactation has been the subject of controversy due to an increased risk of ventricular arrhythmia (VA) and sudden cardiac death of approximately four per 1000 person-years observed among non-lactating adults, including males and females [61,62]. This increased risk relates to the potential of domperidone to prolong the QT interval [63], but the relevance to lactating women has been questioned [64–68]. Domperidone was initially developed as an antiemetic and prokinetic, with previous recommended doses ranging from 30 mg/day to 60 mg/day or greater [66]. Some observational studies suggested that the risk of VA or sudden cardiac death may be increased if daily doses were greater than 30 mg or if patients were male or greater than 60 years old [61,62]. More detail regarding domperidone and QT interval effects among non-lactating adults can be found elsewhere [61,62].

Numerous studies provide reassuring evidence as to the safety of domperidone use in lactating women. A recent randomised double-blind, placebo- and positive-controlled safety study identified that among healthy female volunteers, domperidone at doses up to 80 mg/day did not cause clinically relevant QTc-interval prolongation [69]. Further, all women participating in the trial by Asztalos et al. had an ECG at baseline and at the end of the study. They observed no evidence of prolongation of the QTc-interval with any mother participating in the trial [62]. Perhaps the strongest evidence to guide the clinical practice comes from a recent Canadian population-based cohort study of 320,351 women, of which 45,163 were prescribed domperidone within six months postpartum [70]. The primary outcome consisted of ventricular arrhythmia and data were obtained on whether or not women had a previous history of ventricular arrhythmia. No cases of ventricular arrhythmia were identified among those women with no prior history [71]. All cases of ventricular arrhythmia were observed among women with a previous history. Taken together, these results provide supporting evidence for the limited increased risk of ventricular arrhythmia among women with no previous medical history, and strengthens the support that domperidone should be avoided in those women with a previous history. Notably, among this cohort 90% of women received a dose of domperidone greater than 30 mg/day, suggesting limited impact of dose on ventricular arrhythmia risk. In light of such evidence it appears that domperidone can be used safely, but should be avoided or used with extreme caution in women with established risk factors for QTc-prolongation. This includes women taking medications that inhibit the metabolism of domperidone and/or also prolong the QTc-interval (e.g., fluconazole, erythromycin), and women with a personal or family history of cardiac arrhythmia or family history of unexplained sudden death. Clinical protocols regarding the use of domperidone in clinical practice have been developed and serve to facilitate safer prescribing practices and minimize potential adverse reactions in mothers and their hospitalized premature infants [72].

5.2. Metoclopramide

Similar to domperidone, metoclopramide is a dopamine antagonist and is thought to act as a galactagogue by increasing serum prolactin (Figure 1). In contrast to domperidone, the majority of clinical studies have evaluated the use of metoclopramide among mothers of term infants. Four studies have evaluated a standard dose of 10 mg three times daily, with the treatment duration varying from five to 21 days. One study evaluated three difference metoclopramide doses [39]. Three studies directly measured changes in breast milk volume, one study indirectly evaluated breast milk volume based on infant weight measures before and after a single feed, while the last study evaluated infant weight gain. A dose-response case-cross-over study identified a significant increase from baseline in breast milk yield during a single feed following use of 30 mg/day (42.5 ± 34.7 mL) and 45 mg/day (50.0 ± 35.9 mL) doses, but not 15 mg/day (11.2 ± 28.1 mL) or placebo (4.0 ± 27.5 mL) [39]. Blank et al. identified a greater increase in breast milk volume in those receiving metoclopramide compared with placebo (84 mL/day compared to 29 mL/day, but the difference between the groups was not statistically significantly different [28]. Kauppila et al. identified a significant increase in breast milk supply during the study among those receiving metoclopramide (285 ± 75 mL /day to 530 ± 162 mL /day ($p < 0.01$)), but change in breast milk supply was not reported for those in the placebo arm [38].

Adverse events associated with metoclopramide have been incompletely described. Kauppila et al. reported adverse events occurring among six of eight women, which included tiredness ($n = 6$), headache ($n = 1$), and nausea ($n = 1$) [38]. In the dose-response study by Kauppila et al. seven women complained of side-effects including tiredness ($n = 1$), headache ($n = 1$), anxiety ($n = 1$), hair loss ($n = 1$) and intestinal disorders ($n = 1$), but the relationship between the adverse events and dose were not described [39]. While greater amounts of metoclopramide are secreted into the breast milk compared with domperidone, the reported absolute infant dose ranging from six to 24 micrograms/kg/day is still well below the recommended pediatric dose of 500 micrograms/kg/day [73]. Notably, metoclopramide was detected in the plasma of one infant and while no short-term neonatal adverse events have been identified, whether infants are more susceptible to potential longer-term adverse events of metoclopramide is unknown. This is of concern given the ease with which metoclopramide crosses the blood-brain barrier and therefore the potential to interrupt dopamine signaling in the newborn.

Similar to domperidone, metoclopramide also has the potential to cause serious maternal adverse events. Given that metoclopramide has the ability to cross the blood-brain barrier, it is more likely to cause central nervous system side effects than domperidone. Of particular concern is the increased risk of depression, which is already increased in the postpartum period. In the study by Ingram et al., two of 29 (7%) women receiving open-label metoclopramide after the initial 10-day trial reported experiencing depression [29]. Metoclopramide also has the ability to cause serious and potentially permanent extrapyramidal side effects such as tardive dyskinesia, for which it carries a black box warning in the product information. As such, it is recommended that metoclopramide not be used for longer than five days and that the maximum dose be limited to 10 mg three times daily. A large international internet survey of women who took metoclopramide to increase breast milk supply found that 4.8% of women had either palpitations or racing heart rate, 12% reported depression, and 1 to 7% reported other central nervous system side effects ranging from dizziness and headache to involuntary grimacing and tremors [74]. The overall adverse event profile of metoclopramide makes it less desirable than domperidone to use as a galactagogue, with evidence directly comparing the two outlined below.

5.3. Domperidone Compared To Metoclopramide

While domperidone and metoclopramide are both dopamine antagonists and increase serum prolactin concentrations, they do differ with respect to some pharmacodynamic properties. One of these key properties relates to the ability to cross the blood-brain barrier, with only metoclopramide able to do so in appreciable quantities. This has led to questions regarding potential differences in the efficacy and safety of these two medications. Two studies have directly compared outcomes among women taking domperidone and metoclopramide. Blank et al. identified a slightly greater increase in the daily breast milk volume among nine women receiving domperidone (120 ± 81 mL/day to 239 ± 105 mL/day) than among eleven women receiving metoclopramide (100 ± 53 mL/day to 184 ± 100 mL /day) [28]. The 35 mL/day difference between those receiving domperidone and metoclopramide, however, was not statistically significant. Ingram et al. demonstrated a similar difference in breast milk volume for those receiving domperidone compared with metoclopramide (31.0 mL/day; 95%CI −5.67 to 67.6 mL/day), among a total sample of 65 women [29].

When it comes to maternal adverse events, Blank et al. reported that minor neurobehavioral symptoms of either drowsiness, sleep disturbance, restlessness or dizziness were experienced by two of nine women taking domperidone and two of 11 women taking metoclopramide [28]. In contrast, Ingram et al. reported that seven (20%) of women taking metoclopramide reported experiencing adverse events compared with just three (10%) of women taking domperidone. Notably, one woman taking metoclopramide withdrew from the study due to experiencing bad headaches and dry mouth [29]. Greater incidence of adverse events with metoclopramide is also evident from observational studies, with a 2010 Internet survey of 1990 mothers identifying a 7-fold increased risk of depression, and 4 to 19-fold increased risk of symptoms commonly associated with tardive dyskinesia (e.g., tremors, involuntary grimaces, and jerking) among women taking metoclopramide compared with domperidone [75].

5.4. Sulpiride

Sulpiride is a dopamine antagonist, traditionally used as an antipsychotic, and is thought to act as a galactagogue by increasing serum prolactin (Figure 1). The effects of sulpiride in increasing breast milk supply have been evaluated in two placebo-controlled studies [41,42]. Both utilised a dose of 50 mg three times daily compared with placebo, with the treatment duration ranging from 14 to 28 days. It is unclear from both studies as to the infant gestational age. Both studies demonstrated a greater increase in breast milk volume among those receiving sulpiride, with improvements of 218 mL/day and 315 mL/day and associated reductions in the requirement for supplemental feeds. However, both studies experienced greater than 20% loss to follow-up in the placebo group which was reported to be due to insufficient treatment response, indicating significant potential for bias. While previous clinical trials in lactation have reported few adverse events associated with the use of sulpiride and the doses used in these trials are lower than those thought to produce significant neuroleptic effects in adults, data are still limited and there remains concerns regarding potential maternal or infant effects. Common adverse events include sedation and weight gain, while sulpiride may also cause extrapyramidal side effects similar to metoclopramide. Of greater concern is that the relative infant exposure to sulpiride through breast milk is up to 20% of the maternal weight-adjusted dose [49], which is much greater than that for other galactagogues.

5.5. Growth Hormone

The exact mechanisms of action of growth hormone in lactation remain unclear. The growth hormone is released by the anterior pituitary gland and acts synergistically to prolactin [12]. Administration of the growth hormone has been demonstrated to increase secretion of the insulin-like growth factor, providing evidence that the galactopoietic response to the growth hormone occurs as a result of indirect effects on mammary glands [44]. Two studies have evaluated the administration of the human growth hormone [43,44]. The first placebo-controlled study enrolled mothers of preterm infants (<34 week's gestation) and involved a subcutaneous administration of 0.2 international units/kg/day for seven days. Those receiving growth hormone experienced an increase in breast milk volume (139 ± 49 to 175 ± 46 mL/day; $p < 0.01$), but no difference was evident among those receiving placebo (93 ± 50 to 102 ± 69 mL/day; p = not stated). The second study, by the same authors, enrolled mothers of term infants and investigated three doses ranging from 0.05 to 0.2 international units/kg/day for seven days [44]. Women in the high dose arm (0.2 IU) experienced a greater percentage increase in breast milk volume than those in the low- (0.05 IU) and mid- (0.1 IU) dose arms (36 ± 12.6% compared to. 4.7 ± 9.7%; $p < 0.04$) [44]. Absolute differences in milk volume identified in the two previously reported studies were 25 and 86 mL/day, respectively. No maternal or infant adverse events were reported within either study. Overall, the relatively modest improvements in breast milk supply observed with growth hormone do not seem to justify the high costs and administration challenges associated with the treatment, particularly in comparison to alternative galactagogues such as domperidone or metoclopramide.

5.6. Recombinant Human Prolactin

Given that prolactin is a critical component of establishing lactation, the direct administration of prolactin is viewed as an approach for treating lactation insufficiency that is the result of prolactin deficiency (Figure 1). One study that involved the administration of recombinant human prolactin (r-hPRL) was identified [45]. The placebo-controlled study enrolled mothers of preterm infants (<32 weeks gestation) with prolactin deficiency, and involved subcutaneous administration of either 60 micrograms/kg twice daily or 60 micrograms/kg once daily for seven days. The subgroup analysis revealed that milk volumes were higher in the group treated with twice daily r-hPRL (429 ± 338%) than in the group receiving only placebo (−12 ± 27%; $p < 0.05$). No significant differences were observed with respect to maternal or infant adverse events. While data are limited, this study demonstrates that r-hPRL may be a viable option for the treatment of lactation insufficiency among women with established prolactin deficiency. Given the high costs and administration challenges associated with the treatment, this remains a treatment for rare situations where alternative prolactin stimulating medications such as domperidone are unsuitable, as is the case with the Sheehan syndrome or congenital prolactin deficiency where lactotrophs are absent or reduced in number.

5.7. Thyrotrophin Releasing Hormone

The thyrotrophin-releasing hormone (TRH) acts on the anterior pituitary to increase serum prolactin concentrations (Figure 1). Two studies were identified that involved the administration of TRH, producing mixed results [46,47]. One study randomised women to receive either TRH (n = 5) or placebo (n = 4) orally [47]. Among women treated with TRH, only one woman experienced an increase in serum prolactin. Actual breast milk volumes were not described, with the effect on milk production reported as 'no change' for all women among the treatment and placebo arms of the study. The second study enrolled mothers of term infants and involved the intranasal administration of 1 mg four times daily for 10 days. Those receiving TRH experienced an increase in breast milk volume (142.0 ± 33.9 g/day to 253.0 ± 105.3 g/day; p = 0.014), but no difference was evident among those receiving placebo (150.0 ± 46.2 g/day to 140.6 ± 57.7 g/day; $p = 0.87$). The increase in breast milk volume was accompanied by an observed increase in basal prolactin levels from a mean of 117 (± 45) micrograms/L to 173 (± 56) micrograms/L among women receiving TRH, whereas those receiving placebo experienced a decrease in basal prolactin levels (137 ± 70 micrograms/L to 82 ± 38 micrograms/L). No maternal or infant adverse events were reported in these two studies. However, reported effects on thyroid function have been mixed based on open-label non-controlled studies, with one demonstrating no changes in maternal or infant thyroid function [47], while one study identified two cases of maternal hyperthyroidism among 13 women treated with TRH which required treatment discontinuation [59]. Given the potential severity of adverse events and mixed findings of previous studies, more research on the use of TRH is necessary before its consideration in clinical practice.

5.8. Oxytocin

Despite the important role of oxytocin in regulating milk ejection, no studies have evaluated the use of oxytocin in the setting of established lactation insufficiency. Existing literature is restricted to studies enrolling women soon after delivery, with treatment undertaken in an effort to augment early milk supply and prevent the development of lactation insufficiency [53,54,56–58]. These studies have produced conflicting findings and their relevance to the treatment of women with established lactation insufficiency is uncertain. As it currently stands, there is insufficient evidence whether oxytocin may be an effective treatment for lactation insufficiency. However, given evidence that acute or chronic stress can interfere with oxytocin secretion, inhibiting both milk transfer and mother-infant bonding [20], further research on the potential usefulness of oxytocin in the setting of established lactation insufficiency is warranted, particularly among mothers of preterm infants where high levels of anxiety and depressive symptoms are common.

5.9. Metformin

Metformin is commonly used in the management of type 2 diabetes and is thought to act as a galactagogue through its role as an insulin-sensitising medication (Figure 1). The effects of metformin in increasing breast milk supply have been evaluated in one placebo controlled study [48]. The pilot study enrolled mothers of term infants who were experiencing low breast milk production and who had at least one sign of insulin resistance (e.g., elevated fasting plasma glucose). Women were randomised to either metformin ($n = 10$) or placebo ($n = 5$) with metformin dose increasing from 750 mg/day to 2000 mg/day over a 28-day treatment period. Peak median change in the milk output did not differ between those receiving metformin (+8; IQR –23 to 33 mL/day) or placebo (–58; IQR –62 to –1 mL/day) ($p = 0.31$). Notably, a greater increase in milk output was evident among those who managed to continue taking metformin for the entire treatment period, but only 20% of women assigned to the metformin arm experienced any actual improvement in milk output to day 28. Gastrointestinal adverse events were commonly reported, including diarrhoea (56%), nausea/vomiting (44%), and abdominal pain or cramping (56%), but were mostly of mild or moderate severity. One woman discontinued metformin due to the severity of abdominal cramping. No infant adverse events were reported. Given evidence on metformin is restricted to a small pilot study, its potential role as a galactagogue remains uncertain.

5.10. Summary of Galactagogue Efficacy and Safety

The largest and strongest body of evidence supports the use of domperidone as the first-line medication for the pharmacological management of lactation insufficiency. Domperidone represents a low cost treatment that is well tolerated, with moderate quality evidence that it produces a modest increase in breast milk supply in mothers of preterm infants. Whether results are translatable to mothers of term infants remain unclear as there are a limited number of high quality studies that include this patient population. Domperidone has been associated with serious adverse events and therefore is not suitable for all women, however, following appropriate screening prior to prescribing the risk of serious adverse events appear negligible. There is weak evidence that domperidone may be slightly more effective than metoclopramide and that adverse events are more likely to occur with metoclopramide than domperidone. Metoclopramide may be an appropriate alternative for use in women where domperidone is unsuitable. Antipsychotics do not reflect a viable therapeutic option for the management of lactation insufficiency, largely owing to their side effects such as extrapyramidal reactions and weight gain. Limited evidence and difficulties with respect to access to and administration of medications such as growth hormone, recombinant human prolactin and thyrotrophin-releasing hormone make these less desirable treatment options. Lastly, while an initially promising therapy for improving breast milk supply among women with signs of insulin resistance, initial findings from a pilot study evaluating metformin do not provide compelling evidence to support its use as a therapeutic treatment of lactation insufficiency.

6. Methodological Challenges and Evidence Gaps Regarding Galactagogues

6.1. Previous Study Limitations

Clinical studies evaluating the use of pharmaceutical galactagogues suffer from a number of important methodological limitations. Highlighting these limitations is not only important in considering the quality of existing evidence, but also in informing future research endeavors. The major limitation of the existing evidence lies in the small number of women included in clinical trials, limiting the potential to comprehensively evaluate both the efficacy and safety of these medications. Studies ranged from 7 to 100 participants, with a median of only 24. Many studies were therefore underpowered to detect statistically or clinically significant differences in breast milk supply or adverse events. Further, smaller studies face greater challenges in achieving balance in the distribution of baseline covariates across study groups. For example, in the study by Nommsen-Rivers et al., six out of 10 (60%) women receiving metformin were also taking the herbal galactagogue fenugreek, compared with one out of five (20%) women in the placebo arm [48]. In the study by Knoppert et al., five out of seven (71%) women in the high dose domperidone arm were primiparous, compared with three out of eight (38%) in the low dose domperidone arm [34], whereas in the study by da Silva et al. six out of nine (67%) women in the placebo arm were primiparous, compared with six out of seven (86%) in the domperidone arm [31]. Such differences in the baseline distribution of factors potentially related to breast milk volume or treatment response to galactagogue can significantly bias study findings and highlights the importance of large, appropriately controlled trials.

Many studies suffer from potential selection bias due to the high loss to follow-up in one or both treatment arms. This is particularly important where the loss to follow-up is related to the treatment response and only including data on those successfully completing the study potentially biases the results towards those with more favorable treatment response. Greater than 10% loss to follow-up in one or both intervention arms was evident among nine studies [28,29,31,34,36,37,39,41,42].

Non-pharmacological breastfeeding strategies are the mainstay of treatment for lactation insufficiency, but their implementation and utilisation across clinical trials of galactagogues is often incompletely described. This makes it unclear what strategies have been utilised and whether these have been applied equally across treatment groups.

Two studies reported the encouragement of skin-to-skin contact between the mother and infant [29,37], while some provided advice regarding the minimum number of times women expressed each day, ranging from four to eight however the frequency of pumping was not accounted for despite strong relationships with the milk volume [28,29,34,37,43,48]. All studies where mothers were expressing breast milk reported the use of electric pumps, which have been shown to be more effective in the establishment of milk production than hand expression [75].

Definitions for lactation insufficiency varied widely across studies and involved the use of an absolute volume measure (e.g., <500 mL/day) [27,28,33,34,38,44,45], relative volume measure (e.g.,

< 250 mL/kg/day) [29,35,37,39,41,42,46], requirement for supplemental feeds [30–32,43], insufficient infant weight gain [39], or certain reduction in supply based on previous milk production [14,23,37]. Given that there is no universally accepted definition of lactation insufficiency in the literature, it is unsurprising that definitions used across studies vary greatly. It is uncertain, however, whether the baseline milk volume has any impact on the treatment response to pharmaceutical galactagogues and this could possibly explain some heterogeneity across clinical studies. Linked with this concept is the complete absence of any noted targets regarding the optimal breast milk volume in previously published clinical trials. This is particularly important for mothers of preterm infants where the inclusion of an optimal treatment target could have an impact on the treatment response. For example, if the target is simply defined as meeting the infants' daily feed requirements and the infant is only 1 kg, the target is approximately 150 mL. As the infant grows, however, that target will shift and women whose daily breast milk production remains at 150 mL/day will find it difficult to keep up with their infants' demands and this may lead to eventual cessation of breastfeeding. As an example, Meier and Engstrom have suggested that for mothers of very low birth weight infants a desirable milk volume target is at least 350 mL/day by the end of the second postnatal week [76]. Higher volumes (e.g., 600 to 800 mL/day) are considered even more desirable and are more likely to sustain long-term breastfeeding outcomes [76]. Among the studies included in this review, the median breast milk volume increased from 146 mL/day to 276 mL/day, with only five studies reporting a final mean volume above the proposed target of 350 mL/day [27,30,34,35,38], with all involving the use of either domperidone or metoclopramide. This indicates that the hormonal manipulation utilising existing therapeutic approaches may not be adequate to reverse the poor development of the mammary gland during pregnancy and milk synthesis after it is has been established.

While galactagogues may provide short-term improvements in breast milk supply, their relationship to longer-term breastfeeding outcomes is unknown. Long-term outcomes are particularly challenging to assess among placebo controlled trials as once the trial has completed many women are often able to take the intervention medication off-label. This means long-term outcomes become confounded by maternal behaviors following completion of the trial. With that in mind, few studies have reported long-term breastfeeding outcomes. Asztalos et al. provided the most comprehensive evaluation, with 60% of women still breastfeeding once their infants reached the term gestation corrected age, with 70% requiring additional supplementation [37]. Notably, 67% of women were still using lactation-inducting compounds following completion of the study. A major criticism of previous studies is that the majority only compare differences in supply between groups, not adjusting for baseline milk production. Adjusting for baseline values has been demonstrated to significantly improve statistical power, particularly where there is a strong correlation between baseline and final value [77].

Lastly, while no specific methodological limitation of the studies were included in this review, it is worthwhile to comment on the large number of studies that have investigated the use of pharmaceutical galactagogues in the setting where lactation insufficiency has not been diagnosed [49–58]. We do not feel that one can accurately draw evidence on the efficacy or safety of galactagogues where they are utilised in the absence of a clear established clinical indication. There is currently no evidence to support the prophylactic administration of galactagogues in the absence of the diagnosis of lactation insufficiency, while an associated concern with the use of galactagogues is that they may come at the cost of appropriate utilisation of non-pharmacological treatment strategies.

6.2. Impact of Galactagogues on Breast Milk Composition

Prolactin is thought to mediate changes in breast milk composition occurring during normal lactogenesis [78]. In vitro and animal studies provide evidence that in early lactation prolactin promotes the closure of epithelial tight junctions between alveolar cells and increases the synthesis of α-lactalbumin, which in turn increases breast milk volume [78]. Prolactin has also been demonstrated to influence secretion of immunomodulatory factors in breast milk, with the decreasing concentrations of these factors occurring over the course of lactation mirroring reductions in serum prolactin concentrations also occurring over time [78]. Whether galactagogues, particularly those altering prolactin concentrations, alter the macronutrient composition of the human milk has been examined in just two studies, one involving domperidone [30], and the other recombinant human prolactin [78]. These studies evaluated changes in a number of breast milk constituents including sodium, calcium, lactose, citrate, protein, fat, alpha-lactalbumin, IgA, and oligosaccharides. Such changes in breast milk composition are particularly important for preterm infants, who have higher protein, energy, mineral, and electrolyte requirements than term infants and are at risk of cumulative nutritional deficits and postnatal growth restriction during early life [79]. As such, multi-component breast milk fortifiers (including protein) are routinely added to expressed breast milk to assist preterm infants in meeting their nutrition and growth requirements.

Powe et al. investigated the composition of breast milk among 11 women receiving recombinant human prolactin. The mean time following birth was 12 weeks and ranged from 3.5 to 39 weeks). Overall compositional changes mirrored those in women undergoing normal lactogenesis, with an observed increase in lactose and calcium concentrations, and decreased sodium concentrations that all fell within normal ranges [78]. The observed 200 mL/day increase in breast milk production was thought to be related to the increasing synthesis of alpha-lactalbumin and a corresponding increase in lactose levels, which acts as an osmotic agent to draw additional fluid into milk secretory vesicles [78]. Further, greater maturity was reflected by a decrease in milk sodium concentration, consistent with the closure of tight junctions between the mammary epithelial cells. Lastly, recombinant human prolactin administration led to an increase in overall oligosaccharide concentrations [78]. Oligosaccharide concentrations traditionally decrease during lactogenesis, but observed increases in both neutral and acidic oligosaccharide levels could indicate immunological benefits for the infants. In contrast, no statistically significant differences were observed with respect to changes in milk fat, protein or citrate levels.

In comparison to Powe et al., Campbell-Yeo et al. investigated changes in breast milk composition among a cohort of 45 women randomised to either domperidone or placebo. The mean time following birth was not reported, but 75% of women were randomised prior to four weeks. Compared to those receiving placebo, women administered domperidone experienced a greater improvement in breast milk volume of 163 mL/day, with significant increases in breast milk carbohydrate and calcium, but not sodium. While not statistically significant, mean breast milk protein declined by 9.6% in the domperidone group and increased by 3.6% in the placebo group ($p = 0.16$) [30]. The mechanisms contributing to a possible decrease in the protein concentration are unclear, but are comparable to the 15% reduction in protein observed by Powe et al. following the administration of recombinant human prolactin. It is well known that the milk produced by mothers of preterm infants is higher in protein and this becomes comparable to the breast milk produced by mothers of term infants by 10–12 weeks postpartum [80]. It is possible that the administration of galactagogues altering prolactin levels results in a more rapid transition towards mature breast milk with a resultant reduction in the protein concentration occurring as a natural result of increasing breast milk volume. However, given the limited evidence available, future studies evaluating whether galactagogues alter human breast milk composition are required before a more definitive conclusion can be reached. In order to evaluate this effectively, a comparison of breast milk composition must be made to mothers unaffected by lactation insufficiency in order to determine whether any changes in composition are just reflective of improvements in breast milk supply, or the direct effect of galactagogues. Therefore, a key evidence gap relates to the need for a more comprehensive evaluation of breast milk composition that takes into account postnatal age and breast milk volume.

6.3. Predicting Treatment Response to Galactagogues

There are many factors associated with breastfeeding success, with non-pharmacological treatment strategies clearly first-line in the management of lactation insufficiency. Despite adherence to such strategies (e.g., expressing technique, expressing frequency), the treatment response to galactagogues is noted to be variable in clinical practice. Despite this anecdotal experience, we are unaware of any prospective studies that have examined predictors of treatment response to galactagogues such as domperidone. Most clinical studies simply report the average treatment effect of the galactagogue used, but a small number of studies have reported the individual level data with significant variability in treatment response evident. Wan et al. examined changes in daily breast milk volume in response to treatment with either 30 mg or 60 mg daily doses of domperidone. Among the six women studied, two demonstrated no improvement at all in the daily breast milk volume, representing a non-response rate of 33% [27]. A similar treatment response was identified by Da Silva et al. with two of seven women experiencing less than 10% improvement in daily breast milk volume following the treatment with domperidone, representing a non-response rate of 30% [31].

Given these findings, understanding factors that contribute towards the response to domperidone is important in identifying groups of women most likely to benefit from treatment, while also enabling the potential identification of other important targets for improving breast milk volume. One such factor may be maternal BMI, where it has already been previously reported that overweight/obese mothers of term infants experience a lower prolactin response to infant suckling [81]. The possible impacts of maternal obesity on treatment response to domperidone have been investigated in a recent retrospective cohort study of mothers of preterm infants. In this study, a significant treatment interaction was observed between domperidone and obesity, with obese women less likely to still be breastfeeding their infants at discharge from the Neonatal Unit than non-obese women, despite both receiving domperidone [82]. Obese women were also noted to be more likely to require a prolonged course of domperidone [82]. This may be due to hormonal or inflammatory differences associated with obesity and warrants further investigation. Of particular interest is the role of the toll-like receptor 4 (TLR4) immune signalling pathway, which plays an important role in obesity and the metabolic syndrome [83]. Further, the TLR4 pathway, regardless of whether it is activated by an infection or a sterile inflammatory pathway, is recognised as a key mediator of preterm labor [84]. This is important as TLR4 signalling may be a key factor involved in the delayed secretory activation and development of low milk supply, commonly experienced by many mothers of preterm infants. When exposed to a bacterial by-product in a mouse model of mastitis [85], it is the TLR4 signalling that is responsible for low milk supply, not the bacterium itself [86]. This surprising finding has led to a new paradigm for lactation insufficiency, whereby TLR4 signalling leads to lactocyte cell death and consequently a reduced capacity to produce milk [87]. The presence of asymptomatic breast inflammation could result in unresponsiveness to domperidone (or other galactagogues) whereby the prolactin-mediated signal to stimulate lactocytes to produce more milk is negated by the inflammatory signal. No studies have prospectively evaluated the role of such inflammatory factors in the development of low milk supply or the response to treatment with galactagogues such as domperidone.

In addition to previously described factors, serum prolactin levels could determine the treatment response to galactagogues that alter prolactin concentrations. Kauppila et al. demonstrated that three of five mothers who did not respond to the metoclopramide treatment had the highest basal prolactin levels [72]. Further, while the effects of parity on changes in breast milk supply following treatment with domperidone or metoclopramide have not been previously investigated, there is evidence that nulliparous women are more sensitive to the prolactin stimulatory effects of domperidone/metoclopramide than parous women [88]. This could necessitate the need for different doses or treatment approaches based on parity, but requires further investigation.

Other factors that could influence the treatment response to galactagogues include gestational age at birth and timing of the treatment initiation following birth. Both of these were explored in the recent EMPOWER study, which evaluated the use of domperidone compared to placebo over 14 days [37]. Following the end of the initial treatment period women who received placebo went on to receive domperidone for 14 days. This enabled investigation of the potential effects of delaying the treatment with domperidone for 14 days. At the end of 28 days, the mean daily milk volume was similar between those who immediately started on domperidone compared with those who had treatment with domperidone delayed by 14 days (290 ± 211 mL/day vs. 302 ± 230 mL/day; $p = 0.88$) [37]. Additionally, they observed no significant difference between the number of women who had delivered between 23–26 weeks gestation or 27–29 weeks gestation and achieved a 50% increase in breast milk volume (72.9% compared to 64.2%; $p = 0.38$) [89].

7. Conclusions

The largest and strongest body of evidence supports the use of domperidone as the first-line medication for the pharmacological management of lactation insufficiency but the observed treatment effects are modest at best. There is currently insufficient evidence to determine whether the use of galactagogues has any negative impacts on the micro- and macro-nutrient composition of breast milk and this requires further research. There is emerging evidence of potential differences in the treatment response to galactagogues and further research in this space is critical for improving the therapeutic management of lactation insufficiency. There is a critical need for studies identifying underlying mechanisms contributing towards the development of lactation insufficiency and how these impact the treatment response to various pharmaceutical galactagogues. Such knowledge would facilitate the development and evaluation of novel pharmaceutical treatments for lactation insufficiency while also improve our understanding of individual variability in treatment response to improve utilisation and outcomes of existing treatments.

Author Contributions: Conceptualization, L.E.G.; Writing—Original Draft Preparation, L.E.G.; Writing—Review and Editing, M.E.W., D.T.G.

Funding: This research received no external funding.

Acknowledgments: L.E.G. acknowledges the salary support provided by an Australia National Health and Medical Research Council (NHMRC) Early Career Fellowship (GNT 1070421), Robinson Research Institute Career Development Fellowship and Lloyd Cox Research Fellowship. D.T.G. acknowledges the salary support provided by an unrestricted research grant from Medela AG (Switzerland).

Conflicts of Interest: The authors declare no conflict of interest.

References

1. Gartner, L.M.; Morton, J.; Lawrence, R.A.; Naylor, A.J.; O'Hare, D.; Schanler, R.J.; Eidelman, A.I. Breastfeeding and the use of human milk. *Pediatrics* **2005**, *115*, 496–506.
2. Kramer, M.S.; Kakuma, R. Optimal duration of exclusive breastfeeding. *Cochrane Database Syst. Rev.* **2012**, *8*, CD003517. [CrossRef]
3. Meier, P.; Patel, A.; Esquerra-Zwiers, A. Donor human milk update: Evidence, mechanisms, and priorities for research and practice. *J. Pediatr.* **2017**, *180*, 15–21. [CrossRef]
4. Chung, M.; Raman, G.; Chew, P.; Magula, N.; Trikalinos, T.; Lau, J. Breastfeeding and maternal and infant health outcomes in developed countries. *Evid. Technol. Asses. (Full Rep.)* **2007**, *153*, 1–186.
5. Victora, C.G.; Bahl, R.; Barros, A.J.; França, G.V.; Horton, S.; Krasevec, J.; Murch, S.; Sankar, M.J.; Walker, N.; Rollins, N.C. Breastfeeding in the 21st century: Epidemiology, mechanisms, and lifelong effect. *Lancet* **2016**, *387*, 475–490. [CrossRef]
6. Quigley, M.; Embleton, N.; McGuire, W. Formula versus donor breast milk for feeding preterm or low birth weight infants. *Cochrane Database Syst. Rev.* **2018**, *6*, CD002971. [CrossRef]
7. Vohr, B.R.; Poindexter, B.B.; Dusick, A.M.; McKinley, L.T.; Higgins, R.D.; Langer, J.C.; Poole, W.K. Persistent beneficial effects of breast milk ingested in the neonatal intensive care unit on outcomes of extremely low birth weight infants at 30 months of age. *Pediatrics* **2007**, *120*, e953–e959. [CrossRef]
8. Belfort, M.B.; Anderson, P.J.; Nowak, V.A.; Lee, K.J.; Molesworth, C.; Thompson, D.K.; Doyle, L.W.; Inder, T.E. Breast milk feeding, brain development, and neurocognitive outcomes: A 7-year longitudinal study in infants born at less than 30 weeks' gestation. *J. Pediatr.* **2016**, *177*, 133–139. [CrossRef] [PubMed]
9. Bell, E.H.; Geyer, J.; Jones, L. A structured intervention improves breastfeeding success for ill or preterm infants. *MCN Am. J. Matern. Child. Nurs.* **1995**, *20*, 309–314. [CrossRef] [PubMed]

10. Amir, L.H. Breastfeeding: Managing 'supply' difficulties. *Aust. Fam. Physician* **2006**, *35*, 686–689.

11. Li, R.; Fein, S.B.; Chen, J.; Grummer-Strawn, L.M. Why mothers stop breastfeeding: Mothers' self-reported reasons for stopping during the first year. *Pediatrics* **2008**, *122*, S69–S76. [CrossRef] [PubMed]

12. Lee, S.; Kelleher, S.L. Biological underpinnings of breastfeeding challenges: The role of genetics, diet and environment on lactation physiology. *Am. J. Physiol. Heart Circ. Physiol.* **2016**, *311*, E405–E422. [CrossRef] [PubMed]

13. Neville, M.C. Anatomy and physiology of lactation. *Pediatr. Clin. North. Am.* **2001**, *48*, 13–34. [CrossRef]

14. Crowley, W.R. Neuroendocrine regulation of lactation and milk production. *Compr. Physiol.* **2015**, *5*, 255–291.

15. Forinash, A.B.; Yancey, A.M.; Barnes, K.N.; Myles, T.D. The use of galactogogues in the breastfeeding mother. *Ann. Pharmacother.* **2012**, *46*, 1392–1404. [CrossRef]

16. Stuebe, A.M.; Grewen, K.; Pedersen, C.A.; Propper, C.; Meltzer-Brody, S. Failed lactation and perinatal depression: Common problems with shared neuroendocrine mechanisms? *J. Womens Health (Larchmt)* **2012**, *21*, 264–272. [CrossRef] [PubMed]

17. Amir, L.H. Managing common breastfeeding problems in the community. *BMJ* **2014**, *348*, g2954. [CrossRef] [PubMed]

18. Sultana, A.; Rahman, K.; Manjula, S. Clinical update and treatment of lactation insufficiency. *Med. J. Islamic. World Acad. Sci.* **2013**, *21*, 19–28. [CrossRef]

19. Cregan, M.D.; De Mello, T.R.; Kershaw, D.; McDougall, K.; Hartmann, P.E. Initiation of lactation in women after preterm delivery. *Acta. Obstet. Gynecol. Scand.* **2002**, *81*, 870–877. [CrossRef]

20. Brownell, E.; Howard, C.R.; Lawrence, R.A.; Dozier, A.M. Delayed onset lactogenesis ii predicts the cessation of any or exclusive breastfeeding. *J. Pediatr.* **2012**, *161*, 608–614. [CrossRef]

21. Gilmartin, C.; Amir, L.H.; Ter, M.; Grzeskowiak, L. Using domperidone to increase breast milk supply: A clinical practice survey of australian neonatal units. *J. Pharm. Pract. Res.* **2017**, *47*, 426–430. [CrossRef]

22. Grzeskowiak, L.E.; Hill, M.; Kennedy, D.S. Phone calls to an australian pregnancy and lactation counselling service regarding use of galactagogues during lactation–the Mothersafe experience. *Aust. N. Z. J. Obstet. Gynaecol.* **2018**, *58*, 251–254. [CrossRef] [PubMed]

23. Grzeskowiak, L.E.; Lim, S.W.; Thomas, A.E.; Ritchie, U.; Gordon, A.L. Audit of domperidone use as a galactogogue at an Australian tertiary teaching hospital. *J. Hum. Lact.* **2013**, *29*, 32–37. [CrossRef]

24. Smolina, K.; Morgan, S.G.; Hanley, G.E.; Oberlander, T.F.; Mintzes, B. Postpartum domperidone use in British Columbia: A retrospective cohort study. *CMAJ Open* **2016**, *4*, E13–E19. [CrossRef]

25. Mehrabadi, A.; Reynier, P.; Platt, R.W.; Filion, K.B. Domperidone for insufficient lactation in England 2002-2015: A drug utilization study with interrupted time series analysis. *Pharmacoepidemiol. Drug Saf.* **2018**, *27*, 1316–1324. [CrossRef]

26. Grzeskowiak, L.E.; Dalton, J.A.; Fielder, A.L. Factors associated with domperidone use as a galactogogue at an Australian tertiary teaching hospital. *J. Hum. Lact.* **2015**, *31*, 249–253. [CrossRef] [PubMed]

27. Wan, E.W.; Davey, K.; Page-Sharp, M.; Hartmann, P.E.; Simmer, K.; Ilett, K.F. Dose-effect study of domperidone as a galactagogue in preterm mothers with insufficient milk supply, and its transfer into milk. *Br. J. Clin. Pharmacol.* **2008**, *66*, 283–289. [CrossRef]

28. Blank, C.; Eaton, V.; Bennett, J.; James, S.L. A double blind rct of domperidone and metoclopramide as pro-lactational agents in mothers of preterm infants. Perinatal Society of Australia and New Zealand 5th Annual Conference, Canberra, Australia, 13–16 March 2001; p. 73.

29. Ingram, J.; Taylor, H.; Churchill, C.; Pike, A.; Greenwood, R. Metoclopramide or domperidone for increasing maternal breast milk output: A randomised controlled trial. *Arch. Dis. Child. Fetal. Neonatal. Ed.* **2012**, *97*, F241–F245. [CrossRef] [PubMed]

30. Campbell-Yeo, M.L.; Allen, A.C.; Joseph, K.; Ledwidge, J.M.; Caddell, K.; Allen, V.M.; Dooley, K.C. Effect of domperidone on the composition of preterm human breast milk. *Pediatrics* **2010**, *125*, e107–e114. [CrossRef]

31. Da Silva, O.P.; Knoppert, D.C.; Angelini, M.M.; Forret, P.A. Effect of domperidone on milk production in mothers of premature newborns: A randomized, double-blind, placebo-controlled trial. *Can. Med. Assoc. J.* **2001**, *164*, 17–21.

32. Fazilla, T.E.; Tjipta, G.D.; Ali, M.; Sianturi, P. Domperidone and maternal milk volume in mothers of premature newborns. *Paediatr. Indones.* **2017**, *57*, 18–22. [CrossRef]

33. Inam, I.; Hashmi, A.; Shahid, A. A comparison of efficacy of domperidone and placebo among postnatal women with inadequate breast milk production. *Pak. J. Med. Health Sci.* **2013**, *7*, 314–317.

34. Knoppert, D.C.; Page, A.; Warren, J.; Seabrook, J.A.; Carr, M.; Angelini, M.; Killick, D. The effect of two different domperidone doses on maternal milk production. *J. Hum. Lact.* **2013**, *29*, 38–44. [CrossRef]

35. Petraglia, F.; De Leo, V.; Sardelli, S.; Pieroni, M.L.; D'Antona, N.; Genazzani, A.R. Domperidone in defective and insufficient lactation. *Eur. J. Obstet. Gynaecol. Reprod. Biol.* **1985**, *19*, 281–287. [CrossRef]

36. Rai, R.; Mishra, N.; Singh, D. Effect of domperidone in 2nd week postpartum on milk output in mothers of preterm infants. *Indian J. Pediatr.* **2016**, *83*, 894–895. [CrossRef]

37. Asztalos, E.V.; Campbell-Yeo, M.; da Silva, O.P.; Ito, S.; Kiss, A.; Knoppert, D. Enhancing human milk production with domperidone in mothers of preterm infants: Results from the EMPOWER trial. *J. Hum. Lact.* **2017**, *33*, 181–187. [CrossRef] [PubMed]

38. Kauppila, A.; Anunti, P.; Kivinen, S.; Koivisto, M.; Ruokonen, A. Metoclopramide and breast feeding: Efficacy and anterior pituitary responses of the mother and the child. *Eur. J. Obstet. Gynaecol. Reprod. Biol.* **1985**, *19*, 19–22. [CrossRef]

39. Kauppila, A.; Kivinen, S.; Ylikorkala, O. A dose response relation between improved lactation and metoclopramide. *Lancet* **1981**, *317*, 1175–1177. [CrossRef]

40. Sakha, K.; Behbahan, A. Training for perfect breastfeeding or metoclopramide: Which one can promote lactation in nursing mothers? *Breastfeed. Med.* **2008**, *3*, 120–123. [CrossRef]

41. Ylikorkala, O.; Kauppila, A.; Kivinen, S.; Viinikka, L. Sulpiride improves inadequate lactation. *Br. Med. J.* **1982**, *285*, 249–251. [CrossRef]

42. Ylikorkala, O.; Kauppila, A.; Kivinen, S.; Viinikka, L. Treatment of inadequate lactation with oral sulpiride and buccal oxytocin. *Obstet. Gynecol.* **1984**, *63*, 57–60.

43. Gunn, A.J.; Gunn, T.R.; Rabone, D.L.; Breier, B.H.; Blum, W.F.; Gluckman, P.D. Growth hormone increases breast milk volumes in mothers of preterm infants. *Pediatrics* **1996**, *98*, 279–282.

44. Milsom, S.R.; Rabone, D.L.; Gunn, A.J.; Gluckman, P.D. Potential role for growth hormone in human lactation insufficiency. *Horm. Res. Paediatr.* **1998**, *50*, 147–150. [CrossRef]

45. Powe, C.E.; Allen, M.; Puopolo, K.M.; Merewood, A.; Worden, S.; Johnson, L.C.; Fleischman, A.; Welt, C.K. Recombinant human prolactin for the treatment of lactation insufficiency. *Clin. Endocrinol. (Oxf.)* **2010**, *73*, 645–653. [CrossRef]

46. Peters, F.; Schulze-Tollert, J.; Schuth, W. Thyrotrophin-releasing hormone—A lactation-promoting agent? *BJOG* **1991**, *98*, 880–885. [CrossRef]

47. Zarate, A.; Villalobos, H.; Canales, E.; Soria, J.; Arcovedo, F.; MacGregor, C. The effect of oral administration of thyrotropin-releasing hormone on lactation. *J. Clin. Endocrinol. Metab.* **1976**, *43*, 301–305. [CrossRef]

48. Nommsen-Rivers, L.; Thompson, A.; Riddle, S.; Ward, L.; Wagner, E.; King, E. Feasibility and ccceptability of metformin to augment low milk supply: A pilot randomized controlled trial. *J. Hum. Lact.* **2019**. [CrossRef]

49. Aono, T.; Shioji, T.; Aki, T.; Hirota, K.; Nomura, A.; Kurachi, K. Augmentation of puerperal lactation by oral administration of sulpiride. *J. Clin. Endocrinol. Metab.* **1979**, *48*, 478–482. [CrossRef] [PubMed]

50. De Gezelle, H.; Ooghe, W.; Thiery, M.; Dhont, M. Metoclopramide and breast milk. *Eur J. Obstet. Gynaecol. Reprod. Biol.* **1983**, *15*, 31–36. [CrossRef]

51. Fife, S.; Gill, P.; Hopkins, M.; Angello, C.; Boswell, S.; Nelson, K.M. Metoclopramide to augment lactation, does it work? A randomized trial. *J. Matern. Fetal. Med.* **2011**, *24*, 1317–1320. [CrossRef] [PubMed]

52. Hansen, W.F.; McAndrew, S.; Harris, K.; Zimmerman, M.B. Metoclopramide effect on breastfeeding the preterm infant: A randomized trial. *Obstet. Gynecol.* **2005**, *105*, 383–389. [CrossRef] [PubMed]

53. Huntingford, P.J. Intranasal use of synthetic oxytocin in management of breast-feeding. *Br. Med. J.* **1961**, *1*, 709–711. [CrossRef]

54. Ingerslev, M.; Pinholt, K. Oxytocin treatment during the establishment of lactation. *Acta. Obstet. Gynecol. Scand.* **1962**, *41*, 159–168. [CrossRef]

55. Lewis, P.; Devenish, C.; Kahn, C. Controlled trial of metoclopramide in the initiation of breast feeding. *Br. J. Clin. Pharmacol.* **1980**, *9*, 217–219. [CrossRef] [PubMed]

56. Luhman, L.A. The effect of intranasal oxytocin on lactation. *Obstet. Gynecol.* **1963**, *21*, 713–717. [PubMed]

57. Ruis, H.; Rolland, R.; Doesburg, W.; Broeders, G.; Corbey, R. Oxytocin enhances onset of lactation among mothers delivering prematurely. *Br. Med. J.* **1981**, *283*, 340–342. [CrossRef]

58. Fewtrell, M.S.; Loh, K.; Blake, A.; Ridout, D.; Hawdon, J. Randomised, double blind trial of oxytocin nasal spray in mothers expressing breast milk for preterm infants. *Arch. Dis. Child. Fetal. Neonatal. Ed.* **2006**, *91*, F169–F174. [CrossRef]

59. Tyson, J.; Perez, A.; Zanartu, J. Human lactational response to oral thyrotropin releasing hormone. *J. Clin. Endocrinol. Metab.* **1976**, *43*, 760–768. [CrossRef]

60. Grzeskowiak, L.E.; Smithers, L.G.; Amir, L.H.; Grivell, R.M. Domperidone for increasing breast milk volume in mothers expressing breast milk for their preterm infants – a systematic review and meta-analysis. *BJOG* **2018**, *125*, 1371–1378. [CrossRef]

61. Johannes, C.B.; Varas-Lorenzo, C.; McQuay, L.J.; Midkiff, K.D.; Fife, D. Risk of serious ventricular arrhythmia and sudden cardiac death in a cohort of users of domperidone: A nested case-control study. *Pharmacoepidemiol. Drug Saf.* **2010**, *19*, 881–888. [CrossRef]

62. Van Noord, C.; Dieleman, J.P.; van Herpen, G.; Verhamme, K. Domperidone and ventricular arrhythmia or sudden cardiac death: A population-based case-control study in the netherlands. *Drug Saf.* **2010**, *33*, 1003–1014. [CrossRef]

63. Sewell, C.A.; Chang, C.Y.; Chehab, M.M.; Nguyen, C.P. Domperidone for lactation: What health care providers need to know. *Obstet. Gynecol.* **2017**, *129*, 1054–1058. [CrossRef] [PubMed]

64. Bozzo, P.; Koren, G.; Ito, S. Health canada advisory on domperidone should I avoid prescribing domperidone to women to increase milk production? *Can. Fam. Physician* **2012**, *58*, 952–953. [PubMed]

65. Grzeskowiak, L. Use of domperidone to increase breast milk supply: Are women really dying to breastfeed? *J. Hum. Lact.* **2014**, *30*, 498–499. [CrossRef]

66. Doggrell, S.A.; Hancox, J.C. Cardiac safety concerns for domperidone, an antiemetic and prokinetic, and galactogogue medicine. *Expert Opin. Drug Saf.* **2014**, *13*, 131–138. [CrossRef] [PubMed]

67. Grzeskowiak, L.E. Domperidone for lactation: What health care providers really should know. *Obstet. Gynecol.* **2017**, *130*, 913. [CrossRef] [PubMed]

68. Grzeskowiak, L.E.; Amir, L.H. Pharmacological management of low milk supply with domperidone: Separating fact from fiction. *Med. J. Aust.* **2014**, *201*, 257–258. [CrossRef]

69. Biewenga, J.; Keung, C.; Solanki, B.; Natarajan, J.; Leitz, G.; Deleu, S.; Soons, P. Absence of QTc prolongation with domperidone: A randomized, double-blind, placebo-and positive-controlled thorough QT/QTc study in healthy volunteers. *Clin. Pharmacol. Drug Dev.* **2015**, *4*, 41–48. [CrossRef] [PubMed]

70. Smolina, K.; Mintzes, B.; Hanley, G.E.; Oberlander, T.F.; Morgan, S.G. The association between domperidone and ventricular arrhythmia in the postpartum period. *Pharmacoepidemiol. Drug Saf.* **2016**, *25*, 1210–1214. [CrossRef]

71. Grzeskowiak, L.E.; Smithers, L.G. Use of domperidone and risk of ventricular arrhythmia in the postpartum period: Getting to the heart of the matter. *Pharmacoepidemiol. Drug Saf.* **2017**, *26*, 863–864. [CrossRef]

72. Haase, B.; Taylor, S.N.; Mauldin, J.; Johnson, T.S.; Wagner, C.L. Domperidone for treatment of low milk supply in breast pump–dependent mothers of hospitalized premature infants a clinical protocol. *J. Hum. Lact.* **2016**, *32*, 373–381. [CrossRef]

73. Kauppila, A.; Arvela, P.; Koivisto, M.; Kivinen, S.; Ylikorkala, O.; Pelkonen, O. Metoclopramide and breast feeding: Transfer into milk and the newborn. *Eur J. Clin. Pharmacol.* **1983**, *25*, 819–823. [CrossRef]

74. Hale, T.W.; Kendall-Tackett, K.; Cong, Z. Domperidone versus metoclopramide: Self-reported side effects in a large sample of breastfeeding mothers who used these medications to increase milk production. *Clin. Lact.* **2018**, *9*, 10–17. [CrossRef]

75. Lussier, M.M.; Brownell, E.A.; Proulx, T.A.; Bielecki, D.M.; Marinelli, K.A.; Bellini, S.L.; Hagadorn, J.I. Daily breastmilk volume in mothers of very low birth weight neonates: A repeated-measures randomized trial of hand expression versus electric breast pump expression. *Breastfeed. Med.* **2015**, *10*, 312–317. [CrossRef]

76. Meier, P.P.; Engstrom, J.L. Evidence-based practices to promote exclusive feeding of human milk in very low-birthweight infants. *NeoReviews* **2007**, *8*, e467–e477. [CrossRef]

77. Borm, G.F.; Fransen, J.; Lemmens, W.A. A simple sample size formula for analysis of covariance in randomized clinical trials. *J. Clin. Epidemiol.* **2007**, *60*, 1234–1238. [CrossRef] [PubMed]

78. Powe, C.E.; Puopolo, K.M.; Newburg, D.S.; Lönnerdal, B.; Chen, C.; Allen, M.; Merewood, A.; Worden, S.; Welt, C.K. Effects of recombinant human prolactin on breast milk composition. *Pediatrics* **2011**, *127*, e359. [CrossRef]

79. Radmacher, P.G.; Adamkin, D.H. Fortification of human milk for preterm infants. *Semin. Fetal. Neonatal. Med.* **2017**, *22*, 30–35. [CrossRef]

80. Gidrewicz, D.A.; Fenton, T.R. A systematic review and meta-analysis of the nutrient content of preterm and term breast milk. *BMC Pediatr.* **2014**, *14*, 216. [CrossRef]

81. Rasmussen, K.M.; Kjolhede, C.L. Prepregnant overweight and obesity diminish the prolactin response to suckling in the first week postpartum. *Pediatrics* **2004**, *113*, e465–e471. [CrossRef]

82. Grzeskowiak, L.E.; Amir, L.H.; Smithers, L.G. Longer-term breastfeeding outcomes associated with domperidone use for lactation differs according to maternal weight. *Eur. J. Clin. Pharmacol.* **2018**, *74*, 1071–1075. [CrossRef]

83. Jialal, I.; Kaur, H.; Devaraj, S. Toll-like receptor status in obesity and metabolic syndrome: A translational perspective. *J. Clin. Endocrinol. Metab.* **2014**, *99*, 39–48. [CrossRef]

84. Keelan, J. Pharmacological inhibition of inflammatory pathways for the prevention of preterm birth. *J. Reprod. Immunol.* **2011**, *88*, 176–184. [CrossRef]

85. Ingman, W.V.; Glynn, D.J.; Hutchinson, M.R. Mouse models of mastitis–how physiological are they? *Int. Breastfeed. J.* **2015**, *10*, 12. [CrossRef]

86. Glynn, D.J.; Hutchinson, M.R.; Ingman, W.V. Toll-like receptor 4 regulates lipopolysaccharide-induced inflammation and lactation insufficiency in a mouse model of mastitis. *Biol. Reprod.* **2014**, *90*, 91. [CrossRef] [PubMed]

87. Ingman, W.V.; Glynn, D.J.; Hutchinson, M.R. Inflammatory mediators in mastitis and lactation insufficiency. *J. Mammary Gland Biol. Neoplasia* **2014**, *19*, 161–167. [CrossRef]

88. Brown, T.E.; Fernandes, P.A.; Grant, L.J.; Hutsul, J.-A.; McCoshen, J.A. Effect of parity on pituitary prolactin response to metoclopramide and domperidone: Implications for the enhancement of lactation. *J. Soc. Gynecol. Investig.* **2000**, *7*, 65–69. [CrossRef]

89. Asztalos, E.V.; Kiss, A.; da Silva, O.P.; Campbell-Yeo, M.; Ito, S.; Knoppert, D. Pregnancy gestation at delivery and breast milk production: A secondary analysis from the EMPOWER trial. *Matern. Health Neonatol. Perinatol.* **2018**, *4*, 21. [CrossRef]

MDPI

St. Alban-Anlage 66

4052 Basel

Switzerland

Tel. +41 61 683 77 34

Fax +41 61 302 89 18

www.mdpi.com

Nutrients Editorial Office

E-mail: nutrients@mdpi.com

www.mdpi.com/journal/nutrients